£16.00

*Computational Methods
of
Multivariate Analysis
in
Physical Geography*

# Computational Methods of Multivariate Analysis in Physical Geography

**P. M. Mather**

*Lecturer in Geography,
University of Nottingham*

JOHN WILEY & SONS
London · New York · Sydney · Toronto

Copyright © 1976, by John Wiley & Sons.

All rights reserved.

No part of this book may be reproduced by any means, nor transmitted, nor translated into a machine language without the written permission of the publisher.

**Library of Congress Cataloging in Publication Data:**

Mather, Paul M.
   Computational methods of multivariate analysis in physical geography.

   1. Physical geography — Mathematics. 2. Multivariate analysis. I. Title.
GB21.M37     910'.02'051953     75-23376

ISBN 0 471 57626 3

Typeset in IBM Baskerville by Preface Ltd., Salisbury, Wilts., and printed by The Pitman Press Ltd., Bath

# *Preface*

This book is the outcome of my interest in the use of multivariate analysis in physical geography. One of my earliest impressions of life as a research student in the mid-1960's was the lack of textbooks on advanced statistical and data-processing techniques written by geographers for geographers. Consequently, I had to rely on books such as Rao (1952) and Cattell (1966a) which, although excellent in themselves, required an appreciation of the terminology and research problems of psychometricians and biometricians before the methods themselves could be understood. The appearance of L. J. King's book in 1969 went some way towards relieving this situation, but the level of technical detail was not high. It is still true to say that there is no geographical textbook which adequately sets out the technical basis of the more important multivariate techniques. By 'technical basis' I mean details of the specifications of the statistical models, with information on the assumptions of these models and discussion of the consequences resulting from their violation. I would also include description of alternative methods of estimation when the basic assumptions do not hold, and consideration of the numerical and computational aspects of the particular method, including details of feasible algorithms. This book represents my effort to fill this gap in the literature. The level of mathematical and statistical sophistication of research workers in physical geography is now sufficiently high for a technical book such as this one to be acceptable without the need for large-scale examples and case studies. Such examples have their uses but I feel that, in a book of this nature, specific examples, requiring a reasonable level of knowledge of the research problem involved, would distract the reader from the fundamental points concerning the technical basis of each method. Consequently, the examples are deliberately kept small and simple. In this form, they also provide suitable test data sets for the computer programs and algorithms which are described in the text. It is emphasized that these examples are not meant to be representative applications of the use of multivariate techniques in physical geography.

I anticipate that readers of this book will be postgraduate research workers in physical geography who are actively engaged in the analysis of problems that can be expressed quantitatively. Consequently, this book is not set out in the form of a course in multivariate analysis, but is written in the expectation that it will be used both as a reference book and as a source of computer programs. This approach justifies the concentration on technical and computational details rather than on the examples of the use of various techniques in a research context.

I would like to thank all those who have helped me in the preparation of this book. Professor J. P. Cole, Department of Geography, Univeristy of Nottingham, has been a constant source of ideas and enthusiasm. Mr. R. P. Bradshaw, of the Geography Department, University of Nottingham, has kindly commented on several sections of the manuscript. The staff of the numerical section of the Cripps Computing Centre, University of Nottingham, have provided invaluable advice and assistance. I am also grateful for the comments of an anonymous referee. The editorial staff of John Wiley and Sons, Ltd., have invariable been helpful, and extremely patient. Lastly, it would not have been possible to write this book without the patience and consideration of my wife.

*Nottingham*, 1975. PAUL M. MATHER

# Alphabetical List of Subroutines and Functions

Table numbers are indicated for each entry

ACCINV, A30
ADDCOL, 3.8
ADJUST, 5.7
ALPHA, 5.3
ARBOR, 6.12
BART, 5.19
CANON, 5.7
CANON, 7.8
CHEK, 6.13
CHOLS, A29
CLUMP, 6.13
CLUS, 6.12
CLUSTR, 6.12
COMP, 3.3
COPHGN, 6.12
COPHCR, 6.12
COPY, A4
CROUT, A32
DAPLOT, 3.8
DATIN, 2.1, 7.2
DERIVS, 6.16
DEVIAT, A18
DIST, 6.16
DISTPT, 6.15
DNSAT1, 6.15
DPACC, A7
DT2, A27
EIGVEC, A25
EUCLID, 6.13
FISHER, 3.15
FPROB, 2.1
FSCORE, 5.18
FVAL, 5.7

GJDEF1, A31
GJDEF3, 5.19
GSORTH, A34
HISTGM, A14
HTDQL, A24
IFIND, A19
IDLZD, 5.20
IMAGE, 5.10
IMPORT, 5.7
INN, 6.14
INPT, 6.12, 6.16, 7.7
KALK, A11
LINK, 6.12
MAP, 3.8
MATCH, 6.14
MATPR, A1
MAX, 3.15
MDA, 7.8
MERGE, 6.13
MIN, 6.12
MINIM, 6.12
MINRES, 5.7
MONOT, 6.15
MTMULT, A6
NONLIN, 6.16
NORMLZ, 6.15
NSEIG, A33
ORTHCL, 3.3
OUTPUT, 6.13
OVCHK, A12
PCORD, 6.14
PLACE, 7.8
PLOTIT, 3.3

PLOTS, A8
PREP, 7.8
PREP1, 6.12
PRNTRI, A2
PROMAX, 5.15
RANDOM, A16
RANGE, 6.13
RANK, A13
RATQR, A21
REBAKA, A23
RECOV, 3.3, 3.8
REDUC, A22
REMOVE, 3.15
RESET, 3.15
RESID, 2.1, 3.15
REX, 2.1, 3.15, 4.2
SAMESC, A9
SCALE, A10
SCORE, 4.2
SEARCH, 6.12
SETUP, A15
SIGNIF, 2.1
SIMCAL, 6.14
SMALL, 6.12
SOLVE, A28
SORT, 6.15
SSCP, A3
SWAP, 5.15
SWAP1, 7.7
TARGET, 5.16
TRANS, 2.1
TRBAK1, A26
TRIDI, A20

TRMULT, A5
TSA, 3.8
TSMAIN, 3.8
UPSAT1, 6.15

VARIAN, 3.3
VARMAX, 5.12
VCALC, 3.15
XMOP, 3.3

YDIST, 6.16
ZED, 3.8
ZERO, A18

# Contents

**Introduction** . . . . . . . . . . . . . . . . . . . . . 1
  0.0   Exploratory Data Analysis . . . . . . . . . . . . 1
  0.1   Structure of this Book . . . . . . . . . . . . . . 3
  0.2   Computer Programs . . . . . . . . . . . . . . . 6

PART ONE   ESSENTIALS OF MATRIX ALGEBRA

**1 Mathematical Background** . . . . . . . . . . . . . . 9
  1.0   Introduction . . . . . . . . . . . . . . . . . . . 9
  1.1   Basic Definitions . . . . . . . . . . . . . . . . . 10
  1.2   Further Definitions . . . . . . . . . . . . . . . . 14
  1.3   Vectors . . . . . . . . . . . . . . . . . . . . . 16
  1.4   The Inverse Matrix . . . . . . . . . . . . . . . . 23
  1.5   Eigenvalues and Eigenvectors . . . . . . . . . . . 28
  1.6   Error in Matrix Computations . . . . . . . . . . . 32
  Appendix   Summary of Matrix Theory . . . . . . . . . 35

PART TWO   ANALYSIS OF DEPENDENCE

**2 Regression Analysis: the Linear Model** . . . . . . . . . 39
  2.0   Introduction . . . . . . . . . . . . . . . . . . . 39
  2.1   The Linear Regression Model . . . . . . . . . . . 40
  2.2   Estimation in the Linear Regression Model . . . . . . . 45
  2.3   Computational Aspects of Multiple Regression . . . . . 52
  2.4   Problems in Regression . . . . . . . . . . . . . . 73
      (i)   Multicollinearity and Ridge Regression . . . . . . 73
      (ii)  Heteroscedasticity . . . . . . . . . . . . . 77
      (iii) Autocorrelation . . . . . . . . . . . . . . 82

## 3 Further Aspects of Least-squares Methods . . . . . . . . 94
- 3.0 Introduction . . . . . . . . . . . . . . . . . 94
- 3.1 Least Squares Curve-fitting . . . . . . . . . . . . 95
  - 3.1.0 Introduction . . . . . . . . . . . . . 95
  - 3.1.1 Gram–Schmidt Method: Computation . . . . . 96
  - 3.1.2 Examples . . . . . . . . . . . . . . . . 101
  - 3.1.3 Confidence Intervals for Least-squares Coefficients . 109
- 3.2 Trend Surface Analysis . . . . . . . . . . . . . 116
  - 3.2.0 Introduction . . . . . . . . . . . . . 116
  - 3.2.1 Trend Surface Model . . . . . . . . . . . 117
    - (A) Specification Phase . . . . . . . . . 119
    - (B) Data Point Distributions . . . . . . . . 120
    - (C) Computational Aspects . . . . . . . . 130
    - (D) Inferential Problems . . . . . . . . . 142
  - 3.2.2 Comparison of Trend Surfaces . . . . . . . . 147
  - 3.2.3 Examples . . . . . . . . . . . . . . . . 150
- 3.3 Selection of Explanatory Variables in Multiple Regression . 172
  - 3.3.0 Introduction . . . . . . . . . . . . . 172
  - 3.3.1 The Selection Approach . . . . . . . . . . 173
  - 3.3.2 The Combinational Approach . . . . . . . . 178
- 3.4 Use of Binary Explanatory Variables . . . . . . . . 193
  - 3.4.0 Introduction . . . . . . . . . . . . . 193
  - 3.4.1 Method and Example . . . . . . . . . . . 194
- 3.5 Nonlinear Least Squares Methods . . . . . . . . . 198
  - 3.5.0 Introduction . . . . . . . . . . . . . 198
  - 3.5.1 Methods . . . . . . . . . . . . . . . 198
  - 3.5.2 Example . . . . . . . . . . . . . . . . 207

## PART THREE  ANALYSIS OF INTERDEPENDENCE

## 4 Principal Components Analysis . . . . . . . . . . . . 215
- 4.0 Introduction . . . . . . . . . . . . . . . . . 215
- 4.1 Mathematical Foundations . . . . . . . . . . . . 220
- 4.2 Computation . . . . . . . . . . . . . . . . . 226
- 4.3 Example . . . . . . . . . . . . . . . . . . . 232
- 4.4 Conclusions . . . . . . . . . . . . . . . . . 239

## 5 Factor Analysis . . . . . . . . . . . . . . . . . . 240
- 5.0 Introduction . . . . . . . . . . . . . . . . . 240
- 5.1 Definitions . . . . . . . . . . . . . . . . . 241
- 5.2 Statistical Model . . . . . . . . . . . . . . . 243
- 5.3 Estimation . . . . . . . . . . . . . . . . . 245
  - 5.3.0 Nonstatistical Factor Analysis . . . . . . . . 247
  - 5.3.1 Statistical Factor Analysis . . . . . . . . . 261

|       | 5.3.2 | Image Analysis | 272 |
|---|---|---|---|
| 5.4 | | Rotation of Factors | 275 |
| 5.5 | | Factor Scores | 294 |
| 5.6 | | Summary | 304 |

## PART FOUR  CLASSIFICATION

**6 Classification** . . . . . . . . . . . . . . . . . . . . 309
  6.0  Introduction . . . . . . . . . . . . . . . . . . . 309
  6.1  Cluster Analysis . . . . . . . . . . . . . . . . . . 310
      6.1.0  Types of Cluster Analysis . . . . . . . . . 310
      6.1.1  Measurement of Similarity . . . . . . . . 311
      6.1.2  Hierarchical Clustering Methods . . . . . . . 316
      6.1.3  Display of Hierarchical Relationships . . . . . 321
      6.1.4  Comparison of Hierarchical Clustering Strategies . . 324
      6.1.5  Nonhierarchical Clustering Methods . . . . . 327
  6.2  Ordination . . . . . . . . . . . . . . . . . . . . 329
      6.2.0  Principal Coordinates Analysis . . . . . . . . 330
      6.2.1  Nonmetric Multidimensional Scaling . . . . . 331
      6.2.2  Nonlinear Mapping . . . . . . . . . . . . 339
      6.2.3  Quadratic Loss Functions . . . . . . . . . 340
  6.3  Examples . . . . . . . . . . . . . . . . . . . . . 341
      6.3.0  Hierarchical Clustering . . . . . . . . . . . 342
      6.3.1  Nonhierarchical Cluster Analysis; Nonlinear
             Mapping . . . . . . . . . . . . . . . . 343
      6.3.2  Nonmetric Multidimensional Scaling . . . . . 351
      6.3.3  Principal Coordinates Analysis . . . . . . . . 370
  6.4  Computer Programs . . . . . . . . . . . . . . . . 371

**7 Discriminant Analysis** . . . . . . . . . . . . . . . . . 420
  7.0  Introduction . . . . . . . . . . . . . . . . . . . 420
  7.1  Two-group Discriminant Analysis . . . . . . . . . . 420
  7.2  Multiple Discriminant Analysis . . . . . . . . . . . 432
      7.2.0  Multivariate Analysis of Variance (MANOVA) . . 432
      7.2.1  Multiple Discriminant Analysis . . . . . . . 437
  7.3  Discriminant Analysis based on Binary Attributes . . . 453

**Appendix A**  Fortran Subroutine Library . . . . . . . . . 460

**Bibliography** . . . . . . . . . . . . . . . . . . . . . . 504

**Author Index** . . . . . . . . . . . . . . . . . . . . . 521

**Subject Index** . . . . . . . . . . . . . . . . . . . . . 527

# *Introduction*

## 0.0 Exploratory Data Analysis

The success of modern science in arriving at an understanding of the way in which physical phenomena work and are interrelated is largely due to the examination of supposed causal relationships. In many research situations in pure science, cause and effect have been isolated by the study of relationships between pairs of variables. Experiments carried out under laboratory conditions have made it possible to isolate a few variables of interest, with the other controlling variables being held constant. Physical geography, however, is concerned with complex situations in which many variables are simultaneously related (or even related to their own past values). This results in the obscuring of causal links. Since laboratory studies of the kind mentioned above are generally not feasible, the complex of variables must be studied as a whole. This is due to the fact that changes in one variable may produce corresponding changes in other variables either directly or indirectly, making it very difficult to isolate pairs of strongly-related variables as is possible in many laboratory-oriented sciences.

Chorley and Kennedy (1971) have drawn attention to the advantages of a systems approach in physical geography. Such an approach allows the set of variables under consideration to be viewed as a whole, rather than as the sum of a number of simpler relationships. The behaviour of a complex interacting system of variables cannot easily or efficiently be described in terms of a number of bivariate relationships. In order to understand such systems, it is necessary to use multivariate analysis.

Two opposing views have been taken by users of multivariate methods. Some see the role of these methods as a purely confirmatory one, that is, providing estimates of the parameters of *a priori* models, and allowing significance tests of hypotheses relating to the values taken by these parameters. This attitude is perhaps more relevant to research in the pure sciences, where a firm basis of theory has been established. In other, less well-developed, disciplines multivariate methods are commonly used in an exploratory manner, that is, to suggest, rather than to test, hypotheses. In fact, it could be argued that changes in the orientation of a discipline are more often due to developments in techniques than to changes in theory. New techniques allow the research worker to ask questions of a different, perhaps more complicated, nature. Such initial uses of new techniques are inevitably exploratory rather than confirmatory. (Consider, for example, the state of astronomy after the invention of the telescope.) It is interesting to note that the period 1956 to 1963, during which geography underwent a major internal reorganization, is usually known as the 'quantitative revolution' even though, as Burton (1963) and other writers have shown, the fundamental revolution was in the methodological outlook of geographers. Perhaps the availability of statistical techniques, coupled with the ability to use them, provoked geographers to think of alternative ways of explaining the phenomena and relationships they were scrutinising. However, the 'quantitative revolution' was not necessarily entirely beneficial. Geographers have perhaps been too ready to proclaim themselves to be scientists, and have, as a consequence, eagerly assumed what they thought was 'the' scientific method. Such eagerness in a developing subject can lead to premature regimentation and sterile debate at a time when trial and error methods should be used more freely. Consider a child of three. He does not learn to speak fluently and grammatically by adopting a 'scientific' method of learning; he listens and he experiments with new words and new verbal structures, rejecting some and accepting others on the basis of his success in communicating. In that way, he builds up a body of experience which — at a later stage — allows him to put together words and phrases on the basis of his theoretical knowledge of the language. Despite what St. Paul had to say about casting off childish ways, this analogy has much to commend it.

Exploratory data analysis is, then, a method of examining a set of data from various angles, and of piecing together information about the system of study that is revealed by each analysis. Such information may lead to a subsequent analysis that is more refined and possibly more revealing. The lack of a formal hypothesis may, in some cases, lead to error. However, the possibility of error is not in itself a sufficient reason to reject or to condemn exploratory methods. Knowledge is gained only if the research worker is willing to take the risk of making mistakes resulting in faulty conclusions. Tukey (1969), in a paper entitled *Analysing Data: Sanctification or Detective Work?* makes essentially the same point: '... to concentrate on confirmation, to the exclusion or submergence of exploration, is an obvious mistake. Where does new knowledge come from? How can an undetected criminal be put on trial? ... we learn by taking chances ... no conclusion or inference ever becomes knowledge without risk or error.'

Repetition is important in exploratory work. If it is not possible to take further samples from the population of interest, the given sample can be split into several subsamples, each of which should then be analysed independently. Only results that are consistent across samples should be considered for further investigation. Another aspect of exploratory data analysis that is worthy of comment is the status of uninterpretable results. Such results should not be discarded lightly. Quite often, it is the presence of anomalous results that reveals deficiencies in the form of the model (is the model correctly specified? have important explanatory variables been omitted, or irrelevant ones included?) As Digman (1966, p. 471) remarks: '... there are few things which will stimulate research and thinking as much as a stable but inscrutable phenomenon.' In this context, it could be said that learning comes from the interplay between the investigator, the technique and the data.

It is the author's belief that such an interplay is not possible unless the investigator has a good basic knowledge of the techniques he is using, in addition to a firm general grasp of the system he is studying. Consequently, emphasis is placed in this book on the technical aspects of the specification and assumptions of the multivariate models described, and to computational and other practical difficulties that are often experienced in their application. This is the opposite approach to that favoured by many subject-oriented texts on quantitative methods, the authors of which appear to prefer to work from a case study to the details of the techniques. Since it is assumed that the reader will be actively engaged in research, and will be aware of problems in his own field that are amenable to a multivariate approach, extended case studies are not given. The examples that are included are not meant to be representative applications of the techniques — they are used to show the nature of the results to be expected from a given technique, and to provide test data for the computer programs that are listed in each chapter. References are given to selected applications of the individual techniques, but no comprehensive review of the literature of quantitative physical geography is provided. The reader is recommended to refer to the specialist literature in his own field for details of relevant uses of the various techniques described in this book.

## 0.1 Structure of this book

Readers are assumed to have some familiarity with elementary statistical concepts as laid out by, among others, Agterberg (1974), Kmenta (1971) and Yamane (1964). Since a knowledge of matrix algebra is essential if the material in later chapters is to be intelligible, Part 1 of this book consists of a survey of matrix concepts and operations. No prior knowledge of matrices is assumed. Hand-calculator methods of carrying out the operations described in this part are not given; in all but the most trivial cases the computations will not be performed by hand, and it seems unnecessary to labour the reader with details of outmoded algorithms which, for the most part, are unsuitable for computer implementation. Fortran subroutines to carry out the operations that are mentioned in the text are provided in Appendix A. This Appendix also contains listings of general purpose routines. Attention is given in the closing section of Chapter 1 to the question of computational accuracy in multivariate analysis. This important topic has been generally ignored, and not only in the geographical literature, although it should be self-evident that a computational algorithm should not distort the data. Many readers will, no doubt, be familiar with the phrase 'garbage in, garbage out' but it is true that good data are sometimes transformed into garbage by an inefficient or unstable algorithm. Part 1 closes with an appendix containing a summary of matrix algebra.

The second part of the book consists of a survey of least-squares methods. Chapter 2 deals with the classical ordinary least-squares model and with the strategy to be followed whenever the basic assumptions of the model do not hold. Despite the popularity of regression methods over the last 20 years, many users appear to be unaware of the nature of these assumptions, and few applications of alternative methods of estimation have appeared in the literature of physical geography. In an exploratory analysis it will not normally be known whether or not the disturbance terms in a multiple regression analysis are heteroscedastic or autocorrelated; tests on the residuals from an ordinary least-squares regression provide the opportunity to examine hypotheses about the residuals and allow the investigator to modify the specification of the model either by altering its form or by introducing explanatory variables that were not previously considered. Also, alternative forms of estimation can be employed if such modifications do not produce a model of the classical form. For example, generalized least-squares procedures can be used in the presence of autocorrelated or heteroscedastic disturbances, or ridge regression methods if the explanatory variables are highly correlated.

Extensions of the classical model are dealt with in Chapter 3. Methods of fitting polynomials in one and two explanatory variables (trend surface analysis) are introduced, and the computational and logical problems associated with these techniques are discussed. An important application of regression analysis in exploratory studies is concerned with the selection of important explanatory variables. Two selection methods are discussed – the now traditional stepwise approach and the combinatorial method, which has been made feasible by improvements in the size and speed of modern digital computers. The last two sections in Chapter 3 are concerned with the use of binary explanatory variables and the fitting of nonlinear least-squares models.

Chapters 4 and 5, which form Part 3 of the book, contain a review of methods of analysing the dependence structures within a system of interrelated variables. Principal components analysis is the topic of Chapter 4, while an account of the various methods of factor analysis is given in Chapter 5. Some confusion over the aims of these two techniques is evident from even a casual inspection of the geographical literature. In the author's opinion, principal components analysis is a mathematical transformation of a data matrix to produce a new matrix of scores that has the property that successive orthogonal columns (representing principal components) account for a maximum of the variance or information remaining after the amounts accounted for by preceding columns have been removed. Principal components are thus defined in terms of variance. The principal components score matrix contains the same information as the data matrix, but that information is expressed in a compressed, more economical form. Concepts such as rotation and communality are extraneous to principal components analysis. Factor analysis, on the other hand, is concerned with the structure of the correlations between the variables, and is therefore a suitable technique to use if the pattern of relationships within a system is being scrutinized. Factors are not defined in terms of variance; they are defined in accordance with a hypothesis such as that of simple structure. The methods of factor analysis have recently undergone a considerable revision, and more modern methods — such as those described in this book — have done much to alleviate the criticisms of the technique that have been voiced over the last few years. In particular, the principal axis method has been seen to be a computational expedient rather than a true factoring method.

The fourth and final part of the book is devoted to methods of classification and discrimination. All developing sciences are preoccupied with the classification of the phenomena with which they deal, in order to reduce the number of concepts necessary to understand a particular system. No single classification can be expected to serve all purposes, hence the research worker must specify his own requirements before adopting an existing classification or constructing a new one. Many methods of numerical classification have been described in the literature of various disciplines. Some of the more useful of these are discussed in Chapter 6. Once a suitable classification has been arrived at, it may be necessary to allocate previously unclassified items or observations to their most likely group while minimizing the possibility of error. Chapter 7 is a survey of techniques of discriminant analysis including two-group and multiple-group discriminant analysis as well as a summary of more recent methods of pattern recognition that are suitable for use when the variables are dichotomous rather than continuous in nature.

## 0.2 Computer Programs

Chapters 2 to 7 contain listings of Fortran computer programs to carry out the calculations described in the text. All have been written specifically for this book. Although some of the techniques have previously been programmed (for example, they may be available in one of the standard packages, such as BMD or SPSS) the versions listed here are more flexible, allowing the user to choose between various alternative methods of input and output, for example. Furthermore, because the source listings are provided, and details of the computations are discussed in the text, the reader should be able to modify the programs to suit his own particular requirements. Where alternative estimation procedures are available, for example in factor analysis, individual subroutines are provided. To help readers program the algorithms for which no Fortran listing is included, a library of matrix handling and general purpose subroutines is given in Appendix A. Machine-dependent aspects of Fortran have been avoided as far as possible, and many of the programs have been tested on a small computer (PDP 11), a medium size computer (ICL 1906A) and a large computer (CDC 7600). All of the few machine-dependent instructions are indicated by comments within the listings. The matrix handling routines in Appendix A are based in part on the Algol 60 procedures of Wilkinson and Reinsch (eds.) (1971), to which reference should be made for more detailed computational and algorithmic details.

*Part One*

# Essentials of Matrix Algebra

Two aspects of matrices are dealt with in this Part; (i) the concepts and notation of matrix algebra, and (ii) computational methods for matrices. Matrix algebra provides a concise, abbreviated notation which is well suited to the expression of multivariate relationships. The material covered in the remainder of this book requires a basic knowledge of matrices, including addition, subtraction, multiplication and inversion of matrices; the solution of systems of linear simultaneous equations; the derivation of the eigenvalues and eigenvectors of matrices, and the choice of a computational algorithm to carry out these operations. Geographers, like many other users of multivariate analysis, have tended to ignore the numerical analysis aspects of the topic, possibly due to a belief that the computer can carry out any given operation to the required degree of accuracy, or else to a belief that computational errors will always be of less importance than data measurement errors. Neither of these suppositions is justified, and the purpose of this Part is to show both the utility of matrix methods and the safeguards that are necessary in their use.

*Chapter 1*

# Mathematical Background

## 1.0 Introduction

This chapter contains an introduction to aspects of matrix algebra and a discussion of numerical methods useful in mathematical and statistical applications. An understanding of matrices and vectors will be necessary in order to follow the presentation in later chapters. The use of matrix notation has the advantage of elegance and simplicity over alternative formulations and, by virtue of this, allows a clearer appreciation of the structure of the particular mathematical or statistical model. Secondly, the theorems of matrix algebra can be applied to problems expressed in matrix form, and this may lead to more economical solutions. A solution expressed in matrix form is readily programmed in a high-level language such as Fortran, which allows the use of matrices (or arrays).

A knowledge of the difficulties likely to be encountered in the procedure employed in solving the given problem, together with an appreciation of the characteristics of digital computers, forms the subject-matter of numerical analysis. Many computing procedures still in common use were developed for hand calculation or for desk calculators, and rely on the operator to round intermediate results in a sensible manner. The same algorithm when coded for a digital computer may well produce inaccurate results. Longley (1967), Wampler (1970) and Youngs and Cramer (1971) discuss this point in relation to computer programs for multiple regression analysis, and their results show that even widely-used 'package' programs do not always employ reliable algorithms. Geographers, along with many other users of mathematical and statistical techniques, have generally been unaware of this important aspect of computing.

The level of mathematical expertise required in this chapter is not high. Provided that the reader has mastered elementary algebra and arithmetic he should find little difficulty in understanding the material presented here. A degree of competence in Fortran programming is assumed. Readers who wish to revise their knowledge of programming are referred to Mather (1975B), McCracken (1965), Golden (1965), or to one of the computer manufacturer's programming handbooks. Fortran subroutines are listed in various parts of this book; they are written in standard Fortran-IV, and have been run on one or more of the following computers: the University of Nottingham ICL 1906A; the University of Manchester CDC 7600 and the University of Nottingham Department of Geography DEC PDP-11/05.

A detailed account of matrix algebra and numerical methods is not possible in the space of a single chapter. Results are invariably stated without proofs, and selective accounts of various topics are given. For a more detailed account of matrix algebra see Graybill (1969), Searle (1966) or Bronson (1969). Numerical methods are discussed by Dorn and McCracken (1972), Williams (1972), Carnahan, Luther and Wilkes (1969) and Pennington (1970). Computer applications are specifically discussed in each of these texts, but a fuller coverage of computer methods in matrix algebra is given by Wilkinson and Reinsch (eds.) (1971), whose book contains many Algol-60 procedures, some of which have been translated into Fortran-IV and listed in Appendix A. The two-volume work edited by Ralston and Wilf (1960, 1967) covers in detail the material presented here, and far more besides.

## 1.1 Basic Definitions

A *matrix* is a rectangular arrangement of *elements*, which are usually numbers, in *rows* and *columns*. The matrix is denoted by an upper-case bold face letter, e.g. **A**, while an individual element is given by the same letter in lower case with subscripts denoting the row and column numbers. Thus $a_{51}$ is the element of $A$ lying at the intersection of row 5 and column 1. The row identifier always precedes the column identifier. A matrix with $n$ rows and $p$ columns is said to be an $(n \times p)$ matrix. In a Fortran program, it would have *dimension* $(n \times p)$.

Thus, a $(5 \times 6)$ matrix **Z** can be written as

$$\mathbf{Z} = \begin{bmatrix} z_{11} & z_{12} & z_{13} & z_{14} & z_{15} & z_{16} \\ z_{21} & z_{22} & z_{23} & z_{24} & z_{25} & z_{26} \\ z_{31} & z_{32} & z_{33} & z_{34} & z_{35} & z_{36} \\ z_{41} & z_{42} & z_{43} & z_{44} & z_{45} & z_{46} \\ z_{51} & z_{52} & z_{53} & z_{54} & z_{55} & z_{56} \end{bmatrix}$$

A *vector* is a matrix with only one row or one column. It is denoted by a lower case bold face letter, for example **a, b, c**, while an individual vector element is given by the symbol $a_i$, $b_i$ or $c_i$. Vectors can be either *row vectors* (one row, several columns) or *column vectors* (one column, several rows). For example, **a** is a row vector and **b** is a column vector in the following:

$$a = [1 \quad 2 \quad 3], \qquad b = \begin{bmatrix} 6 \\ 7 \\ 8 \end{bmatrix}$$

A matrix is sometimes considered to be a set of row or column vectors. The matrix Z used above could be thought of as a set of row vectors $z_1$, $z_2$, $z_3$, ..., $z_5$ (with, for example, $z_1 = z_{11}, z_{12}, \ldots, z_{16}$) or a set of column vectors $z_1$, $z_2$, $z_3$, ..., $z_6$ with, for example $z_6 = z_{16}$, $z_{26}, \ldots, z_{56}$. This definition of a matrix is useful when geometrical models of relationships among the elements of matrices and vectors are being developed (Section 1.3).

Matrices having the same dimensions can be *added* and *subtracted* by addition or subtraction of corresponding elements. If $P = A - B$ then $p_{ij} = a_{ij} - b_{ij}$, assuming that A and B have the same dimensions. P will have the same dimensions as A and B. Similarly, if $P = C + Q$ then $p_{ij} = c_{ij} + q_{ij}$. Note that $C + Q = Q + C$.

*Example.* If $A = \begin{bmatrix} 1 & 2 \\ 3 & 4 \end{bmatrix}$ and $B = \begin{bmatrix} 2 & 3 \\ 4 & 5 \end{bmatrix}$, find

(i) C, if $C = A + B$;
(ii) G, if $G = A - B$.

Since $c_{ij} = a_{ij} + b_{ij}$ then, for values of $i$ and $j$ running from 1 to 2 (the number of rows and columns),

$$C = \begin{bmatrix} 3 & 5 \\ 7 & 9 \end{bmatrix} \quad \text{and} \quad G = \begin{bmatrix} -1 & -1 \\ -1 & -1 \end{bmatrix}.$$

The operations of addition and subtraction are thus *defined* only when the matrices involved have the same dimensions.

*Multiplication* of matrices is not so obvious, but is equally simple. As an example, let us take the matrix equation

$$L = JK$$

L is the product matrix, J the 'left-hand matrix' and K the 'right-hand matrix'. In matrix terms K is *premultiplied* by J or alternatively, J is *postmultiplied* by K. Note that multiplication of two matrices is defined only when the right-hand matrix K has the same number of rows as the left-hand matrix J has columns. This can be checked by writing down the order of the two matrices; if K has $m$ rows and $n$ columns and J has $p$ rows and $m$ columns then write down $JK = (p \times m)(m \times n)$. The innermost dimensions match, so the operation is defined. The resulting product matrix L will have $p$ rows and $n$

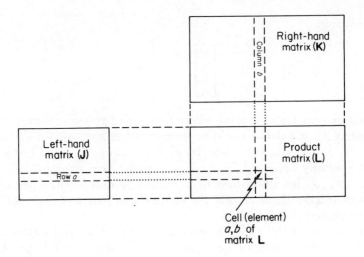

Figure 1.1. Illustrating matrix multiplication (See text for explanation)

columns — the outer dimensions of the left-hand and right-hand matrices.

The idea of matrix multiplication is best explained by reference to Figure 1.1. This shows that the product **L** has the same number of rows as the left-hand matrix and the same number of columns as the right-hand matrix. The element $l_{ab}$ is found by summing the products of the corresponding elements in row $a$ of the left-hand matrix and column $b$ of the right hand matrix. There must, of course, be the same number of elements in the row and the column. For example, let

$$\mathbf{J} = \begin{bmatrix} 1 & 2 & 3 \\ 3 & 2 & 1 \\ 2 & 1 & 3 \end{bmatrix} \quad \text{and} \quad \mathbf{K} = \begin{bmatrix} 1 & 3 & 4 & 7 \\ 2 & 3 & 1 & 6 \\ 3 & 2 & 0 & 5 \end{bmatrix}$$

The matrix **L** (= **JK**) will thus be of order (3 × 4). Element $l_{32}$ is found by multiplying corresponding elements of row 3 of **J** and column 2 of **K** to give $l_{32} = (2 \times 3) + (1 \times 3) + (3 \times 2) = 6 + 3 + 6 = 15$. Element $l_{14}$ is similarly calculated from $l_{14} = (1 \times 7) + (2 \times 6) + (3 \times 5) = 34$. The reader is left to complete the multiplication as an exercise.

The 'row times column' rule explained in the last paragraph is also applicable to the multiplication of two vectors, or to the multiplication of a matrix by a vector. Before such operations are possible the requirement that the left-hand matrix has as many columns as the right-hand matrix has rows must be satisfied.

*Examples.*
(i) Find **m** = **aZ** where

$$\mathbf{a} = [1 \quad 2 \quad 3] \quad \text{and} \quad \mathbf{Z} = \begin{bmatrix} 4 & 3 \\ 2 & 1 \\ 6 & 4 \end{bmatrix}$$

Using the layout of Figure 1.1 as a guide, obtain:

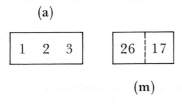

(ii) Find $s = \mathbf{ab}$ where

$$\mathbf{a} = [1 \quad 2 \quad 3], \quad \text{and} \quad \mathbf{b} = \begin{bmatrix} 4 \\ 2 \\ 3 \end{bmatrix}$$

Again using Figure 1.1 as a guide, we find

$$\begin{array}{c} \boxed{\begin{array}{c} 4 \\ 2 \\ 3 \end{array}} \ (\mathbf{b}) \\ \\ (\mathbf{a}) \\ \boxed{1 \quad 2 \quad 3} \qquad \boxed{17} \ (s) \end{array}$$

The reader should verify that the operations **m** = **Za** and $s = \mathbf{ba}$ are not defined, where **Z** and **a** are given in Example (i) and **b** and **a** in Example (ii).

Multiplication of a matrix or a vector by a scalar quantity is straightforward, involving only the multiplication of every matrix (or vector) element by the scalar. Thus, if

**B** = 6.2**A**

where

$$A = \begin{bmatrix} 1 & 4 \\ 2 & 5 \\ 3 & 6 \end{bmatrix} \quad \text{then} \quad B = \begin{bmatrix} 6.2 & 24.8 \\ 12.4 & 31.0 \\ 18.6 & 37.2 \end{bmatrix}$$

A few points concerning the differences between matrix multiplication and ordinary (scalar) multiplication should be noted. (i): The operation $L = JK$ is defined only if the number of rows in $K$ and columns in $J$ are equal. This means that if $L = JK$ then $P = KJ$ is not defined unless $K$ and $J$ have the same number of rows and columns; (ii): Even if both $L = JK$ and $P = KJ$ are defined, the matrices $L$ and $P$ will not be equal, except in special cases; (iii): The following rules carry over from scalar multiplication to matrix multiplication:

$$C(A + B) = CA + CB$$
$$(B + C)D = BD + CD$$
$$P^2 = PP$$

providing, of course, that the operations specified are defined.

## 1.2 Further definitions

A *square* matrix has the same number of rows and columns. This number is the *order* of the matrix. The *principal diagonal* of a square matrix contains the elements of the matrix having the same row and column identifiers. If $A$ is square then $a_{11}, a_{22}, \ldots, a_{nn}$ are the elements on the principal diagonal, given that $n$ is the order of $A$. If all the elements of the principal diagonal equal unity, and all other elements are zero, then the matrix is equal to the *identity matrix*, written $I$. (Note: two matrices are equal only if corresponding elements are equal.) The *null matrix* $O$ has all elements equal to zero. A *diagonal matrix* has all elements equal to zero except those on the principal diagonal, some or all of which are nonzero.

*Examples*

$$I = \begin{bmatrix} 1 & 0 & 0 \\ 0 & 1 & 0 \\ 0 & 0 & 1 \end{bmatrix} \quad \text{(identity matrix of order 3)}$$

$$O = \begin{bmatrix} 0 & 0 & 0 & 0 \\ 0 & 0 & 0 & 0 \\ 0 & 0 & 0 & 0 \\ 0 & 0 & 0 & 0 \end{bmatrix} \quad D = \begin{bmatrix} 4 & 0 & 0 & 0 \\ 0 & 3 & 0 & 0 \\ 0 & 0 & 0 & 0 \\ 0 & 0 & 0 & 2 \end{bmatrix}$$

(null matrix of order 4) (diagonal matrix of order 4)

If a matrix **A** is premultiplied by a diagonal matrix **D** the product matrix **C** is formed by multiplying the $i^{th}$ row of **A** by the $(i,i)^{th}$ element of **D**, i.e. $c_{ij} = d_{ii} \times a_{ij}$. Postmultiplication of **A** by **D** involves multiplying the $i^{th}$ column of **A** by the $i^{th}$ diagonal element of **D**, i.e. $c_{ji} = d_{ii} \times a_{ji}$. The $i^{th}$ diagonal element of a diagonal matrix **D** is frequently written $d_i$ rather than $d_{ii}$. It should be apparent from the context that $d_i$ is an element of a diagonal matrix **D**, not an element of a vector **d**.

*Example.* Find

$$\mathbf{C} = \mathbf{DA}$$

where

$$\mathbf{A} = \begin{bmatrix} 1 & 2 & 3 \\ 2 & 1 & 2 \\ 3 & 3 & 1 \end{bmatrix} \quad \text{and} \quad \mathbf{D} = \begin{bmatrix} 1 & 0 & 0 \\ 0 & 2 & 0 \\ 0 & 0 & 3 \end{bmatrix}$$

Using the rule just formulated, rather than the more cumbersome layout of Figure 1.1, we find that:

$$\mathbf{C} = \begin{bmatrix} 1 & 2 & 3 \\ 4 & 2 & 4 \\ 9 & 9 & 3 \end{bmatrix}$$

The reader should calculate **C** = **AD** as an exercise.

The *transpose* of a matrix **A** is the 'mirror image' of **A**, and is denoted by a superscript prime, **A**′, or a superscript $^T$, for example $\mathbf{A}^T$. The rows of **A** are the columns of **A**′ hence if **A** is a $(p \times q)$ matrix then **A**′ is a $(q \times p)$ matrix.

*Example*

If

$$\mathbf{A} = \begin{bmatrix} 1 & 2 & 3 & 4 & 5 \\ 6 & 7 & 8 & 9 & 8 \\ 7 & 6 & 5 & 4 & 3 \end{bmatrix}, \quad \text{then} \quad \mathbf{A}' = \begin{bmatrix} 1 & 6 & 7 \\ 2 & 7 & 6 \\ 3 & 8 & 5 \\ 4 & 9 & 4 \\ 5 & 8 & 3 \end{bmatrix}$$

A *symmetric matrix* is a square matrix such that the $(i,j)^{th}$ and the $(j,i)^{th}$ elements are equal, that is, the $j^{th}$ element of row $i$ equals the $i^{th}$ element of row $j$. The correlation matrix which is usually denoted by the letter **R** is an example of a symmetric matrix.

$$\mathbf{R} = \begin{bmatrix} 1.0 & 0.6 & 0.1 & 0.5 & 0.4 \\ 0.6 & 1.0 & 0.2 & 0.3 & 0.2 \\ 0.1 & 0.2 & 1.0 & 0.0 & 0.8 \\ 0.5 & 0.3 & 0.0 & 1.0 & 0.5 \\ 0.4 & 0.2 & 0.8 & 0.5 & 1.0 \end{bmatrix}$$

A symmetric matrix and its transpose are equal, i.e. $\mathbf{R} = \mathbf{R}'$. The premultiplication of a matrix by its transpose always gives a symmetric matrix, even if the original matrix is not symmetric, i.e. $\mathbf{C} = \mathbf{A}'\mathbf{A}$ is always a symmetric matrix.

A square matrix with all subdiagonal elements (i.e. elements below the principal diagonal) equal to zero is *upper triangular*. If all elements above the principal diagonal are zero the matrix is *lower triangular*.

$$\begin{bmatrix} 1 & 0 & 0 & 0 \\ 2 & 5 & 0 & 0 \\ 3 & 6 & 8 & 0 \\ 4 & 7 & 9 & 10 \end{bmatrix} \qquad \begin{bmatrix} 1 & 2 & 3 & 4 \\ 0 & 5 & 6 & 7 \\ 0 & 0 & 8 & 9 \\ 0 & 0 & 0 & 10 \end{bmatrix}$$
$$\qquad\quad\text{(a)} \qquad\qquad\qquad\quad \text{(b)}$$

Figure 1.2. Lower (a) and upper (b) triangular matrices.

## 1.3 Vectors

The concept of an algebraic vector was introduced in Section 1.1 where it was defined as a matrix with one row (a row vector) or one column (a column vector). In geometrical terms the set of numbers forming the algebraic vector can be thought of as the coordinates of point P. Geographers are accustomed to thinking of coordinates in terms of a two dimensional space (e.g. map references) or even three-dimensional spaces (location coordinates plus height above Ordnance datum, for example). Thus (6.0, 7.0, 100.0) could fix the position of a point 6 units east, 7 units north and 100 units above a reference point, for example, the origin of the coordinate system. There is no mathematical reason why these ideas should not be extended to 4, 5 or higher dimensional spaces. It is difficult, if not impossible, to visualise a space of more than 3 dimensions, but this should not prevent the acceptance of the concept. Thus, a vector $\mathbf{x} = \{x_1, x_2, x_3, \ldots, x_N\}$ gives the coordinates of a point in an $N$-dimensional space. If the basic axioms of Euclidean geometry are assumed to hold in this $N$-space (and this is usually the case) then the coordinate system is expressed in terms of a set of $N$ axes, all mutually perpendicular, and all passing through the origin $(0,0,0,\ldots,0)$.

The line joining $\mathbf{x}$ to the origin is a geometrical vector, with the properties of magnitude (length) and direction (orientation). Thus, an algebraic vector can be thought of either as representing the position of a point in an $N$-space or else as a line joining that point to the origin of the coordinate system. If we consider the corresponding geometrical vector as a line, then operations on the (algebraic) vector such as addition and multiplication have geometrical meaning also. Thus, multiplication of the algebraic vector x by the scalar $c$ gives:

$$\mathbf{y} = c\mathbf{x} = \{cx_1, cx_2, cx_3, \ldots, cx_N\}$$

while addition of the vectors x and y gives:

$$\mathbf{z} = \mathbf{x} + \mathbf{y} = \{x_1 + y_1, x_2 + y_2, \ldots, x_N + y_N\}$$

assuming these operations to be defined. In geometrical terms, multiplication of the vector x by the scalar $c$ is equivalent to expanding or contracting the corresponding geometrical vector by the factor $c$. If, for example, $c = 0.5$ and $\mathbf{x} = \{1,1\}$ then $\mathbf{y} = c\mathbf{x} = \{0.5, 0.5\}$ (Figure 1.3). The addition of two algebraic vectors can also be shown geometrically. If the vector $\mathbf{x} = \{2,3\}$ gives the position of the point $P_1$, and if $\mathbf{y} = \{1,2\}$ gives the position of the point $P_2$ then the position of $P_3$ is given by $\mathbf{z} = \mathbf{x} + \mathbf{y} = \{3,5\}$ (Figure 1.4).

Using the two concepts of vector addition and multiplication of a

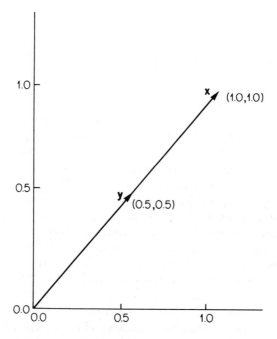

Figure 1.3. Showing multiplication of a vector by a scalar

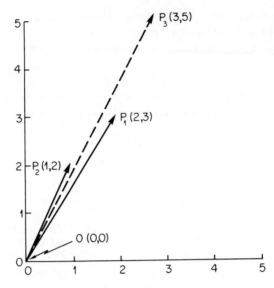

Figure 1.4. Showing vector addition

vector by a scalar a third point can be derived from two given points. In Figure 1.4 the point $P_3(3,5)$ was derived from $P_1(2,3)$ and $P_2(1,2)$ by the formula

$$P_3 = P_1 + P_2$$

In more general terms, given $m$ points we can derive any number of extra points $P_{m+1}, P_{m+2}, \ldots, P_{m+q}$ from a formula of the type

$$P_{m+i} = c_1 P_1 + c_2 P_2 + c_3 P_3 + \ldots + c_m P_m$$

where the $c$'s are scalar multipliers and the P's are the coordinates of the points $P_1$, $P_2$, etc. Such linear combinations ($P_{m+1}, P_{m+2}, \ldots, P_{m+q}$) are *linearly dependent* upon $P_1, P_2, \ldots, P_m$. Conversely, if the coordinates of any one set of points cannot be directly computed from the coordinates of the remaining members of the set then that set of points is *linearly independent*. The number of linearly independent vectors in a set of vectors gives the maximum number of dimensions of the vector space required to contain those vectors. Thus, if the set of vectors $a_1, a_2, a_3, \ldots, a_{10}$ is given and we determine that $a_8$, $a_9$ and $a_{10}$ are linearly dependent on $a_1, a_2, \ldots, a_7$ then the ten vectors lie in a space of not more than seven dimensions. If these ten vectors formed the rows (or columns) of a matrix **A** then that matrix would have *rank* seven. In other words, the rank of a matrix is the number of linearly independent rows or columns it contains. The concept of rank is used in factor analysis (Chapter 5) where, in geometrical terms, it is found necessary to adjust the vectors $a_1, a_2, \ldots, a_n$ so that they lie in a space having the lowest possible dimension, the coordinate system representing the factor axes. This is equivalent to reducing the rank of $A = a_1, a_2, \ldots, a_n$ to a minimum.

In principal components analysis, factor analysis and regression analysis one of the important considerations is whether a given set of points, defined by vectors of observations on a set of variables, is linearly independent. In fact, one aim of principal components analysis is to determine the number of linearly independent vectors in a given set of vectors (Chapter 4).

All the familiar theorems of Euclidean geometry hold in a Euclidean space, no matter what the dimensionality of that space might be. This is of great importance in cluster analysis (Chapter 6) where the distance between two points in an $N$-space is often taken to be a measure of the similarity (or dissimilarity) of the objects they represent, the idea being that if two points are close together then they are relatively similar in terms of the characteristics that the coordinates represent. In two dimensions, Pythagoras's theorem states that $d_{xy}^2 = (x_1 - y_1)^2 + (x_2 - y_2)^2$, that is, the distance $d_{xy}$ between the points defined by vectors $\mathbf{x} = (x_1, x_2)$ and $\mathbf{y} = (y_1, y_2)$ is the square root of the sum of the squared differences between corresponding elements of $\mathbf{x}$ and $\mathbf{y}$. If $\mathbf{x}$ and $\mathbf{y}$ define the same point (i.e. $\mathbf{x}$ and $\mathbf{y}$ are *collinear*) then $d_{xy}$ is the *length* of the geometric vector joining that point and the origin. Note that this formula applies only when their coordinate axes are mutually perpendicular (Figure 1.5). If the coordinate axes are not perpendicular but are separated by an angle $\theta$ that is less than 90° the formula must be modified to take account of this (Figure 1.6). The problem of determining $\cos \theta$ is discussed later in this section.

Measures of similarity other than the Pythagorean (Euclidean) distance can be used in multivariate studies, particularly in classification methods (Chapter 6). Such measures can have a Euclidean representation if they are *metric* measures, that is, if the following conditions

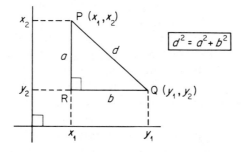

Figure 1.5. Distance between points P and Q in terms of rectangular coordinates

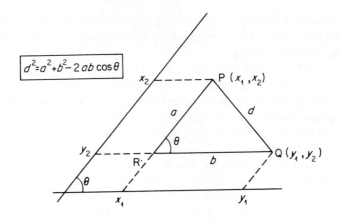

Figure 1.6. Distance between points P and Q in terms of oblique coordinates

hold:

(I) $d_{pp} = 0$
(II) $d_{pq} > 0$ if $q \neq p$
(III) $d_{pq} = d_{qp}$
(IV) $d_{pq} + d_{qr} \geq d_{pr}$

If conditions (I)–(III) hold then the measure is *semi-metric*. If condition (IV) is written as

$$d_{pr} \leq \max(d_{pq}, d_{rq})$$

then the measure is *ultrametric*. The notation max $(x,y)$ means: the larger of the quantities inside the brackets.

The distance $d_{pq}$ between points P and Q is not dependent upon the position of the coordinate axes. One problem that is met with in factor analysis (Chapter 5) is that of *rotating* or *transforming* the coordinate axes of a Euclidean space to new positions in accordance with some specified criterion. This problem can be considered geometrically as in Figure 1.7 in which the distance between points P and Q is seen to be invariant under rotation, that is, the distance PQ is unaltered by the clockwise movement of the axes through an angle $\theta$.

The concept of geometrical representation can be extended to include matrices as well as vectors simply by considering a matrix as a set of column or row vectors. For example, a set of $n$ observations on each of $p$ variables can be represented in the form of $p$ column vectors

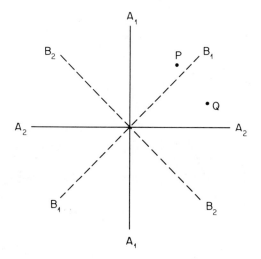

Figure 1.7. Rotation of axes $A_1$ and $A_2$ through angle $\theta$. The positions of the points P and Q are unaffected by the rotation

by the data matrix $X = x_1, x_2, \ldots, x_p$:

$$X = \begin{bmatrix} x_{11} & x_{12} & \cdots & x_{1p} \\ x_{21} & x_{22} & \cdots & x_{2p} \\ \cdots & \cdots & \cdots & \cdots \\ x_{n1} & x_{n2} & \cdots & x_{np} \end{bmatrix}$$

The $p$ *column* vectors $(x_{11}, x_{21}, \ldots x_{n1})$, $(x_{12}, x_{22}, \ldots x_{n2}) \ldots$ can be considered to hold the rectangular cartesian coordinates of $p$ points in a space of $n$ dimensions. Each point represents a variable. The distance of any point $P_j$ from the origin is the length of the (geometric) vector joining the point $P_j$ to he origin. This, as we saw above, is equal to the distance from $P_j = (x_{1j}, x_{2j}, \ldots, x_{nj})$ to the origin $(0, 0, 0, \ldots, 0)$. The distance formula for rectangular coordinates gives $d_{j0}^2 = (x_{1j} - 0)^2 + (x_{2j} - 0)^2 + \ldots + (x_{nj} - 0)^2 = \Sigma_{i=1}^{n} x_{ij}^2$. The variance of a variable $x_j$ is defined as

$$\text{var}(x_j) = \sum_{i=1}^{n} x_{ij}^2 / n$$

assuming that the $x_{ij}$ are in deviation form, hence $d_{j0}^2$ is proportional to the variance of $x_j$. If we take square roots then it is apparent that the distance $d_{j0}$, which is the length of vector $x_j$, is proportional to the standard deviation of $x_j$. The proportionality factor is seen to be $\sqrt{1/n}$.

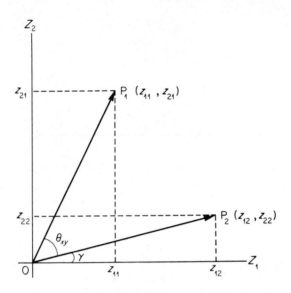

Figure 1.8. Vector representation of two variables in a two-dimensional space

Standardization of X to give each column a mean of zero and unit standard deviation is thus seen to be equivalent to moving the points $P_j (j = 1, p)$ inwards or outwards until the vectors joining the $P_j$ to the origin are all of unit length. The standardized matrix of observed data is usually written Z. The simplest case worth considering $(p = n = 2)$ is shown in Figure 1.8. The angle $\theta_{xy}$ between the vectors $OP_1$ and $OP_2$ can be derived as follows. Assume that the coordinates of $P_1$ are held in the algebraic vector x and those of $P_2$ in the algebraic vector y, that is, $\mathbf{x} = (z_{11}, z_{21})$ and $\mathbf{y} = (z_{12}, z_{22})$. The angle made by the intersection of a vector and a coordinate axis (e.g. the angle $\gamma$ in Figure 1.8) is called the direction angle of the vector. The cosine of this angle is the direction cosine of the vector with respect to the axis. The set of direction cosines $(\lambda_{1i}, \lambda_{2i}, \ldots, \lambda_{ni})$ for all axes are computed from

$$\lambda_i = x_i / L(\mathbf{x}) \tag{1.1}$$

where $x_i$ is the coordinate of vector x on axis $i$ and $L(\mathbf{x})$ is the length of the vector x. Thus, the direction cosines for vector $OP_1$ are $z_{11}/L(OP_1)$ and $z_{21}/L(OP_1)$. The cosine of the angle separating two vectors is the sum of products of corresponding elements of their direction cosines, that is,

$$\cos \theta_{xy} = \sum_{j=1}^{n} \lambda_{jx} \lambda_{jy} \tag{1.2}$$

where $\lambda_x$ is the set of direction cosines for the vector $\mathbf{x}$ (which, in the example, contains the coordinates of $P_1$) and $\lambda_y$ the set of direction cosines for the vector $\mathbf{y}$. Examination of (1.2) shows that it could be written as

$$\cos \theta_{xy} = \frac{\sum_{i=1}^{n} x_i y_i}{L(\mathbf{x})L(\mathbf{y})} \quad (1.3)$$

By definition, $1/\sqrt{n}$ times the length of a geometric vector representing a standardized variable is proportional to the standard deviation of that variable, which is 1.0. For standardized variables only, (1.3) reduces to

$$\cos \theta_{xy} = \sum_{i=1}^{n} x_i y_i / n \quad (1.4)$$

One version of the formula for the correlation between variables $x$ and $y$ is the same as formula (1.4). Thus, an important point is made: the cosine of the angle between two standardized vectors is equal to the product moment correlation coefficient relating the associated variables.

If the angle between all possible pairs of a set of vectors is 90°, the vectors are said to form an *orthogonal set*. If $\mathbf{x}_1, \mathbf{x}_2, \ldots, \mathbf{x}_p$ are an orthogonal set then the matrix $\mathbf{X}$, whose $p$ columns are formed by these vectors, is an *orthogonal matrix*. A set of vectors which, in addition to possessing the property of orthogonality, are also all of unit length is said to be an *orthonormal set*. The normalized eigenvectors of a symmetric positive definite matrix (Section 1.5) form an orthonormal set. (To *normalize* a vector, divide each element of the vector by the vector length.)

## 1.4 The Inverse Matrix

The operations of addition, subtraction and multiplication exist in matrix algebra, as in scalar algebra. Division, however, does not — a matrix cannot be divided by a second matrix. We can achieve the same result by multiplying the matrix by the 'reciprocal' of the second matrix just as, in scalar algebra, $x/y = x(1/y)$. The 'reciprocal' of a matrix is termed its *inverse*, which is denoted by a superscript $-1$. Thus, the inverse of a matrix $\mathbf{A}$ is written $\mathbf{A}^{-1}$. *Matrices that are not square cannot have a true inverse.* Not all square matrices possess inverses, just as some numbers do not have a calculable reciprocal (e.g. 0). A non-invertible square matrix is *singular*. In the same way that a scalar times its reciprocal equals 1, the product of a matrix and its inverse is $\mathbf{I}$, the identity matrix. ($\mathbf{AA}^{-1} = \mathbf{I} = \mathbf{A}^{-1}\mathbf{A}$).

Matrix inversion is used in multiple regression to solve the system of linear equations for $\hat{\boldsymbol{\beta}}$, the vector of estimated regression coefficients.

Assuming that there are three explanatory variables, the equations can be written:

$$\begin{aligned} p_{11}\hat{\beta}_1 + p_{12}\hat{\beta}_2 + p_{13}\hat{\beta}_3 &= h_1 \\ p_{21}\hat{\beta}_1 + p_{22}\hat{\beta}_2 + p_{23}\hat{\beta}_3 &= h_2 \\ p_{31}\hat{\beta}_1 + p_{32}\hat{\beta}_2 + p_{33}\hat{\beta}_3 &= h_3 \end{aligned} \quad (1.5)$$

where **P** is the (3 x 3) matrix of corrected sums of squares and cross-products of the explanatory variables, **h** the (3 x 1) column vector of cross-products of the dependent variable and the explanatory variables, and $\hat{\boldsymbol{\beta}}$ the (3 x 1) column vector of estimated partial regression coefficients. These equations can be written much more concisely in matrix notation as:

$$\mathbf{P}\hat{\boldsymbol{\beta}} = \mathbf{h} \quad (1.6)$$

The reader should verify this by applying the 'row times column' rule of matrix multiplication (Section 1.1 and Figure 1.1).

Premultiplying each side of Equation (1.6) by $\mathbf{P}^{-1}$ gives

$$\mathbf{P}^{-1}\mathbf{P}\hat{\boldsymbol{\beta}} = \mathbf{P}^{-1}\mathbf{h} \quad (1.7)$$

or, since $\mathbf{P}^{-1}\mathbf{P} = \mathbf{I}$ and $\mathbf{I}\hat{\boldsymbol{\beta}} = \hat{\boldsymbol{\beta}}$, Equation (3) reduces to:

$$\hat{\boldsymbol{\beta}} = \mathbf{P}^{-1}\mathbf{h} \quad (1.8)$$

The estimated partial regression coefficients can be obtained by postmultiplying the inverse of the corrected sums of squares and cross-products matrix of the explanatory variables by the vector of cross-products of the dependent variable and the explanatory variables.

Unfortunately, such well-behaved matrices often exist only in theory. It has already been noted that some square matrices do not possess inverses. Other square matrices may be almost singular, that is, their inverses can only be determined accurately if the arithmetic is carried out to a large number of significant digits. The situation can perhaps be clarified by analogy with the problem of determining the exact point of intersection of a pair of straight lines. In Figure 1.9(a) no such point exists since the two lines are parallel. This situation corresponds to the singular matrix. If the lines are almost perpendicular to each other, as in Figure 1.9(b), the point of intersection can be found quite easily. This corresponds to the well-conditioned matrix. In Figure 1.9(c) the point of intersection of the two lines is difficult to determine exactly, and a small increase or decrease in the scale at which the graph is viewed will lead to relatively large changes in the estimated position of the point of intersection. This situation corresponds to the ill-conditioned matrix.

The first feature of an ill-conditioned matrix is the difficulty

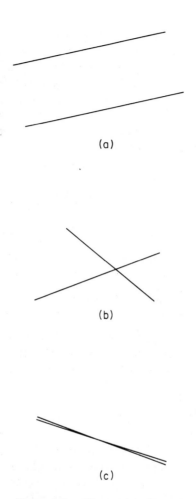

Figure 1.9. Illustrating the concept of matrix condition. See text for explanation

experienced in computing its inverse accurately using floating-point arithmetic and finite word lengths. The second characteristic of an ill-conditioned matrix relates to its use in the solution of a set of simultaneous linear equations, such as those shown in Equation (1.6):

$$P\hat{\beta} = h \qquad (1.6)(bis)$$

The matrix of coefficients in (1.6) is **P**, which — in the context of this equation — is the matrix of sums of squares and cross-products among the explanatory variables in a multiple regression model. If **P** is ill-conditioned, then it is likely that the elements of the estimated regression coefficients vector will change considerably if the elements of the vector **h** of right-hand sides are altered only slightly. This means that the results obtained from the analysis of two separate samples each having exactly the same coefficients matrix **P** but having slightly different right-hand side vectors **h** will differ, perhaps to a considerable extent. Differences in the elements of **h** could result from inaccuracies in recording and measuring the observations, or in roundoff error during computation (see Section 1.7). Fox and Mayers (1968, p. 79) demonstrate the effects of roundoff error on the accuracy of the solution by computing the solution of the same set of simultaneous equations (i) exactly; (ii) rounded to 8 decimals, and (iii) rounded to 4 decimals. Their results were:

|   | (i) | (ii) | (iii) |
|---|---|---|---|
| 1 | 30 | 29.65 | −8.8 |
| 2 | −420 | −416.02 | 90.5 |
| 3 | 1680 | 1666.62 | −196.0 |
| 4 | −2520 | −2502.69 | 59.8 |
| 5 | 1260 | 1253.39 | 75.6 |

Such startling changes would not normally occur with experimental data; this example is included to demonstrate the fact that computational errors can be of great relative magnitude compared to the true value of the solution.

Fox and Mayers term these features of the ill-conditioned matrix *inherent instability*. *Induced instability* is produced by the computing algorithm used. Some algorithms are less reliable than others, and the manner in which the algorithm is programmed can also add to the instability. Of the algorithms for matrix inversion, the best-known is the *Gaussian elimination* algorithm. It is not very reliable when the coefficients matrix is ill-conditioned, as demonstrated in the context of least-squares computations by Longley (1967) and Wampler (1970). To reduce the errors involved in Gaussian elimination, modifications have been made to the method, involving the interchange of the rows and columns of the original matrix to ensure that the calculations are carried out in the most efficient order in terms of roundoff error. This modification is called the *Gauss–Jordan algorithm*, and it is reasonably reliable for most problems. It is described by Bauer and Reinsch in Wilkinson and Reinsch (eds.) (1971); their algorithm GJDEF1 is included in Appendix A.

Other methods of matrix inversion are discussed by the contributors to Wilkinson and Reinsch (eds.) (1971). In his introduction to the volume, Wilkinson suggests that the two most useful algorithms are the Gauss–Jordan algorithm, which is mentioned above, and the *Cholesky method*. This method is slightly more sophisticated than the Gauss–Jordan algorithm, as it involves the factorisation of the matrix **A**, whose inverse is required, into a lower triangular matrix **L** and an upper triangular matrix **L'**. The method can be subjected to a process of iterative refinement to provide a more accurate inverse, if this is required. This is done in subroutine ACCINV which is included in Appendix A. It must be remembered however that if the original data is unreliable, all the computational sophistication in the world will not make up for this fact. Care should be taken, though, with good data to ensure that they are not polluted by the computational methods used in their analysis.

Not all matrices require an elaborate technique such as the Gauss–Jordan or Cholesky methods mentioned above. Diagonal matrices are inverted by replacing the elements of the principal diagonal with the reciprocals of those elements. For example,

$$\mathbf{D} = \begin{bmatrix} 1 & 0 & 0 \\ 0 & 2 & 0 \\ 0 & 0 & 3 \end{bmatrix}, \quad \mathbf{D}^{-1} = \begin{bmatrix} 1 & 0 & 0 \\ 0 & \tfrac{1}{2} & 0 \\ 0 & 0 & \tfrac{1}{2} \end{bmatrix}$$

$$\mathbf{G} = \begin{bmatrix} 7 & 0 & 0 & 0 \\ 0 & 6 & 0 & 0 \\ 0 & 0 & 5 & 0 \\ 0 & 0 & 0 & 4 \end{bmatrix}, \quad \mathbf{G}^{-1} = \begin{bmatrix} \tfrac{1}{7} & 0 & 0 & 0 \\ 0 & \tfrac{1}{6} & 0 & 0 \\ 0 & 0 & \tfrac{1}{5} & 0 \\ 0 & 0 & 0 & \tfrac{1}{4} \end{bmatrix}$$

So far, the solution of the linear simultaneous equations $\mathbf{P}\hat{\boldsymbol{\beta}} = \mathbf{h}$ has been described in terms of matrix inversion, getting $\hat{\boldsymbol{\beta}} = \mathbf{P}^{-1}\mathbf{h}$ and then finding $\hat{\boldsymbol{\beta}}$ by postmultiplying $\mathbf{P}^{-1}$ by $\mathbf{h}$. This is very inefficient if we only require the vector $\hat{\boldsymbol{\beta}}$. Methods of equation solution, based on the Gauss–Jordan or Cholesky algorithms, can be used to determine $\hat{\boldsymbol{\beta}}$ much more quickly than by the matrix inversion and multiplication method outlined above. In most applications considered in later chapters we are, however, interested either in an inverse matrix itself, or in quantities derived from the inverse matrix. Thus, in some kinds of factor analysis (Chapter 5) the most popular estimate of communality is the square of the mutiple correlation coefficient (SMC) of each variable with the remaining variables. If **R** is the correlation matrix, and if $r^{ii}$ represents the $i^{\text{th}}$ diagonal element of $\mathbf{R}^{-1}$ then $\text{SMC}_i = 1.0/r^{ii}$. In multiple regression analysis the solution to

$$\hat{\boldsymbol{\beta}}_{(s)} = \mathbf{R}^{-1}\mathbf{g}$$

is required where $\hat{\boldsymbol{\beta}}_{(s)}$ is the vector of standardized regression coefficients, **R** is the correlation matrix for the explanatory variables, and **g** is the vector of correlations between the dependent and explanatory variables. If we require the vector $\hat{\boldsymbol{\beta}}_{(s)}$ of standardized regression coefficients alone, then the method using matrix inversion is less efficient than equation-solving algorithms such as that used in subroutine SOLVE (Appendix A). However, the elements of $\mathbf{R}^{-1}$ are usually required in other computations in regression (for example, in the determination of standard errors) so inversion techniques are usually preferred. The method of orthogonal polynomials (Chapter 3) can reduce some of the problems associated with badly-conditioned matrices.

A quantity associated with the inverse matrix is the *determinant* of a matrix. The determinant of matrix $\mathbf{A}$ is written $\det(\mathbf{A})$ or $|\mathbf{A}|$. The determinant is a scalar ranging from 0 (in the case of a singular matrix) to an unbounded upper limit. The determinant of the correlation matrix $\mathbf{R}$ is, however, constrained to lie in the range $0 \leq \det(\mathbf{R}) \leq 1$. There are several ways of computing determinants; some methods are completely impracticable for any but the smallest matrices. The method used in subroutine DT2 (Appendix A) is based on the Cholesky factorization described above in connection with the inverse matrix. The determinant can be used as a measure of the condition of a matrix. The topic of matrix condition is discussed above. A matrix with a near-zero determinant is badly conditioned, and computational error can be expected in results based on its inverse. A matrix with a large determinant (near 1 in the case of $\mathbf{R}$) is well conditioned. Other uses of determinants are in the statistical tests associated with multivariate analysis of variance and multiple discriminant analysis (Chapter 7). Further discussion of the topic of matrix condition can be found in Section 3.2.

## 1.5 Eigenvalues and Eigenvectors

The eigenvalues and eigenvectors of a matrix are numerical quantities associated with, or derived from, that matrix. They are also known as the proper, latent, or characteristic roots and vectors of a matrix. Eigenvalues and eigenvectors can be given specific meaning in certain statistical techniques (for example, in principal components analysis (Chapter 4), factor analysis (Chapter 5), principal coordinates analysis (Chapter 6) and in multiple discriminant analysis (Chapter 7)). Other uses of eigenvalues and eigenvectors are described by Gould (1967), who also gives a geometrical explanation of these quantities. In algebraic terms, the eigenvalues of an $(n \times n)$ matrix $\mathbf{A}$ are held in a diagonal matrix $\Lambda$ with elements $\lambda_1, \lambda_2, \ldots, \lambda_n$. It is conventional to arrange the eigenvalues $\lambda_i$ in descending order. Associated with each eigenvalue is a column vector of eigenvectors, x; the $n$ column vectors make up the matrix of eigenvectors, $\mathbf{X}$. The eigenvalues and eigenvectors are related to $\mathbf{A}$ by the characteristic equation of $\mathbf{A}$, which is:

$$(\mathbf{A} - \lambda \mathbf{I})\mathbf{x} = 0 \tag{1.9}$$

or, alternatively

$$\mathbf{Ax} = \lambda \mathbf{x} \tag{1.10}$$

given that the elements of **x** are not all zero. In this equation $\lambda$ is a scalar which can take on any of the $n$ values of the solution of (1.9). This equation can be written out as a set of homogeneous linear equations

$$\left.\begin{array}{l}(a_{11} - \lambda)x_1 + a_{12}x_2 + \ldots + a_{1n}x_n = 0 \\ a_{21}x_1 + (a_{22} - \lambda)x_2 + \ldots + a_{2n}x_n = 0 \\ a_{n1}x_1 + a_{n2}x_2 + \ldots + (a_{nn} - \lambda)x_n = 0\end{array}\right\} \tag{1.11}$$

The equations (1.11) have a solution only if the determinant of the coefficients matrix is zero, that is, if

$$\det(\mathbf{A} - \lambda \mathbf{I}) = 0 \tag{1.12}$$

This determinant can be expanded to form an $n^{th}$ degree polynomial in $\lambda$, the $n$ roots of which are the eigenvalues of $A$. For example, if $n = 2$ and $A$ is of the conventional form

$$\mathbf{A} = \begin{bmatrix} a_{11} & a_{12} \\ a_{21} & a_{22} \end{bmatrix}$$

then (1.12) can be written as

$$(a_{11} - \lambda)(a_{22} - \lambda) - a_{12} \cdot a_{21} = 0$$

or:

$$a_{11} \cdot a_{22} - a_{22}\lambda + \lambda^2 - a_{11}\lambda - a_{12} \cdot a_{21} = 0$$

or:

$$\lambda^2 - (a_{22} + a_{11})\lambda - a_{12} \cdot a_{21} + a_{11} \cdot a_{22} = 0 \tag{1.13}$$

This is a second-degree polynomial in $\lambda$. Its solutions are given by the usual method of solving quadratic equations, that is, if $ax^2 + bx + c = 0$ then $x = (-b \pm (b^2 - 4ac))/2a$. A simple example will help to explain the procedure. If

$$\mathbf{A} = \begin{bmatrix} 1 & \tfrac{1}{2} \\ \tfrac{1}{2} & 1 \end{bmatrix}$$

then the characteristic equation of **A** can be written in accordance with (1.13) as:

$$\lambda^2 - 2\lambda - \tfrac{1}{4} + 1 = 0$$

that is,

$$\lambda^2 - 2\lambda + 3/4 = 0$$

The solutions of this equation are found by substituting $a = 1$, $b = -2$ and $c = 3/4$ into the formula for the roots of a quadratic equation; this formula is given above. It yields the values $\lambda = 1\frac{1}{2}$ or $\frac{1}{2}$. The eigenvalues of $A$ are therefore $1\frac{1}{2}$ and $\frac{1}{2}$. These eigenvalues are often considered as the diagonal elements of a matrix $\Lambda$ which in this case is:

$$\Lambda = \begin{bmatrix} 1\frac{1}{2} & 0 \\ 0 & \frac{1}{2} \end{bmatrix}$$

It is not possible to determine unique values of the elements of each eigenvector associated with these eigenvalues, but the ratios of the elements can be found using Equation (1.11). For the first eigenvalue $\lambda_1$ the ratio of the elements $x_{11}$ and $x_{12}$ of the associated eigenvector $\mathbf{x}_1$ are given by:

$$(1 - 1\frac{1}{2})x_{11} + \frac{1}{2}x_{12} = 0$$

$$\frac{1}{2}x_{11} + (1 - 1\frac{1}{2})x_{12} = 0$$

It is apparent from the first of these simultaneous equations that $x_{11} = x_{12}$ so that any pair of positive of negative real numbers would serve as the first eigenvector of $A$ so long as the relationship $x_{11} = x_{12}$ was satisfied. A similar procedure using $\lambda_2 = \frac{1}{2}$ gives the ratios $(x_{21}/x_{22}) = (-1/1)$. The fact that unique numerical values cannot be given to the elements of $\mathbf{x}_1$ and $\mathbf{x}_2$ is very important, for it implies that the arbitrary values given to these elements must be scaled in some way. The scaling process is termed the *normalization* of the eigenvectors, and it may take one of several forms, such as

(1) the division of each element of the arbitrary eigenvector by the largest element in the vector, so that the largest element of the normalized eigenvector has the value 1;

(2) The division of each element of the arbitrary eigenvector by the square root of the sum of the squares of the elements (that is, by the length of the arbitrary eigenvector) so that the length of the normalized eigenvector is 1. Such eigenvectors are said to be unit length or just unit eigenvectors;

(3) the division of each element of the arbitrary eigenvector by the first element of that vector so that the first element of the normalized vector is unity.

A unit length eigenvector can be further scaled so that its length is equal to any quantity $c$ simply by multiplying each element of the vector by $c$. The computer subroutines HTDQL and EIGVEC/TRBAKI in Appendix A return unit-length eigenvectors.

Although the method described above for the extraction of eigenvalues and eigenvectors could in theory be used for any matrix A, in practice it would be found to be inordinately slow. Several numerical methods which are much faster than the method of expanding the characteristic equation are available. The choice of method depends on whether A is symmetric. If it is, then the calculations are easier to perform. Nonsymmetric matrices do not necessarily have orthogonal eigenvectors, and their eigenvalues may be either real or complex. The nonsymmetric matrices that we shall encounter in Chapter 7 of this book are of the form $A = B^{-1}C$, which can be reduced to form such that the methods for symmetric matrices can be used, by the application of subroutines REDUC and REBAKA (Appendix A). We will therefore not give further consideration to the eigenproblem when the matrix is not symmetric.

The oldest method of finding the eigenvalues and eigenvectors of a symmetric matrix is due to Jacobi. It is a relatively simple procedure, and is accurate even when some of the eigenvalues are equal or nearly equal. (This is a major difficulty with some methods.) If the $(n \times n)$ matrix A has exactly $n$ nonzero eigenvalues then the 'JK' process — a modification by Kaiser (1973) of the Jacobi technique — can be used. The general Jacobi method is described by Kuo (1972).

Recent developments in numerical analysis have led to the introduction of a number of highly efficient methods of eigenvalue extraction. (Mulaik, 1972; Wilkinson and Reinsch, (eds.) 1971). If the size of the matrix exceeds approximately (10 x 10) then the QL algorithm is faster than the Jacobi method. The QL algorithm (programmed as subroutine HTDQL in Appendix A) can be modified to compute a selected number of eigenvalues (subroutines TRIDI and RATQR) and the corresponding eigenvectors (subroutines EIGVEC and TRBAK1). If more than 25 per cent of the eigenvalues and eigenvectors are required it is more efficient to calculate them all using HTDQL. Finally, if the matrix is of the form $A = B^{-1}C$ (as in multiple discriminant analysis, Chapter 7) then use subroutine REDUC to reduce the problem to standard form, compute the eigenvalues and eigenvectors of this reduced form using HTDQL or TRIDI/RATQR, then convert the eigenvectors of the reduced form into the eigenvectors of A by a call to subroutine REBAKA. Details of the calculations involved are beyond the scope of this chapter; readers wishing to pursue the topic should follow up the references given above.

Some definitions relating to, and properties of eigenvalues and eigenvectors will be briefly considered. A matrix with no zero or negative eigenvalues is termed *positive definite*. If all the eigenvalues of a matrix are nonnegative, that matrix is *positive semidefinite*. The terms *negative definite* and *negative semidefinite* can be defined analogously. The number of nonzero eigenvalues of a matrix is equal to the *rank* of that matrix, defined as the number of linearly independent rows (and columns) it possesses. The determinant of a matrix can be computed as the product of the eigenvalues. It follows that a matrix that has any zero eigenvalues is singular. The sum of the diagonal elements of a symmetric matrix **A**, termed the *trace* of **A** (written tr(**A**) or less frequently, Spur(**A**)) is equal to the sum of the eigenvalues. The *basic structure* of a symmetric matrix **A** is given by $\mathbf{A} = \mathbf{X\Lambda X'}$ where $\mathbf{\Lambda}$ is the diagonal matrix of eigenvalues of **A** and **X** the matrix of eigenvectors.

## 1.6 Error in Matrix Computations

The statistical and mathematical techniques that form the subject-matter of the remainder of this book frequently require the computation of a matrix of some kind from the observational data, or involve numerical manipulation of these matrices. The computer subroutines described in Appendix A have been written so as to minimize computational error. This is pointless if the matrix on which the particular operations are to be carried out has been computed inaccurately. As noted in the introduction to this chapter, many of the methods of computing these initial matrices, such as correlation matrices or variance–covariance matrices, are based upon the antiquated technology of the desk calculator. Good algorithms should be developed with the characteristics of the computer in mind. Pennington (1970) and Dorn and McCracken (1972) give very readable accounts of such errors, of which there are three main types, due to inaccuracy in data preparation, errors of machine representation of floating point numbers, and arithmetic errors. These could be termed physical, mathematical and computational errors. The extent and importance of the last two types of error depends upon the length and internal structure of the computer word — that is, the total number of binary digits making up one word, and the length of the exponent and mantissa, and on the manner in which the arithmetical operations are performed. If the structure of the computer word is unknown an algorithm described by Malcolm (1972) can be used to determine the number of bits in the exponent and mantissa, and also to discover whether the computer 'truncates' or 'rounds'.

Errors in data preparation are inevitable and it is logical therefore to check all input data carefully, More serious, because they cannot be observed, are errors of representation and of arithmetic. The exact representation of all real numbers is not possible in any single number system. Thus, if the base 10 number system is used, quantities such as $\frac{1}{3}$, $\frac{1}{6}$, $\pi$ and e cannot be expressed accurately. Similarly, a digital computer cannot represent or store all real numbers beyond a certain level of accuracy, for example 7–8 decimal digits (IBM System/360), or

10–11 (ICL 1900 series). Accuracy can be increased by using double precision representation. Integer values are held exactly providing that they lie in a specified range, which varies from one machine to another.

Errors in floating point arithmetic are the most serious of all. Malcom (1971) discusses errors in floating point addition while Pennington (1970) shows how subtraction of quantities of almost equal magnitude causes considerable loss of accuracy in the result. Because a program may have produced satisfactory answers on previous occasions, it does not follow that it will work for all data sets. (It is said of a famous numerical analyst that he was nervous when he boarded a plane because he knew it had been designed using floating point arithmetic.) The program should, therefore, embody algorithms designed to avoid so far as possible those situations with which the computer cannot deal adequately. A good example of the development of a computing algorithm is the work of Youngs and Cramer (1971) who have compared several methods of finding the elements $p_{ij}$ of the corrected sums of squares and cross-products matrix, **P**. The commonly used formula is

$$p_{ij} = \sum_{k=1}^{n} x_{ki}x_{kj} - \left(\sum_{k=1}^{n} x_{ki}\right)\left(\sum_{k=1}^{n} x_{kj}\right)$$

(See, for example, Hemmerle, 1967, p. 78.) Youngs and Cramer found that this formula '... can lead to quite erroneous results ...' and that this situation does not necessarily improve if double-precision representation is used, because the error is due to subtraction of nearly equal quantities. Since $\mathbf{P}^{-1}$ is required in multiple regression analysis, the vector of estimates $\hat{\boldsymbol{\beta}}$ could be considerably in error if **P** is ill-conditioned and the above formula used to compute its elements. (We noted earlier that the elements of $\hat{\boldsymbol{\beta}}$ vary considerably in response to slight changes in the elements of $\hat{\mathbf{P}}$ if **P** is ill-conditioned.) Longley (1967) gives this as one of the reasons for the remarkably poor performance of several generally available regression programs.

A possible alternative to the short-cut formula discussed above is the definitional formula

$$p_{ij} = \sum_{k=1}^{n} (x_{ki} - \bar{x}_i)(x_{kj} - \bar{x}_k)$$

as recommended by Davis (1973, p. 422). This is a two-stage procedure; the means $\bar{x}_i$ and $\bar{x}_j$ must be computed first, then the deviations $(x_{ki} - \bar{x}_i)$. Thus, the entire data-matrix **X** must be retained in core or else transferred to and later recovered from backing store. Either method involves added overheads in computing time, memory requirements or both. (This is the reason the 'short-cut' formula is so popular.)

A highly accurate method, involving only a single pass through the data, is described by Youngs and Cramer (1971). The procedure is set out below.

Step (i): Set $V_1 = 0$
Step (ii): Define $f(x_{ki}) = f(x_{k-1,i}) + x_{ki}$, where $f(x_{0i}) = 0$ for $i = 1, 2, \ldots p$, and $x_{ki}$ is the $k^{th}$ observation on the $i^{th}$ variable.
Step (iii): Let $V_k = V_{k-1} + (kx_{ki} - f(x_{ki}))(kx_{kj} - f(x_{kj}))/k(k-1)$ for $k = 2, 3, \ldots n, i = 1, 2, \ldots, p$ and $j = i, i+1, \ldots p$.
Step (iv): Set $p_{ij} = V_n$

Although it appears formidable this algorithm is easily programmed (see subroutine SSCP, Appendix A). A simple example with $p = 2$ variables and $n = 4$ observations per variable will serve to illustrate the procedure.

$$\text{Given } \mathbf{X}' = \begin{bmatrix} 1 & 2 & 3 & 4 \\ 5 & 4 & 3 & 2 \end{bmatrix}$$

Firstly, set up $f(x_{k1})$ and $f(x_{k2})$ for $k = 0, 1, \ldots, 4$ using step (ii) and having previously set $V_1 = 0$

| | | | | | |
|---|---|---|---|---|---|
| $f(x_{k1})$ | 0 | 1 | 3 | 6 | 10 |
| $f(x_{k2})$ | 0 | 5 | 9 | 12 | 14 |
| $k$ | 0 | 1 | 2 | 3 | 4 |

Now apply the formula of step (iii):

$V_1 = 0$ (from step (i))
$V_2 = 0 + (2 \times 2 - 3).(2 \times 4 - 9)/2 = -\frac{1}{2}$
$V_3 = -\frac{1}{2} + (3 \times 3 - 6).(3 \times 3 - 12)/6 = -2$
$V_4 = -2 + (4 \times 4 - 10).(4 \times 2 - 14)/12 = -5$

Hence $p_{12} = -5$. This can be confirmed by the use of either of the alternative formulae given in the preceding paragraph, e.g.

$$\sum_{k=1}^{n} x_{k1} x_{k2} = 30; \quad \sum_{k=1}^{n} x_{k1} = 10; \quad \sum_{k=1}^{n} x_{k2} = 14,$$

therefore

$$p_{12} = 30 - \frac{10 \times 14}{4} = -5.$$

The 'short cut' formula is much easier for hand computation, for which it was designed. The Youngs and Cramer algorithm involves no subtractions of what may be nearly equal quantities and hence is suited to the characteristics of computer arithmetic.

The purpose of this chapter has been to introduce to the reader some of the concepts of matrices and vectors, and to point out numerical methods of performing operations on matrices on matrices and vectors. No attempt has been made to provide proof or even justification of some of the formulae quoted. The reader is strongly recommended to do some follow-up reading; the books by Searle (1966), Gibson and Mayatt (1965) and Liebeck (1971) are recommended along with the references given earlier in the chapter. Other useful references are: Walsh (1966), Acton (1970), Branfield and Bell (1970), and Hemmerle (1967). To conclude this chapter, a summary of matrix theory is provided as an appendix.

### Appendix: Summary of Matrix Theory
(It is assumed that all operations mentioned are defined given the matrices concerned)

1. $A \pm B = B \pm A$
2. $A + (B + C) = A + B + C$
3. $A - (B - C) = A - B + C$
4. $A(B + C) = AB + AC$
5. $A(BC) = (AB)C = ABC$
6. $AB \neq BA$ (in general).
7. $A + A = 2A$
8. $A^3 = AAA$
9. $BA^3 B \neq B^2 A^3$ or $A^3 B^2$
10. $AO = A = OA$
11. $AB = O$ does not imply either that $A = O$ or $B = O$ or that both $A$ and $B$ equal $O$.
12. $C = AB$ is not necessarily symmetrical even if both $A$ and $B$ are.
13. $A' = A$ if $A$ is a symmetric matrix.
14. $A'A$ is always a symmetric matrix.
15. $AI = IA = A$
16. $B'AB$ is symmetric if $A$ is symmetric.
17. $(A + B)' = A' + B'$
18. $(A')' = A$
19. $A^{-1} A = I$. $A^{-1}$ is the inverse of $A$, which must be a square matrix.
20. $(A + B)^{-1} \neq A^{-1} + B^{-1}$ (in general).
21. $(A^{-1})^{-1} = A$
22. The inverse of a diagonal matrix is found by replacing the values in the diagonal cells by their reciprocals.
23. Premultiplication of a matrix $A$ by a diagonal matrix $D$ (i.e. $DA$) is accomplished by multiplying the elements in row $i$ of $A$ by $d_i$. Post-multiplication of $A$ by $D$ (i.e. $AD$) is accomplished by multiplying the elements in column $j$ of $A$ by $d_j$.

24. Orthogonal matrices are square. If $\mathbf{A}$ is orthogonal $\mathbf{AA'} = \mathbf{I} = \mathbf{A'A}$ and $\mathbf{A}^{-1} = \mathbf{A'}$. $\mathbf{A}^{-1}$ and $\mathbf{A'}$ are also orthogonal.
25. The eigenvectors of a real symmetric matrix are orthogonal. If the associated eigenvalues are equal or very close it may be difficult if not impossible to compute the eigenvectors, however. Only square matrices have eigenvalues and eigenvectors.
26. If $\mathbf{\Lambda}$ is the diagonal matrix of eigenvalues of $\mathbf{A}$ then

$$\sum_{i=1}^{n} \lambda_i = \text{tr}(\mathbf{A}) \quad \text{and} \quad \prod_{i=1}^{n} \lambda_i = \det(\mathbf{A})$$

27. If $\mathbf{X}$ is the matrix of column eigenvectors associated with the eigenvalues $\mathbf{\Lambda}$ of a real symmetric matrix $\mathbf{A}$ then

$$\mathbf{A} = \mathbf{X\Lambda X'}$$

($\mathbf{\Lambda}$ is a *diagonal* matrix of eigenvalues)
28. The rank of a matrix equals the number of nonzero eigenvalues it possesses.
29. Only square matrices possess determinants.
30. $\det(\mathbf{I}) = 1$.
31. $\det(\mathbf{A}) = 0$ if any columns (or rows) of $\mathbf{A}$ are linear combinations of other columns (or rows). $\mathbf{A}$ is then said to be singular. This implies (i) $\mathbf{A}^{-1}$ does not exist, though a generalized inverse $\mathbf{A}^-$ may be computable; (ii) the rank of $\mathbf{A}$ is less than its order.
32. $\det(\mathbf{D})$, the determinant of a diagonal matrix, equals the product of the diagonal elements of $\mathbf{D}$.
33. The rank $p$ of $(\mathbf{A} + \mathbf{B})$ is less than or equal to the sum of the ranks of $\mathbf{A}$ and $\mathbf{B}$, i.e. $p(\mathbf{A} + \mathbf{B}) \leq p(\mathbf{A}) + p(\mathbf{B})$.
34. $p(\mathbf{AB})$ is less than or equal to the smaller of $p(\mathbf{A})$ and $p(\mathbf{B})$.
35. The square root of a diagonal matrix $\mathbf{D}$, written $\mathbf{D}^{\frac{1}{2}}$, is another diagonal matrix with diagonal elements equal to the positive square roots of the diagonal elements of $\mathbf{D}$. E.g. $\mathbf{F} = \mathbf{D}^{\frac{1}{2}}$ then $f_{ii} = +\sqrt{d_{ii}}$. If $\mathbf{G} = \mathbf{D}^{-\frac{1}{2}}$ then $g_{ii} = 1.0/+\sqrt{d_{ii}}$.

Part Two

# Analysis of Dependence

The determination of the relationship between a variable of interest and a set of explanatory or controlling variables has always been a prime objective of exploratory investigations. The apparent simplicity of the methods of multiple regression has led some to disregard the practical difficulties encountered in their application. Computational problems have, in the main, been ignored although such problems can be severe, especially when regression methods are extended to deal with spatial variation, as in trend surface analysis. This Part consists of two chapters. In the first of these the classical regression model is described, particular attention being paid to (a) the specification of the model, including details of the classical assumptions; (b) the strategies that can be followed whenever the classical assumptions do not hold, and (c) computational algorithms. The second chapter of this Part is concerned with extensions of the Linear Model, including polynomial curve fitting, trend surface analysis, selection and combinatorial methods of finding the 'best regression', and the use of binary explanatory variables. The chapter closes with an introduction to the use of nonlinear least squares techniques.

*Chapter 2*

# Regression Analysis: the Linear Model

## 2.0 Introduction

Regression analysis involves the specification and identification of the type and nature of the dependence of a single variable upon a set of controlling, predictor or explanatory variables. The relationship may be used in interpolating or predicting values of the dependent variable (often known as curve-fitting) or it may form part of the process of scientific explanation, in which the variation in the dependent variable is accounted for in terms of the explanatory variables. The basic postulate of regression analysis is that the variation in the dependent variable is made up of two parts — one part which is deterministically related to the explanatory variables, and a second part which appears to be random. This apparently random variation is usually attributed to the effects of minor explanatory variables not included in the regression model, plus the effects of measurement error and, possibly, truly random variation. For this reason, the random term (usually called the disturbance term if the regression model is applied to a population, or the residual term if the regression model is being fitted to a sample) is often assumed to be normally distributed. The deterministic part of the relationship may take any form but the linear model is used most frequently in exploratory studies, partly because the linear model is computationally least complex, and partly because a linear relationship is the easiest to comprehend. It may be difficult in the early stages of an investigation to interpret complex multivariable relationships, especially if these relationships are nonlinear in nature. When a body of theory has been built up, nonlinear relationships may be postulated and alternative regression models employed. In exploratory studies, it is probably wise to err on the side of conservatism, and the linear model is the least complex.

Economists and other social scientists have, in the main, been the most active users of regression techniques; so much so that the subject appears almost to have become identified with econometrics (Johnston, 1972; Kmenta, 1971; Huang, 1970; Aigner, 1970). Applications of regression analysis in geography are also numerous, and a bibliographical guide has been prepared by Green-Wootten (1972). Papers by Potter (1953), Melton (1957), Krumbein (1959), Benson (1962), Morisawa (1962), Wong (1963, 1971) and Rodda (1967) illustrate the use of the technique in physical geography. Other applications of regression analysis to research problems in physical geography are reported by Carson and Sutton (1971), Clarke (1973), Jansen and Painter (1974), Lavalle (1968), Mackay (1967), Stephenson (1971), Thomas and Benson (1970), Woodruff and Hewlett (1970) and Yevyevitch (1972). Computational aspects of multiple regression are covered by Anderssen and Osborne (1969, 1970). Discussion of individual topics mentioned in this chapter are to be found in Hodges and Moore (1972), Cochrane and Orcutt (1949), Farrar and Glauber (1967) and Silvey (1969). Lastly, an application of the principle of least total deviations rather than least sum of squared deviations is given by Davies (1967).

In this chapter, emphasis is placed upon the requirements or assumptions of the model, but it should be understood at the outset that this is a counsel of perfection. Normal distributions, for example, are rarely, if ever, found in nature — though they may be closely approximated. One should attempt to maintain a balance between the two extreme approaches of completely disregarding all assumptions and requirements on the one hand and of insisting on their absolute satisfaction on the other. Special consideration is also given to computational problems. It is well-known that least-squares solutions are prone to serious computational error. Longley (1967) gives several examples of regression coefficients produced by widely-available computer programs that are inaccurate in every digit! The cause of this regrettable state of affairs is often a lack of awareness of the characteristics of computer arithmetic allied to a disregard of methods of numerical analysis. Thus, computing formulae devised originally for use with hand-calculating machines are frequently incorporated in computer programs, the results being at best inaccurate and at worst completely misleading. Many presentations of regression analysis concentrate on theory; although the formulae so obtained help in acquiring an understanding of the meaning of the concepts they are unsuitable for use in computer algorithms.

## 2.1 The Linear Regression Model
The linear regression model describes the linear relationship between a random vector variable y and a set of explanatory variables $x_1$, $x_2, \ldots, x_k$ which can be regarded as forming the columns of a matrix

X. These explanatory variables are sometimes known as independent variables, predictor variables, or controlling variables. The term 'linear model' is understood to mean that the model is linear in the parameters, not in the actual terms themselves. Thus $y = \beta_0 + \beta_1 x^2$ is a linear model but $y = \beta_0 x^{\beta_1}$ is not.

The general form of the regression relationship is:

$$y_i = \beta_0 + \beta_1 x_{i1} + \beta_2 x_{i2} + \ldots + \beta_k x_{ik} + \epsilon_i \quad (i = 1, 2, \ldots, n) \quad (2.1)$$

where $\epsilon_i$ is the disturbance term associated with the ith observed value of $y$. If $\mathbf{x_0}$ is a unit vector then (2.1) can be rewritten as

$$y_i = \beta_0 x_{i0} + \beta_1 x_{i1} + \beta_2 x_{i2} + \ldots + \beta_k x_{ik} + \epsilon_i$$
$$(i = 1, 2, \ldots, n) \quad (2.2)$$

which is equivalent to

$$y_i = \sum_{j=1}^{k} \beta_j x_{ij} + \epsilon_i \quad (i = 1, 2, \ldots, n) \quad (2.3)$$

or, in matrix notation,

$$\mathbf{y} = \mathbf{X}\boldsymbol{\beta} + \boldsymbol{\epsilon} \quad (2.4)$$

In equation (2.4) $\mathbf{X}$ is the matrix with columns $\mathbf{x}_0, \mathbf{x}_1, \mathbf{x}_2, \ldots, \mathbf{x}_k$. To show that this matrix of explanatory variables is augmented on the left by a column of ones, the subscript (1) will be added thus: $\mathbf{X_{(1)}}$.

The $\beta$'s are the parameters of the model. It can be shown that they are linear functions of the $y_i$. The term $\beta_0$ is the constant term or intercept, and gives the value of $y$ when all the $x$'s are zero. It is the level of $y$ in the absence of any control by the $x$'s. The remaining $\beta_i$ give the change in the corresponding $x$ when $y$ is increased by one unit, independent of the level of the other $x$'s. The $\beta_i$ are therefore termed *partial regression coefficients*. Since the $x$'s may be measured on different scales, the values of the $\beta_i$ are not directly comparable; to obtain an idea of the relative influence of the $x$'s each $\beta_i$ is standardized by multiplying it by the ratio of the standard deviation of the corresponding $x$ to the standard deviation of $y$. To avoid confusion, the partial regression coefficients are written as $\beta_i$ and their standarized equivalents as $\beta_{(s)i}$. The relationship between the two is:

$$\beta_{(s)i} = \beta_i \cdot s_i/s_y \quad (2.5)$$

where $s_i$ is the standard deviation of $x_i$ and $s_y$ the standard deviation of $y$.

The model has so far been presented in terms of a population of $x$'s and $y$'s. Usually one has to deal with a sample, and the population parameters (the $\beta$'s in Equation (2.1)) have to be estimated. In this text,

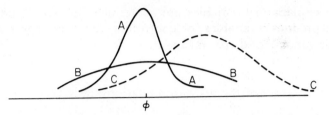

Figure 2.1. Showing bias and minimum variance properties of estimators. The sampling distributions of three estimators of the parameter $\phi$ are shown. Estimators $A$ and $B$ are unbiased but $A$ has a smaller variance than $B$. Estimator $C$ is biased

estimators will be indicated by a circumflex; for example $\hat{\beta}_i$ is an estimator of $\beta_i$. Several methods of estimation are possible (see Yamane, 1964; Kmenta, 1971) but some are applicable only when special conditions, such as normality, obtain. Each method of estimation is based on a formula, called an estimator. When actual numerical values are inserted into the formula the result is an estimate. Statisticians have shown that the most preferable types of estimators of a particular kind (see Figure 2.1). In some cases, as mentioned below in Section 2.2, these properties may not be relevant. In the general case, however, these properties are accepted as necessary. An unbiased estimator is one which produces a sampling distribution with a mean equal to the value of the parameter being estimated. For example, an estimator of the population mean $\mu_x$ could be defined as

$$\hat{\mu}_x = \left( \sum_{i=1}^{n} x_i \right) / n \tag{2.6}$$

and a large number of random samples of size $n$ drawn from the population. The values of $\hat{\mu}_x$ will not be equal for each sample; in fact, the values of $\hat{\mu}_x$ will have a frequency distribution, called the sampling distribution of $\hat{\mu}_x$. The mean of the sampling distribution should equal the population mean (in this example) if $\hat{\mu}_x$ is unbiased. Several estimators may be unbiased, and the criterion used to select the 'best estimator' is that of minimum variance. In other words, the sampling distribution of the chosen estimator has a smaller variance than any other sampling distribution in the class of unbiased estimators. Since comparison of all possible estimators is usually impossible, attention is restricted to estimators that are linear in the observations. The best linear unbiased estimator (BLUE) of $\mu_x$ is, in fact, given by (2.6) which can be rewritten as

$$\hat{\mu}_x = \frac{1}{n} x_1 + \frac{1}{n} x_2 + \ldots + \frac{1}{n} x_n \tag{2.7}$$

to show that $\hat{\mu}_x$ is a linear function of the $x_i$. A good introduction to the topic of parameter estimation is given by Kmenta (1971).

In the regression model the parameters $\beta$ are estimated by the method of least-squares which produces best linear unbiased estimates, given that certain assumptions are satisfied. The quantity minimized in least-squares estimation is

$$\sum_{i=1}^{n} e_i^2 = \sum_{i=1}^{n} (y_i - \hat{y}_i)^2 \tag{2.8}$$

which is the sum of squares of the residuals. Here, $y_i$ is the $i^{th}$ observation on the dependent variable and $\hat{y}_i$ is the computed or predicted value, obtained from

$$\hat{y}_i = \sum_{j=0}^{k} \hat{\beta}_j x_{(1)ij} \tag{2.9}$$

The $e_i$ are residuals from the regression, and are estimates of the $\epsilon_i$ in (2.1). The derivation of the least-squares estimators of the $\beta_i$ will not be given here, as some knowledge of calculus is required. A full description of their derivation can be found in Johnston (1972). The formula is:

$$\hat{\beta} = (X'_{(1)} X_{(1)})^{-1} X'_{(1)} y \tag{2.10}$$

Equation (2.10) is the best linear unbiased estimator (BLUE) of $\beta$ provided that the following assumptions are met:

(I) The mean of $\epsilon$ is zero. This implies that no important explanatory variable has been omitted in the specification of the regression model, and that the chosen $x$'s represent the major controls on the variability of $y$.

(II) The variance of $\epsilon$, written $\sigma^2$, is constant at each level of the $x$'s. This assumption is equivalent to requiring that the variance of $y$ is constant over all $x$ values. This statement is explained most clearly when only one explanatory variable is involved. Firstly, reference to Equation (2.3) or (2.4) shows that the value $y_i$ is made up of two parts: one determined by the $x$'s, while the other is a random or stochastic term $\epsilon_i$. Equation (2.3) could be combined with (2.9) to give

$$y_i = \hat{y}_i + \epsilon_i \tag{2.11}$$

when the population is considered. The term $\epsilon_i$ supposedly includes the joint effects of small independent explanatory variables not included in the model, plus the effects of natural randomness and measurement error in the determination of $y_i$. Thus, if several samples of $y$ are drawn at the same value of $x$ then the difference between the values will be due entirely to the $\epsilon$ term, for Equation (2.9) shows that $\hat{y}$ will be constant if $x$ does not change. Hence, $\epsilon$ has the same distribution form as $y$. Secondly, assume that a large number of measurements of $y$ have been made at each of a set of $x$ values. The variability in $\epsilon$ will ensure

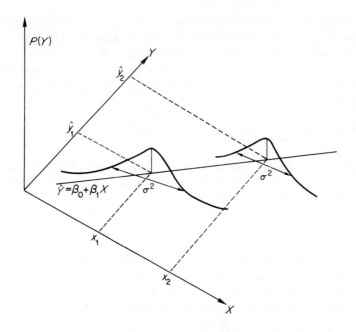

Figure 2.2. Conditional distributions of $Y$ given $X_1$ and $X_2$. Both distributions have the same variance, $\sigma^2$

that $y$ will have a frequency distribution at each point $x$. This is the conditional frequency distribution of $y$ given $x$, written $P(y \mid x)$. Assumption II requires that the variance of each conditional distribution of $y$ given $x$ is identical. It is known as the assumption of homoscedasticity. Figure 2.2 illustrates the point.

(III) The explanatory variables are nonrandom and are measured without error. The $x$'s are considered to be mathematical variables; each realization is unique and is not a sample drawn from an underlying frequency distribution. Thus, the values of location coordinates vary from place but their values at any particular place are fixed, and can, in theory, be measured without error. At each fixed point we might observe a value, such as temperature, which is a sample from the population of all temperature values that could have been observed at that place. Temperature is therefore a random variable whereas location is not.

(IV) The explanatory variables are not perfectly linearly related. The concept of linear dependence is explained in Chapter 1. If a perfect linear relationship exists between two or more of the $x_i$ then Equation (2.10) cannot be solved by normal methods, for $(\mathbf{X}'_{(1)}\mathbf{X}_{(1)})^{-1}$ will not exist. Problems may be encountered if this matrix is nearly singular, for the inverse will be difficult to compute accurately. The problem is not purely a computational one, for if two explanatory variables are closely related it will be difficult to sort out their individual contributions to the variance of $y$. If the $x$'s are linearly related they are said to be collinear or multicollinear.

(V) The number of observations, $n$, exceeds the number of explanatory variables, $k$. This is a necessary requirement for the existence of $(X'_{(1)}X_{(1)})^{-1}$.

(VI) The values $\epsilon_i$ should be independent of each other, that is, the variance-covariance matrix of the $\epsilon_i$ is $\sigma^2 I$. Dependence shown by similarity of adjacent values of $\epsilon_i$ is known as autocorrelation. A regression of mean daily temperature on time over a period of 20 years may show no overall trend yet the residuals will be grouped into positive and negative sets, corresponding to summer and winter. There will, then, be a strong positive autocorrelation among the residuals. Spatial autocorrelation may also exist in data that are collected for areal sampling units.

(VII) If the statistical tests of significance are to be used the conditional distribution of $y$ given $x$ should be approximately normal. Provided that the model is correctly specified, i.e. no important explanatory variables have been omitted, and that assumptions I—VI are satisfied then it is reasonable to assume that $\epsilon$ is normally distributed with a mean of zero and variance $\sigma^2 I$, since the sum of a number of small independent variables is approximately normally distributed according to the Central Limit Theorem.

Violation of assumptions II, IV and V will not affect the property of unbiasedness, but the estimator will not be the minimum-variance estimator. If the aim of the study is to find the best-fitting (in the least-squares sense) straight line or other function, then the coefficients given by these estimators will be acceptable, and only assumptions IV and V will be relevant. Providing Equation (2.10) is soluble, the resulting line or surface will be the 'least-squares' function. On the other hand, if the aim is to estimate the parameters of a population regression model then all assumptions except VII are necessary. Assumption VII is required when interval estimates are to be computed, or whenever tests of significance are to be carried out.

Besides the papers and textbooks mentioned elsewhere in this chapter, the following are recommended sources of further information on the regression model and its assumptions: Bottenberg and Ward (1960), Daniel and Wood (1971), Darlington (1968), Dickinson (1968), Draper and Smith (1966) and Poole and O'Farrell (1971).

## 2.2 Estimation in the linear regression model

To avoid unnecessary confusion in the presentation of this topic, the terminology will be defined at the outset. Readers will find that terminology varies from author to author and also that the same term is occasionally used by different authors to represent separate concepts. However, we can at least try to be internally consistent.

$\mathbf{X}$     ($n \times k$) matrix made up of $n$ observations on $k$ explanatory variables;

$\mathbf{X_{(1)}}$     ($n \times q$) matrix made up of the $k$ columns of $\mathbf{X}$ plus an additional column on the left, consisting entirely of 1's;

$\mathbf{y}$     ($n \times 1$) column vector of observations on the dependent variable;

$\mathbf{R}$     ($k \times k$) matrix of correlations among the explanatory variables. Do not confuse with $R^2$, the coefficient of determination;

$\mathbf{P}$     ($k \times k$) matrix of sums of squares and cross-products of the explanatory variables, corrected for the means;

$\mathbf{g}$     ($k \times 1$) column vector of correlations between the dependent variable and the $k$ explanatory variables;

$\mathbf{h}$     ($k \times 1$) column vector of cross-products of the dependent variable and the $k$ explanatory variables, corrected for the means;

$\boldsymbol{\beta}$     ($q \times 1$) vector of regression parameters. $\beta_0$ is the regression constant and $\beta_i (i = 1, 2, \ldots, k)$ are the partial regression coefficients;

$\boldsymbol{\beta}_{(s)}$     ($k \times 1$) vector of standardized regression coefficients;

$\hat{}$     indicates an estimator;

$\boldsymbol{\epsilon}$     ($n \times 1$) column vector of disturbances;

$\mathbf{e}$     estimator of $\boldsymbol{\epsilon}$;

$s_i^2$     sample variance of the $i^{th}$ explanatory variable;

$s_y^2$     estimator of the variance of the dependent variable;

$s^2$     estimator of the variance of the disturbances, $\sigma^2$.

$\mathbf{y}^*$     ($n \times 1$) column vector of observations on the dependent variable expressed as deviation from the mean;

$\mathbf{X}^*$     ($n \times k$) matrix of observations on the explanatory variables expressed as deviations from the means;

$n$     number of observations per variable;

$k$     number of explanatory variables;

$q$     $k + 1$;

$\hat{\boldsymbol{\beta}}_{(R)}$     ($k \times 1$) column vector of ridge estimates;

$\hat{\mathbf{b}}$     generalized least squares (GLS) estimator of $\boldsymbol{\beta}$.

To begin this section, a summary of the relationships presented earlier is provided. The (sample) regression equation is:

$$\hat{y}_i = \hat{\beta}_0 x_{(1)i0} + \hat{\beta}_1 x_{(1)i1} + \ldots + \hat{\beta}_k x_{(1)ik} \quad (i = 1, 2, \ldots, n) \quad (2.12)$$

where $x_{(1)ij}$ is the $ij^{th}$ element of $\mathbf{X_{(1)}}$ and $\hat{y}_i$ is the $i^{th}$ predicted value of the dependent variable. The residual at the $i^{th}$ data point is:

$$e_i = y_i - \hat{y}_i = y_i - \sum_{j=0}^{k} \hat{\beta}_j x_{(1)ij} \quad (2.13)$$

In matrix notation, Equation (2.12) becomes:
$$\hat{y} = X_{(1)}\hat{\beta} \tag{2.14}$$
in which

$$\hat{y} = \begin{bmatrix} \hat{y}_1 \\ \hat{y}_2 \\ \hat{y}_3 \\ . \\ . \\ . \\ \hat{y}_n \end{bmatrix}, \quad \hat{\beta} = \begin{bmatrix} \hat{\beta}_0 \\ \hat{\beta}_1 \\ \hat{\beta}_2 \\ \hat{\beta}_3 \\ . \\ . \\ . \\ \hat{\beta}_k \end{bmatrix}, \quad \text{and}$$

$$X_{(1)} = \begin{bmatrix} x_{(1)10} & x_{(1)11} & \cdots & x_{(1)1k} \\ x_{(1)20} & x_{(1)21} & \cdots & x_{(1)2k} \\ . & & & . \\ . & & & . \\ . & & & . \\ x_{(1)n0} & x_{(1)n1} & \cdots & x_{(1)nk} \end{bmatrix}$$

Remember that the first column of $X_{(1)}$, that is, the elements $x_{(1)10}$, $x_{(1)20}, \ldots, x_{(1)n0}$, consists entirely of 1's. Since
$$y = X_{(1)}\hat{\beta}$$
then, premultiplying both sides by $X'_{(1)}$
$$X'_{(1)}y = (X'_{(1)}X_{(1)})\hat{\beta}$$
Premultiplying both sides by $(X'_{(1)}X_{(1)})^{-1}$,
$$(X'_{(1)}X_{(1)})^{-1}X'_{(1)}y = (X'_{(1)}X_{(1)})^{-1}(X'_{(1)}X_{(1)})\hat{\beta}$$
$$= \hat{\beta}$$
because
$$(X'_{(1)}X_{(1)})^{-1}(X'_{(1)}X_{(1)}) = I$$
and
$$I\hat{\beta} = \hat{\beta}$$
Hence the vector of estimated regression parameters $\hat{\beta}$ is found from
$$\hat{\beta} = (X'_{(1)}X_{(1)})^{-1}X'_{(1)}y \tag{2.15}$$

The matrix $(X'_{(1)}X_{(1)})$ is the matrix of raw sums of squares and cross-products of the explanatory variables, and $X'_{(1)}y$ is the vector of cross-products of the explanatory variables and the dependent variable. The elements of $(X'_{(1)}X_{(1)})$ and of $X'_{(1)}y$ can become very large, and are often of widely varying magnitudes, hence the inverse of $(X'_{(1)}X_{(1)})$ may be difficult to compute accurately. To minimize this difficulty the matrix $P$ of corrected sums of squares and cross-products is often used in place of the raw sums of squares and cross-products matrix. The corresponding regression equation is:

$$y^* = X^*\hat{\beta} + \epsilon \tag{2.16}$$

The superscript * indicates that the $x_{ij}$ and the $y_i$ values have been converted to deviations from their respective means. To convert the $y_i^*$ back to the original units, $\hat{\beta}_0$ is obtained from

$$\hat{\beta}_0 = \bar{y} - \sum_{i=1}^{k} \bar{x}\hat{\beta}_i \tag{2.17}$$

In this context, $\hat{\beta}$ is the vector of estimated partial regression coefficients excluding the $\beta_0$ term. The vector $\hat{\beta}$ is derived in a manner similar to that used above [Equation (2.15)], thus

$$\hat{\beta} = P^{-1}h \tag{2.18}$$

since

$$P = (X^{*'}X^*)$$

and

$$h = X^{*'}y^*$$

The partial regression coefficients are converted to standardized form by

$$\hat{\beta}_{(s)i} = \hat{\beta}_i \cdot \frac{s_i}{s_y} \tag{2.19}$$

However, the $\hat{\beta}_{(s)i}$ can be obtained directly from

$$\hat{\beta}_{(s)} = R^{-1}g \tag{2.20}$$

and the partial regression coefficients can likewise be found from the converse of Equation (2.19):

$$\hat{\beta}_i = \hat{\beta}_{(s)i} \cdot \frac{s_y}{s_i} \tag{2.21}$$

Direct calculation of the standardized coefficients is preferable for two reasons; (i) the $\hat{\beta}_{(s)}$ are dimensionless, and their relative magnitudes can be used as an indication of the relative importance of the corresponding $x_i$; (ii) the elements of **R** range in value between $-1$ and $+1$, consequently $\mathbf{R}^{-1}$ is more easily obtained than $\mathbf{P}^{-1}$ or $(\mathbf{X}'_{(1)}\mathbf{X}'_{(1)})^{-1}$. The determinant of **R** lies in the range 0 (singular) to 1 (orthogonal) and hence it can be used as an indication of the condition of **R**.

Miesch and Connor (1968) use a measure of ill-conditioning which they attribute to Booth (1957). This measure is the determinant of the correlation matrix **R** after dividing each row element by the sum of the squared elements in that row. The resulting determinant ranges from 0 (ideally conditioned or orthogonal) to $\pm 1$ (singular). A third measure is the $P$-condition number of the matrix, which is the ratio of the largest to the smallest eigenvalues of the matrix. The $P$-condition number can range from 1 (orthogonal) to infinity (singular).

An indication of the precision of the estimators $\hat{\boldsymbol{\beta}}$ is given by their standard errors, written $s_{\hat{\beta}_i}$. These standard errors are simply the square roots of the variances of the estimators, $\mathrm{var}(\hat{\beta}_i)$, which are defined by:

$$\mathrm{var}(\hat{\beta}_i) = \sigma^2 \, [\mathbf{P}^{-1}]_{ii} \tag{2.22}$$

In this equation, $\sigma^2$ is the variance of the disturbances, assumed constant for all values of the dependent variable. It is estimated by $s^2$, the variance of the residuals (Equation 2.23).

$$s^2 = \frac{1}{n-k-1} \cdot \sum_{i=1}^{n} e_i^2$$

$$= \frac{1}{n-k-1} \cdot \mathbf{e}'\mathbf{e} \tag{2.23}$$

The value of $s^2$ can be found easily even if the vector $\hat{\boldsymbol{\beta}}_{(s)}$ has been computed directly using Equation (2.20). However, an initial step — the computation of the coefficient of determination, $R^2$ — is necessary. $R^2$ is a measure of the proportion of variation in the dependent variable that is accounted for by the explanatory variables. It is calculated from

$$R^2 = \hat{\boldsymbol{\beta}}'_{(s)}\mathbf{g} \tag{2.24a}$$

If $R^2$ is being used as a measure of the goodness of fit of a set of regression functions, it should be remembered that the value of $R^2$ will never fall as additional explanatory variables are included. In addition, the sample $R^2$ given by Equation (2.24a) is a biased estimator of the population $R^2$ if $n$ is small. An unbiased estimator is $\bar{R}^2$, termed '$R^2$ corrected for degrees of freedom'. It is:

$$\bar{R}^2 = \frac{n-1}{n-k-1} \cdot (1-R^2) \tag{2.24b}$$

$\bar{R}^2$ can be used as a criterion for the selection of the best regression equation from a set of such equations (see Section 3.1). However, there is no statistical theory to support comparisons based on $\bar{R}^2$ (or $R^2$) except when they are estimating a population $R^2$ of zero. Nevertheless, on intuitive grounds alone, $R^2$ is one of the most widely used regression statistics. The value of $s^2$ is given by:

$$s^2 = \left(1 - \frac{R^2}{n-k-1}\right) \cdot s_y^2 \qquad (2.25)$$

The standard errors of the $\hat{\beta}_i$ can be found even if $\mathbf{P}^{-1}$ is not available:

$$s_{\hat{\beta}_i}^2 = s^2 \, [\mathbf{R}^{-1}]_{ii} \cdot \frac{s_y}{s_i} \qquad (2.26)$$

The estimated partial regression coefficients should always be accompanied by a statement of their standard errors. Usually the standard error is bracketed below the numerical value of the regression coefficient.

The scatter of the $y_i$ values is postulated to be made up of (i) a linear trend and (ii) random fluctuations about this trend. As stated above, the strength of the linear trend is measured by $R^2$. On the assumption that the conditional distributions of the $y_i$ are normal one can test the null hypothesis that the apparent linear trend is a result of sampling fluctuations, thus the null hypothesis is $H_0 : \beta_1 = \beta_2 = \ldots = \beta_k = 0$, which is equivalent to $H_0 : R^2 = 0$. The alternative hypothesis is that at least one of the $\beta_i (i = 1, k)$ is nonzero in the population.

The test statistic for this null hypothesis is:

$$F = \left(\frac{R^2}{k}\right) \bigg/ \left(\frac{1-R^2}{n-k-1}\right) \qquad (2.27)$$

with $k$ and $(n - k - 1)$ degrees of freedom. The alternative hypothesis is accepted if $F$ exceeds the tabled value of the $F$-ratio with the given degrees of freedom and significance level $\alpha$. Thus, if $k = 3$, and $n = 10$ and $\alpha = 5\%$ then the tabled value of the $F$-ratio is $F_{(3,6,5\%)} = 4.7571$.

It is also possible to test whether the individual partial regression coefficients are significantly different from zero. Here, $H_0$ is: $\beta_i = 0$ $(i = 1, k)$ and $H_A$ is $\beta_i \neq 0$ $(i = 1, k)$. Again, the assumption that the conditional distributions of the $y_i$ are normal must be satisfied for the significance test to be accurate (though this assumption may be relaxed when $n$ is large). The test statistic in this case is $t$, computed from:

$$t = \frac{\hat{\beta}_i}{s_{\hat{\beta}_i}} \qquad (2.28)$$

where $s_{\hat{\beta}_i}$ is, as previously, the standard error of the $i^{th}$ partial regression coefficient. The same numerical value of $t$ is obtained from:

$$t = \frac{\hat{\beta}_{(s)i}}{\sqrt{s^2 [\mathbf{R}^{-1}]_{ii}}} \qquad (2.29)$$

when standardized partial regression coefficients have been computed. Since the test is two-tailed, we look up tables of one-tailed Student's $t$ with $(n - k - 1)$ degrees of freedom and $p = \alpha/2$. Hence, for 6 degrees of freedom and $\alpha = 5\%$ the tabled value of $t$ is 2.447. (Some tables of Student's $t$ are designed for two-tailed tests, in which case $p = \alpha$.)

The square of $t$ can be compared with the $F$-ratio with 1 and $(n - k - 1)$ degrees of freedom. The result will be identical since the two-tailed $t$-test is equivalent to the one-tailed $F$-test. The $t$-statistic is useful in that it can be used to test $H_0 : \beta_i = z$ where $z$ is any number, not necessarily zero, or it can be used as a one tailed test (e.g. where $H_A$ is $\beta_i > 0$) or it can be used to construct confidence intervals around $\beta_i$. Occasionally one finds values of $t$ which are negative; in this case the absolute value $|t|$ is used.

If the explanatory variables show a strong linear relation to one another a paradoxical situation will often arise. The $t$-tests of the individual hypotheses $H_0 : \beta_i = 0$ are accepted, while the $F$-test of $H_0 : \beta_1 = \beta_2 = \ldots = \beta_k = 0$ is not accepted. For example, Doornkamp and King (1971) related drainage area $(y)$ to the total length of streams $(x_1)$ and total number of streams $(x_2)$ in the given basin and found that

$$y = -1.81 + 0.42 \, x_1 + 0.13 \, x_2$$
$$\phantom{y = -1.81 + 0}(0.22) \quad\ \ (0.26)$$

In this example, $n = 10$. The computed $t$ values for the tests of $H_0 : \beta_1 = 0$ and $H_0 : \beta_2 = 0$ are 1.89 and 0.51 respectively. The tabled value of $t$ with 7 degrees of freedom and $\alpha = 5\%$ is 2.365, so $H_0$ is accepted in both cases. However the test of $H_0 : \beta_1 = \beta_2 = 0$ produces an $F$-value of 12.38. The tabled $F$-ratio with 2 and 7 degrees of freedom at $\alpha = 5\%$ is 4.74, hence $H_0$ was rejected, and it can be concluded that one or both of the partial regression coefficients probably differ from zero! This apparent contradiction is due to the fact that the correlation between $x_1$ and $x_2$ is 0.88. This close relationship effectively prevents the investigator from estimating the individual effect of the explanatory variables. Furthermore, strong linear relationships among the explanatory variables make $\mathbf{R}$ an ill-conditioned matrix. These are the consequences of multicollinearity, which is discussed in more detail in a later section.

Other anomalous relationships between the results of individual $t$-tests and the $F$-ratio test are discussed by Geary and Leser (1968). These relationships should cast doubt on the validity of the entire regression.

We now turn to the calculation of confidence limits around the $\hat{\beta}_i$. We assume that standardized partial regression coefficients $\hat{\beta}_{(s)}$ have been computed from Equation (2.20), and that $s^2$ is available. (Equation 2.25). The 95 per cent confidence intervals are found from the equation

$$\hat{\beta}_{(s)i} - t_{0.25}(s^2[\mathbf{R}^{-1}]_{ii})^{1/2}d \leqslant \beta_i \leqslant \hat{\beta}_{(s)i} + t_{0.25}(s^2[\mathbf{R}^{-1}]_{ii})^{1/2}d$$

in which

$$d = \frac{s_y}{s_i} \tag{2.30}$$

and $t_{0.25}$ is the 2.5 per cent point of Student's $t$-distribution with $(n-k-1)$ degrees of freedom.

The effects of departures from the basic assumptions of the regression model as set out in Section 2.1 on the estimates of $\beta$ and on the outcomes of the $t$ and $F$-tests are considered in detail by Bohrnstedt and Carter (1971). These authors conclude that the $t$-tests are little affected by non-normality of the conditional distributions of the $y_i$ when $n$ is 60 or more. The $F$-ratio test is also only slightly affected by heteroscedasticity of the variances of the $y_i$ unless this is marked. However, if the variances are heteroscedastic, $\hat{\beta}$ is no longer the minimum variance estimator of $\beta$, though it is still unbiased. Linear relationships among the independent variables can be equally upsetting, and it is not infrequently found under these conditions that some of the $\hat{\beta}_i$ have the wrong sign. The effect of violations of the basic assumptions of the regression model are considered in more detail in a later section.

## 2.3 Computational Aspects of Multiple Regression

Many computing formulae presently in general use were not designed with the characteristics of digital computers in mind. It is therefore not unexpected to find that a surprisingly high proportion of regression analysis programs are inaccurate, some grossly so (Freund (1963), Longley (1967) and Wampler (1970)). The main reasons for this lack of accuracy are errors generated in the computation and inversion of the matrix $\mathbf{R}$ (or $\mathbf{P}$ or $(\mathbf{X}'_{(1)}\mathbf{X}_{(1)})$ as the case may be). Methods which were appropriate for hand calculation have been adapted for computer use and, generally speaking, this has led to a level of accuracy which is entirely unacceptable. (The sceptic should refer to the appendix to Longley's 1967 paper.) In this section some programming considerations are mentioned first then details are given of methods of testing an available program to check its accuracy.

The first major step in multiple regression calculations is the computation of the elements of $\mathbf{R}$. In Section 1.6 it was shown that the most common formula for this calculation is:

$$r_{ij} = \frac{\sum_{k=1}^{n} x_{ki} x_{kj} - \frac{\sum_{k=1}^{n} x_{ki} \sum_{k=1}^{n} x_{kj}}{n}}{s_i s_j}$$

This formula is usually applied in two steps — firstly, the numerator is calculated, giving the elements of the variance–covariance matrix S. Next, the $(i,j)^{th}$ element of S is divided by $\sqrt{[S]_{ii}[S]_{jj}}$, the product of the standard deviations of variables $x_i$ and $x_j$, to form the $(i,j)^{th}$ element of R.

This formula has one major advantage — the sums ($\Sigma_{k=1}^{n} x_{ki}$ and cross-products ($\Sigma_{k=1}^{n} x_{ki} x_{kj}$) can be cumulated as each row of the data matrix is read in. If the number of variables is $p$, then only $p$ store locations are used and successive rows of the data matrix overwrite preceding ones. Unfortunately, the formula has also a major defect — it tends to produce inaccurate results. Youngs and Cramer (1971) found it the least accurate of seven formulae they tested. Accumulation of sums and cross-products in double-precision may help, but store requirements are increased by a factor of 2, and the result is often less accurate than the single precision version of the alternative algorithm provided by Youngs and Cramer (1971) which was described in Section 1.6.

Since R is symmetric, we need only compute the elements of its upper triangle. (The upper triangle is computed rather than the lower triangle because subroutine ACCINV given in Appendix A requires this.) Subroutine SSCP (Appendix A) can be used for this purpose. In the computer program (Table 2.1) the vector **g** of correlations between the explanatory variables and the dependent variable is stored in the last column of the array SS, which, after the call to SSCP, contains,

$r_{11}$ $r_{21}$ $r_{31}$ ........... $r_{k1}$ $g_1$
... $r_{22}$ $r_{32}$ ............ $r_{k2}$ $g_2$
....... $r_{33}$ ............ $r_{k3}$ $g_3$
............................
............................
.........                    $r_{kk}$ $g_k$
.........                         ... $g_{k+1}$

where elements below the diagonal are undefined and $g_{k+1}$ is the self-correlation of the dependent variable. Next, the inverse of the matrix R (held in rows $2 - k + 1$ and columns $2 - k + 1$ of array SS) is calculated using subroutine ACCINV (Appendix A). This routine ensures that $R^{-1}$ is found to within machine accuracy by an iterative

## Table 2.1
### Computer program for multiple linear regression

```
C REGRESSION ANALYSIS PROGRAM BY PAUL M. MATHER, DEPT. OF GEOGRAPHY,
C UNIVERSITY OF NOTTINGHAM.
C
C******************************  NOTES ON PROGRAM USE  *****************
C REQUIRES I/O UNITS AS FOLLOWS: LINEPRINTER (CHANNEL 6),
C CARD READER (CHANNEL 5). TAPE OR DISC FILE (CHANNEL 3).
C IF DATA TO BE READ FROM FILE, CHANNEL 9 SHOULD BE ALLOCATED
C TO THE APPROPRIATE STORAGE DEVICE.
C CORE REQUIREMENTS ARE (NV*NV+1) + (NV*NV)+4*NV+2*NOBS
C NV AND NOBS ARE DEFINED BELOW.
C DON'T FORGET TO ALTER DIMENSION OF X AND IX IN MAIN PROGRAM FOR
C LARGE DATA SETS.  ALSO NUM SHOULD BE SET TO THE DIMENSION OF X.
C
C DATA IN ORDER
C
C CARD 1: TITLE CARD IN 20A4 FORMAT.  IF STOP PUNCHED IN COLS 1-4 THE
C PROGRAM IS TERMINATED.
C
C CARD 2: NV  NOBS
C NV: NUMBER OF VARIABLES. PUNCH IN COLS 1 - 3.
C NOBS: NUMBER OF OBSERVATIONS PER VARIABLE. PUNCH IN COLS 4-6 OF CARD.
C
C CARD 3 FMT, THE DATA INPUT FORMAT, IN 20A4 FORMAT. INCLUDE OPENING AND
C CLOSING BRACKETS.
C IF DATA TO BE READ FROM FILE CARD 3 IS BLANK.
C
C CARD 4: IND, THE NUMERIC IDENTIFIER OF THE DEPENDENT VARIABLE
C PUNCH IN COLS 1-2.
C
C CARD 5: TRANSFORMATION CODES. ONE PER VARIABLE. CODES ARE: 1: NO
C TRANSFORM, 2: LOG(BASE 10) TRANSFORM.  PUNCH IN (80I1) FORMAT,
C I.E. CODE FOR VARIABLE 1 IS PUNCHED IN COLUMN 1.
C CARD 6: DATA CAN BE READ FROM CARDS OR FROM DISC/MAGTAPE FILE.
C IF DATA IS READ FROM CARDS THE WORD CARD SHOULD BE PUNCHED
C IN COLS. 1-4. IF DATA IS TO BE READ FROM FILE PUNCHED THE WORD FILE
C IN COLS. 1-4. FILE DATA IS ASSUMED TO BE UNFORMATTED AND TO BE LOCATED
C ON INPUT CHANNEL 9.
C
C CARD 7 ET SEQ. : DATA CASE BY CASE IN FORMAT SPECIFIED ON CARD 3.
C EACH CASE SHOULD BEGIN ON A NEW CARD.
C IF DATA IS TO BE READ FROM FILE THIS SET OF CARDS IS OBVIOUSLY
C OMITTED.
C
C CARD 8: FMT, THE OUTPUT FORMAT FOR USE IN SUBROUTINE*MATPR*.
C SEE APPENDIX A.
C
C RETURN TO CARD 1 (TITLE) FOR ANALYSIS OF SUBSEQUENT DATA SETS OR PUNCH
C STOP IN COLS 1-4 OF CARD FOLLOWING FMT.
C
C SUBROUTINES REQUIRED: ACCINV, CHOLS,PLOTS,SCALE,KALK,OVCHK,
C MATPR AND PRNTRI.   ALL ARE LISTED IN APPENDIX A.
C
C REFER TO CHAPTER 2 FOR MORE INFORMATION.
C
      DIMENSION X(2000),IX(2000)
      EQUIVALENCE (X,IX)
      DATA STOP/4HSTOP/
C IF YOU ALTER THE DIMENSION STATEMENT DON'T FORGET TO CHANGE *NUM* ALSO
      NUM=2000
      WRITE(6,200)
  200 FORMAT('1MULTIPLE LINEAR REGRESSION PROGRAM'/
     1' PAUL M. MATHER'/
     2' DEPARTMENT OF GEOGRAPHY'/
     3' UNIVERSITY OF NOTTINGHAM'/
     4' VERSION DATED AUGUST 19TH,1974'/)
      J=0
```

Table 2.1  *continued*

```
C READ TITLE CARD.  IF *STOP* IS PUNCHED, TERMINATE THE PROGRAM.
    1 READ(5,101) (X(I),I=1,20)
      IF(X(1).EQ.STOP) STOP
      WRITE(6,102) J, (X(I),I=1,20)
  101 FORMAT(20A4)
  102 FORMAT(I1,' TITLE OF JOB    ',20A4)
      READ(5,100) NV,NOBS
  100 FORMAT(2I3)
      WRITE(6,250) NV, NOBS
  250 FORMAT('0 INPUT SPECIFIES',I6,' VARIABLES'/
     + 13X,' AND',I6,' CASES'///)
C SET UP BASE ADDRESSES OF ARRAYS
      NV1=NV+1
      N1=NOBS+NOBS+1
      N2=N1+NV
      N3=N2+NV1*NV
      N4=N3+NV
      N5=N4+NV*NV
      N6=N5+NV
      NCV=N6+NV*NV
      NVC=NV*NOBS+NOBS*4+1
C CHECK THAT DIMENSION OF X AND IX IS SUFFICIENT
      IF(NCV.LT.NUM.AND.NVC.LT.NUM) GOTO 2
      NCV=MAX0(NCV,NVC)
      WRITE(6,300) NCV
  300 FORMAT(' *** DIMENSION OF A IN MASTER PROGRAM SHOULD BE AT LEAST'
     1,I6,/' EXECUTION TERMINATED.  CHANGE THIS PARAMETER AND RE-INPUT')
      STOP
    2 NUM=NV*NOBS+1
C CALL *DATIN* TO READ IN THE DATA
      CALL DATIN(X,NOBS,NV,IX(NUM))
C CALL MAIN ROUTINE *REX*
      CALL REX(X(1),X(N1),X(N2),X(N3),X(N4),X(N5),X(N6),NV,NOBS,NV1)
C CALL *PLOT* TO PRODUCE SCATTER DIAGRAM OF RESIDUALS V. PREDICTED Y.
      CALL PLOTS(X,1,2,NOBS,NOBS,2,0,IX(N1),NOBS)
      WRITE(6,400)
  400 FORMAT('0PLOT OF RESIDUALS V PREDICTED Y'//
     1          'OVERTICAL AXIS IS RESIDUAL'// ' HORIZONTAL AXIS IS PREDICTED
     1 Y')
      J=1
      GO TO 1
      END



      SUBROUTINE REX(PLT,X,SS,S,B,PBC,BB,NV,NOBS,NV1)
C *REX* IS THE CONTROLLING ROUTINE.
      DIMENSION X(NV),S(NV),SS(NV1,NV),FMT(20),B(NV),PBC(NV),
     + BB(NV,NV),PLT(NOBS,2)
C THE FOLLOWING PARAMETER IS MACHINE DEPENDENT.  SEE NOTES ON *ACCINV* IN
C APPENDIX A.
      EPS=2.0**(-37)
C READ OUTPUT FORMAT SPECIFICATION FOR USE IN *MATPR*.  SEE APPENDIX A.
      READ(5,8) FMT
      IND=-2
C COMPUTE CORRELATIONS AMONG X'S AND BETWEEN Y AND THE X'S IN THE UPPER
C TRIANGLE OF ARRAY *SS*
      CALL SSCP(S,SS,X,NV,NOBS,NV1,IND,3,6,FMT,1)
      IP=1
      NN=NV1*NV
C OUTPUT THE CORRELATION MATRIX USING *PRNTRI* -- SEE APPENDIX A.
      CALL PRNTRI(SS,NN,NV1,24HCORRELATION MATRIX        ,3,IP,NV,1,8,6)
      CON=S(NV)
      REWIND 3
      K=NV-1
    8 FORMAT(20A4)
```

Table 2.1  *continued*

```
C INVERT MATRIX OF CORRELATIONS AMONG THE X'S USING ITERATIVE
C IMPROVEMENT ALGORITHM*ACCINV* ---SEE APPENDIX A.
      CALL ACCINV(SS,BB,PBC,NV,K,L,NV1,NV,EPS)
      WRITE(6,150) L
  150 FORMAT('0',I4,' ITERATIONS IN ACCINV')
      IF(K) 9,10,9
   10 WRITE(6,100)
  100 FORMAT('OCORRELATION MATRIX IS SINGULAR --- THIS PROBLEM HAS BEEN
     +ABANDONED')
      GOTO 11
    9 IF(NV) 12,13,12
   13 WRITE(6,200)
  200 FORMAT('OITERATIONS NOT CONVERGING IN *ACCINV*'// MAYBE EPS IS INC
     +ORRECTLY SET'// THIS PROBLEM HAS BEEN ABANDONED')
   12 IP=1
C PUT INVERSE INTO *BB* TO FACILITATE HANDLING IN LATER PROCEDURES.
C (SEE FIG. A1 FOR THE REASONS FOR THIS!)
      DO 14 I=1,K
      DO 14 J=1,I
      BB(I,J)=SS(I+1,J)
   14 BB(J,I)=BB(I,J)
      NN=NV*NV
      CALL MATPR(32HINVERSE  OF  CORRELATION  MATRIX,4,FMT,6,BB,NN,NV,
     + K,K,IP,1,8)
C COMPUTE STANDARDIZED REGRESSION ESTIMATES
      DO 15 I=1,K
      B(I)=0.0
      DO 15 J=1,K
   15 B(I)=B(I)+BB(I,J)*SS(J,NV)
      WRITE(6,3)
      WRITE(6,5) (I,B(I),I=1,K)
      WRITE(6,2)
C COMPUTE PARTIAL REGRESSION COEFFICIENTS
      XNV=X(NV)
      DO 4 I=1,K
      A=B(I)*XNV/X(I)
      PBC(I)=A
      CON=CON-A*S(I)
    4 WRITE(6,5) I,A
    2 FORMAT('OPARTIAL REGRESSION COEFFICIENTS : '//)
    3 FORMAT('OSTANDARDISED PARTIAL REGRESSION COEFFICIENTS : '//)
    5 FORMAT(' ',I6,G25.12)
      WRITE(6,7) CON
    7 FORMAT('OREGRESSION CONSTANT =  ',E16.7)
C CALL *SIGNIF* TO CARRY OUT T AND F TESTS AND TO COMPUTE R-SQUARED
      CALL SIGNIF(B,K,SS,S,X,BB,NV,NOBS,NV1)
C CALL *RESID* TO COMPUTE RESIDUALS FROM REGRESSION
      CALL RESID(B,K,NOBS,X,NV,PBC,CON,PLT)
   11 RETURN
      END



      SUBROUTINE DATIN(X,NOBS,NV,IV)
C SUBROUTINE TO READ IN RAW DATA
      DIMENSION X(NOBS,NV),ITRAN(80),FMT(20),TR(2),IV(NOBS,4)
      DATA FILE,TR(1),TR(2)/4HFILE,4HNONE,4HLOG /
      REWIND 3
C READ INPUT FORMAT INTO ARRAY *FMT*. THIS MAY BE A DUMMY IF DATA TO
C BE READ FROM FILE AND NOT FROM CARDS.
      READ(5,2) FMT
    2 FORMAT(20A4)
C *IND* GIVES THE NUMERIC IDENTIFIER OF THE DEPENDENT VARIABLE.
      READ(5,3) IND
    3 FORMAT(I2)
```

Table 2.1  *continued*

```
      WRITE(6,4) IND
    4 FORMAT('0DEPENDENT VARIABLE IS NO. ',I4)
C *ITRAN* = TRANSFORMATION CODES
      READ(5,5) ITRAN
    5 FORMAT(80I1)
      WRITE(6,10)
      DO 11 I=1,NV
      J=ITRAN(I)
   11 WRITE(6,12) I, TR(J)
   10 FORMAT('0    TRANSFORMATIONS SPECIFIED ARE'//' VAR NO    TRANSFORMATI
     +ON'//)
   12 FORMAT(' ',I4,10X,A4)
C *SOURCE* HOLDS EITHER THE WORD FILE OR THE WORD CARDS DEPENDING ON
C THE DATA INPUT TYPE.  DATA FROM FILE SHOULD BE UNFORMATTED.
      READ(5,2) SOURCE
      INPUT=1
      IF(SOURCE.EQ.FILE) INPUT = 2
      IF(INPUT.EQ.2) REWIND 9
      WRITE(6,8)
    8 FORMAT('0       DATA MATRIX'//)
      DO 1 I=1,NOBS
C READ DATA FROM FILE OR CARDS
      GOTO (50,60),INPUT
   50 READ(5,FMT) (X(I,J),J=1,NV)
      GOTO 70
   60 READ(9) (X(I,J),J=1,NV)
C CALL TRANSFORMATION ROUTINE
   70 CALL TRANS(X,NV,NOBS,ITRAN,I)
      WRITE(6,7) I, (X(I,J),J=1,NV)
    7 FORMAT(' ',I3,2X,(10G11.5))
C DEPENDENT VARIABLE MUST BE LAST -- IF IT ISN'T, PUT IT THERE
      IF(IND.EQ.NV) GOTO 1
      N1=IND+1
      T=X(I,IND)
      DO 18 M=N1,NV
   18 X(I,M-1)=X(I,M)
      X(I,NV)=T
    1 WRITE(3) (X(I,J),J=1,NV)
      ENDFILE 3
      REWIND 3
      IF(IND.NE.NV) WRITE(6,19) NV
   19 FORMAT('0**DEPENDENT VARIABLE HAS BEEN MOVED TO COLUMN',I4,' BUT T
     1HE ORDER OF THE EXPLANATORY VARIABLES IS UNALTERED.'///)
      WRITE(6,13)
   13 FORMAT('0PLOTS OF EXPLANATORY VARIABLES V. DEPENDENT VARIABLE'///)
      CALL PLOTS(X,1,-NV,NOBS,-NOBS,NV,0,IV,NOBS)
      RETURN
      END



      SUBROUTINE RESID(B,K,NOBS,X,NV,PRC,CON,PLT)
C THIS SUBROUTINE COMPUTES RESIDUALS FROM THE REGRESSION
      DIMENSION PRC(NV),B(K),X(NV),PLT(NOBS,2)
      F=NOBS-K-1
      WRITE(6,1)
    1 FORMAT('1RESIDUALS FROM REGRESSION'///'          OBS. Y          CALC.
     1Y          RESIDUAL'//)
    2 FORMAT(' ',3G15.5)
      DO 3 I=1,NOBS
      READ(3) X
      RR=CON
      DO 4 J=1,K
    4 RR=RR+PRC(J)*X(J)
      PLT(I,2)=RR
      PP=X(NV)
```

Table 2.1  *continued*

```
      PQ=PP-RR
      PLT(I,1)=PQ
    3 WRITE(6,2) PP,RR,PQ
      WRITE(6,5)
    5 FORMAT('1FREQUENCY HISTOGRAM FOR RESIDUALS')
C CALL HISTGM FROM LIBRARY (APPENDIX A) TO PRODUCE A FREQUENCY PLOT
C OF THE RESIDUALS.
      CALL HISTGM(PLT,20,NOBS)
      RETURN
      END




      SUBROUTINE SIGNIF(B,K,SS,SE,S,BB,NV,NOBS,NV1)
C THIS SUBROUTINE CARRIES OUT THE STANDARD SIGNIFICANCE TESTS AS
C DESCRIBED IN CHAPTER 2.
      DIMENSION B(K),SE(NV),S(NV),SS(NV1,NV),BB(NV,NV)
C SIGNIFICANCE LEVELS ARE COMPUTED VIA A CALL TO SUBROUTINE FPROB
C (APPENDIX A).
      SNV=S(NV)
      RSS=0.0
      DO 8 I=1,K
    8 RSS=RSS+SS(I,NV)*B(I)
      SIG=1.0-RSS
      G=K
      LL=NOBS-K-1
      H=LL
      S2=SIG/H
      WRITE(6,2)
    2 FORMAT('1STANDARD ERRORS OF PARTIAL REGRESSION COEFFICIENTS:'//)
      DO 1 I=1,K
      SE(I)=SQRT(S2*BB(I,I))
      D=SE(I)*SNV/S(I)
    1 WRITE(6,5) I, D
      WRITE(6,4) LL
      DO 3 I=1,K
      A=B(I)/SE(I)
      A=ABS(A)
      SIGF=1.0-SQRT(FPROB(A,1,LL))
    3 WRITE(6,5) I, A, SIGF
    4 FORMAT(////' T VALUES FOR TEST OF H0: B=0. DEGREES OF FREEDOM = '
     *,I6//'  INDEP VAR.       T VALUE      SIGNIFICANCE LEVEL'//)
    5 FORMAT(' ',I4,6X,2G17.6)
    6 FORMAT('1ANALYSIS OF VARIANCE TEST OF REGRESSION'//
     1' TOTAL SS = ',7X,G17.6//
     2' REGRESSION SS = ',2X,G17.6//
     3' RESIDUAL SS = ',4X,G17.6//
     4' F-RATIO = ',8X,G17.6//
     5' DEGREES OF FREEDOM ARE: ',2I6/
     6' SIGNIFICANCE LEVEL = ',G17.6)
      F=(RSS/G)/(SIG/H)
      SIGF=1.0-FPROB(F,K,LL)
      WRITE(6,6) SS(NV,NV), RSS, SIG, F, K, LL, SIGF
      R2=RSS
      R2COR=R2-G/H*(1.0-R2)
      WRITE(6,7) R2, R2COR
    7 FORMAT(/////' R SQUARED = ',G16.7///'OR SQUARED CORRECTED FOR DEGREE
     *S OF FREEDOM = ',G16.7)
      RETURN
      END




      SUBROUTINE TRANS(X,NV,NOBS,IT,K)
      DIMENSION X(NOBS,NV),IT(NV)
```

Table 2.1    *continued*

```
      DO 1 I=1,NV
      J=IT(I)
      GOTO (1,2), J
    2 IF(X(K,I).LE.0.0) GOTO 3
      X(K,I)=ALOG10(X(K,I))
    1 CONTINUE
      RETURN
    3 WRITE(6,4) I, X(K,I)
    4 FORMAT(' *** ATTEMPT TO FORM LOG OF NEGATIVE OR ZERO NUMBER, ITEM'
     1,I6,'  OF LAST CARD   - ',F12.5)
      STOP
      END




      FUNCTION FPROB(FR,M,N)
      PI=3.1415926535
      FPROB=0.0
      CON=1.0
      FM=M
      FN=N
      IF(MOD(M,2)) 70,80,70
   70 IF(MOD(N,2)) 40,60,40
   40 IF(N-1) 5,30,5
   30 THETA=ATAN(SQRT(FN/(FM*FR)))
      J=M/2
      GOTO 7
    5 THETA=ATAN(SQRT(FM*FR/FN))
      J=N/2
    7 SINE=SIN(THETA)
      SINSQ=SINE*SINE
      COSQ=1.0-SINSQ
      COSN=SQRT(COSQ)
      IF((M.EQ.1).AND.(N.EQ.1)) GOTO 50
      DO 10 I=1,J
      FPROB=FPROB+CON
      TWI=2*I
   10 CON=CON*TWI*COSQ/(TWI+1.0)
   50 FPROB=1.0-2.0*(FPROB*SINE*COSN+THETA)/PI
      IF(N-1) 110,1000,110
  110 FPROB=1.0-FPROB
      IF(M-1) 120,1000,120
  120 FCTR=CON
      CON=1.0
      PEP=0.0
      FNM1=N-1
      J=M/2
      DO 20 I=1,J
      PEP=PEP+CON
      TWI=2*I
   20 CON=CON*(FNM1+TWI)*SINSQ/(TWI-1.0)
      FPROB=FPROB-2.0*FN*FCTR*SINE*COSN*PEP/PI
      GOTO 1000
   60 X=FN/(FN+FM*FR)
      J=N/2
      FMS2=M-2
      GOTO 85
   80 X=FM*FR/(FN+FM*FR)
      J=M/2
      FMS2=N-2
   85 OWX=1.0-X
      DO 90 I=1,J
      FPROB=FPROB+CON
      TWI=2*I
      CON=CON*(FMS2+TWI)*X/TWI
      IF(CON.LT.0.000001) GOTO 91
```

Table 2.1  *continued*

```
  90 CONTINUE
  91 IF(MOD(M,2)) 100,95,100
  95 FPROB=1.0-OWX**(FN/2.0)*FPROB
     GOTO 1000
 100 FPROB=OWX**(FM/2.0)*FPROB
1000 RETURN
     END
```

procedure which is described in Appendix A. If **R** is nearly singular the use of ACCINV will lead to distinctly improved results, while if the matrix **R** is well-behaved the number of iterations in ACCINV will be low, giving a computing time requirement similar to that of straightforward matrix inversion routines. The use of accurate computational methods cannot, however, compensate for bad data. From this point on the program is straightforward. Equation (2.20) is used to calculate $\hat{\vec{\beta}}_{(s)}$ and the partial regression coefficients are found from (2.21). Next, (2.24a) and (2.24b) are used to find $R^2$ and $\bar{R}^2$. The sequence (2.25) and (2.26) gives $s^2$ and the $s^2_{\hat{\beta}_i}$, and the $t$-values for $H_0 : \beta_i = 0$ are derived from (2.29). The $F$-test of $H_0 : R^2 = 0$ is given by (2.27). These calculations are performed in subroutine SIGNIF. The next subroutine, RESID, computes the residuals from the regression using (2.9) and (2.13). Subroutine PLOTS (Appendix A) is used to plot the dependent variable against the explanatory variables, and also to plot $\hat{y}$ against $\hat{e}$. Finally, subroutine HISTGM produces a frequency distribution histogram for the residuals.

The use of the program is illustrated with reference to a set of data taken from the paper by Nash and Shaw (1966). The data matrix is shown in Table 2.2. The three columns represent measurements of drainage basin area [$A$] in square miles, rainfall [$R$] in inches and mean annual discharge [$Q$] in cubic feet per second for 57 drainage basins in the United Kingdom. In this example, $Q$ is the dependent variable and $A$ and $R$ are the explanatory variables. The logarithms (base 10) of these variables are taken as this gave better linear functional relationships (Figure 2.3). The plot of log $Q$ against log $A$ is reasonably linear, though the relationship between log $Q$ and log $R$ is much less so. This is borne out by the entries in the correlation matrix (Table 2.3) which show that log $A$ and log $Q$ have a correlation of 0.7456 while log $R$ and log $Q$ have a correlation of only 0.3598. The two explanatory variables are themselves related ($r = -0.2726$). Only one iteration was required in ACCINV. The standardized partial regression coefficients are: $\hat{\beta}_{(s)1} = 0.911$, $\hat{\beta}_{(s)2} = 0.608$, showing that basin area is more important than rainfall in explaining the variation in mean annual discharge for the sample of 57 basins. The partial regression coefficients are: $\hat{\beta}_0 = -4.072$, $\hat{\beta}_1 = 0.838$ and $\hat{\beta}_2 = 2.156$. (Note that inferences about the relative importance of $x_1$ and $x_2$ would have been quite wrong had

Table 2.2
Hydrological data for multiple regression example (Reproduced with permission from J. E. Nash and B. L. Shaw, *Inst. Civil Engrs., Proc. Sympos. River Flood Hydrol.*, London, 115—136 (1966))

|    | X1      | X2     | Y        |
|----|---------|--------|----------|
| 1  | 28.00   | 953.00 | 2595.00  |
| 2  | 328.00  | 845.00 | 10741.00 |
| 3  | 11.00   | 665.00 | 544.00   |
| 4  | 151.00  | 822.00 | 11176.00 |
| 5  | 528.00  | 451.00 | 15562.00 |
| 6  | 6.00    | 507.00 | 269.00   |
| 7  | 45.00   | 367.00 | 1632.00  |
| 8  | 85.00   | 264.00 | 699.00   |
| 9  | 81.00   | 297.00 | 249.00   |
| 10 | 485.00  | 314.00 | 5690.00  |
| 11 | 187.00  | 391.00 | 4275.00  |
| 12 | 36.00   | 237.00 | 178.00   |
| 13 | 27.00   | 243.00 | 244.00   |
| 14 | 75.00   | 255.00 | 447.00   |
| 15 | 90.00   | 258.00 | 597.00   |
| 16 | 86.00   | 263.00 | 478.00   |
| 17 | 41.00   | 265.00 | 166.00   |
| 18 | 651.00  | 247.00 | 2764.00  |
| 19 | 1170.00 | 242.00 | 3689.00  |
| 20 | 565.00  | 256.00 | 3229.00  |
| 21 | 44.00   | 240.00 | 99.00    |
| 22 | 180.00  | 244.00 | 389.00   |
| 23 | 117.00  | 250.00 | 793.00   |
| 24 | 30.00   | 256.00 | 223.00   |
| 25 | 8.00    | 252.00 | 180.00   |
| 26 | 400.00  | 256.00 | 1298.00  |
| 27 | 140.00  | 309.00 | 495.00   |
| 28 | 3810.00 | 289.00 | 11832.00 |
| 29 | 137.00  | 283.00 | 658.00   |
| 30 | 153.00  | 342.00 | 1263.00  |
| 31 | 8.00    | 338.00 | 226.00   |
| 32 | 257.00  | 321.00 | 3218.00  |
| 33 | 617.00  | 339.00 | 6269.00  |
| 34 | 102.00  | 279.00 | 1185.00  |
| 35 | 125.00  | 284.00 | 1007.00  |
| 36 | 859.00  | 269.00 | 4708.00  |
| 37 | 1670.00 | 372.00 | 15687.00 |
| 38 | 782.00  | 464.00 | 10841.00 |
| 39 | 5.00    | 964.00 | 442.00   |
| 40 | 1560.00 | 409.00 | 19475.00 |
| 41 | 28.00   | 769.00 | 1747.00  |
| 42 | 342.00  | 329.00 | 1673.00  |
| 43 | 138.00  | 391.00 | 5169.00  |
| 44 | 732.00  | 499.00 | 16545.00 |
| 45 | 495.00  | 561.00 | 17458.00 |
| 46 | 64.00   | 655.00 | 4296.00  |
| 47 | 4.00    | 997.00 | 787.00   |
| 48 | 44.00   | 580.00 | 924.00   |
| 49 | 401.00  | 569.00 | 9252.00  |
| 50 | 8.00    | 515.00 | 420.00   |
| 51 | 61.00   | 297.00 | 573.00   |
| 52 | 160.00  | 344.00 | 1967.00  |
| 53 | 58.00   | 407.00 | 1931.00  |
| 54 | 235.00  | 304.00 | 2634.00  |
| 55 | 78.00   | 304.00 | 763.00   |
| 56 | 262.00  | 440.00 | 6340.00  |
| 57 | 216.00  | 498.00 | 10655.00 |

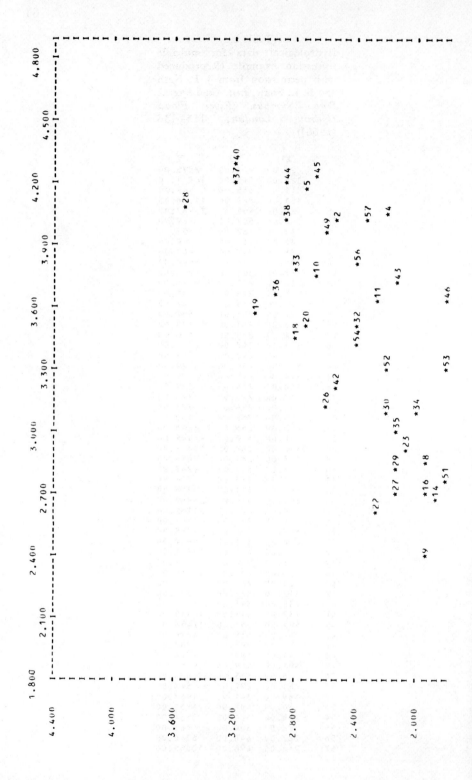

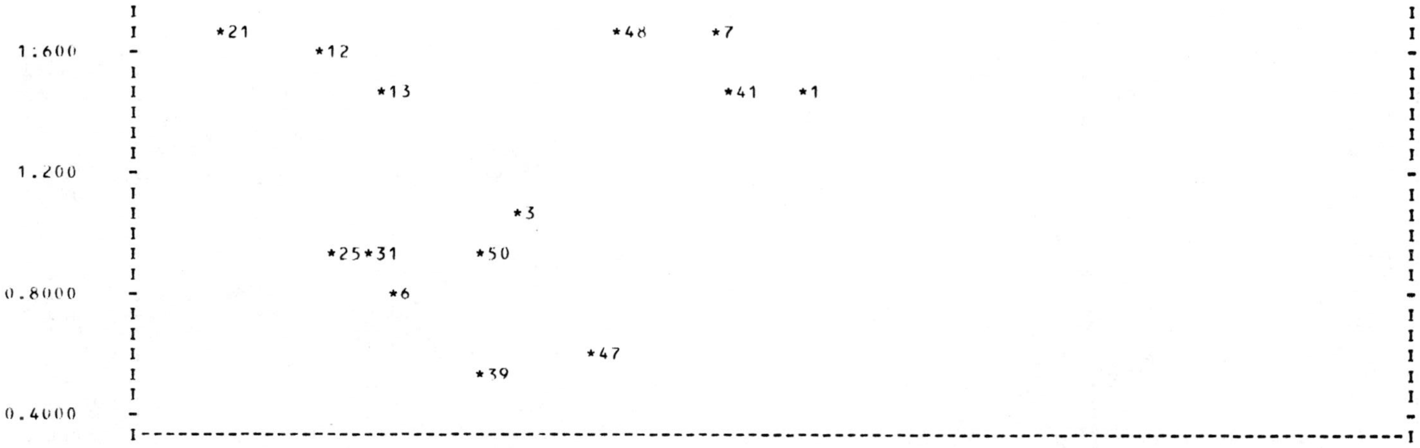

```
                                    I---I-------I-------I-------I-------I
                                    I                                   I
                                    I                                   I
1.600                               I                                   I
                                    I       *48    *7                   I
                                    I                                   I
                                    I                          *41  *1  I
                                    I                                   I
1.200      *21    *12               I                                   I
                          *13       I                                   I
                                    I                                   I
                                    I               *3                  I
                                    I        *25*31  *50                I
                                    I                                   I
0.8000                              I              *6                   I
                                    I                                   I
                                    I                        *47        I
                                    I                                   I
                                    I                   *39             I
                                    I                                   I
0.4000                              I                                   I
                                    I---I-------I-------I-------I-------I
```

PLOT OF COL.   1 (VERTICAL AXIS) AGAINST COL.   3 (HORIZONTAL AXIS)

NUMBER OF OVERPRINTS =    4

OVERPRINT TABLE
ITEM NO.  ALSO LOCATION OF ITEM NO.

8           15
8           55
12          17
13          24

Figure 2.3. (a) Scatter plot of log basin area v. log mean annual discharge

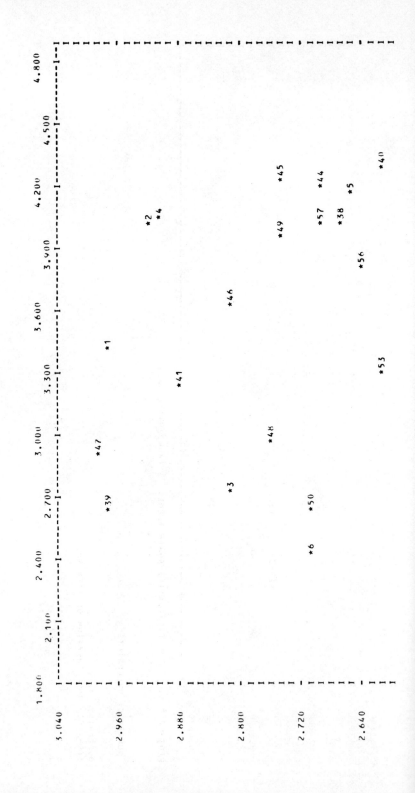

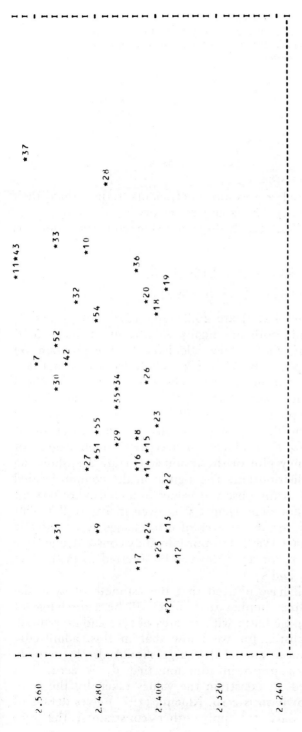

Figure 2.3. (b) Scatter plot of log rainfall v. log mean annual discharge

Table 2.3
(a) Matrix of correlations among hydrological variables; (b) inverse of matrix of correlations among explanatory variables

(a)

|   | 1 | 2 | 3 |
|---|---|---|---|
| 1 | 1.0000 | −0.2726 | 0.7456 |
| 2 |  | 1.0000 | 0.3598 |
| 3 |  |  | 1.0000 |

(b)

|   | 1 | 2 |
|---|---|---|
| 1 | 1.0803 |  |
| 2 | 0.2945 | 1.0803 |

we relied on the partial regression coefficients rather than their standardized counterparts.) The standard errors of $\hat{\beta}_1$ and $\hat{\beta}_2$ are $s_{\hat{\beta}_1} = 0.0414$, $s_{\hat{\beta}_2} = 0.1598$. Thus, the regression equation could be written out in full as:

$$\log Q = -4.072 + 0.838 \log A + 2.156 \log R.$$
$$\quad\quad\quad\quad\quad\quad (0.0414) \quad\quad (0.1598)$$

The $t$-values derived from (2.29) are 20.21 and 13.49 for $\beta_1$ and $\beta_2$ respectively. With 54 d.f. both are highly significant ($t_{0.001} = 3.46$ with 60 d.f.). The significance values calculated by the program are 0.00002 and 0.0003 respectively. We can conclude from this that it is highly probable that $\beta_1 \neq 0$ and $\beta_2 \neq 0$. The $F$-value (2.28) is 238.8 which again is significant at less than 0.001. In fact, the significance level printed out by the program is 0.000000. $R^2$ is 89.8 per cent and $\bar{R}^2$ is 89.5 per cent. From this we may conclude that the regression of $\log Q$ on $\log A$ and $\log R$ produces an acceptably high degree of explanation. The frequency plot of the residuals (Figure 2.4) shows an approximately normal distribution. The figures in the column headed 'number' could be used as the observed values in a chi-square test for normality. Lastly, the plot of e against $\hat{y}$ is given in Figure 2.5. This plot shows no tendency for the scatter of e to change systematically with $\hat{y}$, thus indicating that (i) var($e_i$) is approximately constant, and (ii) e is not related either to $x_1$ or $x_2$. (This can be inferred from the fact that $\hat{y}$ is a function of $x_1$ and $x_2$.)

Perceptive readers will have noticed that the estimate of $\beta_0$ in the example is $-4.072$, which implies that there will be a discharge of $10^{-4.072}$ c.f.s. in a drainage basin with an area of zero and no rainfall. This is obviously unrealistic, for we know that in this, admittedly unlikely, situation the mean annual discharge should also be zero. In other words, there is a priori information that $\beta_0$ is zero. This information can be used to constrain the values taken by the $\beta_i$ to ensure that the intercept term is zero. Kmenta (1971) gives details of the computations necessary to apply other constraints; the zero intercept constraint is the most common, at least in physical geography, so it will be described here. Let X be the matrix of observations on the explanatory variables (without a leading column of 1's), y the ($n \times 1$) column vector of observations on the dependent variable and $\hat{\beta}_i$ the estimator of the $i^{th}$ parameter in the model:

$$y_i = \beta_1 x_{i1} + \beta_2 x_{i2} + \ldots + \beta_k x_{ik} + \epsilon_i$$

Then

$$\hat{\beta} = (X'X)^{-1} X'y$$

Note that the elements of X are not corrected for the mean — they are the raw observations on the explanatory variables. Applying this technique to the Nash and Shaw data used in the preceding example, the regression equation becomes:

$$y_i = 0.667 x_{i1} + 0.715 x_{i2}$$

The coefficient of $x_1$ drops from 0.838 to 0.667, but the second coefficient shows a far greater reduction — from 2.156 to 0.715. Since the constrained regression is no longer the best fit line, in the least squares sense, $R^2$ can be expected to fall, which it does, from 0.898 to 0.739. This is because the regression line is now forced to pass through the origin.

The program listed in Table 2.1 has proved remarkably accurate on the ICL 1906A and DEC PDP 11/05 computers. Longley (1967) provides test data (Table 2.4) and results computed by hand to 14 figures. (The ICL 1906A works to 10–11 significant figures in single precision). Longley found that '... many programs tested with identical inputs produced results which differed from each other in every digit .. ' (1967, p. 827). There is no indication on the printout to indicate error, and the fact that the results are interpretable should not be taken to imply that they are accurate. Longley's results and those achieved using the ICL 1906A computer are compared in Table 2.5. Two iterations were necessary in ACCINV to produce these results.

Checking a program with reference to one data set is clearly not sufficient, for it may be better conditioned than the data set to be analysed. Mullett and Murray (1971) describe a means of examining rounding error in any given data set. The steps are:

(1) Compute the full regression $y = \hat{\beta}_0 + \hat{\beta}_1 x_1 + \hat{\beta}_2 x_2 + \ldots + \hat{\beta}_k x_k$.
(2) Regress $y + \alpha x_i$ on the same set of explanatory variables. The parameter $\alpha$ must be nonzero. Repeat this step using different $\alpha$'s and different $x_i$.

If there is only insignificant rounding error the following results must

```
FREQUENCY HISTOGRAM FOR RESIDUALS

SCALING FACTOR =    14.2857

NUMBER  PERCENT
        MIDPOINT

   3    5.7632    |********
        0.3836    |********
                  |********
                  |********
   1    1.7544    |****
        0.5405    |****
                  |
   0    0.0000    |
        0.2970    |
   2    3.5088    |*******
        0.2537    |*******
                  |*******
                  |*******
                  |*******
                  |*******
                  |*******
   4    7.0175    |***************
        0.2104    |***************
                  |***************
                  |***************
                  |***************
                  |***************
                  |***************
                  |***************
                  |***************
   3    5.7632    |***************
        0.1671    |***************
                  |***************
                  |***************
   3    5.7632    |***************
        0.1238    |***************
                  |***************
                  |***************
                  |***************
                  |***************
   6   10.5263    |***************
        0.0805    |***************
                  |***************
                  |***************
                  |***************
                  |***************
                  |***************
   7   12.2807    |***************
        0.0372    |***************
```

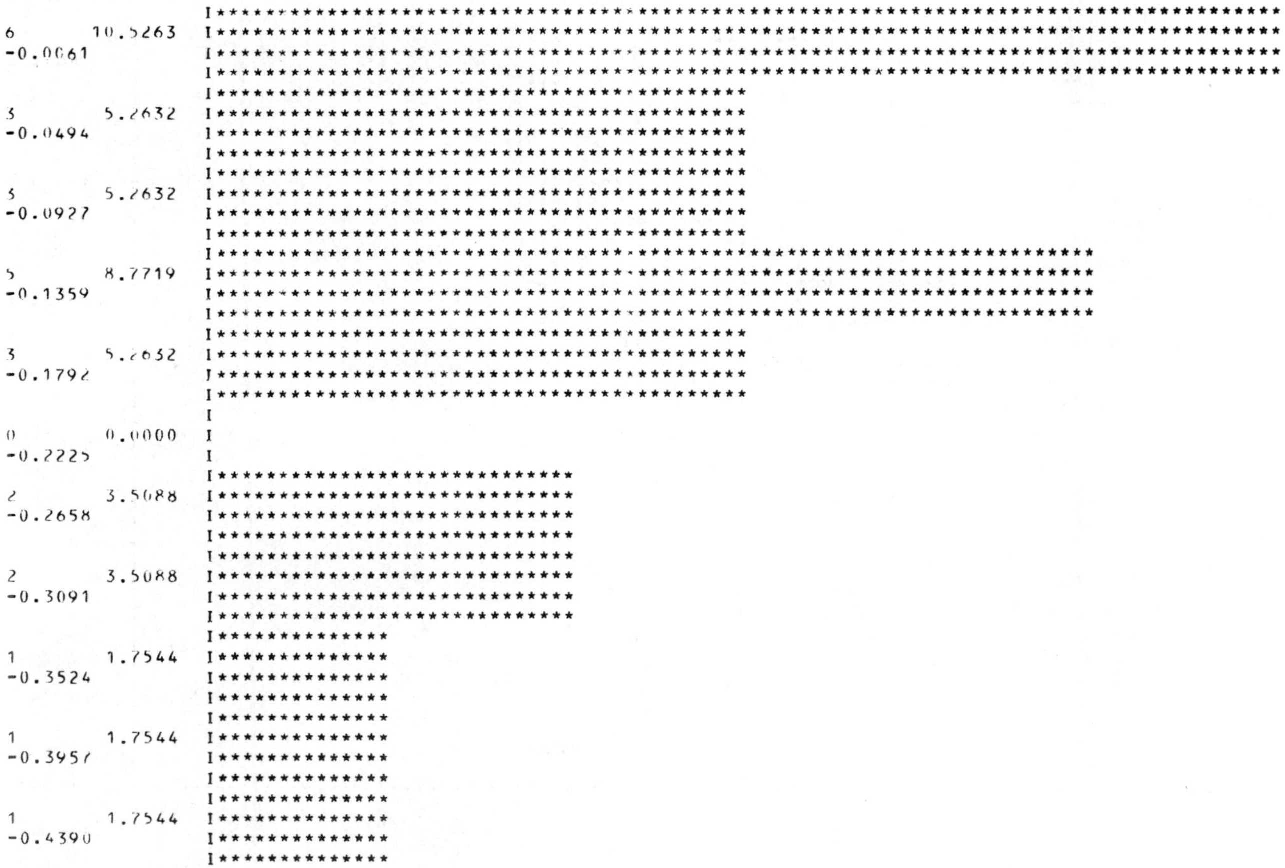

Figure 2.4. Frequency distribution of residuals

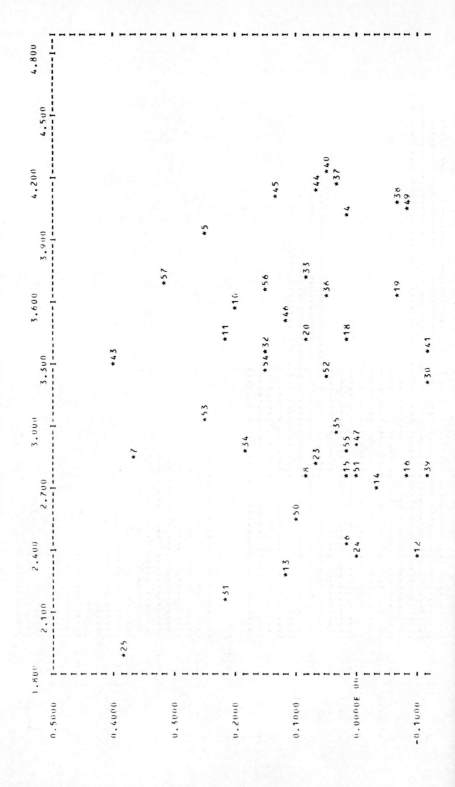

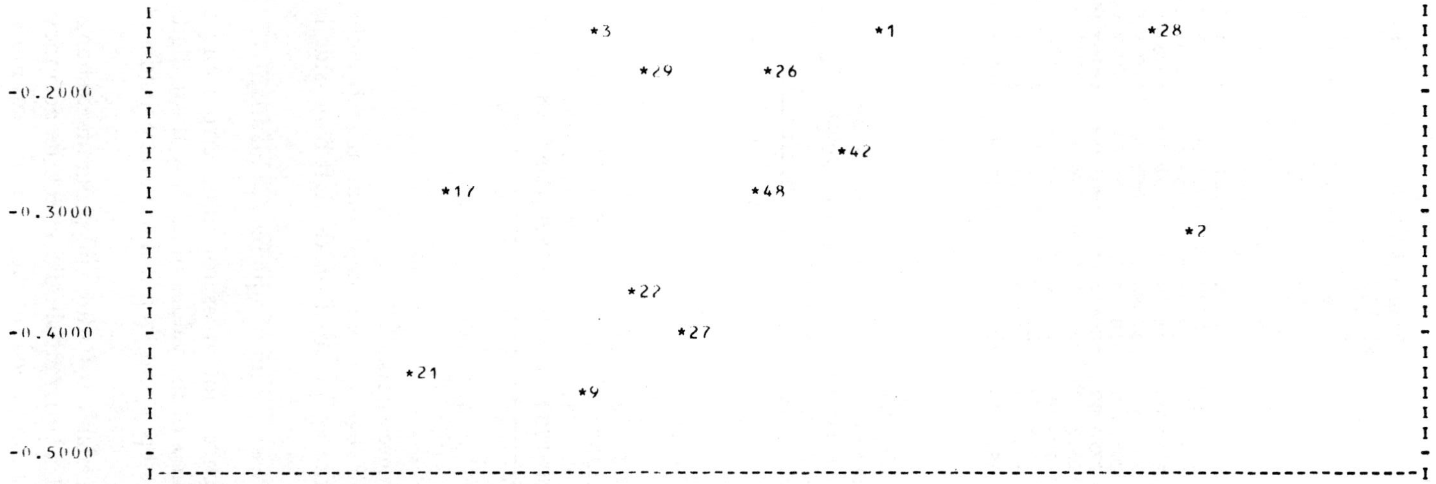

```
                                                                        *28


                                              *3            *1
                                       *19   *26
                                                    *42
-0.2000                                      *48

           *17                         *22
-0.3000                                              *27
                                              *9
-0.4000    *21

-0.5000
```

PLOT OF COL. 1 (VERTICAL AXIS) AGAINST COL. 2 (HORIZONTAL AXIS)

NUMBER OF OVERPRINTS =    0

PLOT OF RESIDUALS V PREDICTED Y

VERTICAL AXIS IS RESIDUAL
HORIZONTAL AXIS IS PREDICTED Y

Figure 2.5. Plot of $e_i$ versus $\hat{y}_i$ for regression example

Table 2.4
Longley's test data (Reproduced with permission from J. M. Longley, *Journ. Amer. Statist. Assoc.*, 62, 819–829 (1967))

|    | 1      | 2         | 3       | 4       | 5         | 6       | 7        |
|----|--------|-----------|---------|---------|-----------|---------|----------|
| 1  | 83.00  | 234289.00 | 2356.00 | 1590.00 | 107608.00 | 1947.00 | 60323.00 |
| 2  | 88.50  | 259426.00 | 2325.00 | 1456.00 | 108632.00 | 1948.00 | 61122.00 |
| 3  | 88.20  | 258054.00 | 3682.00 | 1616.00 | 109773.00 | 1949.00 | 60171.00 |
| 4  | 89.50  | 284599.00 | 3351.00 | 1650.00 | 110929.00 | 1950.00 | 61187.00 |
| 5  | 96.20  | 328975.00 | 2099.00 | 3099.00 | 112075.00 | 1951.00 | 63221.00 |
| 6  | 98.10  | 346999.00 | 1932.00 | 3594.00 | 113270.00 | 1952.00 | 63639.00 |
| 7  | 99.00  | 365385.00 | 1870.00 | 3547.00 | 115094.00 | 1953.00 | 64989.00 |
| 8  | 100.00 | 363112.00 | 3578.00 | 3350.00 | 116219.00 | 1954.00 | 63761.00 |
| 9  | 101.20 | 397469.00 | 2904.00 | 3048.00 | 117388.00 | 1955.00 | 66019.00 |
| 10 | 104.60 | 419180.00 | 2822.00 | 2857.00 | 118734.00 | 1956.00 | 67857.00 |
| 11 | 108.40 | 442769.00 | 2936.00 | 2798.00 | 120445.00 | 1957.00 | 68169.00 |
| 12 | 110.80 | 444546.00 | 4681.00 | 2637.00 | 121950.00 | 1958.00 | 66513.00 |
| 13 | 112.60 | 482704.00 | 3813.00 | 2552.00 | 123366.00 | 1959.00 | 68655.00 |
| 14 | 114.20 | 502601.00 | 3931.00 | 2514.00 | 125368.00 | 1960.00 | 69564.00 |
| 15 | 115.70 | 518173.00 | 4806.00 | 2572.00 | 127852.00 | 1961.00 | 69331.00 |
| 16 | 116.90 | 554894.00 | 4007.00 | 2827.00 | 130081.00 | 1962.00 | 70551.00 |

Table 2.5
Comparison of results of analysis of Longley's test data (Table 2.4). (Reproduced with permission from J. M. Longley, *Journ. Amer. Statist. Assoc.*, 62, 819–829 (1967))

| Coefficient | Longley     | ICL 1906A      |
|-------------|-------------|----------------|
| intercept   | −3482259.0  | −3482256.0     |
| 1           | 15.061929   | 15.06187138    |
| 2           | 0.03581923  | −0.03581908    |
| 3           | −2.02023050 | −2.0202349     |
| 4           | −1.03322720 | −1.03322750    |
| 5           | −0.05110374 | −0.05110532    |
| 6           | 1829.151900 | 1829.15009238  |

hold:

(a) $\hat{\beta}_0$ is the same for all the regressions.
(b) The $\hat{\beta}_i$ are the same in all regressions except that at step 2 the regression coefficient of the $x_i$ included on the left-hand side is increased by $\alpha$.
(c) The residuals, $R^2$ and regression sum of squares are invariant.

This method provides both a check for gross rounding errors and a means of determining the accuracy of the values of $\hat{\beta}$. Only those digits which remain constant through step 1 and step 2 should be considered as potentially meaningful.

In the present era of freely-available software and program exchange it is imperative that the user of a program should check its accuracy, especially if the program was not developed on the particular computer available to him. If this precaution is omitted the results of any subsequent analysis could be inaccurate.

## 2.4 Problems in regression
### (i) Multicollinearity and ridge regression

The problem of linear relationships among the explanatory variables has been mentioned above in the discussion of tests of significance of the regression coefficients. If the explanatory variables are linearly related it becomes difficult to sort out their separate contributions to the variance (sums of squares) of the dependent variable and, furthermore, since such linear relationships imply that $R$ will be badly conditioned, the determination of $R^{-1}$ may involve computational error. The diagonal elements of $R^{-1}$ will tend to be large, which in turn implies that the standard errors of the coefficients will be large also. Though the $\hat{\beta}_i$, being least squares estimators, will be unbiased, a large standard error indicates that the probability of getting an individual $\hat{\beta}_i$ that is far from the population $\beta_i$ is increased. This is not critical if the regression is being used in a descriptive (curve fitting) manner, for by definition the least squares line or surface is 'best fit'. However, if the regression model is being fitted, so that the parameters do have a physical interpretation, then the ordinary least squares (OLS) estimators may be seriously misleading. As Hoerl and Kennard (1970a) point out, '... the results of these errors are critical when $X\beta$ is a true model. Then the least squares estimates often do not make sense when put into the context of the physics, chemistry and engineering of the process which is generating the data.' (p. 55).

The presence of linear relationships among the explanatory variables is termed multicollinearity. It is a matter of degree. There is every variation possible, from perfect multicollinearity, when $R$ is singular, to complete absence of multicollinearity, in which case $R$ is an orthogonal matrix. The measurement of the degree of multicollinearity is difficult. If there are two explanatory variables, the degree of linear relationship between them can be inferred from their correlation. For more than two explanatory variables, '... the presence of perfect multicollinearity does not necessarily mean that the correlation between any two explanatory variables must be perfect, or even particularly high...' (Kmenta, 1971, p. 383). However, if two explanatory variables have a correlation of ± 1.0 then perfect multicollinearity necessarily follows.

In the past, one could do very little when faced with the problem of multicollinearity. Geographers have traditionally looked to principal components analysis (Chapter 4) for a solution, since the aim of principal components analysis is to replace the data matrix by a matrix which, in addition to other properties, is orthogonal. The drawback in this approach is that one builds an artificial barrier between the regression equation (based on principal components) and the system which is described by the variables. Another possibility is to eliminate some of the explanatory variables from the regression, but this involves throwing away what may well be useful information.

One way out of the problem is to use the technique known as ridge regression. (Hoerl and Kennard, 1970a,b; Marquardt, 1970; Jones, 1972; Stone and Conniffe, 1973; Smith and Goldstein, 1975). Rather than minimize $e'e$ these authors suggest that we should try to minimize

$$L^2 = (\hat{\boldsymbol{\beta}} - \boldsymbol{\beta})' \cdot (\hat{\boldsymbol{\beta}} - \boldsymbol{\beta}) \qquad (2.31)$$

to give a vector of estimates which have a smaller mean square error than the least-squares estimates. (Mean-square error is a measure of the dispersion of an estimate around the true value of the parameter.) The expected value of $L^2$, written $E(L^2)$ is

$$E(L^2) = \sigma^2 \operatorname{tr}(\mathbf{R}^{-1})$$
$$= \sigma^2 \sum_{i=1}^{k} (1/\lambda_i) \qquad (2.32)$$

where $\lambda_i$ is the $i^{\text{th}}$ eigenvalue of $\mathbf{R}$. If $\mathbf{R}$ is an identity matrix then the $\lambda_i$ will all be equal to 1.0. However, as linear relationships among the independent variables become more pronounced, the smallest eigenvalue $\lambda_{\min}$ of $\mathbf{R}$ approaches zero, so $1/\lambda_{\min}$ and hence $\Sigma (1/\lambda_i)$ will become very large. This indicates that $E(L^2)$ is also very large. It follows that the ordinary least squares estimates derived under these conditions by the methods of Sections 2.3 and 2.4 will be too large in absolute value and may even have the wrong sign.

Hoerl (1962) showed that this instability of the least-squares estimators can be reduced if the following estimator of $\boldsymbol{\beta}$ is used:

$$\hat{\boldsymbol{\beta}}_{(\mathbf{R})} = (\mathbf{R} + k\mathbf{I})^{-1} \mathbf{g} \qquad (2.33)$$

The vector $\hat{\boldsymbol{\beta}}_{(\mathbf{R})}$ contains the 'ridge estimates' of $\boldsymbol{\beta}$. The term 'ridge' is used because Equation (2.33) is mathematically similar to a quadratic response function when $k \geq 0$. The parameter $k$ is a nonnegative constant, usually falling within the range $0 \leq k \leq 1$.

The estimates $\hat{\boldsymbol{\beta}}_{(\mathbf{R})}$ are not true least-squares estimates unless $k = 0$. Consequently, the error sum of squares associated with $\hat{\boldsymbol{\beta}}_{(\mathbf{R})}$ will be larger than the error sum of squares associated with the ordinary least-squares estimates $\hat{\boldsymbol{\beta}}$. However, the parameter $L^2$ will be reduced in magnitude.

The choice of a value for $k$ is usually accomplished by inspection of a plot of $k$ against $\hat{\boldsymbol{\beta}}^k_{(\mathbf{R})}$, where $\hat{\boldsymbol{\beta}}^k_{(\mathbf{R})}$ are the values of the ridge estimates for a particular value of $k$. This plot is usually referred to as a *ridge trace*. An example is given in Figure 2.6. Usually, $\hat{\boldsymbol{\beta}}^k_{(\mathbf{R})}$ initially changes very rapidly as $k$ increases, but the values begin to stabilize at a point generally between $k = 0.1$ and $k = 0.3$. The values of $\hat{\boldsymbol{\beta}}^k_{(\mathbf{R})}$ at or near the point of stabilization are taken to be the ridge estimates of $\boldsymbol{\beta}$. Hoerl and

Kennard (1970b) give two examples of the use of the ridge trace. Their first example involves a 10 x 10 matrix of correlations among the explanatory variables. Values of $\hat{\beta}^k_{(R)}$ are found for 15 values of $k$ between 0 and 1, six of these values lying in the interval $0 \leqslant k \leqslant 0.1$. The ridge trace shows that the initial values of $\hat{\beta}^k_{(R)}$ alter rapidly as $k$ increases. Their variables $x_5$ and $x_6$ show a particularly rapid decline in absolute value. The system stabilizes between $k = 0.2$ and $k = 0.3$, and a $\hat{\beta}^k_{(R)}$ selected from within this range will be likely to give a closer estimate of $\beta$ than the least-squares estimates. The ridge trace also shows that the coefficients of $x_5$ and $x_7$ move towards zero as $k$ increases, and so, in Hoerl and Kennard's terminology, they 'do not hold their predictive power'. Hence, they could be removed from the regression equation. This procedure of screening variables may give more meaningful results than the more usual stepwise regression method discussed in Chapter 3, and may also be preferable to the use of principal components analysis (Chapter 4).

As an example of the use of ridge regression, a correlation matrix from the paper by Jones (1972) is used. These correlations are descriptive measures only, since the $x_i$ are assumed to be fixed constants. Hence, tests of significance applied to these correlation coefficients would be meaningless. Four of the six explanatory variables measure properties of the surface material of Clark Street Beach, Evanston, Illinois, and the remaining two explanatory variables measure geographical position of the sampling points, in terms of $(u,v)$ coordinates. The 30 sampling points were the points of intersection of a sampling grid. In detail, the $x_i$ are:

$x_1$ : $U$ coordinate (across-beach distance);
$x_2$ : $V$ coordinate (along-beach distance);
$x_3$ : sorting coefficient, $\phi$;
$x_4$ : geometric mean of the beach sand;
$x_5$ : phi mean, $\bar{\phi}$, of the beach sand.

The dependent variable was the slope of the beach measured at each of the 30 sampling points. The correlations between the beach variables are given in Table 2.6.

The eigenvalues of **R** are 3.111, 1.008, 0.715, 0.164 and 0.003, giving a ratio of about 980:1 between the values of the maximum and minimum eigenvalues. This ratio is sometimes used as a measure of the 'condition' of a matrix. (If all the $x_i$ were uncorrelated, **R** would be an identity matrix with all eigenvalues equal to 1.0.)

The computed inverse of **R** is given in Table 2.7. Multiplying the rows of $\mathbf{R}^{-1}$ by g in turn gives $\hat{\beta}_{(s)}$, the estimated ordinary least-squares standardized partial regression coefficients. These differ somewhat from

Table 2.6
Matrix **R** and vector **g** for beach variables (Reproduced with permission from T. A. Jones, *Journ. Intern. Assoc. Mathem. Geol.*, 4, 203–218 (1972) Table 1)

|   | $x_1$ | $x_2$ | $x_3$ | $x_4$ | $x_5$ |   |
|---|---|---|---|---|---|---|
| $x_1$ | 1.000 |  |  |  |  |   |
| $x_2$ | 0.000 | 1.000 |  |  |  |   |
| $x_3$ | −0.490 | 0.039 | 1.000 |  |  | **R** |
| $x_4$ | 0.595 | 0.071 | −0.866 | 1.000 |  |   |
| $x_5$ | −0.376 | 0.002 | 0.986 | −0.807 | 1.000 |   |
| $y$ | 0.802 | 0.165 | −0.291 | 0.505 | −0.240 | **g** |

those shown in Jones's Figure 3 presumably because his estimates were computed from a more precise form of Table 2.6. This illustrates the instability of least-squares estimators in a situation of high multicollinearity.

The vector of least-squares estimates is $\hat{\beta}^0_{(R)}$ { =1.528, −0.241, 8.307, 1.340, −6.775}. The extremely high values associated with $x_3$ and $x_5$ seem to indicate that beach slope varies directly with phi sorting and inversely with the phi mean of the surface material. However, if we invert $(R + kI)$ with $k = 0.1$ we find that the elements of the inverse matrix are much closer in value (Table 2.8).

Table 2.7
Inverse of matrix **R** shown in Table 2.6

| | | | | |
|---|---|---|---|---|
| 3.224 | | | | |
| −0.692 | 1.416 | | | |
| 16.926 | −8.433 | 180.404 | | |
| 0.860 | −1.416 | 23.969 | 7.274 | |
| −14.782 | 6.909 | −152.154 | −17.437 | 131.381 |

Table 2.8
Inverse of $(R + 0.1I)$

| | | | | |
|---|---|---|---|---|
| 1.338 | | | | |
| 0.030 | 0.935 | | | |
| 0.471 | −0.295 | 5.943 | | |
| −0.712 | −0.235 | 1.415 | 2.847 | |
| −0.487 | 0.100 | −4.128 | 0.578 | 4.866 |

The vector $\hat{\beta}^0_{(R)^1}$ derived from this matrix is {0.698, 0.121, 0.305, 0.277, −0.050}. Variable $x_5$ has not held its predictive power, its coefficient being virtually zero. The most important controls of beach slope new seem to be position across the beach, sorting and geometric mean of the beach sand, all acting in a direct relationship. The situation is shown better in the ridge trace (Figure 2.6).

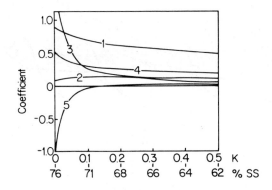

Figure 2.6. Ridge trace for Jones data (Reproduced with permission from T. A. Jones, *Journ. Intern. Assoc. Mathem. Geol.*, 4, 203–218 (1972), Figure 3)

*(ii) Heteroscedasticity*

A second assumption of the linear regression model is that the variances of the residuals are equal. If this is not so, then the least-squares estimators $\hat{\beta}$ do not possess the property of minimum variance of all linear estimators, though they are still unbiased. Variances will be heteroscedastic if they increase or decrease with any of the $x_i$ (see Figure 2.7). Since the predicted values of $y$ are linear functions of the $x_i$ we can plot $e_i$ against $\hat{y}_i$ and observe the resulting scatter. If the plot shows a random distribution, the variance of the conditional distributions of $y_i$ will tend to be homoscedastic. If a systematic change occurs, variances are probably heteroscedastic.

Draper and Smith (1966) and Daniel and Wood (1971) give a full discussion of the use and interpretation of residual plots. Reference should also be made to Anscombe (1960, 1961), Kruskal (1960), Anscombe and Tukey (1963), and Larsen and McCleary (1972). The last of these papers proposes a rather different method of plotting residuals. Larsen and McCleary (1972) define the $i^{th}$ 'partial residual vector' as 'the dependent variable vector corrected for all independent variables except the $i^{th}$ variable' (p. 781). This is claimed to give a clearer picture of nonlinear relationships, outliers, and inhomogeneous variances.

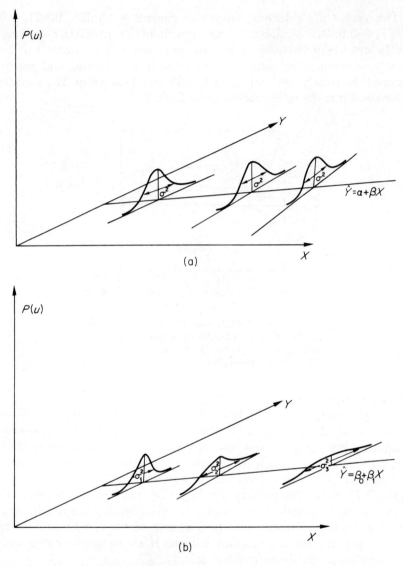

Figure 2.7. (a) Homoscedastic and (b) heteroscedastic variances (Reproduced with permission from J. Johnston, *Econometric Methods, Second Edition*, McGraw-Hill, New York, 1972, Figures 2.1 and 7.1)

If the assumption of homoscedastic variances does not apply then, as noted previously, the estimators $\hat{\beta}$ are still unbiased but are not the best linear estimators since they no longer have the property that their variances are the smallest of all linear estimators. Moreover, the variances of the $\hat{\beta}$ computed from the formulae given in Section 2.2 are biased, so that the significance tests described in that same section will produce inaccurate and possibly misleading results. However, the least-squares relationship will always be the 'best-fit' in that $e'e$ is minimized.

One method of tackling the problem is to use the method of generalized least-squares, which can be used whenever, we know, or can postulate, the form of the relationship between var $(e_i)$ and the individual explanatory variables. The assumption that var $(e_i) = \sigma^2 I$ is replaced by the assumption that var $(e_i) = \sigma^2 G$ where $G$ is a diagonal matrix of weights.

$$G = \begin{bmatrix} g_{11} & 0 & \ldots & 0 \\ 0 & g_{22} & \ldots & 0 \\ \vdots & \vdots & & \vdots \\ 0 & 0 & \ldots & g_{nn} \end{bmatrix} \qquad (2.34)$$

The $g_{ii}$ are used to weight the $x_i$ to produce generalized least-squares estimators $\hat{b}$:

$$\hat{b} = (X'G^{-1}X)^{-1} X'G^{-1} y \qquad (2.35)$$

The inverse of $G$ can be determined readily by defining a second matrix $P$ where

$$P^{-1} = \begin{bmatrix} \sqrt{\lambda_1} & 0 & \ldots & 0 \\ 0 & \sqrt{\lambda_2} & \ldots & 0 \\ \vdots & \vdots & & \vdots \\ 0 & 0 & & \sqrt{\lambda_n} \end{bmatrix} \qquad (2.36)$$

where

$$g_{ii} = 1/\lambda_i.$$

It follows that

$$G = PP'$$

and

$$P^{-1} G (P^{-1})' = I$$

therefore

$$(P^{-1})' P^{-1} = G^{-1} \qquad (2.37)$$

The variances of the $\hat{b}$ are the diagonal elements of $(X'G^{-1}X)^{-1}$

multiplied by $\sigma^2$ which is itself found from

$$\sigma^2 = \frac{1}{n-k-1} \cdot \mathbf{e}'\mathbf{G}^{-1}\mathbf{e} \qquad (2.38)$$

and

$$\mathbf{e} = \mathbf{y} - \mathbf{X}\hat{\mathbf{b}} \qquad (2.39)$$

As an example, consider the relationship $y = \beta_0 + \beta_1 x + e_i$ where the variance of the $e_i$ is proportional to $x_i^2$, that is,

$$\text{var}(e_i) = \sigma^2 x_i^2 \qquad (2.40)$$

The diagonal elements of **G** are thus equal to $x_i^2$ and the elements $\sqrt{\lambda_i}$ of the diagonal of **P** are equal to $1/x_i$ since $g_{ii} = 1/\lambda_i$. The inverse of **G** is found from Equation (2.37), and the GLS estimators $\hat{\mathbf{b}}$ are computed using Equation (2.35).

The situation is more complicated when the variance of the residuals is related to the value of more than one of the explanatory variables. One possibility is the construction of a linear function of the explanatory variables involved, and its use in the equations given above.

Often one is not in a position to make specific statements about the relationship between $\sigma_i^2$ and the $x_i$. In this case, a logarithmic transformation is often applied to the dependent variable so as to reduce the scatter as $y_i$ increases. Ordinary least squares estimators are then found. If this is done, it should be remembered that the quantity $(\log \mathbf{e})'(\log \mathbf{e})$ is minimized rather than $\mathbf{e}'\mathbf{e}$, so that true OLS estimators for the original variables cannot be found simply by taking the antilogarithms of the computed $\hat{\beta}_i$. Unfortunately, it is probable that by taking the logarithm of $y$ the distribution form of $y$ will be altered and it is necessary to assume that $y$ is normally distributed at each of the $n$ observation points if the significance tests are to be applicable. The use of transformations of the dependent variable therefore raises some difficulties.

Apart from purely visual checks of scatter plots of $e_i$ against $y_i$, and against each of the x's, there are several methods of testing the validity of the assumption of homoscedasticity, though none of them is universally accepted. Johnston (1972) discusses these methods, two of which are now described. The first is applicable when a large number of observations is available. The dependent variable is split into $M$ classes in order of the magnitude of $y$ and a value $\lambda$ is computed from

$$\lambda = \prod_{i=1}^{M} \left(\frac{s_i}{n_i}\right)^{n_i/2} (\Sigma s_i / \Sigma n_i)^{\Sigma n_i/2} \qquad (2.41)$$

In this equation, $M$ is the number of classes and $n_i$ is the number of observations in the $i^{th}$ class, i.e. $\Sigma_{i=1}^{M} n_i = n$, and $s_i$ is found from

$$s_i = \sum_{j=1}^{n_i} (y_{ij} - \bar{y}_i)^2 \qquad (2.42)$$

The test statistic $\mu$ is equal to $-2 \log_e \lambda$ and is distributed as chisquare with $M-1$ degrees of freedom under the assumption of homogeneous variances. The hypothesis of homogeneity is not accepted if $\mu > \chi^2_{m-1}$ at a given significance level.

An alternative test of homogeneity of variances is simply to compute the Spearman rank correlation coefficient between the absolute values of the residuals, $|e_i|$, and each of the independent variables with which the variance might be associated. This test involves replacing $|e_i|$ and $x_i$ by their ranks, and computing

$$r = 1 - \frac{6 \sum_{i=1}^{n} d_i^2}{n^3 - n} \qquad (2.43)$$

where $d_i^2$ is the difference between the ranks of the $i^{th}$ values of $|e_i|$ and $x_i$. If $r$ is significantly different from zero, the hypothesis of homogeneity is not accepted.

A simple example will serve to illustrate the fact that the variance of OLS estimators is increased in the presence of heteroscedasticity. The equation $y = 1.5 + 2x_1 + 3x_2 + e$ was used to generate sets of artificial data. Both $x_1$ and $x_2$ were drawn randomly from a distribution uniform on the range $(1,10)$. The residual term $e_i$ was drawn randomly from a normal distribution with zero mean and unit variance. It was then weighted by the corresponding value of $x_1^2$. The variance of the residuals is thus made proportional to $x_1^2$. Thirty samples of size 150 were generated by this method, using the same values of $x_1$ and $x_2$ each time. The results are shown in Table 2.9.

Table 2.9 shows that the mean values of the OLS and GLS estimators

Table 2.9
Comparison of OLS and GLS estimators

| Coefficient | OLS | | | GLS | | | True |
|---|---|---|---|---|---|---|---|
| | Mean | Max | Min | Mean | Max | Min | |
| $\beta_1$ | 1.97 | 14.78 | −15.51 | 1.92 | 5.02 | −1.69 | 1.5 |
| $\beta_2$ | 2.08 | 6.20 | −0.55 | 1.83 | 3.97 | −0.09 | 2.0 |
| $\beta_3$ | 2.74 | 5.62 | 0.20 | 2.98 | 3.59 | 2.39 | 3.0 |

over 30 trials are very similar, the GLS estimator being closer to the true value except in the case of $\hat{\beta}_2$. The ranges of the values of $\hat{\beta}_i$ are markedly wider in the OLS case, however. For example, one set of OLS estimators gave $y = -15.51 + 2.85x_1 + 5.37x_2$. With the same data the GLS method produced $y = 1.28 + 2.05x_1 + 3.02x_2$. The conclusion to be drawn is that the probability of getting estimates $\hat{\beta}$ that are very far from the true parameters $\beta$ increases considerably in the presence of heteroscedasticity if OLS methods are used. The fact that OLS estimators are still unbiased is of little comfort when only one sample is available. Unfortunately, a priori knowledge of the relationships between the variance of the residuals and the independent variables is required before GLS methods can be usefully employed.

*(iii) Autocorrelation*

Assumptions II and IV of the classical regression model require that the disturbance terms [$\epsilon_i$ in Equation (2.1)] are independently distributed random variables with expectation zero and identical variances, $\sigma^2$. The last of these requirements, homogeneity of variances, was considered in the preceding section. Independence of the population disturbance terms implies that the magnitude and sign of a particular disturbance have no effect on the magnitudes or signs of the following disturbance terms. For example, the $i^{th}$ observation in a random sample of roundness measures of pebbles taken from a river bed may depart considerably from any of the other values, possibly due to excessive measurement error, to a mistake in recording the observed value, or to the fact that the pebble concerned may be of foreign origin. The roundness values of pebbles $i + 1$, $i + 2$ and so on will not be affected by the anomalous value recorded for pebble $i$. In contrast, consider the following situation. The dependent variable in the regression model $y = \beta_0 + \beta_1 x + \epsilon$ is the peak discharge recorded during a single storm at a series of stations located along a river. The explanatory variable is the distance from the river source to the recording station. The aim of the regression is to estimate the rate at which the flood peak increases downstream, and the residuals from this regression (estimating the population disturbances) can be expected to reflect errors in the measurement of the discharge, as well as the influence of local conditions. The measurement errors can be assumed to be unrelated. However, the flood peak at station $i$ will be affected by such factors as the entry of tributaries, the discharge of storm sewers and the effects of flood protection works. These factors will affect all stations downstream. Thus, a series of positive residuals could be expected below a tributary junction or an urban area. There would thus be a marked tendency for positive residuals to group together; the same would be true for negative residuals. In other words, the value of the $i^{th}$ residual would be partially predictable from the values of the $(i-1)^{th}$ residual.

This relationship can be expressed as

$$\epsilon_j = \rho \epsilon_{j-1} + u_j$$

where $u_j$ is a normal, independently distributed random variable and $\rho$ is the first order serial (autocorrelation) coefficient, measuring the degree of linear relationship between adjacent values of $\epsilon$. This model is said to have a first order autoregressive structure. Second and higher order autoregressive models are possible but they are obviously more complex than the first order model.

In the flood discharge example above, and in all time series, the direction of dependence can be stated explicitly. The concepts of forwards and backwards along a line, or of before and after in time, allow the specification of the direction of dependence. These concepts cannot easily be transferred to the two dimensional or spatial situation. In such cases, dependence may exist in all directions around point $i$. Taking the example of a regular pattern of points, the disturbance at the point with coordinates $(p,q)$ could be influenced by, and in turn be an influence upon, the disturbances at points $(p+1,q)$, $(p-1,q)$, $(p,q+1)$ and $(p,q-1)$. Assuming only a first order dependence, as in the flood model discussed above, the statistical model for the autoregressive structure of the disturbances is

$$\epsilon_{pq} = a\epsilon_{p+1,q} + b\epsilon_{p-1,q} + c\epsilon_{p,q+1} + d\epsilon_{p,q-1} + u_{pq}$$

An example of a simple situation involving two dimensional dependence is the pattern resulting from throwing a stone into the calm waters of a lake or pond. The shock waves extend outwards in all directions from the point of impact, and the amplitude of the wave at a given point $(p,q)$ is a function of the effects of local circumstances such as water depth, or presence of vegetation, and the strength of the initial disturbance. Adjacent points will exhibit a similar pattern of disturbance, and the nature of the dependence can be inferred from the generating process, in this case the impact of the stone on the water surface. If the generating process is known, the direction of the relationship can be stated, but in some circumstances knowledge of the generating processes is inadequate.

The phenomena that are discussed above are referred to as autocorrelation. In regression analysis it is autocorrelation of the disturbance terms that is of interest, though in other situations involving, for example, the study of spatial processes, autocorrelation structures in observed variables will be of interest. The situations given above as examples both involve first order autoregressive disturbances. The analysis of autocorrelation structures of observed variables can be carried out using the methods of spectral analysis (Rayner, 1971; Yevyevitch, 1972; Agterberg, 1974). However, adequate procedures exist to deal with the problem of autocorrelated residuals in regression.

The effects of autocorrelated disturbances on the OLS estimates are the same whether or not the pattern of observations is one- or two-dimensional. These effects are: (a) OLS estimators of are neither efficient nor asymptotically efficient; (b) the variances of the $\hat{\beta}_i$ and $e_i$ are biased, and (c) predictions based on the OLS estimates will be inefficient. The direction of bias in the variances of the parameters will be downwards if the autocorrelation is positive, resulting in confidence intervals that are too narrow and in incorrect values for the significance tests described earlier in this chapter. The OLS estimators of the parameters are still unbiased and consistent, even if the disturbances are autocorrelated, although it is likely that their numerical values for a given sample will be incorrect.

Before the alternative procedures are described, tests for autocorrelated disturbances are discussed. In the case of one-dimensional series, several tests have been put forward by econometricians. The best-known of these is the Durbin-Watson test (Kmenta, 1971, p. 295). Work on statistical tests for the presence of autocorrelation among regression residuals in the two-dimensional situation is of relatively recent origin (Cliff and Ord, 1973) and their methods have not, as yet, been widely applied. The Durbin-Watson test involves the calculation of a statistic $d$, defined as

$$d = \frac{\sum_{i=2}^{n}(e_i - e_{i-1})^2}{\sum_{i=1}^{n} e_i^2}$$

The $e_i$ are the residuals calculated from a particular sample, and are estimates of the population disturbances, the $\epsilon_i$. To test the null hypothesis that there is no autocorrelation among the disturbances, refer to tables of the critical points of $d$ (these tables can be found in Durbin and Watson, 1951, reproduced as Table D-5A in Kmenta, 1971). The table gives a lower limit, $d_L$ and an upper limit, $d_U$. The decision rules are as follows: (a): reject $H_0$ if $d < d_L$; (b) do not reject $H_0$ if $d < d_U$ (c) the test is inconclusive if $d_L \leq d \leq d_U$. These rules should be used if the alternative hypotheses is $H_1 : \rho > 0$. For a two-tailed test, with $H_1 : \rho \neq 0$ the decision rules are: (a) reject $H_0$ if $d < d_L$ or if $d > 4 - d_L$; (b) do not reject $H_0$ if $d_U < d < 4 - d_U$; (c) the test is inconclusive if $d_L \leq d \leq d_U$ or if $4 - d_U \leq d \leq 4 - d_L$.

The test statistic for use in the case of two-dimensional autocorrelation has been derived by Cliff and Ord (1973). Readers are referred to this monograph for full details and extensive worked examples. The distribution for the test statistic, termed $I$, has been evaluated under two different sets of assumptions: (i) assumption $N$, which requires that the $\epsilon_i$ are a random sample of size $n$ drawn from a normally-distributed population, and (ii) assumption $R$, randomization. Irrespective of the underlying distributional form of the population, the observed value of $I$ is considered relative to the set of all possible values $I$ could have taken if the $\epsilon_i$ had been randomly permuted around the system of point locations or areas. Assumption $N$ is the more appropriate if the regression model is being used, though assumption $R$ is valuable if the sampling points are fixed locations and do not represent a true random sample. Under assumption $N$, the distribution of $I$ is defined (a) for the case of a single explanatory variable, and (b) for several explanatory variables.

In the case of a single explanatory variable, $I$ is derived from the residuals $e_i$ calculated for a sample of size $n$ as follows:

$$I = \frac{n \sum_{i=1}^{n} \sum_{j=1}^{n} w_{ij} e_i e_j}{W \sum_{i=1}^{n} e_i^2} \quad (i \neq j)$$

where $w_{ij}$ is the $(i,j)^{\text{th}}$ element of the weight matrix $\mathbf{W}$, the derivation of which is described below. The scalar $W$, not to be confused with the matrix $\mathbf{W}$, is the sum of the $w_{ij}$, that is,

$$W = \sum_{\substack{i=1 \\ (i \neq j)}}^{n} \sum_{j=1}^{n} w_{ij}$$

The expected value of $I$ under assumption $N$ is:

$$E(I) = -\frac{1 + I_{1x}}{(n-2)}$$

$I_{1x}$ is the first order autocorrelation coefficient for $x$, given by:

$$I_{1x} = n\mathbf{x}^{*\prime}\mathbf{W}\mathbf{x}^*/W\mathbf{x}^{*\prime}\mathbf{x}^*$$

Lastly, the variance of $I$, $\text{var}(I)$, is computed from:

$$\text{var}(I) = \frac{1}{n(n-2)} \left[ \frac{n^2 S_1 - n S_2 + 3W^2}{W^2} + 2I_{1x}^2 - \left(\frac{n}{w}\right)^2 \frac{\sum_{i=1}^{n}(v_i - \bar{v})^2}{\mathbf{x}^{*\prime}\mathbf{x}^*} \right] - \frac{1}{(n-2)^2}$$

where

$$S_1 = \tfrac{1}{2} \sum_{\substack{i=1 \\ (i \neq j)}}^{n} \sum_{j=1}^{n} (w_{ij} + w_{ji})^2,$$

$$S_2 = \sum_{i=1}^{n} \left[ \sum_{j=1}^{n} w_{ij} + \sum_{j=1}^{n} w_{ji} \right]^2,$$

$$v_i = \sum_{j=1}^{n} (w_{ij} + w_{ji}) x_j$$

and

$$x_i^* = x_i - \bar{x}.$$

The distribution of $I$ is asymptotically normal, so that the ratio $(I - E(I))/(\text{var}(I))^{\frac{1}{2}}$ is approximately normally distributed with zero mean and variance 1. Thus, by reference to tables of the standard normal distribution, an appropriate percentage point can be read off from the upper or the lower tail. Note that a negative value of $I$ indicates negative spatial autocorrelation.

To illustrate the use of the $I$ statistic in regressions involving one explanatory variable, the following simple example has been chosen. The dependent variable, $y$, is the rainfall in millimetres recorded at each of 12 raingauges in the Dove and Manifold valleys, Derbyshire/Staffordshire, on 15 July 1973. The explanatory variable $x$, is the altitude of the raingauge in metres. Logarithms of both variables were used, as this increased $R^2$ from approximately 0.58 to 0.68. The regression equation was:

$$y = 0.549 + 0.473x$$
$$(0.244)\ (0.102)$$

The $t$-test of the slope coefficient (0.473) was significant at the 95 per cent level, so a positive relationship between rainfall amount and altitude can be shown to be likely. However, the estimates are unreliable if there is autocorrelation among the residuals, so the $I$ statistic was computed using $w_{ij} = 1$ if the $i^{\text{th}}$ and $j^{\text{th}}$ raingauge sites were first nearest neighbours, and $w_{ij} = 0$ otherwise. This gave $I = -0.0958$, $I_{1x} = -0.3998$, $E(I) = -0.0600$, and the standard error of $I$ (defined as the square root of var $(I)$) was 0.2279. The resulting standard normal deviate was $-0.1571$, which is well below the 95 per cent point of the lower tail of the standard normal distribution, $-1.96$. The null hypothesis of zero spatial autocorrelation among the regression residuals is therefore accepted.

Before going on to the case of several variables, the derivation of the matrix **W** deserves some attention. The form of **W** should reflect the investigator's knowledge of the generating process, that is, the process producing the autocorrelation. In the absence of such knowledge, empirical values of $w_{ij}$ have to be used. In these cases, several alternative forms of weight should be tried. In theory, the $w_{ij}$ should reflect the interaction between points $i$ and $j$. Thus, $w_{ij}$ may be the reciprocal of the distance (or of the square root of the distance) between points $i$ and $j$. If both $i$ and $j$ are areas and not points, then $w_{ij}$ could measure the proportion of the boundary of area $i$ that is in contact with area $j$, excluding any sections of boundary that form the edge of the study area. A third possibility is to allow $w_{ij}$ to reflect both distance and shared boundary length. A simpler form of weighting is to give $w_{ij}$ the value 1 if points $i$ and $j$ are first (or second nearest neighbours, and set $x_{ij}$ to zero otherwise. Notice that, in general, $w_{ij}$ need not be equal to $w_{ji}$. Also, $w_{ii}$ is not defined. The choice of $w_{ij}$'s is important, for the values of $E(I)$ and var $(I)$ are dependent upon the values of the weights. It appears that forms of **W** involving distances and their reciprocals emphasise the spatial autocorrelation among residuals more strongly than binary (1/0) weights.

The procedure for evaluating $I$ can now be extended to the case of several explanatory variables. As before, $\mathbf{X}_{(1)}$ is the matrix containing the column vectors $\mathbf{x}_1, \mathbf{x}_2, \ldots, \mathbf{x}_k$ with $\mathbf{x}_1$ being a unit vector and $\mathbf{x}_2$ to $\mathbf{x}_k$ the observations on the $(k-1)$ explanatory variables. The value of $I$ for the sample is

$$I = \frac{n \sum_{\substack{i=1 \\ (i \neq j)}}^{n} \sum_{j=1}^{n} w_{ij} e_i e_j}{W \sum_{i=1}^{n} e_i^2}$$

and

$$E(I) = -I_{1x}/(n-k)$$

with

$$I_{1x} = (n/W) \sum_{\substack{i=1 \\ (i \neq j)}}^{n} \sum_{j=1}^{n} w_{ij} d_{ij}$$

and

$$d_{ij} = \mathbf{x}_i^{\circledR} (\mathbf{X}_{(1)}' \mathbf{X}_{(1)})^{-1} \mathbf{x}_j^{\circledR}$$

The symbol $\circledR$ indicates that the $(1 \times k)$ vectors $\mathbf{x}_i^{\circledR}$ and $\mathbf{x}_j^{\circledR}$ are *rows* of $\mathbf{X}_{(1)}$ and not columns, as in normal usage. The variance of $I$ is calculated by the cumbersome formula given below (Cliff and Ord, 1973).

$$\operatorname{var}(I) = \frac{n}{W^2(n-k)} \left\{ \frac{n^2 S_1 - n S_2 + 3 W^2}{n^2} \right.$$

$$\left. + \frac{1}{n} \sum_{\substack{i=1 \\ (i \neq j)}}^{n} \sum_{j=1}^{n} \left( \sum_{m=1}^{n} w_{im} + \sum_{m=1}^{n} w_{mi} \right) \left( \sum_{m=1}^{n} w_{jm} \right) \right.$$

$$+ \sum_{m=1}^{n} w_{mj} \Bigg) d_{ij} + 2 \left( \sum_{\substack{i=1 \\ (i \neq j)}}^{n} \sum_{j=1}^{n} w_{ij} d_{ij} \right)^{2}$$

$$- \Bigg[ \sum_{\substack{i=1 \\ (i \neq j \neq k)}}^{n} \sum_{j=1}^{n} \sum_{k=1}^{n} (w_{ik} + w_{ki})(w_{jk} + w_{kj}) d_{ij}$$

$$+ \sum_{\substack{i=1 \\ (i \neq j)}}^{n} \sum_{j=1}^{n} (w_{ij} + w_{ji})^{2} d_{ii} \Bigg] + \frac{1}{n} \sum_{\substack{i=1 \\ (i \neq j \neq k)}}^{n} \sum_{j=1}^{n} \sum_{k=1}^{n} (w_{ij} + w_{ji})(w_{ik}$$

$$+ w_{ki})(d_{ii} d_{jk} - |d_{ij} d_{ik}|) \Bigg\} - (n-k)^{-2}$$

The transformation $(I - E(I))/(\text{var}(I))^{\frac{1}{2}}$ is again used to give a standard normal deviate, for large $n$.

If these tests show that it is likely that the disturbance terms are autocorrelated, than one or more of three conclusions can be drawn: (a) the form of the model is incorrect: (b) one or more relevant explanatory variables have been omitted, or (c) the disturbances are truly autocorrelated, that is, they have an autoregressive structure. The form of the regression model is usually assumed to be linear; a simple example will show how apparent autocorrelation can be produced by the use of a linear model when the true model is curvilinear or nonlinear. In Figure 2.8 the regression line (a) has positive residuals at

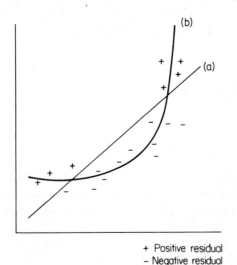

+ Positive residual
− Negative residual
[with respect to curve (a)]

Figure 2.8. Apparently autocorrelated residuals (See text for explanation)

low $x$ values, negative residuals at medium $x$ values and positive residuals again at high $x$ values. Visual examination of this diagram shows that a closer fit, giving apparently non-autocorrelated residuals, is achieved if regression curve (b) is fitted. Nonlinear models, or transformation of the terms in a linear model, may therefore be appropriate if it is thought that the form of the model is incorrectly specified.

Omission of one or more explanatory variables can leave some systematic variation remaining in the residuals after the variation due to the included explanatory variables has been removed. Plotting the computed residuals against possible candidates for inclusion as explanatory variables may lead to the identification of the missing $x$'s. If the data points are spatially distributed, a map of the residual values may lead to similar conclusions. This approach is followed by Benson (1962).

If the form of the model, and the number of explanatory variables have been correctly specified, the third possibility is that the disturbances have an autoregressive structure. This implies that the variance-covariance matrix of the disturbances ($E(\epsilon_i \epsilon_j)$) is not of the form assumed by the classical model, which requires that

$$E(\epsilon_i \epsilon_j) = \sigma^2 I \quad (i,j = 1,2,\ldots,n)$$

As in the case of heteroscedastic variances, the method of generalized least squares can be used, but only if the value of $\rho$, the serial correlation coefficient, is known. This is unusual in practice, so details of the GLS approach are not given here. Details can be found in Johnston (1972, p. 259). Several methods of estimating $\rho$ and the regression parameters have been suggested. Two of the most widely used are the Cochrane–Orcutt iterative procedure (CO) and the Durbin two-stage algorithm (D-2). The CO and the D-2 methods use the following transformation to achieve a situation in which the disturbances are not autocorrelated. OLS estimators can then be used. The transformation is:

$$(y_i - \rho y_{i-1}) = \beta_0(1-\rho) + \beta_1(x_i - \rho x_{i-1}) + u_i \qquad (2.44)$$

assuming that

$$\epsilon_i = \rho \epsilon_{i-1} + u_i$$

It will be shown later that both schemes can be easily modified to allow for (i) higher-order autoregressive structures, and (ii) more than one explanatory variable. The CO procedure is as follows:

(a) use OLS to find estimates of the parameters of the model

$$y = \hat{\beta}_0 + \hat{\beta}_1 x + e_i$$

(b) compute the residuals $e_i$ for the OLS regression and from these residuals obtain

$$\hat{\rho} = \sum_{i=2}^{n} e_i e_{i-1} \bigg/ \sum_{i=2}^{n} e_{i-1}$$

(c) now transform the $x$ and $y$ terms in the model as suggested in Equation (2.44) using $\hat{\rho}$ as an estimator of $\rho$. This will reduce the number of variables from $n$ to $(n-1)$. Now find estimates $\hat{\beta}_0^*$ and $\hat{\beta}_1$ in the equation

$$(y_i - \hat{\rho} y_{i-1}) = \hat{\beta}_0^* + \hat{\beta}_1 (x_i - \hat{\rho} x_{i-1}) + u_i$$

The $u_i$ are the new error terms. If the estimator $\hat{\rho}$ was accurate and if the $\epsilon_i$ were related by a first order autoregressive relationship then a Durbin-Watson test on the $u_i$ should result in acceptance of the hypothesis that there is no autocorrelation amongst them. If this is so, then the process has converged. All that remains is to backtransform $\hat{\beta}_0^*$ to give an estimator $\hat{\beta}_0$ of $\beta_0$. This is described below. If the null hypothesis is not accepted, then:

(d) Find revised estimates of the $e_i$ from

$$e_i = (y_i - \hat{\rho} y_{i-1}) - (\hat{\beta}_0^*/(1-\hat{\rho}) + \hat{\beta}_1 x_i)$$

and calculate a new estimate of $\rho$, written $\hat{\hat{\rho}}$. Now repeat step (c) and, if necessary, step (d) until either (i) the estimators of $\rho$ converge or (ii) the null hypothesis of zero autocorrelation among the $u_i$ is accepted.

The estimator of $\beta_0$ is obtained from $\hat{\beta}_0^*$ as follows:

$$\hat{\beta}_0 = \hat{\beta}_0^*/(1-\tilde{\rho})$$

where $\tilde{\rho}$ is the most recent estimate of $\rho$.

The CO method is discussed in several textbooks on econometrics (Johnston, 1972; Kmenta, 1971). It seems to be agreed that there is little necessity to go beyond two or three iterations. Convergence of the procedure is assured, but the terminating point may represent a local rather than a global minimum (see Chapter 3). However, there is no evidence that multiple minima are common (Johnston, 1972).

The CO procedure can be extended simply to deal with several explanatory variables and to provide for second and higher order autoregressive structures. If there are several $x$'s, simply transform each one using $(x_{ij} = x_{ij} - \hat{\rho} x_{i,j-1})$ and compute estimates of the $\beta_i$ as usual. Remember to backtransform $\hat{\beta}_0^*$. For higher order autoregressive structures, the transformation is $(x_{ij} - \hat{\rho}_1 x_{i,j-1} - \hat{\rho}_2 x_{i,j-2} \ldots)$. Some authors have suggested that, when estimating the $\rho_i$ in the autoregressive model

$$\epsilon_i = \rho_1 \epsilon_{i-1} + \rho_2 \epsilon_{i-2} + \ldots + u_i$$

the standard errors of the coefficients $\hat{\rho}_i$ are obtained in the usual manner and $t$-tests are carried out to determine the probability that each $\hat{\rho}_i$ is estimating a population $\rho$ that is significantly different from zero. Only those $\hat{\rho}_i$ that are significant should be used in the transformation of $y$ and the $x$'s.

An alternative to the CO method is the D-2 method. This is a two-stage estimation procedure and involves

(i) rewriting the regression model in the form

$$y_i = \beta_0^* + \rho y_{i-1} + \beta_1 x_i + \beta_2 x_{i-1} + u_i$$

(ii) Use OLS to obtain estimates of the parameters $\beta_0^*$, $\beta_1$ and $\beta_2$. The estimator of $\rho$, denoted by $\hat{\rho}$, is then used to construct new variables $(y_i - \hat{\rho} y_{i-1})$ and $(x_i - \hat{\rho} x_{i-1})$.

(iii) Estimate the parameters of the model

$$(y_i - \hat{\rho} y_{i-1}) = \beta_0^* + \beta_1 (x_i - \hat{\rho} x_{i-1}) + u_i$$

As in the CO procedure, the estimator $\hat{\beta}_0^*$ must be backtransformed to give an estimator of $\hat{\beta}_0$. The relationship is:

$$\hat{\beta}_0 = \hat{\beta}_0^* / (1.0 - \hat{\rho})$$

The D-2 procedure can be extended to deal with the case of more than one explanatory variable, and with higher-order autoregressive structures, in the same way that the CO procedure was modified. Details of the necessary modifictions are given above.

The use of the CO procedure is illustrated using a data set consisting of 31 daily measurements of mean sediment concentration (y) and mean discharge (x) for October, 1966 for the Smoky Hill River, Kansas. The observations are listed in Table 2.10. OLS estimation gave the following regression equation:

$$y = 33.31 + 0.3087x$$

(As in the case of the discharge/area example above, the parameter $\beta_1$ could have been estimated subject to the constraint $\beta_0 = 0$, but — since this example is meant to be purely illustrative — the unconstrained regression equation is evaluated.) The value of the Durbin–Watson statistic $d$ computed from the residuals from this regression was 0.9574. Using the alternative hypothesis of positive autocorrelation, the significance points at the 5 per cent level are $d_L = 1.36$ and $d_U = 1.50$ for $n = 31$. Since $d$ is less than $d_L$ the null hypothesis that there is zero autocorrelation among the disturbances is

Table 2.10
Data for autocorrelation example: mean discharge and suspended sediment concentration, Smoky Hill River, Kansas, October 1966 (Reproduced with permission from Kansas Geol. Survey, *Spec. Distr. Publicn. 48*, 1970, Figure 8)

| Day | Mean discharge (cfs) | Mean suspended sediment concentration (mg/l) |
| --- | --- | --- |
| 1 | 197 | 86 |
| 2 | 194 | 95 |
| 3 | 197 | 85 |
| 4 | 190 | 92 |
| 5 | 178 | 72 |
| 6 | 178 | 64 |
| 7 | 178 | 68 |
| 8 | 178 | 73 |
| 9 | 171 | 69 |
| 10 | 171 | 66 |
| 11 | 164 | 48 |
| 12 | 152 | 49 |
| 13 | 146 | 49 |
| 14 | 140 | 62 |
| 15 | 328 | 66 |
| 16 | 566 | 62 |
| 17 | 730 | 280 |
| 18 | 800 | 270 |
| 19 | 840 | 280 |
| 20 | 638 | 310 |
| 21 | 315 | 310 |
| 22 | 220 | 290 |
| 23 | 186 | 66 |
| 24 | 171 | 65 |
| 25 | 167 | 66 |
| 26 | 164 | 96 |
| 27 | 161 | 82 |
| 28 | 161 | 82 |
| 29 | 152 | 100 |
| 30 | 143 | 98 |
| 31 | 143 | 100 |

not accepted at the 5 per cent level. The regression of $e_i$ on $e_{i-1}$ gave

$$e_i = 0.5211 e_{i-1} \quad (i = 2, 3, \ldots, n)$$

and new variables $(y_i - 0.5211 y_i{-}1)$ and $(x_i - 0.5211 x_{i-1})$ were calculated. OLS estimation of the regression equation involving these new variables produced the equation

$$(y_i - 0.5211 y_{i-1}) = 24.1843 + 0.2499(x_i - 0.5211 x_{i-1})$$

The Durbin–Watson $d$ was computed for the residuals from this regression; its value was $d = 1.7470$. With $n = 30$ (since one observation was lost in the transformation) the upper and lower significance points at the 5 per cent level are 1.49 and 1.35. The computed value of $d$ exceeds $d_U$ so the alternative hypothesis of positive autocorrelation is rejected in favour of the null hypothesis. The best predictions of sediment concentration from mean daily discharge are therefore to be achieved by the equation

$$y_i = 24.1843 + 0.2499(x_i - 0.5211x_{i-1}) + 0.5211y_{i-1}$$

The value of $\hat{\beta}_0$ can be transformed to give the regression in terms of the original variables. This is:

$$y_i = 50.4980 + 0.2499x_i$$

It should be noted that the use of the correction for autocorrelation results in a reduction in the $R^2$ value; thus, for the OLS equation $R^2 = 0.5290$ whereas the $R^2$ value for the CO equation is 0.5096. This is to be expected, as the OLS regression line is, by definition, the best-fit line.

An example of the use of the Cliff and Ord I statistic is given in Section 3.2.2 in connection with trend surface analysis. The use of methods similar to the CO and D-2 procedures with spatial data is complex. Cliff and Ord (1973) provide details of the computational method.

Chapter 3

# Further Aspects of the Least-squares Methods

## 3.0 Introduction

The multiple linear regression model, described in Chapter 2, is only one application of the general linear model. In the present chapter other aspects of this general model are considered, as well as its extension to nonlinear problems. The problem of fitting a polynomial function of one explanatory variable is described in Section 3.1. Such functions are generally used in curve-fitting rather than hypothesis-testing situations; for example, an empirical relationship between stage and discharge is often required in applied hydrological research. An extension of the polynomial function to include two explanatory variables is treated in Section 3.2. In trend surface analysis the two explanatory variables are the map coordinates of a set of data points. The resulting function is therefore a two-dimensional surface or trend surface. Trend surface analysis has been widely used — and widely criticized — by physical geographers. It has often been confused with the mathematical technique of surface fitting. Although the same numerical techniques can be employed in both cases, the aims and assumptions of the two approaches differ quite fundamentally. Furthermore, the requirements of the technique are not always fully appreciated; it is unwise to expect an apparently complex technique of analysis to compensate for faults in experimental design.

Methods of selecting a subset of important explanatory variables for use in a prediction equation are covered in Section 3.3. Many methods are available, the most widely-used being the stepwise regression technique. Other selection procedures have been made feasible by the availability of large and fast computers. Some of these methods are discussed, with examples of their application.

The third section of this chapter is concerned with an aspect of regression analysis that is not, perhaps, as widely appreciated as it might be, namely, the use of dummy explanatory variables. By the use of this straightforward modification, the multiple regression model can be extended to include (in a more flexible form) the analysis of variance. (See Darlington, 1968, and Searle, 1971).

The closing section deals with the nonlinear least-squares problem. In the introduction to Chapter 2 it was stated that the linear regression model is generally only an approximation (and sometimes not a very good approximation) to the true structure of the data. Nonlinear minimization techniques have been developed to deal efficiently with the estimation of the parameters of the nonlinear least-squares model. The estimates of the parameters of the nonlinear model are arrived at by an iterative procedure which requires initial approximations of their correct values. This, together with the necessity to state the form of the model in advance, means that nonlinear models are used only when a body of experience and theory is available. This has limited their application in physical geography to a large extent.

## 3.1 Least Squares Curve-Fitting
### 3.1.0 Introduction

The visual interpolation of a 'best-fit' curve through a bivariate scatter diagram is one of the most common starting points to the analysis of experimental data. The aim may be to construct a curve from which the values of $y$ for any value of $x$ in a given range can be read. An example of this is the construction of stage-discharge curves by hydrologists. Alternatively, the aim may be to identify a trend, which may or may not be linear, in a set of data relating to measurement of a single variable over time or along a transect. A third possibility is the fitting of a statistical model to the data. The relationship between a dependent variable, $y$, and a set of explanatory variables consisting of the successive nonnegative integral powers of $x$, may be postulated *a priori* and — if the assumptions of the regression model (Chapter 2) are satisfied — the normal procedures of the linear regression model can be used. Usually it is not possible to specify a polynomial model in $x$ beyond the second or third degree.

Curve-fitting or interpolation should be used with a combination of caution and common sense. If a polynomial function approximates the form of relationship between $y$ and $x$ it does not follow that the *true* relationship is of this form, for it can be shown that *any* single-valued function can be represented to a required degree of accuracy by a polynomial function of sufficiently high degree. In interpolation the aim is to find the best fitting curve, not to identify the mechanism which generates the observed data values. This distinction between approximation and model-fitting is a vital one.

The computation of the least-squares coefficients in polynomial approximation is often inaccurate and, in some cases, grossly so. This is due to the fact that the matrix of sums of squares and cross-products of the powers of $x$ becomes progressively more ill-conditioned as the degree of the polynomial is increased. Consequently, the solution of the equations $\hat{\boldsymbol{\beta}} = (\mathbf{X}'\mathbf{X})^{-1}\mathbf{X}'\mathbf{y}$ is progressively more inaccurate. Wampler (1970) has carried out a comparative study of methods of solving for $\hat{\boldsymbol{\beta}}$ in this situation, and his results indicate that the method of matrix inversion using Gaussian elimination is the least successful, and — in some cases — is disastrously inaccurate. For example, in Wampler's first test ($y1$) (see below) one program using elimination methods produced the vector (2991622, −6065892, 2218821, −296194.5, 16462.20, −322.5731) when the true solution was (1, 1, 1, 1, 1, 1). A more accurate method is Gram—Schmidt orthogonalization, using the method of Bjork (1967). This is the method described here. It is relatively simple to program and use, and — even in badly-conditioned problems — is surprisingly accurate, though the degree of accuracy is affected by the size of the residuals (Mather, 1975A).

The method can also be used to estimate parameters in multiple regression or trend-surface analysis. In multiple regression there are advantages in keeping to the established procedures, especially if an iterative refinement inversion subroutine is used, as in Chapter 2. However, the Gram—Schmidt method is accurate, rapid, and does not require matrix inversion. Hence, it will be described in a general fashion so that readers can adapt it to multiple regression computations if they so desire. The use of the method in trend-surface analysis is described in the next section. Thus, $\mathbf{x}_i$ can be interpreted as (i) the $i$th explanatory variable in a multiple regression equation; or (ii) $x^i$ in a polynomial least-squares equation, or (iii) $u^p v^q$ in a trend-surface equation.

*3.1.1 Gram—Schmidt Method: Computation*

If a variable $y$ is approximated by a polynomial function in a single predictor variable $x$, the relationship can be expressed as:

$$y = a_0 + a_1 x + a_2 x^2 + \ldots + a_k x^k \tag{3.1}$$

The successive terms $1, x, x^2, \ldots, x^k$ in this polynomial function are correlated, i.e. they are non-orthogonal. In the terminology of regression, the explanatory variables are highly collinear and so the matrix $\mathbf{X}'\mathbf{X}$ (or its derivatives $\mathbf{P}$ and $\mathbf{R}$) will be ill-conditioned. An orthogonal set of variables, $\boldsymbol{\Phi}$, can be derived from the $x$'s; they are called *orthogonal polynomials*. The $\boldsymbol{\Phi}$ contain the same information as the $x$'s; they can be thought of as the projections of the $\mathbf{x}_i$ on an arbitrary orthogonal coordinate system. In this respect, they resemble superficially the principal components of $\mathbf{X}'\mathbf{X}$ (see Chapter 4). Equation (3.1) can be rewritten in terms of $\phi$ as:

$$y = b_0 \phi_0 + b_1 \phi_1 + b_2 \phi_2 + \ldots + b_k \phi_k \tag{3.2}$$

There are two steps to be performed — (i) conversion of the $x_i$, to the $\phi_i$, and (ii) determination of the orthogonal coefficients $b_i$ and their conversion to the polynomial coefficients $a_i$.

Several alternative methods of computing the elements of matrix $\Phi$ are available. Forsythe (1957) uses a recurrence relationship which is economical in terms of computer storage. This method is also described by Fox and Mayers (1968) and by Carnahan, Luther and Wilkes (1969). As described in these texts, the method can be used only for polynomials in a single predictor variable. An older method, (Rushton, 1951) which can also be used for the case of several predictors (thus allowing its application to multiple regression and trend surface problems) is based on the Cholesky decomposition of the symmetric matrix $X'X$ into a lower triangular matrix $L$ and an upper triangular matrix $L'$ (see Chapter 1). An even older method, dating back to the Nineteenth Century, is the Gram—Schmidt process. The 'classical' form of this algorithm is well known and is described in most numerical analysis texts (e.g. Ralston, 1965). It is not particularly accurate since the calculation of each orthogonal vector $\phi_i$ is based on the computed values of the preceding $\phi_i$, therefore if errors occur then they accumulate and the system may become unstable, that is, successive columns of $\Phi$ may no longer be orthogonal. A modified form of the Gram—Schmidt procedure is due to Bjork (1967) and is usually highly accurate. (Accuracy tends to decline as the goodness of fit of the regression line or trend surface approaches zero. It is also a function of the word-length of the computer used. See Mather, 1975A.) The modified Gram—Schmidt procedure does not require the matrix $(X'X)$ but works instead upon the matrix $X$ of raw data.

Since the position of the $\phi_i$ in the vector space spanned by the $x_i$ is arbitrary, it is customary to align $\phi_1$ along the line of $x_1$. The component of $x_2$ in the dimension defined by $\phi_1$ is calculated and subtracted from $x_2$ to give the second orthogonal vector, $\phi_2$, and so on.* This is, in fact, classical Gram—Schmidt and it can be written as:

$$\left.\begin{aligned}\phi_1 &= x_1 \\ \phi_2 &= x_2 - \frac{x_2' \cdot \phi_1}{\phi_1' \cdot \phi_1} \cdot \phi_1 \\ &\quad \vdots \\ \phi_k &= x_k - \sum_{i=1}^{k-1} \frac{x_k \cdot \phi_i}{\phi_i' \cdot \phi_i} \cdot \phi_i\end{aligned}\right\} \quad (3.3)$$

---

*From this point on, $k$ is used to indicate the total number of terms in the equation (=degree of polynomial + 1). This is done in order to allow subscripts to run from 1 to $k$.

It is usual to re-express this set of equations as

$$\left.\begin{array}{l} x_1 = c_{11}\phi_1 \\ x_2 = c_{21}\phi_1 + c_{22}\phi_2 \\ \quad \cdot \quad \cdot \\ \quad \cdot \quad \cdot \\ \quad \cdot \quad \cdot \\ x_k = c_{k1}\phi_1 + \ldots\ldots\ldots\ldots c_{kk}\phi_k \end{array}\right\} \quad (3.4)$$

where

$$\begin{array}{ll} c_{ij} = x_i'\phi_j/\phi_j'\phi_j & \text{for } i > j \\ c_{ij} = 1 & \text{for } i = j \\ c_{ij} = 0 & \text{for } i < j \end{array}$$

The matrix **C** is therefore lower triangular with a unit diagonal. If the $\phi_i$ are normalized to unit length and written $\phi_i^*$ these equations become:

$$\left.\begin{array}{l} x_1 = l_{11}\phi_1^* \\ x_2 = l_{21}\phi_1^* + l_{22}\phi_2^* \\ \quad \cdot \quad \cdot \\ \quad \cdot \quad \cdot \\ \quad \cdot \quad \cdot \\ x_k = l_{k1}\phi_1^* + \ldots\ldots\ldots\ldots + l_{kk}\phi_k^* \end{array}\right\} \quad (3.5)$$

with

$$\begin{array}{ll} l_{ij} = (\phi_j'\phi_j)^{1/2} \cdot c_{ij} & (i > j) \\ l_{ij} = (\phi_j'\phi_j)^{1/2} & (i = j) \\ l_{ij} = 0 & (i < j) \end{array}$$

The matrix **L** is also a lower triangular matrix and is the matrix **L** in the Cholesky factorization $\mathbf{A} = \mathbf{LL'}$. This is the link between Rushton's (1951) method and the Gram–Schmidt process.

Bjork (1967) has modified the above equations to minimize the incidence of rounding error. Instead of computing $\phi_i$ in one step, the following scheme is employed:

$$\left.\begin{array}{l} x_i^{(1)} = x_i - \dfrac{x_i'\phi_1}{\phi'\phi} \cdot \phi_1 \\[1em] x_i^{(2)} = x_i^{(1)} - \dfrac{x_i^{(1)'}\phi_2}{\phi_2'\phi_2} \cdot \phi_2 \\[1em] \phi_i = x_i^{(i-2)} - \dfrac{x_i^{(i-2)'}\phi_{i-1}}{\phi_{i-1}'\phi_{i-1}} \cdot \phi_{i-1} \end{array}\right\} \quad (3.6)$$

This method is used in Subroutine GSORTH (Appendix A).

The elements of **L** can be interpreted as the coordinates of the $x_i$ with reference to an orthonormal basis, $\phi_i^*$. The projection of $x_i$ onto the subspace containing $x_1, x_2, x_3, \ldots, x_{i-1}$ is the vector

$$l_{i1}\phi_1^* + l_{i2}\phi_2^* + \ldots + l_{i,i-1}\phi_{i-1}^*$$

Since the $\phi_i^*$ have unit length, the square of this vector is

$$l_{i1}^2 + l_{i2}^2 + \ldots + l_{i,i-1}^2$$

The perpendicular distance of $x_i$ from the sub-space spanned by $x_j$ ($j = 1, i - 1$) is $l_{ii}$ ($= \phi_i'\phi_i$). If the angle between $x_i$ and the subspace is $\Psi_i$ then it is possible to verify that $\sin^2 \Psi_i = l_{ii}^2/(x_i'x_i)$. Low values of $\Psi_i$ mean that $x_i$ is almost collinear with the preceding $x_j$ ($j = 1, i - 1$) and hence nearly linearly dependent on them. The vectors $x_i$ and $\phi_i^*$ can thus be eliminated if $\sin^2 \Psi_i$ is small ($<1.0E-5$). Such a procedure could reduce the effects of multicollinearity or provide a method of selecting linearly independent subsets of regressors, from a larger set, as in stepwise regression (Section 3.3).

If the modified Gram—Schmidt process is carried out on the partitioned matrix (**X** | **y**), which contains an initial column of ones, the values of the $k$ predictor variables in columns 2 to $k + 1$, and the values of the dependent variable in column $k + 2$, the following results are derived:

(i) the coefficients $b_i$ ($i = 1, k$) of the orthogonal polynomials (Equation 3.2) are contained in the last row of **L**, excluding the diagonal term;

(ii) the residuals from the least-squares curve are held in the column vector $\phi_{k+2}$;

(iii) the residual sum of squares is $(\phi_{k+2}'\phi_{k+2})$;

(iv) the relative contributions of the $x_i$ to the explained sum of squares is $l_{k+2,k+2}^2$.

Recovery of the $a_i$ of Equation (3.1) from the $b_i$ of Equation (3.2) is straightforward; firstly, $a_k$ is derived from

$$a_k = b_k/l_{kk}$$

and the remaining $a$'s from

$$a_i = \left\{ b_i - \sum_{j=i+1}^{k} (l_{ji} \cdot a_j) /l_{ii} \right\} \quad (i = k-1, k-2, \ldots, 1) \quad (3.7)$$

In subroutine GSORTH (Appendix A) it is possible to add extra terms to the least-squares equation without recalculating previous values. If the approach used in Chapter 2 had been adopted then the equation $a = (X'X)^{-1}X'y$ would need to be solved for every combination of $x$'s. However, in using orthogonal polynomials the first $p$

column vectors $\phi_i^*$ $(i = 1, p)$ remain the same even when additional terms are added. Consequently, only the $\phi_i^*$ corresponding to the newly-added terms need to be calculated. Since many polynomial curve-fitting exercises are sequential, beginning with $y = a_0 + a_1 x$, then adding $x^2$, $x^3$, and so on, this feature is of considerable significance.

Comments within subroutines GSORTH and RECOV give a sufficiently detailed explanation of the workings of those routines. Double-precision accumulation of inner products is used in both routines; this is probably wise if the computer on which the routine is implemented has a single-precision word length of 48 bits, and essential if the word length is 36 or 32 bits. The performance of GSORTH in the five test problems of Wampler (1970) is quite acceptable. These test problems are given by the equations

$$y1 = 1 + x + x^2 + x^3 + x^4 + x^5 \quad (x = 0, 1, \ldots, 20)$$
$$y2 = 1 + 0.1x + 0.01x^2 + 0.001x^3 + 0.0001x^4 + 0.00001x^5$$
$$(x = 0, 1, \ldots, 20)$$
$$y3 = y1 + \Delta$$
$$y4 = y1 + 100 \Delta$$
$$y5 = y1 + 1000 \Delta$$

where $\Delta'$ is the vector (759, −2048, 2048, −2048, 2523, −2048, 2048, −2048, 1838, −2048, 2048, −2048, 1838, −2048, 2048, −2048, 2523, −2048, 2048, −2048, 759).

A measure of the condition of a set of linear simultaneous equations is given by the ratio of the maximum and minimum eigenvalues of the associated coefficients matrix. This ratio is termed the P condition number of the coefficients matrix. It can range from 1 for an orthogonal matrix to infinity for a singular matrix. The P condition number for the coefficients matrix for the second test problem is $4.095 \times 10^{13}$, and for test problems 1, 3, 4, and 5 it is $6.829 \times 10^{13}$. All five test problems have poorly conditioned coefficients matrices, and considerable error can therefore be expected in the solution of the equations. Wampler (1970) describes a method of determining the number of correct digits attained by a particular solution. Negative values can occur, and their absolute value indicates the number of orders of magnitude by which the result is in error. A summary of the performance of GSORTH relative to the algorithms tested by Wampler is given in Table 3.1.

The entries in Table 3.1 show average number of correct digits. Program ORTHO is from the Omnitab package and uses classical Gram–Schmidt, with reorthonormalizing, which increases accuracy. It was run on a Univac 1108 (D = 8 decimal digits in single precision) with double precision accumulation of inner products (DD = 18 decimal digits). BMD 03R is from the BMD package, uses Gaussian elimination

Table 3.1
Comparison of GSORTH with other algorithms. (For explanation see text.)

| Algorithm | Problems | | | | |
|---|---|---|---|---|---|
| | y1 | y2 | y3 | y4 | y5 |
| GSORTH | 7.23 | 9.08 | 7.42 | 5.70 | 3.70 |
| ORTHO | 6.46 | 3.90 | 3.90 | 3.23 | 1.34 |
| BMD 03R | 2.29 | −0.12 | 1.04 | 1.04 | 1.21 |
| BJORK | 5.78 | 3.90 | 3.89 | 2.00 | 0.02 |
| BJORK-GOLUB | 6.23 | 8.00 | 8.00 | 8.00 | 6.91 |

based on the correlation matrix, and was run on a Univac 1108 in single precision (D = 8). BJORK uses an algorithm similar to GSORTH. The results quoted were obtained on a Univac 1108 in single-precision (D = 8). BJORK-GOLUB is a routine implementing the method of orthogonal Householder transformations. It was also run on an 1108 with double-precision accumulation of inner products. GSORTH was run on an 1CL 1906A (D = 11, DD = 21). It produced results not markedly inferior to BJORK-GOLUB on y3, superior on y1 and y2, and somewhat poorer on y4 and y5 (though still better than other algorithms).

## 3.1.2 Examples

The construction of stage-discharge or rating curves is used here as an example of curve-fitting and interpolation. Given a series of measurements $(h_i, q_i)$ where $h_i$ is the height of the water surface and $q_i$ the discharge then an interpolating formula of the type $q_i = f(h_i)$ is required so that a value of $Q$ can be read off for any value of $H$ (Bruce and Clarke, 1966, pp. 87–90; Ward, 1967, p. 387). The data for this particular example comes from Yevyevitch (1972, p. 175) and are listed in Table 3.2 together with the residuals from the first-degree and second-degree polynomial curves. The equation of the least-squares linear (first-order) curve is:

$$q_i = -24.349 + 3.0141 h_i$$

The pattern of residuals from this line shows a distinct systematic variation, the signs being positive in the case of observations 1–5, negative for 6–10 and positive again for 11–13. The graph of the function (Figure 3.1) confirms the diagnosis that a more flexible curve should be fitted. However, out of the total sums of squares of $0.17556 \times 10^7$ the linear fit explains $0.15816 \times 10^7$ or 90.1 per cent.

The second-degree polynomial (quadratic) is

$$q_i = 29.687 + 0.56897 h_i + 0.71988 \times 10^{-2} h_i^2$$

Table 3.2

Data for curve-fitting example. The residuals from the first and second order curves (Figures 3.1 and 3.2) are also shown. (Reproduced with permission from V. Yevyevich, *Probability and Statistics in Hydrology*, Water Resources Publications, Ft. Collins, 1972, p. 175)

|    | $H$ (stage) cm | $Q$ (discharge) cu. metres per second | residual $(Q - \bar{Q})$ (i) Linear | (ii) Quadratic |
|----|------|---------|--------|-------|
| 1  | −23  | 15.55   | 4.86   | 0.43  |
| 2  | −22  | 15.46   | 5.19   | 1.04  |
| 3  | −16  | 20.07   | 2.36   | −0.22 |
| 4  | −16  | 21.99   | 0.43   | −2.14 |
| 5  | 14   | 36.11   | 2.95   | 6.02  |
| 6  | 33   | 59.82   | −3.52  | 1.41  |
| 7  | 46   | 86.58   | −15.40 | −9.96 |
| 8  | 69   | 110.96  | −7.74  | −2.31 |
| 9  | 88   | 136.52  | −1.02  | 3.37  |
| 10 | 120  | 204.40  | −2.77  | −1.63 |
| 11 | 136  | 232.87  | 7.35   | 6.37  |
| 12 | 220  | 492.50  | 10.78  | −2.60 |
| 13 | 400  | 1412.48 | 3.40   | 0.22  |

Figure 3.2 shows this to be a much closer fit to the data than the first-order curve. The explained sum of squares is 99.97 per cent of the total. Hence, this curve is as close to the data-points as we would normally require, and from it the value of $q_i$ for any $h_i$ in the range −25 to 400 cm can be derived. One should be careful in using approximations such as this one, for the control points are unequally spaced — in fact, only two points lie in the range $140 < h_i < 400$. Computational sophistication cannot compensate for the inadequacy of an experimental design.

A second example of least-square polynomial curve fitting is the work of Shideler (1973) who took 45 equally spaced samples of sediment from a coastal barrier chain between Cape Henry, Virginia and Cape Hatteras, N. Carolina, over a length of 90 nautical miles. The barrier chain has three well-defined environments — beach, dune and aeolian flat. At each sampling site, sediments were taken separately from each environmental type and the four moment measures (mean, standard deviation, skewness and kurtosis) were calculated for the sand-size material (0.0 $\phi$ to 4.0 $\phi$). For each of the three environmental types, first-degree curves of the form $y = a_0 + a_1 x$ were fitted to establish the presence of any large-scale linear trend, subsequent analysis was carried out on the residuals from this trend. This involved the fitting of polynomials up to degree 10.

Results showed a linear increase towards the south in the mean size of sediments from the foreshore and berm populations, interpreted as a response to regional wave refraction patterns established by coastal morphology and to a southwards reduction in shelf width. Both of these factors combined to produce a southward increase in average wave energy. The strength of this trend is very small — 8.92 per cent (berm) and 5.95 per cent (foreshore) of the sum of squares. The tenth-degree polynomial fitted to the residuals from the linear trend accounted for 35.26 per cent of the sum of squares in the case of the foreshore sediments, and is thought to represent changes in the texture of source materials. The explanation of the tenth-degree polynomial rises to 59 per cent in the case of the berm population and 37.3 per cent in the dune population. Both patterns are given a physical interpretation by Shideler, who also deals with the trends evident in the other moment measures. Figure 3.3 shows the trend (1st order curve), and the 10th-order polynomial fitted to each of the three populations. The upper part of each diagram shows the trend (dashed line). Below this is the 10th-order function for the residuals from this trend.

The third example is taken from the work of Abrahams (1972) who is concerned with the relationship between drainage density and precipitation in Eastern Australia. Drainage density is used as a surrogate for sediment yield since both measure topographic dissection to a certain extent, and both are functions of climate and lithology. Abrahams suggests that the general pattern of change in drainage density and sediment yield should be similar, though the magnitude of the change would be expected to vary. He took a sample of 52 basins and fitted a least-squares curve to the logarithms of drainage density and mean annual precipitation, though the numerical procedure he used was not described. The percentage of variance explained by curves of order 1 to 4 are: 0.14 per cent, 33.55 per cent, 52.5 per cent and 53.22 per cent. The third-order curve was therefore selected.

The question of when to stop is a difficult one. If approximation is the aim, the answer is: stop when the unexplained variation is considered to be negligible. If, on the other hand, trend isolation is the aim the answer is paradoxical; it is: when the trend is isolated. This is a circular argument, for the purpose of the exercise is to find a trend! A trend is usually considered to be represented by a smooth curve, so fitting of trend lines could cease when the curve begins to break up. This question recurs later in the chapter in connection with trend surface analysis.

Abrahams (1972) decided that a third order curve isolates the trend in his data. This curve is shown in Figure 3.4; it indicates a minimum drainage density at around 20″ mean annual precipitation, rising to a maximum at 100″. On the assumption that drainage density is zero when there is no rainfall, a second maximum at 0–7″ can be inferred. Apart from this maximum the curve is in broad agreement with the

```
1552.18    1384.54    1216.90    1049.25    881.61    713.97
```

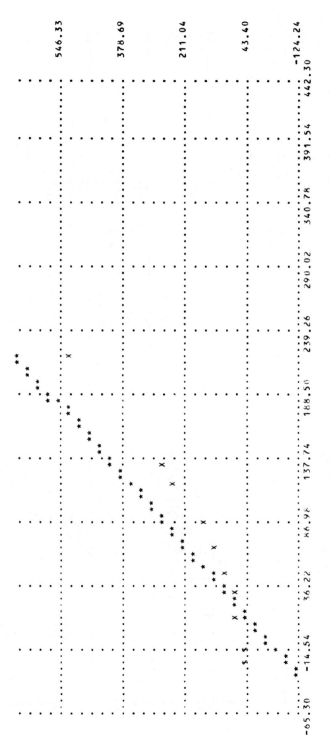

Figure 3.1. Linear fit to Yevyevitch data

1552.18  1384.54  1216.90  1049.25  881.61  713.97

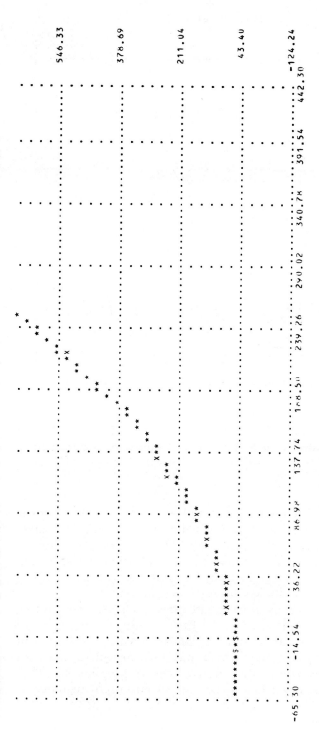

Figure 3.2. Quadratic fit to Yevyevitch data

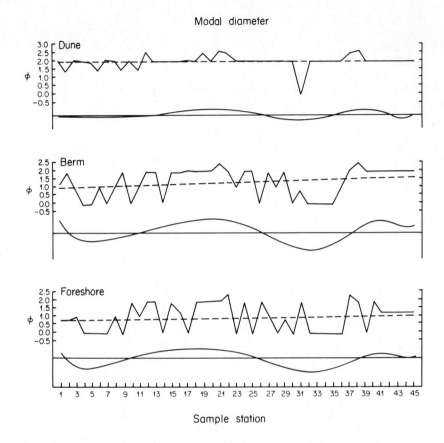

Figure 3.3. Comparative trend analysis of modal diameter (Reproduced with permission from G. L. Schideler, *Sedim. Geology*, 9, 195–220 (1973) Figure 3 by permission of Elsevier Scientific Publishing Co., Amsterdam)

generalized findings of climatic geomorphologists, though Fournier (1960) postulated a second peak in the sediment yield: rainfall curve. Douglas (1967), using a small sample ($n = 21$) from Eastern Australia, found peaks at 1.6–2.0″ and at about 78″ in the sediment-yield/mean annual runoff curves. It is argued that this second peak occurs in response to 'acutely seasonal, high intensity cyclonic storms'.

Polynomial curve-fitting by least-squares has, then, two contrasting uses; the first is interpolation, as in the first example (rating curve in hydrology). The second and third examples illustrate the use of polynomials in identifying trends which may then provide a means of making generalizations about the mechanisms responsible for their production. One must, nevertheless, be careful not to fit high-degree polynomial curves and give the parameters physical meaning, for the true relationship may well be quite different.

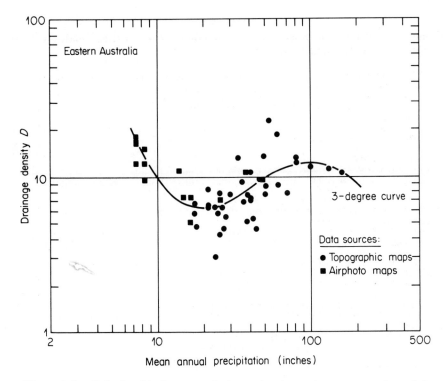

Figure 3.4. Relationship between drainage density and mean annual precipitation (Reproduced with permission from A. D. Abrahams, *Australian Geog. Studies*, 10, 19–41 (1972), Figure 2A)

### 3.1.3 Confidence Limits on Least-Squares Coefficients

This section closes with a brief discussion of a specialized point which will not affect most users of polynomial curve-fitting programs. However, if a model involving a polynomial in $x$ can be specified, and if the assumptions of the regression model are satisfied, then the linear regression model

$$y = \hat{\beta}_0 + \hat{\beta}_1 x + \hat{\beta}_2 x^2 + \ldots + \hat{\beta}_k x^k$$

is quite valid. Rather than use the methods of Chapter 2 we can make use of orthogonal polynomials to provide the estimates of the $\beta_i$ in the above equation. In regression analysis we are likely to be interested in the variances of the $\hat{\beta}_i$ which are required in the $t$-tests of the null hypotheses that $\beta_i = 0$.

The inverse of matrix **L** is required in order to compute the variances of the $\hat{\beta}_i$. Since **L** is lower triangular, the usual methods of matrix inversion described in Chapter 1 are inapplicable. The following scheme is adapted from Rushton (1951, p. 97) and is included as subroutine VARIAN in the accompanying computer program (Table 3.3). If

Table 3.3
Computer program for least squares polynomial curve fitting, multiple linear regression and polynomial regression

```
C#################### NOTES ON PROGRAM USE ####################################
C THE PROGRAM WILL CARRY OUT (A) MULTIPLE LINEAR REGRESSION;
C (B) POLYNOMIAL REGRESSION FOR ANY GIVEN DEGREE POLYNOMIAL;
C (C) POLYNOMIAL CURVE FITTING FROM  DEGREE 1 TO SPECIFIED
C UPPER LIMIT.
C I/O UNITS REQUIRED: CARD READER (LOGICAL UNIT NO 5) AND
C LINEPRINTER (LOGICAL UNIT NO 6).
C
C DATA IN ORDER:
C [1] JOB NAME - USE ANY OR ALL OF COLS 1-80
C [2] COLS 1-3: NUMBER OF OBSERVATIONS (INDIVIDUALS, UNITS, AREAS)
C      COLS 4-6: NUMBER OF VARIABLES PLUS 1. IN POLYNOMIAL OPTIONS THE
C      NUMBER OF VARIABLES IS COUNTED AS DEGREE OF POLYNOMIAL +1.
C      E. G. POLYNOMIAL REGRESSION OPTION, DEGREE 2, ENTER 004 HERE.
C [3] DATA INPUT FORMAT -COLS 1-80.
C [4] OUTPUT FORMAT FOR USE IN *MATPR*. COLS 1-80. SEE APPENDIX A.
C [5] COL 1: ENTER 1 (MULTIPLE REGRESSION), 2 (POLYNOMIAL REGRESSION)
C      OR 3 (POLYNOMIAL CURVE FITTING).
C [6] - [ ]: DATA MATRIX, READ ROW BY ROW IN SPECIFIED FORMAT
C ENTER THE EXPLANATORY (PREDICTOR) VARIABLES FIRST, FOLLOWED BY THE
C VALUE OF THE DEPENDENT VARIABLE.
C
C DIMENSION OF X IN MAIN PROGRAM MUST BE AT LEAST 2*N+2*N*M+2*M+LD
C WHERE N IS THE NUMBER OF OBSERVATIONS, M IS THE NUMBER OF VARIABLES
C +1 AND LD=M*(M+1)/2.
C IF DIMENSION OF X IS ALTERED, CHANGE NDIM ACCORDINGLY.
C NEEDS SUBROUTINE GSORTH FROM APPENDIX A.
C
      DIMENSION X(3000),TITLE(20)
      NDIM=3000
      READ(5,11) TITLE
   11 FORMAT(20A4)
      WRITE(6,12)    TITLE
   12 FORMAT('0      JOB NAME'/'0',20A4///)
      READ(5,1) N,M
    1 FORMAT(2I3)
      NM=N*M
      LD=M*(M+1)/2
      N1=NM+1
      N2=N1+LD
      N3=N2+M
      N4=N3+M
      N5=N4+N
      N6=N5+NM
      NTOP=N6+N
      IF(NTOP.GT.NDIM) GOTO 100
      CALL ORTHCL(X(1),X(N1),X(N2),X(N3),X(N4),X(N5),X(N6),N,M,NM,LD)
      GOTO 102
  100 WRITE(6,101) NTOP
  101 FORMAT('0DIMENSION OF X IN MASTER PROGRAM SHOULD BE INCREASED TO',
     1 I6)
  102 WRITE(6,103) TITLE
  103 FORMAT('0END OF JOB FOR ',20A4)
      STOP
      END


      SUBROUTINE ORTHCL(X,C,D,COEFFS,Y,V,XX,N,M,NM,LD)
      DIMENSION X(N,M),C(LD),D(M),V(N,M),COEFFS(M),Y(N),XX(N)
      DIMENSION FMT(20),FMT1(20)
      COMMON /PLOTS/ XMAX,XMIN,A,B,Q,ZZ
      NUO=0
      ETA=1.0E-5
```

Table 3.3   *continued*

```
      IC=0
      WRITE(6,13)
   13 FORMAT('1PROGRAM TO COMPUTE MULTIPLE LINEAR REGRESSION AND'/
     +' POLYNOMIAL REGRESSION   USING BJORK''S(1967) PROCEDURE'//)
      READ(5,40) FMT, FMT1
   40 FORMAT(20A4)
      READ(5,100) IOPT
  100 FORMAT(I1)
      GOTO (101,105,102), IOPT
  101 WRITE(6,103)
  103 FORMAT('0              MULTIPLE LINEAR REGRESSION'///)
      DO 104 I=1,N
      READ(5,FMT) (X(I,J),J=2,M)
  104 X(I,1)=1.
      IP=0
      MM=N*M
      CALL MATPR(16HDATA   MATRIX    ,2,FMT1,6,X,MM,N,N,M,IP,2,8)
      L1=1
      L2=M
      LOOP1=1
      LOOP2=1
      GOTO 99
  105 WRITE(6,106)
  106 FORMAT('0POLYNOMIAL LEAST SQUARES REGRESSION OPTION'/)
      WRITE(6,108)
  108 FORMAT('0 DATA MATRIX - X AND Y'///)
      DO 107 I=1,N
      X(I,1)=1.0
      READ(5,FMT) X(I,2),X(I,M)
  107 WRITE(6,10) X(I,2),X(I,M)
      LOOP1=M-2
      LOOP2=M-2
      L1=1
      L2=M
      IF(M.EQ.3) GOTO 112
      MM=M-1
      DO 111 I=1,N
      DO 111 J=3,MM
  111 X(I,J)=X(I,J-1)*X(I,2)
  112 GOTO 99
  102 WRITE(6,11)
   11 FORMAT('0LEAST-SQUARES POLYNOMIAL FIT'//' DATA MATRIX'//       X
     1                Y'/)
      L=1
      DO 12 I=1,N
      READ(5,FMT) X(I,2),Y(I)
      X(I,1)=1.
      X(I,3)=Y(I)
   12 WRITE(6,10) (X(I,J),J=2,3)
   10 FORMAT(' ',2E16.8)
      L1=1
      L2=3
      LOOP1=1
      LOOP2=M-2
   99 DO 210 KJ=LOOP1,LOOP2
      IF(IOPT.NE.1) WRITE(6,14) KJ
   14 FORMAT('1             POLYNOMIAL OF DEGREE',I2//)
      IF(KJ.EQ.LOOP1) GOTO 999
      L2=KJ+2
      L1=L2-1
      DO 20 I=1,N
      X(I,L2)=Y(I)
   20 X(I,L1)=X(I,KJ)*X(I,2)
  999 CALL GSORTH(X,V,C,D,N,M,LD,NU,L1,L2,IC,ETA)
      NUO=NUO+NU
      WRITE(6,3) NU,ETA,NUO
```

Table 3.3   *continued*

```
    3 FORMAT('ONUMBER OF LINEAR DEPENDENCIES IN DATA SET AT THIS STAGE =
     1 ',I4/' WITH ETA = ',E16.8/ ' CUMULATIVE NUMBER OF LINEAR DEPENDE
     2NCIES = ',I6)
      WRITE(6,4)
    4 FORMAT('0       RESIDUALS'//)
      WRITE(6,55) (I,V(I,L2),I=1,N)
   55 FORMAT(' ',I4,E20.8)
      CALL RECOV(C,D,LD,M,N,L1,L2,COEFFS,RSS)
      RES=D(L2)
      TSS=RES+RSS
      FPC=RSS/TSS*100.0
      WRITE(6,50) TSS,RSS,RES,FPC
   50 FORMAT('0 TOTAL SS = ',G20.5//'  EXPLAINED SS = ',G20.5/
     +' RESIDUAL SS = ',G20.5/' EXPLAINED SS AS PERCENT OF TOTAL = ',
     + G20.5//)
      S2=RES/FLOAT(N-L2+1)
      WRITE(6,31) S2
   31 FORMAT('OVARIANCE OF RESIDUALS = ',E16.7)
      NDF1=M-2
      NDF2=N-NDF1
      FRAT=(RSS/FLOAT(NDF1))/(RES/FLOAT(NDF2))
      WRITE(6,52) FRAT,NDF1,NDF2
   52 FORMAT('OF-TEST OF ENTIRE REGRESSION'// F RATIO = ',E16.7,
     +' WITH',I4,' AND',I4,' DEGREES OF FREEDOM')
      IF(IOPT.EQ.3) GOTO 96
      M=M-1
      CALL VARIAN(C,LD,M,S2,D,COEFFS)
      IF(IOPT.EQ.1) GOTO 210
   96 DO 97 I=1,N
   97 XX(I)=X(I,2)
      CALL PLOTIT(COEFFS,M ,XX,    X(1,L2),N,KJ,LOOP1)
  210 CONTINUE
      RETURN
      END




      SUBROUTINE VARIAN(C,LD,N,S2,VAR,COEFFS)
      INTEGER R
      DOUBLE PRECISION D,E,F
      DIMENSION C(LD),VAR(N),COEFFS(N)
      J(I,K)=I*(I-1)/2+K
C INVERT LOWER TRIANGULAR MATRIX STORED LINEARLY ROWWISE
      DO 2 I=1,N
      L1=J(I,I)
      C(L1)=1.0/C(L1)
      IF(I.EQ.N) GOTO 2
      I1=I+1
      DO 3 K=I1,N
      R=K-I
      D=0.D0
      DO 4 IM=1,R
      M=IM-1
      L1=J(I+M,I)
      L2=J(K,I+M)
      E=DBLE(C(L1))
      F=DBLE(C(L2))
    4 D=D+E*F
      L1=J(K,I)
      L2=J(K,K)
      E=DBLE(C(L2))
      F=-D/E
    3 C(L1)=SNGL(F)
    2 CONTINUE
      WRITE(6,10)
   10 FORMAT('0 COEFFICIENT            STD. ERROR         T VALUE'/)
```

Table 3.3   *continued*

```
      DO 5 I=1,N
      D=0.D0
      DO 6 K=I,N
      L1=J(K,I)
      E=DBLE(C(L1))
    6 D=D+E*E
      F=D*DBLE(S2)
      VAR(I)=DSQRT(F)
      TVAL=ABS(COEFFS(I)/VAR(I))
    5 WRITE(6,7) I,COEFFS(I),VAR(I),TVAL
    7 FORMAT(' ',I2,3E16.7)
      RETURN
      END




      SUBROUTINE RECOV(C,D,LD,M,N,L1,L2,COEFFS,RSS)
      DOUBLE PRECISION G,H,P
      DIMENSION C(LD),D(M),COEFFS(M)
      RSS=0.0
      WRITE(6,6)
    6 FORMAT('1POLYNOMIAL COEFFICIENTS'///)
      IE=L2*(L2+1)/2-L2
      L21=L2-1
      DO 1 I=1,L21
      IF(I.GT.1) RSS=RSS+C(IE+I)**2
    1 WRITE(6,7) I,C(IE+I)
    7 FORMAT(' ',I6,2X,E20.11)
      K=L2
      DO 10 J=1,L21
      K1=K
      K=K-1
      P=C(IE+K)
      G=0.D0
      L=K*(K-1)/2+K
      H=C(L)
      IF(H-0.D0) 20,10,20
   20 IF(K.EQ.L21) GOTO 11
      DO 12 L=K1,L21
      LL=L*(L-1)/2+K
   12 G=G+C(LL)*COEFFS(L)
   11 COEFFS(K)=(P-G)/H
   10 CONTINUE
      WRITE(6,5)
    5 FORMAT('0 LEAST SQUARES COEFFICIENTS'///)
      DO 3 I=1,L21
      IF(D(I).EQ.0.0) GOTO 4
      WRITE(6,21) I, COEFFS(I)
      GOTO 3
   21 FORMAT(' ',I6,2X,E20.11)
    4 WRITE(6,2) I
    3 CONTINUE
    2 FORMAT(' ',I6,2X,'COEFFICIENT ELIMINATED')
      RETURN
      END




      SUBROUTINE PLOTIT(COEFS,M,X,Y,N,KJ,LOOP1)
      DIMENSION COEFS(M),X(N),Y(N),SL(100),E(100),F(100),OV(11)
      COMMON /PLOTS/ XMAX,XMIN,A,B,C,Z
      DATA DOT/'.'/,STAR/'*'/,BLANK/' '/
      DATA OV/'X','$','"','#','^','=','A','B','C','D','E'/
      CALL SORT(X,Y,N)
```

Table 3.3    *continued*

```
      IF(KJ.GT.LOOP1) GOTO 14
      CALL XMOP(X,N,XMAX,XMIN)
      C=XMAX-XMIN
      XMAX=XMAX+0.1*C
      XMIN=XMIN-0.1*C
      C=XMAX-XMIN
   14 XSCA=C/100.
      CALL CALCF(COEFS,XMAX,XSCA,E,F,M)
      CALL SORT(E,F,100)
      IF(KJ.GT.LOOP1) GOTO 11
      Z=Y(1)-Y(N)
      A=Y(1)+0.1*Z
      B=Y(N)-0.1*Z
      Z=A-B
   11 DO 12 I=1,N
    2 X(I)=((X(I)-XMIN)/C)* 99.0+1.5
   12 Y(I)=((Y(I)-B)/Z)*59.0+1.5
      ZK=Z/10.
      DO 3 I=1,100
      F(I)=((F(I)-B)/Z)*59.0+1.5
    3 E(I)=((E(I)-XMIN)/C)* 99.0+1.5
      IN2=1
      IN3=1
      J=61
      WRITE(6,1)
    1 FORMAT('1')
      YY=A+ZK
      DO 99 JK=1,60
      J=J-1
      IF( MOD(J,6 ).EQ.0) GOTO 4
      IF(J.EQ.1) GOTO 4
      DO 6 I=1,100
    6 SL(I)=BLANK
      DO 5 I=1,100,10
    5 SL(I)=DOT
      SL(100)=DOT
      GOTO 7
    4 DO 8 I=1,100
    8 SL(I)=DOT
    7 IF(IN2.GT.N) GOTO 36
      DO 13 IN=IN2,N
      K=INT(Y(IN))
      IF (J-K) 13,15,16
   15 L=INT(X(IN))
      IP=1
      IF(SL(L).EQ.BLANK.OR.SL(L).EQ.DOT) GOTO 19
      CALL COMP(OV,L,SL(L),IP)
   19 SL(L)=OV(IP)
   13 CONTINUE
   16 IN2=IN
   36 IF(IN3.GT.100) GOTO 37
      DO 9 IN=IN3,100
      K=INT(F(IN))
      IF (J-K) 9,18,17
   18 L=INT(E(IN))
      IF(L.LT.1) GOTO 9
      IF(SL(L).NE.BLANK.AND.SL(L).NE.DOT) GOTO 9
      SL(L)=STAR
    9 CONTINUE
   17 IN3=IN
  200 FORMAT(6X,100A1)
   37 IF(MOD(J,6).EQ.0) GOTO 98
      IF(J.EQ.1) GOTO 98
      WRITE(6,200) SL
      GOTO 99
   98 YY=YY-ZK
      WRITE(6,97) SL,YY
```

Table 3.3  *continued*

```
   97 FORMAT(6X,100A1,1X,F8.2)
   99 CONTINUE
      YY=C/10.
      XX=XMIN-YY
      DO 96 I=1,11
      XX=XX+YY
   96 E(I)=XX
      WRITE(6,95) (E(I),I=1,11)
   95 FORMAT(1X,11(F8.2,2X))
      WRITE(6,10)
   10 FORMAT(////' PLOT OF FUNCTION VALUES AND OBSERVED VALUES OF THE DEP
     +ENDENT VARIABLE VERSUS X'//
     +' FUNCTION VALUES PLOTTED AS *'//' OBSERVED VALUES PLOTTED AS SYMBO
     +L FROM OVERPRINT TABLE'///'        OVERPRINT TABLE'/
     +' SYMBOL  NO OF DATA PTS'/ '      X      1'//'   $      2'//
     +'    "       3'/'    #      4'//'   ^      5'//'   =      6'//
     +'    A       7'/'    B      8'//'   C      9'//'   D     10'//
     +'    E      11 OR MORE')
      RETURN
      END




      SUBROUTINE COMP(A,L,S,IP)
      DIMENSION A(11)
      DO 1 I=1,10
      IF(A(I).NE.S) GOTO 1
      IP=I+1
      GOTO 2
    1 CONTINUE
      IP=11
    2 RETURN
      END




      SUBROUTINE XMOP(X,N,XA,XB)
      DIMENSION X(N)
      XA=X(1)
      XB=XA
      DO 1 I=2,N
      IF(X(I).GT.XA) XA=X(I)
      IF(X(I).LT.XB) XB=X(I)
    1 CONTINUE
      RETURN
      END




      SUBROUTINE SORT(X,Y,N)
      DIMENSION X(N),Y(N)
      N1=N-1
      DO 2 J=1,N1
      A=Y(J)
      K=J
      J1=J+1
      DO 1 I=J1,N
      IF(Y(I).LE.A) GOTO 1
      A=Y(I)
      K=I
    1 CONTINUE
      IF(K.EQ.J) CONTINUE
      Y(K)=Y(J)
      Y(J)=A
```

Table 3.3    *continued*

```
        A=X(K)
        X(K)=X(J)
        X(J)=A
      2 CONTINUE
        RETURN
        END



        SUBROUTINE CALCF(COEFS,XMAX,XSCA,E,F,M)
        DIMENSION COEFS(M),E(100),F(100)
        Z=XMAX+XSCA
        DO 1 I=1,100
        Z=Z-XSCA
        Q=COEFS(1)+Z*COEFS(2)
        IF(M.EQ.2) GOTO 3
        Y=Z*Z
        DO 2 J=3,M
        Q=Q+COEFS(J)*Y
      2 Y=Y*Z
      3 E(101-I)=Z
      1 F(101-I)=Q
        RETURN
        END
```

$U = L^{-1}$ then

$$u_{jj} = 1.0/l_{jj} \quad (j = 1,2,\ldots,k)$$

and

$$u_{j,j+r} = -\sum_{m=0}^{r-1} u_{j,j+m}\, l_{j+m,j+r} \quad (j = 1,2,\ldots,k; r = j-1)$$

The variance of the residuals is found as before by dividing the residual sum of squares by the degrees of freedom, $n - k - 1$. The variance of the $j$th estimator is

$$\text{var}(\hat{\beta}_j) = s^2 \sum_{i=j+1}^{k} u_{j+1,i}^2$$

The tests described in Chapter 2 can be carried out using these computed variances.

## 3.2 Trend Surface Analysis
### 3.2.0 Introduction

Geographers have traditionally begun analyses of spatially-distributed variables by mapping their areal distributions. Generalizations concerning the variation in a variable of interest have then been arrived at by visual analysis or by the use of a method such as that of generalized contours. Trend surface analysis (TSA) is an attempt to handle the same problem by the use of a statistical model. However, traditional methods of analysis — such as plotting the locations of the sample points and the values at these points on a map — should not be disregarded, for a knowledge of the data point distribution can help to avoid both logical and practical difficulties.

The earliest uses of TSA were in geophysics (DeLury, 1950; Oldham and Sutherland, 1955), but it soon found applications in geology (Miller, 1956). It was not until the publication of the work of Chorley and Haggett (1965) that TSA became widely used in geography; however, in the last few years several examples have appeared in the geographical literature (C. A. M. King, 1969; Doornkamp, 1972; Rodda, 1970; Unwin, 1973). These applications of TSA have met with some criticism, which is based on both philosophical and practical grounds. As in other disciplines and with other techniques, the dichotomy between the exploratory and the confirmatory use of the method has led to considerable discussion. Practical difficulties concern the choice of algorithm to be used in computing the solution to the trend equations. It was pointed out in Section 3.1 that sums of squares and cross-products matrices derived from power series expansions are usually ill-conditioned, so that traditional methods such as those used in Chapter 2 in connection with multiple regression are unreliable. A further difficulty concerns the form of the data-point distribution which can add to the problem of ill-conditioning.

The literature on TSA is considerable and will not be extensively reviewed here. Davis (1973), Harbaugh and Merriam (1968), Whitten (1975) and Norcliffe (1969) supply good bibliographies. In the following sections attention will be given to the logical problems encountered in trend-surface analysis, the practical difficulties involved in the solution of trend-surface equations, the comparison of trend-surface maps, and the programming of trend-surface calculations.

### 3.2.1 Trend Surface Model

In TSA it is assumed that the value of each observation $(z_i)$ at point $i$ consists of contributions from (i): a large-scale process which can be considered to generate the overall general pattern of spatial variation, or trend; (ii): from a small-scale process acting locally, and (iii): from random fluctuations, including errors of measurement. Thus,

$$z_i = t_i + l_i + e_i \tag{3.8}$$

where $t_i$ and $l_i$ are the trend and local components. It is virtually impossible in practice to separate the last two components, hence this equation is generally written as

$$z_i = t_i + e_i \tag{3.9}$$

where $e_i$ represents the deviation from the trend surface at point $i$.

Parsley (1971) gives useful definitions of these terms as follows:

(i) regional: a response to a process (or processes) which operates over an area greater than that under consideration and without repetition in the control area.

(ii) local: a response to a process (or processes) which operates over an area smaller than the control area but larger than the mean area.

(iii) residual: a response to a process (or processes) which operates over an area smaller than the mean area' (Parsley, 1971, p. 282).

In these definitions the term 'control area' means the smallest area containing all the sample points, while the mean area is the control area divided by the number of points.

The regional response, or trend component, can take any form. With nonperiodic data it is usual to fit a power-series polynomial function of the location $(u, v)$ coordinates of the sample points, since any single-valued function can be approximated by a polynomial of sufficiently high degree. Thus,

$$\begin{aligned} t_i &= f(u_i, v_i) \\ &= \alpha_{00} + \alpha_{10} u_i + \alpha_{01} v_i + \alpha_{20} u_i^2 \\ &\quad + \alpha_{11} u_i v_i + \alpha_{02} v_i^2 + \ldots + \alpha_{0q} v_i^q \end{aligned} \quad (3.10)$$

This equation is similar to the standard multiple regression equation (Chapter 2); the $\alpha_{ij}$ are analogous to the $\beta_i$, but generally they have no physical meaning. Like partial regression coefficients their estimates are computed by the method of least squares which ensures (in the case of random samples) that they have the desirable properties outlined in Chapter 2, subject to the same conditions, which can be summarized as:

I: the mean of the disturbances $(e_i)$ is zero;

II: the variance of the $e_i$, and hence of the $z_i$, is constant over the control area;

III: the $(u_i, v_i)$ are measured without error, and their powers and cross-products are also calculated without error;

IV: the $(u_i, v_i)$ and their powers and cross-products are not perfectly linearly related;

V: the number of sample points, $n$, is large relative to the number of coefficients ($\alpha$'s) to be estimated, and the distribution of these points over the control area is randon;

VI: the error terms are not spatially autocorrelated;

VII: the conditional distribution of the $z_i$ given the $(u_i, v_i)$ is normal.

Assumption VII is necessary whenever statistical tests of significance are to be applied. The other assumptions are made in order that the $\hat{\alpha}_{ij}$ (trend coefficients estimated from the data) are good estimates of the true trend coefficients, $\alpha_{ij}$. In practical instances, where the aim is to describe a given data distribution rather than to fit a model, these assumptions can be relaxed for — whatever the nature of the data — the trend surface is always the best-fitting polynomial surface (in the least-squares sense) of the given order.

*(A) Specification Phase*   The use of TSA in exploratory studies has been remarked upon in a previous paragraph. Much needless confusion has been generated by a failure to realize the distinction between 'model-fitting' or confirmatory analyses which are statistical in nature in the strict sense of the word, and exploratory or descriptive studies, which do not involve statistical inference but which may use 'subject-matter reasoning'. Norcliffe (1969) puts the view of the school of thought which holds that the form of the spatial response to a given process should be specified in advance and TSA used in a hypothesis-testing capacity. The alternative view is more widely held, perhaps because it is less restrictive; it allows the researcher to fit progressively more complex surfaces to his data in an attempt to find a reasonable specification of the spatial components of variance. This approach is costly in terms of computer time and is often the first resort of those who 'let the data speak for themselves'. The main difficulty concerns the evaluation of the significance of the fit of the trend surface to the data. Would a trend surface of the same order, fitted to data points having the same spatial distribution but taking on random values, give as good a fit? Statistical (or quasi-statistical) tests are often invalid, both on technical and logical grounds. If sequential fitting is used then the tests of the significance of the fit of successive surfaces are not independent and so the distribution of the $F$-ratio (the test statistic employed) is uncertain. Often, too, the true degrees of freedom are impossible to ascertain because of poor data point distributions and spatial autocorrelation among the residuals. Furthermore, it would seem illogical to suggest that, because a given surface is a poor fit to the data, a more flexible surface should not be considered.

Even if a trend surface is judged to fit the data sufficiently well, some problems still remain. The temptation to build *a posteriori* theories may prove irresistible, and such theorizing is rarely fruitful; witness the 'Cretaceous cover' hypotheses in British geomorphology (Chorley, 1965, p. 149). The human mind has an extraordinary capacity to perceive pattern and order, even when none exists (Vernon, 1962; Hanson, 1958). Curry's remarks, quoted by Norcliffe (1969, p. 340) are apt in this context: '... for statistical results to be meaningful there needs to be a close wedding between procedures and theory — with theory at the head of the household.'

On the other hand it can be argued that a restrictive, denying approach may prove as damaging in the long run as the indiscriminate or thoughtless use of a technique. Gould (1970), for example, seems to argue in favour of a less inhibited approach to quantitative methods when he exhorts us to make mistakes — 'good mistakes' — from which we can learn. In the absence of theory, this kind of probing may produce vital evidence (Tukey, 1969). Charles Darwin, for example, observed in his autobiography that '... without any theory (I) collected facts on a wholesale scale ... when I see the list of books of all kinds that I read and abstracted ... I am surprised at my industry.' The exploratory approach to trend surface fitting may at least provide the opportunity to make a generalization or put forward a tentative hypothesis, even if it may later be proved to be incorrect.

*(B) Data Point Distributions*   The spatial pattern of the data points is an often-overlooked though vital aspect of any geographical study, for the validity of the conclusions depends to a large extent on the pattern of the data points. It would be illogical to attempt to infer the shape of the trend surface in a given area if few data points were available, or if the distribution was such that large sections of the area were not represented. Furthermore, the spatial arrangement of the data points can have profound effects on the shape of a computed trend surface. Doveton and Parsley (1970) recommend that the following criteria be adhered to:

(i) there should be at least as many, preferably many more, points as there are coefficients in the trend equation. (The number of coefficients is $(k + 1)(k + 2)/2$ where $k$ is the order of the surface.)
(ii) the points should be evenly distributed;
(iii) the map area should be approximately square.

Since in many cases the data-point distribution is fixed, interest centres on the distortion that can be produced if these requirements are not satisfied. Doveton and Parsley (1970) consider criteria (ii) and (iii) in detail. Criterion (i) is a statistical and computational requirement, for if the number of coefficients to be estimated exceeds the sample size then the number of degrees of freedom will be negative, thus inhibiting any subsequent statistical hypothesis-testing, while the matrix of sums of squares and cross-products will be singular, preventing the calculation of the $\hat{\alpha}$'s by the usual methods.

Ideally, the data points should be located so as to give full coverage of the area of interest, thus allowing reasonable generalizations to be drawn. The pattern of data points also has an effect on the accuracy of calculations, as Doveton and Parsley (1970) found. They examined (i) a pattern consisting of data points lying in a linear belt across the control area; (ii) a pattern in which the majority of data points were restricted to a narrow strip across the map, with the rest scattered randomly, and (iii) a clustered pattern in which data points were restricted to well-defined groups. The experiments showed that a type (i) distribution produced considerable distortion whereas a type (ii) distribution apparently had little distorting effect. The results of the study are summarized in Figure 3.5.

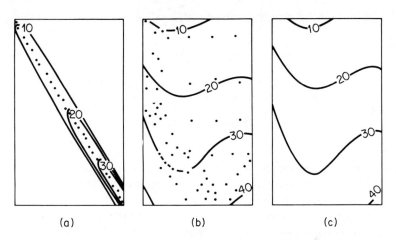

Figure 3.5. Effect of data point distributions on trend surfaces: (a) distribution type (i); (b) distribution type (ii); (c) original surface (Reproduced with permission from J. M. Doveton and A. J. Parsley, *Trans. Inst. Min. Metall.*, B, 79, B197–B207 (1970), Figures 2(f), (d) and (i))

Clustering of data points (type (iii)) had a lesser distorting effect than either of types (i) and (ii). This led to the conclusion that '... clustering can be accepted where simple regional patterns are the only interest'. (Doveton and Parsley, 1970, p. B200.) The results of this experiment are shown in Figure 3.6. A second aspect of type (iii) data point distributions is the effect on the outcome of standard statistical tests. This is considered in Section (D).

The experiments of Doveton and Parsley used data for which the fit of the computed trend surface was high – in excess of 90 per cent explained variance. At lower levels of explanation the influence of the data point distribution may be more marked. For high levels of explanation, trend surface analysis is thought to be '... robust to all forms of data point unevenness except type (i) linearity'. (Doveton and Parsley, 1970.) There may, however, be computational difficulties; these are discussed at a later point in this section.

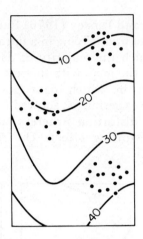

Figure 3.6. Effect of clustering of data points on trend surface (Reproduced with permission from J. M. Doveton and A. J. Parsley, *Trans. Inst. Min. Metall.*, B, 79, B197–B207 (1970), Figure 2(a))

A measure of the degree of clustering (or absence of randomness) in a spatial point pattern is the nearest neighbour statistic, which was originally developed by Clark and Evans (1954) for use in ecological work. It has since been widely used in human geography, and its application is discussed by L. J. King (1969), Haggett (1965), and Harvey (1969). A brief summary is given here.

The expected mean distance between a given point and its nearest neighbour in a uniform distribution is

$$\bar{d} = 0.5\, p^{0.5}$$

where $p$ is the density of points per unit area and distance and area are measured in the same basic units, e.g. miles and square miles. The determination of $p$ is somewhat subjective, for the boundaries of the area are rarely defined accurately. One can either take the area of the smallest polygon enclosing the point pattern or the area of smallest rectangle which could be fitted to the distribution. The observed mean nearest neighbour distance $\bar{D}$ is calculated next; this is

$$\bar{D} = \frac{1}{n} \sum_{i=1}^{n} D_i$$

where $n$ is the number of data points and $D_i$ the distance from point $i$ to the point closest to it. The ratio $R$ of $\bar{D}$ to $\bar{d}$ is the nearest-neighbour statistic. It ranges in value from 0 (if all the points are coincident) to 2.1491, when the points lie form a regular hexagonal pattern. A value of $R = 1$ indicates randomness. Thus, for $R < 1$ the pattern is more clustered than random and for $R > 1$ the pattern is more regular than random. It is possible to test statistically the null hypothesis that the value of $R$ is 1.0 in the population from which the points were drawn. (L. J. King, 1969.) This assumes that the nearest-neighbour distances are normally distributed. Other references to the use of nearest neighbour analysis in geography are to be found in Dacey (1960, 1962, 1966), Tinkler (1971) and Curry (1964); in plant ecology in Greig-Smith (1964), and in geology in Miller and Kahn (1962) and Davis (1973). The technique has been extended to deal with 2nd, 3rd, ..., $j$th order nearest neighbours by Dacey (1963).

Kershaw (1973, Chapter 7) discusses at length an alternative approach to the detection of nonrandomness in spatial point patterns. This involves the subdivision of the control area into rectangular subareas, or quadrats, and the calculation of the numbers of quadrats with $0, 1, \ldots, n$ individuals. The quantity $m = \Sigma_{i=0}^{n} a_i f_i / k$ is then derived, where $a_i$ and $f_i$ are respectively the number of individuals per quadrat and the frequency count for the $i$th class (see example below). The value $k$ is the total number of quadrats. The expected frequencies $f_i^*$ can be derived by multiplying the Poisson distribution ($e^{-m}$, $me^{-m}$, $(m^2/2!)e^{-m}$ $(m^3/3!)e^{-m}$, $(m^4/4!)e^{-m}$ ...) by $k$, where $m$ is the mean density of individuals as derived above. The chi-square test can be applied to determine the probability of the observed distribution having been derived by random sampling from a Poisson distribution with parameter $m$. The procedure is illustrated using an example from Kershaw (1973). One hundred quadrats were located randomly in an area containing a random point pattern. The classes ($a_i$) and frequencies ($f_i$) were:

| $a_i$ | 0 | 1 | 2 | 3 |
|---|---|---|---|---|
| $f_i$ | 46 | 34 | 14 | 6 |

hence $m = \Sigma_{i=0}^{3} a_i f_i / k = 0.8$. The successive terms of the Poisson series with $m = 0.8$ are:

$$e^{-0.8} = 0.4493$$

$$0.8 \, e^{-0.8} = 0.3594$$

$$\frac{0.8^2}{2!} e^{-0.8} = 0.1438$$

$$\frac{0.8^3}{3!} e^{-0.8} = 0.0383$$

which, when multiplied by $k = 100$, give the $f_i^*$ ($i = 0,3$). Table 3.4 shows the layout for the $\chi^2$ test.

The value of chi-square is the sum of the elements of the bottom

Table 3.4
Chi-square calculations, quadrat example

| $a_i$ | 0 | 1 | 2 | 3 |
|---|---|---|---|---|
| $f_i^*$ | 44.93 | 35.94 | 14.38 | 3.83 |
| $f_i$ | 46 | 34 | 14 | 6 |
| $(f_i^* - f_i)^2 / f_i^*$ | 0.0269 | 0.1006 | 0.0111 | 1.2737 |

row, giving $\chi^2 = 1.4123$ with degrees of freedom equal to two less than the number of classes, in this case $(4 - 2) = 2$. The calculated value, 1.4123, has a probability of only 0.5 or so ($\chi^2_{0.5,2} = 1.39$) and so the chances of the difference between the $f_i$ and the $f_i^*$ arising fortuitously are not high. The observed distribution can therefore be safely considered to have a random distribution.

A difficulty in the use of quadrat analysis is determination of the size of the quadrat. For example, with a distribution that is badly clustered the null hypothesis of non-randomness can often be rejected if the quadrat size is large in relation to the control area. In addition, the chi-square test is not very powerful when the number of observations in a particular class is less than 5. Harvey (1969), L. J. King (1969) and Greig-Smith (1964), in addition to Kershaw (1973), provide further details of this method of analysis.

The use of measures of randomness in the distribution of sampling points in trend surface analysis is thus open to some doubt. The effects of clustering or linearity in the pattern depend not only upon the degree to which they are present but also on the goodness of fit of the trend surface, so no generalization about 'acceptable' values of $\chi^2$ or the nearest neighbour statistic are possible. However, on grounds of common sense, one would not attempt to fit a (continuous) trend surface when data were available only for a small part of the total area.

The effect of data-point distribution upon the condition of the sums of squares and cross-products matrix **P** has been considered briefly by Miesch and Connor (1968). (The concept of the condition of a matrix is discussed in Chapter 1.) In order to allow comparison with the results of Miesch and Connor we will deal with the correlation matrix **R** rather than the matrix **P**. (Refer to Chapter 2 for details of the effects of ill-conditioning of **R** on the resulting regression — in this case trend — coefficients.) Two measures of condition are in general use; the first is the determinant, det(**R**), which, in the case of the correlation matrix varies from 1.0 (perfect condition) to 0.0 (singular). The second is termed the $P$-condition number of the matrix and is the ratio of the absolute values of the highest to the lowest eigenvalues of **R**. $P$ equals 1.0 for a perfectly conditioned (orthogonal) matrix and infinity if the

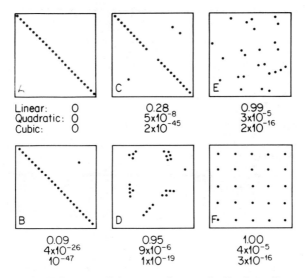

Figure 3.7. Condition numbers of **R** for first, second and third order trend surfaces for various data point distributions (Reproduced with permission from A.T. Miesch and J.J. Connor, *Computer Contribution*, 27, Figure 3 (1968), Kansas Geol. Survey)

matrix is singular. Miesch and Connor use as a measure of condition the determinant of **R** after it has been normalized by dividing the elements of row $j$ by the sums of squares of the elements of the row. The range of this measure is from 0 (singular) to ±1 (perfect condition). The results of Miesch and Connor are reproduced here as Figure 3.7.

Three further examples are studied here. In the first, the control points lie in a band running NW—SE (Figure 3.8). The $(u, v)$ coordinates of this spatial distribution were generated by randomly choosing $u_i$ from a uniform distribution between 1.0 and 10.0. The corresponding $v_i$ was computed by adding a quantity $d_i$ to the given $u_i$, where $d$ was selected randomly from a distribution uniform on the range $(-2.5, 2.5)$. The correlation between $u_i$ and $v_i$ is 0.9334. Results for the correlation matrices computed for trend surfaces of order 1 to 5 inclusive are shown in Table 3.5.

Case 2, shown in Figure 3.9, is a random distribution, with both $u_i$ and $v_i$ drawn from a distribution uniform in the range (1, 10). The correlation between $u_i$ and $v_i$ is 0.0612. The condition numbers and determinants of the matrices of correlations among the $u_i$ and $v_i$ and their powers and cross-products are shown in Table 3.6. The improvement in matrix condition over case 1 is marked for the first and second order but problems might be anticipated in dealing with the quartic and quintic surfaces by techniques employing matrix inversion.

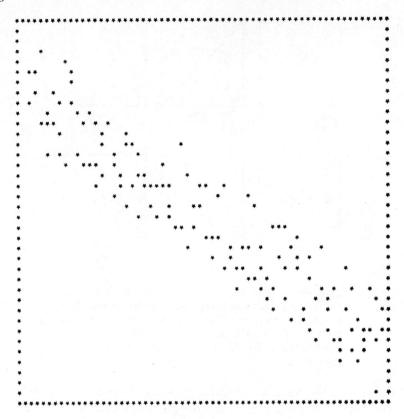

Figure 3.8. Near linear data point distribution

Table 3.5
Condition of **R** for near-linear data point distribution (Figure 3.8)

| Order | Computed determinant | $P$ condition number |
|---|---|---|
| 1 | 0.129 | $0.290 \times 10^2$ |
| 2 | $0.132 \times 10^{-6}$ | $0.185 \times 10^5$ |
| 3 | $0.993 \times 10^{-21}$ | $0.101 \times 10^8$ |
| 4 | $0.305 \times 10^{-46}$ | $0.516 \times 10^{10}$ |
| 5 | 0.000 | $0.359 \times 10^{12}$ |

The third distribution considered is a regular pattern of data points equally spaced over the control area (Figure 3.10). The improvement over case 2 is not great until the 4th and 5th orders are reached (Table 3.7).

The conclusion to be drawn from these experiments is that methods of determining the trend coefficients using matrix inversion techniques are likely to be inaccurate for all but linear and quadratic surfaces. The

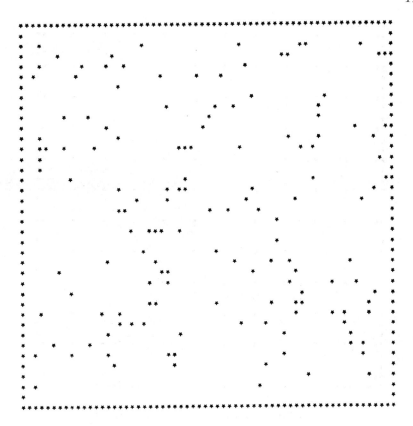

Figure 3.9. Random data point distribution

Table 3.6
Condition of **R** for random data point distribution (Figure 3.9)

| Order | Computed determinant | $P$ condition number |
|---|---|---|
| 1 | 0.996 | $0.113 \times 10^1$ |
| 2 | $0.143 \times 10^{-3}$ | $0.208 \times 10^3$ |
| 3 | $0.133 \times 10^{-13}$ | $0.176 \times 10^5$ |
| 4 | $0.221 \times 10^{-32}$ | $0.146 \times 10^7$ |
| 5 | $0.118 \times 10^{-62}$ | $0.103 \times 10^9$ |

distribution of data points may enhance the difficulty, the best results being achieved when the sample points lie on a regular grid. A linear pattern produces the worst results from a numerical point of view, and conclusions drawn from an analysis of such a distribution are, in any case, illogical since information exists only for part of the map. Lack of control on the position of the surface outside the linear strip containing the data points will result in wild variations in the predicted values as

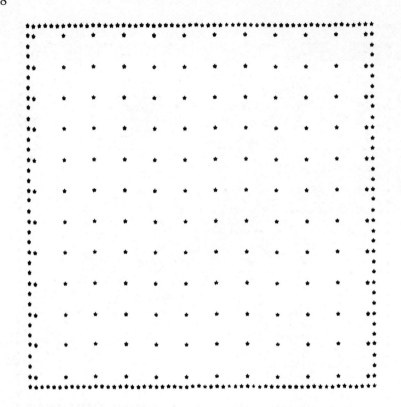

Figure 3.10. Regular data point distribution

Table 3.7
Condition of **R** for regular data point distribution (Figure 3.10)

| Order | Computed determinant | $P$ condition number |
|---|---|---|
| 1 | 0.985 | $0.128 \times 10^1$ |
| 2 | $0.537 \times 10^{-3}$ | $0.125 \times 10^3$ |
| 3 | $0.958 \times 10^{-12}$ | $0.843 \times 10^4$ |
| 4 | $0.219 \times 10^{-28}$ | $0.483 \times 10^6$ |
| 5 | $0.219 \times 10^{-55}$ | $0.269 \times 10^8$ |

the order of the surface is increased. The computational and inferential aspects of trend surface analysis are considered in more detail in later sections.

In their analysis of the distortion of trend surfaces resulting from uneven data point distributions, Doveton and Parsley (1970) point out that one frequently-overlooked aspect concerns the shape of the control area. Elongated control areas give rise to prominent 'edge effects' which are shown by the tendency of the contours of trend surfaces of order 2 or higher to lie subparallel to the long axis of the control area. Such a tendency is imposed upon the trend surface by the shape of the control area rather than by any variation in the data values. An example of the use of an elongate control area is provided by Chorley (1969), who related the elevation of the Lower Greensand ridge in S.E. England to the thickness of the formation, the projection of its upper contact and the percent silt and clay content of the soils by comparing trend-surface maps of the four variables. These four maps are shown in Figure 3.11, and evidence of the influence of map shape is clearly seen.

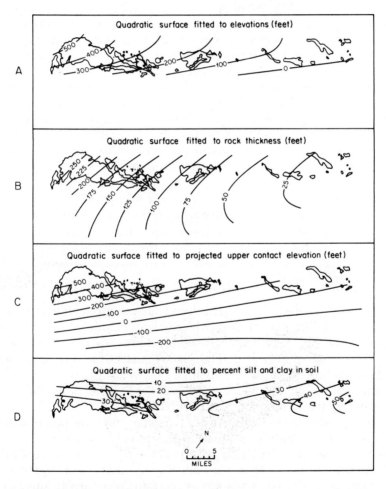

Figure 3.11. Trend surfaces based on elongate control area (Reproduced with permission from R. J. Chorley, *Geol. Mag.*, **106**, 231–248 (1969), Figure 6, by permission of Cambridge University Press)

No mention has been made so far of two related problems: the method of sampling to be used and the size of sample. These points are often overlooked or ignored for, in many instances the data exist only at a given set of locations. In these cases the question is whether such a data set can be said to constitute a random sample from any population. It could be argued that the population is the given set of points, in which case statistical inference would be unnecessary. Some workers have maintained that there is a theoretical population (for example, the population of corrie floor heights which could have existed if the surface topography and altitude had been different) but it is doubtful whether inferences about such populations have any particular interest. The trend surface is a description of one aspect of the particular distribution and the question of statistical inference does not arise. Other studies (e.g. Chorley, Haggett, Stoddart and Slaymaker, 1966) deal with spatial distributions which are available for sampling at a large number of points. In such a situation a balance must be maintained between randomness of the sampling procedure (to permit valid inferences) and the constraints discussed earlier in this section. A random sampling process can leave some areas under-represented, with potentially serious consequences for the accuracy of the resulting trend-surface calculations. A possible answer is to use spatially stratified random sampling, taking randomly a set number of points from each of a given number of quadrats. The number of points and the size of the quadrats will depend upon the scale of the study. At least two points per unit area are required in order to differentiate systematic variation from random noise. The question of sampling in the context of trend surface analysis has not been given a great deal of consideration. However, Berry and Marble (eds.) (1968) provide a comprehensive account of spatial sampling in general terms.

*(C) Computational Aspects* The trend-surface model is a special case of the General Linear Model, which also includes multiple linear regression and polynomial regression. However, there are several practical differences between trend-surface analysis and multiple linear regression, the most important of which involve:

(i) the generation of the powers and cross-products of the $(u_i, v_i)$ coordinates of the sampling points in their correct order;

(ii) the production of lineprinter maps of the computed surfaces (and, if desired, of confidence surfaces and residual values);

(iii) the accurate calculation of estimates of the parameters of the trend equation, given that the matrix of sums and squares and cross-products of the $(u_i, v_i)$ is almost certain to be nearly singular for surfaces of order 3 or more.

*Generation of the polynomial terms* in the trend equation is usually performed sequentially, that is, the first-order or linear terms are

| Order | Terms | | | | |
|---|---|---|---|---|---|
| 0 | 1 | | | | |
| 1 | $u$ | $v$ | | | |
| 2 | $u^2$ | $uv$ | $v^2$ | | |
| 3 | $u^3$ | $u^2v$ | $uv^2$ | $v^3$ | |
| 4 | $u^4$ | $u^3v$ | $u^2v^2$ | $uv^3$ | $v^4$ |
| ... | ... | ... | ... | ... | ... |
| ... | ... | ... | ... | ... | ... |

Figure 3.12 Terms associated with increasing order of trend surface.

calculated first, then the terms for each successively higher order. Even where all the terms in the highest-order equation are desired, it is convenient to calculate them sequentially (Figure 3.12).

The terms shown in Figure 3.12, apart from the unit vector corresponding to the zero order surface, are obtained as follows: the $k$th order surface has $(k + 1)$ associated terms. The first $k$ of these are found by multiplying by $u_i$ the $k$ terms associated with the surface of order $(k - 1)$. The $(k + 1)$th term is equal to the $k$th term on the previous line multiplied by $v_i$. Thus, if $u_i = 3$ and $v_i = 7$ the three second-order terms would be (3 x 3), (7 x 3) and (7 x 7), i.e. 9, 21 and 49. The four third-order terms are 27, 63, 147 and 343. This procedure is easily programmed.

The number of digits that can be represented in one computer word is limited. The IBM 360 range holds floating point numbers to 7—8 decimal digits, while the ICL 1900 series can manage 10—11 decimal digits. This means that only the first $n$ significant digits of the number are correctly represented, where $n$ is the number of digits that the machine uses. Thus, if the coordinate system on which the location coordinates $(u_i, v_i)$ are measured is badly chosen, then the powers and cross-products of the location coordinates will become very large and loss of precision will result. It is for this reason that the origin of the coordinate system should be placed in the centre of the control area. Large values of $u$ or $v$ should be avoided.

*Lineprinter maps* provide a quick and reasonably accurate method of displaying trend surfaces. Their production entails the evaluation of the trend equation at each of a set of points distributed according to the grid pattern. The grid points correspond to lineprinter character positions. Thus, an $(n \times m)$ grid is formed by $n$ printer lines each capable of holding $m$ characters. Elaborate methods of shading are possible, using overprinting of characters (such a method is described and programmed in Fortran by Howarth, 1971). See also Davis and McCullagh (eds.) (1975). A simpler method, which is adequate for most circumstances, is described here. No matter how they are produced, lineprinter maps have certain inherent limitations, especially of size and fineness of detail. The grid mesh cannot be smaller in size than one lineprinter character, which is usually $\frac{1}{10}$th inch wide and $\frac{1}{6}$th inch long. The map itself must be an integral number of characters in size. Its maximum dimensions are limited by the type of printer used. Most lineprinters produce 60 lines per page, each line capable of holding 120, 132 or 160 characters. It is generally best to program for the 120-character line in order to minimize the problems of transferring programs to other computers. The 'page-throw' mechanism of the lineprinters can be switched off, either by hardware or software, hence the map could have unlimited vertical extent. A map greater than 120 character positions wide could be accommodated by printing in sections.

Since there does not appear to be an established convention for the location of the origin, it will be assumed to lie at the bottom left-hand corner of the map. The number of characters per line, selected when the subroutine is called, will be denoted by NC and the number of lines by NL. The top line of the map therefore contains NC character positions the centres of which have coordinates $(u_{min}, v_{max})$, $(u_{min} + x\text{axis}, v_{max})$, ..., $(u_{max}, v_{max})$. The symbols $u_{min}$, $v_{min}$, $u_{max}$, $v_{max}$ denote the minimum and maximum values of the location coordinates, while $x$ axis is the number of units of $u$ corresponding to one character position along the horizontal axis; this is found from $x$-axis $= (u_{max} - u_{min})/$NC. Similarly, the second line of the map is a series of points with coordinates $(u_{min}, v_{max} - y\text{axis})$, $(u_{min} + x\text{axis}, u_{max} - y\text{axis})$, ..., $(u_{max}, v_{max} - y\text{axis})$. Here, $y$axis is the number of units of $v$ corresponding to one line-width.

Beginning with character position 1, the values of $z_i$ are computed for each character position of line 1, using the same algorithm as that described above under the heading 'generation of polynomial terms'. On this occasion, the polynomial expansion to the desired order of the $(u, v)$ coordinates of the particular character position is followed by evaluation of the trend equation. (It is assumed that the parameters of the equation have been estimated by some method. The choice of algorithm for such estimation is discussed below.) Thus, for coordinates $(u_{min}, v_{max})$, i.e. the last character position of line 1, we find

$$z_i = \hat{\alpha}_0 + \hat{\alpha}_1 u_{min} + \hat{\alpha}_2 v_{max} + \hat{\alpha}_3 u_{min} v_{max} + \ldots + \hat{\alpha}_k v_{max}^m$$

To determine the contour interval containing $z_i$ two further parameters are required, REFCON and CONT. The former is the reference contour, which should be chosen to lie near the mean of the $z$'s. An integral value is preferable. CONT is the contour interval. Again this should be well-chosen value, avoiding the extremes of too many or too few contour intervals. The subroutine listed in Table 3.8 allows for up to 20 contour intervals. The reference contour is the 11th interval, and is

Table 3.8

Computer program for trend surface analysis by orthogonal polynomials

```
C
C TREND SURFACE ANALYSIS USING ORTHOGONAL POLYNOMIALS.   PROGRAM BY
C P.M.MATHER, SEPT. 1973.    SEE CHAPTER 3 SECTION 3.2 FOR DETAILS.
C
C NOTE: ALL INTEGERS SHOULD BE RIGHT JUSTIFIED, AND THE DECIMAL POINT
C SHOULD BE PUNCHED FOR FLOATING POINT NUMBERS IN THE INPUT SEQUENCE
C DESCRIBED BELOW.
C
C THE FOLLOWING I/O UNITS ARE REQUIRED: CARD READER (UNIT 5), LINE-
C PRINTER (UNIT 6) AND -OPTIONALLY- A TAPE OR DISC FILE (UNIT 9) IF THE
C DATA SET IS TO BE READ FROM FILE -- SEE BELOW.
C
C *****************DIMENSION OF X MUST BE AT LEAST 4M+2(N*M)+M*(M+1)/2**
C *****************SET N TO CURRENT DIMENSION OF X**********************
C DATA IN ORDER:
C 1]:          N   M   INPUT
C COLS 1-3:   N, THE NUMBER OF POINTS.
C COLS 4-6:   M, THE MAX. ORDER SURFACE TO BE FITTED.
C COL 7:      ENTER 0 IF DATA TO BE READ FROM CARDS (CHANNEL 5) OR 1
C                 IF DATA TO BE READ IN UNFORMATTED FORM FROM A TAPE OR
C                 DISC FILE (CHANNEL 9). SEE BELOW
C 2]         FN   FC   REFCON   CONT   INIT
C COLS 1-10: LENGTH OF TREND MAP AS PRINTED ON OUTPUT, INCHES.
C COLS 11-20:WIDTH OF PRINTED TREND MAP, IN INCHES.
C COLS 21-30: VALUE OF REFERENCE CONTOUR IN TERMS OF Z UNITS.  IF ZERO,
C              A DEFAULT VALUE, NEAR TO THE AVERAGE OF Z, IS USED.
C COLS 31-40: CONTOUR INTERVAL. USUALLY, ABOUT 10 CONTOUR INTERVALS
C              SHOULD COVER THE RANGE OF Z.  IF 0 ENTERED A DEFAULT
C              VALUE IS USED.
C COL 41:     ENTER 0 IF THE COORDINATES OF THE CORNERS OF THE TREND
C              MAP ARE TO BE COMPUTED, OR 1 IF THEY ARE TO BE READ IN.
C              IF THE LATTER, SEE BELOW [3].
C 3]: IF INIT = 1 UMAX,UMIN,VMAX,VMIN   IN FORMAT(4F10.0).
C          UMAX,UMIN = MAXIMUM AND MINIMUM ALONG HORIZONTAL AXIS
C          VMAX,VMIN = MAXIMUM AND MINIMUM ALONG VERTICAL AXIS
C***** CARD 3 REQUIRED ONLY IF INIT = 1*******************
C 4]: DATA INPUT FORMAT FOR CARD INPUT.  LEAVE THIS CARD BLANK IF THE
C INPUT DATA IS READ FROM FILE (INPUT = 1 ON CARD 1).
C 5]: DATA IN FORM U(I),V(I),Z(I).
C          U(I) IS HORIZONTAL COORDINATE
C          V(I) IS VERTICAL COORDINATE
C          Z(I) IS VALUE OF DEPENDENT VARIABLE AT (U(I),V(I))
C          IF READ FROM CARDS START EACH SET OF (U,V,Z) ON A FRESH CARD
C          IN FORMAT SPECIFIED ON CARD 4.
C          IF DATA ON FILE, EACH (UVZ) SET SHOULD FORM ONE RECORD.
C
C
      DIMENSION X(5000)
      N=5000
      CALL TSMAIN(X,N)
      STOP
      END




      SUBROUTINE TSMAIN(X,NU)
C SET UP BASE ADRESSES
      DIMENSION X(NU)
      READ(5,1) N,M,INPUT
      KO=M
    1 FORMAT(2I3,I1)
```

Table 3.8   *continued*

```
      M=((M+1)*(M+2))/2+1
      NM=N*M
      LD=M*(M+1)/2
      N1=NM+1
      N2=N1+LD
      N3=N2+M
      N4=N3+NM
      N5=N4+M
      N6=N5+M
      N7=N6+M
      N8=N7+M
    4 FORMAT(' ACTUAL CORE NEEDED =',I7)
      IF(N8.GT.NU) GOTO 2
      CALL TSA(X(1),X(N1),X(N2),X(N3),N,M,NM,LD,X(N4),X(N5),KO,X(N6),X(N
     17),INPUT)
      RETURN
    2 WRITE(6,3) N8
    3 FORMAT('O DIMENSION OF X TOO SMALL - INCREASE TO',I6)
      RETURN
      END




      SUBROUTINE TSA(X,C,D,V,N,M,NM,LD,DD,MEANS,KO,COEFFS,WORK,INPUT)
C MAIN SUBROUTINE
      REAL MEANS(M),X(N,M),V(N,M),C(LD),DD(M),D(M),COEFFS(M),WORK(M),
     * FMT(20)
      S=0.0
      UMAX=1.E-10
      VMAX=UMAX
      UMIN=1.E10
      VMIN=UMIN
      ZMIN=UMIN
      ZMAX=UMAX
      WRITE(6,930)
  930 FORMAT('1 TREND SURFACE ANALYSIS PROGRAM USING ORTHOGONAL POLYNOMI
     *ALS'// P. M. MATHER,GEOGRAPHY DEPT., UNIV. OF NOTTINGHAM'/
     *' JULY 1973')
      READ(5,903) FN,FC,REFCON,CONT ,INIT
C INIT=1 MEANS COORDINATES OF MAP CORNERS ARE SUPPLIED.
      IF(INIT.EQ.1) READ(5,903) UMAX,UMIN,VMAX,VMIN
      READ(5,8825) FMT
 8825 FORMAT(20A4)
      WRITE(6,940)
  940 FORMAT(///'O                   DATA MATRIX'//
     +'          U              V           Z'///)
      DO 902 I=1,N
      X(I,1)=1.0
      IF(INPUT.EQ.1) GOTO 1
      READ(5,FMT) (X(I,J),J=2,4)
      GOTO 2
    1 READ(9) (X(I,J),J=2,4)
    2 S=S+X(I,4)
      IF(INIT.EQ.1) GOTO 902
      IF(X(I,2).GT.UMAX) UMAX=X(I,2)
      IF(X(I,2).LT.UMIN) UMIN=X(I,2)
      IF(X(I,3).GT.VMAX) VMAX=X(I,3)
      IF(X(I,3).LT.VMIN) VMIN=X(I,3)
      IF(X(I,4).GT.ZMAX) ZMAX=X(I,4)
      IF(X(I,4).LT.ZMIN) ZMIN=X(I,4)
  902 WRITE(6,900) (X(I,J),J=2,4)
  900 FORMAT(' ',3F12.5)
  903 FORMAT(4F10.0,I1)
      SS=S/FLOAT(N)
      IF(REFCON.GT.SS-SS*0.25.AND.REFCON.LT.SS+SS*0.25) GOTO 899
      REFCON=AINT(SS)
```

Table 3.8 *continued*

```
         CONT=(ZMAX-ZMIN)/15.0
         WRITE(6,898) REFCON, CONT
 898 FORMAT('0 *** REFCON AND CONT RESET TO ',G12.4,'   AND',G12.4)
 899 NL=INT(FN*6.0+0.5)
         NC=INT(FC*10.0+0.5)
         XAXIS=(UMAX-UMIN)/FLOAT(NC-1)
         YAXIS=(VMAX-VMIN)/FLOAT(NL-1)
         XAXIS=XAXIS+0.000001*XAXIS
         YAXIS=YAXIS+0.000001*YAXIS
         CALL DAPLOT(X,N,XAXIS,YAXIS,UMAX,VMAX,UMIN,VMIN,NL,NC)
         EPS=1.0E-7
C EPS IS THE TEST CRITERION USED IN THE ORTHOGONAL POLYNOMIAL ROUTINE.
C RESET AS NECESSARY. SEE CHAPTER 3 FOR DETAILS.
         DO 905 L=1,KO
C MAIN LOOP.
         WRITE(6,911) L
         IF(L.GT.1) GOTO 906
         L2=4
         L3=1
         IC=0
         GOTO 908
 906 L3=L2
         L2=(L+2)*(L+1)/2+1
         DO 907 I=1,N
 907 X(I,L2)=X(I,L3)
         CALL ADDCOL(X,N,M,L,MEANS)
 908 CALL GSORTH(X,V,C,D,N,M,LD,NU,L3,L2,IC,EPS)
         CALL RECOV(C,D,LD,M,DD,L3,L2,COEFFS,WORK,L,RSS)
         TSS=RSS+D(L2)
         WRITE(6,922) TSS
 922 FORMAT('0 TOTAL SUMS OF SQUARES = ',F20.4)
         WRITE(6,910) D(L2)
 910 FORMAT('0RESIDUAL SUM OF SQUARES = ',F20.4)
         RSS=RSS/TSS*100.
         WRITE(6,912) RSS
 912 FORMAT('0EXPLAINED SUM OF SQUARES =',F12.4,'%')
 911 FORMAT('1             TREND SURFACE OF ORDER',I4//)
         WRITE(6,821)
 821 FORMAT('0',5X,'U',9X,'V',9X,'Z',5X,'PRED Z',8X,'RES'//)
         DO 941 I=1,N
         ZZ=X(I,L2)-V(I,L2)
 941 WRITE(6,921) X(I,2),X(I,3),X(I,L2),ZZ,V(I,L2)
 921 FORMAT(' ',5F10.4)
         LV=L2-1
         CALL MAP(NL,NC,COEFFS,WORK, REFCON,CONT,XAXIS,YAXIS,LV,L,VMAX,UMIN
        +)
         WRITE(6,13) UMIN,VMAX,UMAX,VMAX,UMIN,VMIN,UMAX,VMIN
  13 FORMAT('0COORDINATES OF MAP CORNERS ARE:'/
        :5X,' TOP LEFT',5X,2F12.4/
        :5X,' TOP RIGHT     ',2F12.4/
        :5X,' BOTTOM LEFT   ',2F12.4/
        :5X,' BOTTOM RIGHT  ',2F12.4)
 905 CONTINUE
         RETURN
         END




         SUBROUTINE RECOV(C,D,LD,M,DD,K1,K2,COEFFS,WORK,IORD,RSS)
C CONVERTS ORTHOGONAL COEFFICIENTS TO LEAST-SQUARES COEFFICIENTS AND
C COMPUTES Z-SQUARE ARRAY.
         DOUBLE PRECISION G,H,P
         DIMENSION C(LD),D(M),DD(M),COEFFS(M),WORK(M)
         LE=K2*(K2+1)/2-K2
         WRITE(6,4)
    4 FORMAT('0ORTHOGONAL COEFFICIENTS'///)
```

Table 3.8  *continued*

```
      RSS=0.0
      K21=K2-1
      DO 5 I=1,K21
      IF(I.GT.1) RSS=RSS+C(LE+I)**2
    5 WRITE(6,7) I,C(LE+I)
      K=K2
      DO 10 J=1,K21
      K1=K
      K=K-1
      P=C(LE+K)
      G=0.D0
      L=K*(K-1)/2+K
      H=C(L)
      IF(K.EQ.K21) GOTO 11
      DO 12 L=K1,K21
      LL=L*(L-1)/2+K
   12 G=G+C(LL)*COEFFS(L)
   11 COEFFS(K)=(P-G)/H
   10 CONTINUE
      WRITE(6,9)
    9 FORMAT('0     COEFFICIENTS OF TREND EQUATION ARE'///)
   77 FORMAT(' ',E25.11)
      WRITE(6,7) (I,COEFFS(I),I=1,K21)
    7 FORMAT(' ',I6,2X,E20.8)
      WRITE(6,20)
   20 FORMAT('0 Z SQUARE ARRAY'///)
      L=IORD+1
      LL=L
      IP=1
      IS=1
      INCBIG=1
      INC=2
   23 DO 21 I=1,L
      WORK(I)=C(LE+IS)**2
      IS=IS+INC
   21 INC=INC+1
      IF(L.EQ.LL) GOTO 25
      L1=L+1
      DO 26 I=L1,LL
   26 WORK (I)=0.0
   25 WRITE(6,22) (WORK(I),I=1,LL)
      L=L-1
      IS=IP+INCBIG
      IP=IS
      INCBIG=INCBIG+1
      INC=INCBIG+1
      IF(IS.LT.K2-1) GOTO 23
   22 FORMAT(' ',10G11.4)
      RETURN
      END


      SUBROUTINE ADDCOL(X,N,M,K,MEANS)
C ADDS COLUMNS TO DATA MATRIX AS SURFACE ORDER INCREASES
      REAL MEANS(M),X(N,M)
      L=K+1
      L1=L*(L-1)/2+1
      L0=L1-K-1
      L2=L1+K-1
      DO 2 I=L1,L2
      L0=L0+1
      DO 1 J=1,N
    1 X(J,I)=X(J,L0)*X(J,2)
    2 CONTINUE
```

Table 3.8  *continued*

```
      DO 3 J=1,N
      X(J,L2+1)=X(J,L0)*X(J,3)
    3 CONTINUE
      RETURN
      END



      SUBROUTINE MAP(NL,NC,ALPHA,VEC,REFCON,CONT,XAXIS,YAXIS,NO,IORD,
     *VMAX,UMIN)
C PRINTS LINEPRINTER MAP OF TREND SURFACE
      DIMENSION ALPHA(NO),VEC(NO),SLINE(120),SYMB(23)
      DATA SYMB/'1',' ','2',' ','3',' ','4',' ','5',' ','$',' ','6',
     * ' ','7',' ','8',' ','9',' ','0','I','-'/
    1 FORMAT('1        TREND SURFACE MAP FOR ORDER',I8/' PART',I4//)
    2 FORMAT(' ',120A1)
      KMAX=0
      KMIN=20
      NO2=110
      NSTR=NC/110
      LEFT=NC-110*NSTR
      IF(LEFT.GT.0) NSTR=NSTR+1
      DO 96 KOUNT=1,NSTR
      IF(KOUNT.EQ.NSTR.AND.LEFT.GT.0) NO2=LEFT
      WRITE(6,1) IORD,KOUNT
      NO22=NO2+2
      WRITE(6,2) (SYMB(23),I=1,NO22)
      V=VMAX+YAXIS
      DO 12 L=1,NL
      V=V-YAXIS
      U=UMIN-XAXIS+(KOUNT-1)*110*XAXIS
      DO 6 I=1,NO2
      U=U+XAXIS
      Z=ZED(ALPHA,VEC,NO,IORD,U,V)
      K=11+INT((Z-REFCON)/CONT)
      IF(Z.LT.REFCON) K=K-1
      IF(K.LT.1) K=1
      IF(K.GT.20) K=20
      IF(K.GT.KMAX) KMAX=K
      IF(K.LT.KMIN) KMIN=K
    6 SLINE(I)=SYMB(K)
   12 WRITE(6,2) SYMB(22),(SLINE(I),I=1,NO2),SYMB(22)
      WRITE(6,2) (SYMB(23),I=1,NO22)
   96 CONTINUE
      WRITE(6,4) REFCON,CONT
    4 FORMAT('0       REFERENCE CONTOUR =',F12.4/
     *'        CONTOUR INTERVAL   =',F11.4)
      WRITE(6,3)
    3 FORMAT('0  KEY TO CONTOUR VALUES'//)
      DO 5 I=KMIN,KMAX
      J=11-I
      P=REFCON-J*CONT
      Q=P+CONT
      IF(I.EQ.1) GOTO 7
      IF(I.EQ.20) GOTO 8
      WRITE(6,9) (SYMB(I),J=1,6),P,Q
      GOTO 5
    7 WRITE(6,10) (SYMB(I),J=1,6),P
   10 FORMAT(' ',6A1,2X,F10.4,' AND BELOW')
      GOTO 5
    8 WRITE(6,11) (SYMB(I),J=1,6),P
   11 FORMAT(' ',6A1,2X,F10.4,'  AND ABOVE')
    9 FORMAT(' ',6A1,2X,F10.4,'  TO  ',F10.4)
    5 CONTINUE
      RETURN
      END
```

Table 3.8 *continued*

```
      FUNCTION ZED(ALPHA,VEC,NO,N,U,V)
C COMPUTES VALUE ON SURFACE AT POINT (U,V)
      DIMENSION ALPHA(NO),VEC(NO)
      ZED=ALPHA(1)+ALPHA(2)*U+ALPHA(3)*V
      IF(N.LT.2) RETURN
      VEC(2)=U
      VEC(3)=V
      DO 1 M=2,N
      L1=M-1
      L2=M*(M+1)/2
      DO 2 I=1,M
      L2=L2+1
      VEC(L2)=VEC(L2-M)*U
    2 ZED=ZED+ALPHA(L2  )*VEC(L2)
      VEC(L2+1)=VEC(L2  -M)*V
    1 ZED=ZED+ALPHA(L2+1)*VEC(L2+1)
      RETURN
      END




      SUBROUTINE DAPLOT(X,N,XAXIS,YAXIS,UMAX,VMAX,UMIN,VMIN,NL,NC)
C PRODUCES MAP OF LOCATION OF DATA POINTS
      DIMENSION X(N,3),SLINE(113),SYMB(10)
      DATA BL,ST,BAR,DOT/1H ,1H*,1H-,1H./
      DATA SYMB/1H1,1H2,1H3,1H4,1H5,1H6,1H7,1H8,1H9,1H0/
    1 FORMAT('1PLOT OF DATA POINT DISTRIBUTION'// SCALE IS THE SAME AS T
     +HE TREND MAPS'///)
    2 FORMAT(' ',120A1)
      WRITE(6,1)
      NSTR=NC/110
      NO2=113
      LEFT=NC-110*NSTR
      IF(LEFT.GT.0) NSTR=NSTR+1
      DO 5 KOUNT=1,NSTR
      IF(KOUNT.EQ.NSTR.AND.LEFT.GT.0) NO2=LEFT+3
      NO22=NO2+2
      WRITE(6,4) KOUNT
    4 FORMAT(' PART',I6//)
      WRITE(6,2) (BAR,I=1,NO22)
      V=VMAX+YAXIS
      IF(KOUNT-1) 9,7,9
    9 U=U2
      GOTO 8
    7 U=UMIN
    8 U2=U+NO2*XAXIS
      DO 3 JK=1,NL
      IJ=0
      V=V-YAXIS
      VV=V-YAXIS
      DO 10 I=1,NO2
   10 SLINE(I)=BL
      DO 6 I=1,N
      X1=X(I,3)
      X2=X(I,2)
      IF(X1.LT.VV.OR.X1.GT.V) GOTO 6
      IF(X2.LT.U.OR.X2.GT.U2) GOTO 6
      K=INT((X2-U)/XAXIS)+1
      IF(K.LT.IJ) GOTO 6
      SLINE(K)=ST
      J=I
      IJ=I+3
      IF(J.LT.100) GOTO 11
      SLINE(K+1)=SYMB(J/100)
      J=J-(J/100)*100
      GOTO 12
```

Table 3.8   *continued*

```
   11 SLINE(K+1)=SYMB(10)
      IF(J.LT.10) GOTO 13
   12 L=J/10
      IF(L.EQ.0) L=10
      SLINE(K+2)=SYMB(L)
      IF(L.EQ.10) GOTO 14
      J=J-L*10
      GOTO 14
   13 SLINE(K+2)=SYMB(10)
   14 IF(J.EQ.0) J=10
      SLINE(K+3)=SYMB(J)
    6 CONTINUE
    3 WRITE(6,2) DOT,(SLINE(I),I=1,NO2),DOT
    5 WRITE(6,2) (BAR,I=1,NO22)
      END
```

represented by the $ sign. The value $z_i$ lies in the $q$th contour interval, where $q$ is calculated from

$$q = 11 + \text{INT}(z_i - \text{REFCON})/\text{CONT}$$

The value 11 refers to the fact that the reference contour lies at the lower end of the 11th interval. Thus, the $q$th printing character is to be output at $(u_{\min}, v_{\max})$. Rather than print each character individually they are stored in a vector SLINE, which can hold up to 120 characters. When the last printing character of line 1 has been placed in SLINE this vector is printed out, using (120A1) format. If desired, the character I can be printed before and after the vector SLINE, and a line of '-' characters can be printed above and below the map to produce the appearance of a border. The process of filling and printing SLINE is repeated for all NL lines.

*The numerical method of estimating* $\alpha$, the vector of trend coefficients, is one of the more controversial aspect of trend-surface calculations. Most textbooks and the majority of published computer programs make use of the ordinary least-square (OLS) approach. This involves the calculation — for each order of trend-surface — the *basic matrix* $X'_{(1)}X_{(1)}$, **P** or **R** depending on the approach adopted. The vector $\hat{\boldsymbol{\alpha}}$ or its standardized equivalent $\hat{\boldsymbol{\alpha}}_{(s)}$ is then found from one of the following equations, using as far as possible the terminology of Chapter 2:

$$\left.\begin{array}{l}\text{(i) } \hat{\alpha} = (X'_{(1)}X_{(1)})^{-1} X'_{(1)}y; \\ \text{(ii) } \hat{\alpha} = P^{-1}h \\ \text{or } \text{(iii) } \hat{\alpha}_{(s)} = R_{uv}^{-1}g_{uvz}\end{array}\right\} \quad (3.11)$$

Method (ii) requires that $\hat{\alpha}_0$ be estimated separately by

$$\hat{\alpha}_0 = \bar{z} - \hat{\alpha}_1 \bar{u} - \hat{\alpha}_2 \bar{v} - \hat{\alpha}_3 \overline{uv} \ldots - \hat{\alpha}_k \bar{v}^m \quad (3.12)$$

Method (iii) gives the vector of standardized trend coefficients, which have to be converted to raw form by

$$\hat{\alpha}_i = \hat{\alpha}_{(s)i} \cdot \frac{s_i}{s_z} \quad (i = 1, 2, \ldots, k) \tag{3.13}$$

where $s_i$ is the standard deviation of the $i$th term in the equation and $s_z$ the standard deviation of the dependent variable. Again, $\hat{\alpha}_0$ must be recovered separately, from Equation (3.12).

The OLS approach usually results in several difficulties, specifically:

(i) the matrices $X'_{(1)}X_{(1)}$, $P$ or $R_{uv}$ are likely to be nearly singular especially as $m$, the surface order increases. A poor data-point distribution (Subsection (B)) may exaggerate the problem.

(ii) whichever of the three basic matrices is used a poor choice of algorithm may lead to considerable loss of accuracy especially if the method is implemented on a computer using short word formats (32 or less bits per real number). Rounding error may result in a matrix which is in fact singular being 'inverted' or it may cause the computer to detect singularity where none exists. The choice of algorithm in this context is discussed in Chapter 1.

(iii) as mentioned above, the magnitude of the numbers generated in the evaluation of high-order polynomial terms may exceed machine capacity if the coordinate origin is not well-chosen. For example, the true value of $333^5$ is 4 094 691 316 893. An ICL 1900 series computer (48 bit words) represents this number by 4 094 691 316 000 while an IBM 360 computer (32 bit words) would store this number as 4 094 691 000 000. No indication of such rounding or chopping will be apparent to the user. These problems can be minimized by the choice of a sensible origin for the coordinate system and, if traditional computing methods are to be applied, by the use of $R_{uv}$ as the basic matrix.

(iv) the traditional methods of finding $\hat{\alpha}$ are inefficient. The inverse of the basic matrix must be found for each separate order of matrix. Since $m$, the number of terms, increases rapidly as $k$, the order of the surface, rises then computational problems can be expected. All the elements of $\hat{\alpha}$ must be computed every time the order of the surface is increased.

To minimize these problems the use of the method of orthogonal polynomials is recommended. No matter what technique of determining $\hat{\alpha}$ is used, the consequences of the near-singularity of $R_{uv}$ (or whichever basic matrix is employed) cannot be eliminated. However, the added accuracy of orthogonal polynomials, combined with the fact that only the additional orthogonal coefficients need be computed as the order of the surface rises, make the technique attractive in most trend surface applications. The use and calculation of orthogonal polynomials has already been described in an earlier section (3.1.1). Modifications necessary to use the same basic techniques in a trend-surface program are described below. Early examples of the use of orthogonal polynomials are provided by Oldham and Sutherland (1955) and Grant (1957). Krumbein and Graybill (1965) present Grant's formulation for the case of regularly gridded data. However, the extension of the method to deal with randomly spaced data points has not been accomplished until recently (Spitz, 1966: Dixon, 1969; Dixon and Spackman, 1970; Dixon et al., 1972; Whitten, 1970, 1972). Other recent uses of orthogonal polynomials have been in the context of mathematical surface-fitting prior to automatic contour interpolation (Crain and Bhattacharyya, 1967; Crain, 1970; Czegledy, 1972). A good general review of the subject is contained in the paper by Berztiss (1964).

In Section 3.1.1 it was concluded that both the classical Gram—Schmidt process as used by Whitten (1970) and Rushton's Cholesky-based method generally produce unacceptable error. Bjork's (1967) modified Gram—Schmidt procedure was chosen as the most suitable computing method. Forsythe's (1957) method of recurrence relationships is extended to two dimensions by Spitz (1966) and Dixon and Spackman (1970). However, the method of Bjork is simple and accurate, and it readily allows conversion of the estimates of the parameters from the orthogonal basis to the observed $(u, v)$ polynomial system. (Subroutine GSORTH, Appendix A.)

One of the advantages of the use of orthogonal polynomials is provision of information concerning the importance (in terms of contribution to the explained sum of squares) of each orthogonal coefficient (Whitten, 1970, 1972). The use of this information in the determination of the most appropriate trend surface is illustrated in the example below. Usually, the information is displayed in the form of a table, termed the $Z^2$ array. The subscript of the $Z_{ij}^2$ refers to the order of the corresponding term — either $u$ or $v$ — in the polynomial expansion. Thus, $Z_{12}$ corresponds to the term in $uv^2$, while $Z_{32}$ corresponds to the term in $u^3v^2$. The layout of the $Z^2$ array is shown in Table 3.9.

Table 3.9
Layout of $Z^2$ array

| | | | |
|---|---|---|---|
| — | $Z_{01}^2$ | $Z_{02}^2$ | $Z_{03}^2 \ldots \ldots \ldots Z_{0m}^2$ |
| $Z_{10}^2$ | $Z_{11}^2$ | $Z_{12}^2$ | $Z_{13}^2 \ldots \ldots Z_{1,m-1}^2$ |
| $Z_{20}^2$ | $Z_{21}^2$ | $Z_{22}^2$ | $Z_{23}^2 \ldots . Z_{2,m-2}^2$ |
| $\ldots\ldots\ldots\ldots\ldots\ldots\ldots\ldots\ldots\ldots$ | | | |
| $\ldots\ldots\ldots\ldots\ldots\ldots\ldots\ldots$ | | | |
| $\ldots\ldots\ldots\ldots\ldots\ldots$ | | | |
| $Z_{m0}^2$ | | | |

Grant (1957) and Oldham and Sutherland (1955) discuss and illustrate the use of the $Z^2$ array in differentiating between the trend or regional component and the residual component. Grant (1957) describes statistical tests which can be carried out on the elements of the $Z^2$ array. These tests have rarely been applied in practice. The use of the $Z^2$ array is illustrated in Section 3.2.2.

The use of orthogonal polynomials does not avoid the problem of multicollinearity, the effects of which were discussed in Section 2.4(i). It was suggested in that section that the method of ridge regression, while giving estimates that are biased, may nevertheless produce more realistic sample values whenever the explanatory variables are closely linearly related, a situation to be expected in trend surface analysis. Jones (1972) gives a brief example of such an approach. However, the technique has not yet been evaluated in simulation experiments, in which the nature of the trend is known exactly. Preliminary investigations indicate that in most, if not all, situations ridge estimators are inefficient compared with standard OLS estimators. Further work is required before the method of ridge regression can be accepted as a standard approach in trend surface analysis in a situation of high multicollinearity. An alternative approach could involve the calculation of the principal components of $(X'X)$ or $R_{uv}$ and the subsequent use of the principal components scores matrix in place of the powers and cross-products of $u$ and $v$.

One drawback inherent in trend-surface analysis is that only a single dependent variable can be used. Davis (1973) describes a method by which the technique can be extended from 3 to 4 dimensions. For more than one dependent variable, the use of canonical correlation methods has been proposed by Lee (1969A, 1969B). Another approach that might be adopted is to use scores on a principal component or factor as the dependent variable. Whichever method is used, the problem of identifying the canonical vector, factor or principal component remains.

*(D) Inferential Problems*  Under this heading will be outlined the difficulties raised by:

(i) the choice of null hypothesis;
(ii) the use of standard statistical procedures, and
(iii) the place of such procedures in exploratory studies.

Three null hypotheses can be postulated in the context of a trend-surface analysis. Firstly there is the hypothesis that the variance regarded as being due to the 'trend' in the sample is the result of chance alone, or, in other words, a variance reduction as great or greater than that actually achieved could have been obtained by random sampling of a population in which there was no trend. Secondly, one could set up the null hypothesis that the increase in explained variance resulting from raising the order of the trend surface from $k$ to $k + 1$ is again zero in the population. Finally, it is possible to test each individual coefficient in the trend equation against the null hypothesis that it is zero in the population. This approach is used in the stepwise fitting of trend surfaces (Miesch and Connor, 1968) in which the aim is not the identification of a statistical trend but the modelling of a particular mathematical surface. Hence, Miesch and Connor use not only polynomial terms in $u$ and $v$ but also incorporate trigonometric and logarithmic functions in $u$ and $v$. More economical procedures for surface fitting are available, for example, the use of bicubic spline functions (Whitten and Koelling, 1973; Koelling and Whitten, 1973). Generally speaking, the trend coefficients are not given a physical meaning and consequently the third hypothesis is rarely applied. However, it can be compared to the use of the $Z^2$ array, discussed in the previous section, in which the significance of individual $Z_{ij}^2$ is assessed subjectively.

The first two hypotheses can be tested using analysis of variance methods, given certain assumptions which are discussed below. The calculations are best shown in tabular form, as in Table 3.10, in which $RSS_p$ indicates the residual sum of squares for the $p$th order trend surface, and $ESS_p$ is the corresponding explained sum of squares. The

Table 3.10
Analysis of variance layout for trend surface analysis (See text for explanation)

| (i) Source of variation | (ii) Sum of squares | (iii) D.F. | (iv) Mean square | (v) $F$ |
|---|---|---|---|---|
| Trend surface of order $p + 1$ | $ESS_{p+1}$ | $NC_{p+1} - 1$ | $EMS_{p+1}$ | $\dfrac{EMS_{p+1}}{RMS_{p+1}}$ * |
| Deviation from $(p + 1)^{th}$ order trend surface | $RSS_{p+1}$ | $N - NC_{p+1}$ | $RMS_{p+1}$ | |
| Trend surface of order $p$ | $ESS_p$ | $NC_p - 1$ | $EMS_p$ | $\dfrac{EMS_p}{RMS_p}$ ** |
| Deviation from $p^{th}$ order trend surface | $RSS_p$ | $N - NC_p$ | $RMS_p$ | |
| Additional terms in the $(p + 1)^{th}$ order trend surface | $ESS_{inc}$ | $NC_{p+1} - NC_p$ | $EMS_{inc}$ | $\dfrac{EMS_{inc}}{RMS_{p+1}}$ *** |

number of coefficients (including the constant term) in the $p$th order trend equation is $NC_p$, while the number of observations is $N$. The table shows: (i) the source of variation; (ii) the sum of squares attributable to that source; (iii) the corresponding degrees of freedom; (iv) the mean square, found by dividing (ii) by (iii), and (v) the $F$-ratio. The first such $F$-ratio, marked * in Table 3.10, is the computed value of $F$ used in the testing of the null hypothesis that the $(p + 1)$th order trend surface does not account for a statistically significant proportion of the total sum of squares. If this value of $F$ is less than the tabled value of $F$ with the degrees of freedom given in Table 3.10 and at a suitable level of significance then the null hypothesis is accepted.

The second $F$-ratio in column (v) of Table 3.10, marked **, is used to test the null hypothesis that the $p$th order trend surface does not account for a statistically significant proportion of the total sum of squares of the dependent variable. A procedure similar to that described above is followed. The third entry in column (v) of Table 3.10, marked ***, is the $F$ ratio that is used to test the null hypothesis that the increase in the explained sum of squares that results from the raising of the order of the fitted surface from $p$ to $p + 1$ could have arisen by chance. This is equivalent to postulating that the trend coefficients associated with the $(p + 1)$th order surface are zero in the population, that is:

$$H_0 : \alpha_{NC_{(p)}+1} = \alpha_{NC_{(p)}+2} = \ldots = \alpha_{NC_{(p+1)}-1}$$

where $NC_{(p)}$ is the number of coefficients of order $p$ and less. For example, if $p = 3$ then $NC_{(p)} = 10$ and the extra coefficients associated with the $(p + 1) = 4$th order surface are $\alpha_{11}, \alpha_{12}, \ldots, \alpha_{14}$. The testing of the null hypothesis is carried out as before.

The blind application of these tests of significance is not to be recommended, as several assumptions must be fulfilled if they are to be reliable (Parsley and Doveton, 1969). These are:

(a) the observations on the dependent variable represent a random sample the size of which is considerably greater than the number of coefficients defining the trend surface;

(b) the conditional distribution of the dependent variable is normal;

(c) the residuals from the trend surface are not significantly spatially autocorrelated, and have homoscedastic variances.

Satisfaction of the first assumption frequently requires mental sleights-of-hand. To use an example mentioned earlier, given a fixed population of $n$ corries one could argue that they are really a sample of all possible corries that might have formed had topographic conditions been different. This does not imply, however, that the present configuration is random. The hypothesis could be, however, that the observed configuration is but one example drawn from a population comprising an infinite number of corrie distribution patterns in a given area, and might lead to questions concerning the relationship of the observed distribution to the mean pattern in the population. In most instances, one is concerned with *describing* the given pattern in terms of generalizations at various scales, and relating these generalized patterns to the distributions of other, possibly causally linked, variables. In such circumstances, significance tests are both unnecessary and inapplicable. Whenever the observations do represent a sample, for example, annual rainfall totals measured at a fixed number of points, then the statistical tests described above are strictly valid if the distribution of the sample points is random. Hence, clustered and linear data point distributions — which also affect the fit of the trend surface — do not allow the correct use of standard inferential tests.

A second requirement is normality of the conditional distribution of the dependent variable. This assumption, together with the third assumption which concerns the residuals from the trend-surface, are discussed in detail in Chapter 2 in the context of multiple linear regression. The tests hold approximately in the case of large samples even if distributions are non-normal, though the degree to which the significance levels are valid will obviously depend on the nature and extent of the non-normality. Spatial autocorrelation among the residuals is a more serious problem. Firstly, it is difficult to measure, secondly the least-squares estimates of the trend coefficients lose their minimum-variance property, and thirdly standard statistical tests become invalid, partly due to the affect of autocorrelation on the degrees of freedom of the $F$-ratio tests. Agterberg (1967) quotes an example of this with reference to a one-dimensional trend. Using standard formulae, the degrees of freedom were computed as 19 and 39, and with these degrees of freedom the computed $F$ ratio was high enough to allow rejection of the null hypothesis under consideration. However, assuming that the residuals followed a first-order Markov scheme of the form

$$R_{k+1} = \rho R_k + \epsilon_{k+1}$$

where $R_i$ is the $i$th residual and $\rho$ the first-order serial correlation coefficient, Agterberg was able to show that the true degrees of freedom were 19 and 16, at which point the computed $F$-ratio was non-significant. Testing for spatial (two-dimensional) autocorrelation is discussed in Section 2.4(iii). An example of the use of such tests in the context of trend surface analysis is given below (Section 3.2.2).

In some cases the apparent autocorrelation is due to interaction between the dependent variables and explanatory variables other than location. One possibility is to employ a two-stage least-squares estimation procedure as described by Agterberg and Cabillo (1969). A two-stage estimation procedure has also been proposed by Parsley (1971) whose method, however, requires gridded data. This necessitates the use of two-dimensional interpolation algorithms prior to the trend analysis. Autocorrelation may also result if the functional form of the trend surface model is incorrect. We have considered only linear models, whereas there is no reason to suppose a linear relationship between the dependent variable and the $(u, v)$ coordinates and their powers. Very little has been published on the application of nonlinear models in trend surface analysis. James (1967, 1970) gives an example of a nonlinear surface-fitting procedure using Marquardt's algorithm (see Section 3.5) but other practical examples are nonexistent. This is an obvious area for further research.

The consequences of heteroscedasticity of residual variances are similar to those resulting from autocorrelation. A brief discussion is given in Chapter 2. It must be remembered, however, that these assumptions are necessary only where inferential tests are being applied. Irrespective of the form of the distributions the least-squares surface is always the 'best-fit' surface of that order by definition. Thus, if the investigation is concerned not with estimating the properties of the population trend function but with describing the variation exhibited by a spatially-distributed phenomenon the questions of normality, spatial autocorrelation and heteroscedasticity do not arise. However, even in 'descriptive' applications resort is often made to $F$-ratio tests in order to decide on the order of surface which best represents a regional trend. In these cases, the assumptions must be satisfied if the tests are to be meaningful. Often these tests are applied sequentially, as in stepwise regression (Section 3.3.1). Again, this distorts the true significance levels.

The above remarks point towards the conclusion that $F$-ratio tests, or other measures based on sums of squares explained, such as $R^2$, should not be accepted uncritically in the determination of the statistical significance of a sample trend function in the manner advocated by Chayes (1970). The levels associated with the tests are dubious on statistical grounds, while they can be faulted also on logical grounds. As the order of the trend surface increases so it becomes more flexible. A complex pattern cannot be described by a low order surface simply because the surface cannot 'bend' sufficiently. A linear surface fitted to a corrugated pattern might well prove to be nonsignificant, and the fitting procedure terminated. This would assume that the increment to the explained sums of squares decreases monotonically as the order of the fitted surface increases.

How, then, are sample trend-surfaces to be examined for significance other than on purely subjective grounds? This is a difficult question, and no immediate answer can be forthcoming. Work by Howarth (1967), Norcliffe (1969), Tinkler (1969) and Unwin (1970) indicates that simulation experiments with random data provide a partial answer. Howarth (1967) describes a series of experiments carried out with a sample of size 100. The explained sum of squares (ESS) for the linear surface was at least 6 per cent in 95 per cent of the trials, rising to 12.0 and 16.2 per cent respectively for the quadratic and cubic surfaces. In all cases the true value of ESS was 0. Unwin (1970) reported further work on the values of ESS derived from clustered data point distributions. He concluded that '... it is possible that imperfect distributions may yield (explained sums of squares) values that are sufficiently high to give an incorrect assessment when using standard tests of significance based on the assumption of a random distribution of control.' Significance levels produced by this simulation method would have to provide both for various types of data-point distribution and different sample sizes — clearly an enormous task.

An alternative procedure is to sample randomly from the existing set of data points and compare the resulting trend surfaces. Confirmation of a hypothesis can come only from repetition, not from significance testing. Hence, if replicate samples are unobtainable, the given sample can be split in several random ways and trend surfaces derived from each subsample. (See Doornkamp, 1972, for an example.) If the same overall pattern emerges one can be more confident of its real existence. This method presupposes an original sample that is large enough to enable realistic subsamples to be drawn from it. Furthermore, replication cannot help directly in the decision as to which order of trend-surface 'fits' the data.

These conclusions seem both to be pessimistic and to confirm the suspicion that statistical techniques merely delay a subjective decision rather than replace it. However, until the methods of statistical inference are further refined, it is better to be suspicious than to follow blindly contemporary trends.

### 3.2.2 Comparison of trend surfaces

The relative lack of research work in the field of map comparison is surprising in view of the importance of the topic. When geographical variables are mapped it is logical to ask in what respects the resulting patterns agree or disagree. This is especially true when there is some supposed causal or relational link between the two variables being compared. A simple correlation measure does not provide much geographical or spatial information; the degree of overall association can be inferred from the correlation coefficient, but the spatial distribution of resemblance cannot be estimated from a single number. In this section procedures for the comparison of trend maps will be reviewed, but it will become clear that much more work in this area is necessary before useful and reliable methods become available.

The first group of procedures can be termed direct measures because they are based on the predicted trend values as represented by the trend surface maps. The simplest of these measures is the correlation of vectors of predicted (trend) values for two variables measured at the same data points. The product moment correlation coefficient is computed for the vectors of predicted values at the $n$ data points, and the degree of overall resemblance is inferred from the value of the coefficient (which ranges from +1, indicating perfect positive association, through zero, which implies no relationship, to $-1$, showing a reciprocal relationship between the two variables). The first drawback of this simple scheme is the fact that it is a generalization of what may in fact be a complex pattern. One could easily think of a situation in which the two variables were spatially related in a positive sense in one sector of the study area and in a negative sense in another sector. These two relationships would be averaged by the correlation measure, leading to a coefficient whose value might lead the observer to conclude that there was no significant spatial relationship between the two variables. In other words, it will be difficult to say what a value of $r = 0.75$ actually means. It certainly does not mean that 75 per cent of the areas of the two maps are similar, or that the values at each point have a similarity of 0.75. Since the value 0.75 is a composite one, it could be made up of any number of different patterns. The second criticism that can be levelled against this method is that the continuous trend surfaces are being compared at a finite number of discrete points. The behaviour of the surface between the points is not taken into account. This may prove to be misleading, especially in cases in which the distribution of data points is not regular.

The same concept has therefore been applied to vectors of predicted values measured on a grid superimposed on the trend surface. Although the surface is now better represented, this is by no means an acceptable method since the behaviour of the trend surface between data points is to some extent dependent on the distribution of those data points. The surface may be completely unreliable in an area of the map that contains only a few of the data points.

A further modification, which is subject to the comments made in the previous paragraphs, involves the calculation of a correlation measure (the product moment correlation coefficient has normally been used) at each of a set of grid points over the map area (Robinson, 1962; Robinson and Caroe, 1967). At each corresponding pair of grid points on the two maps to be compared a vector of predicted values derived for a number of adjacent points is computed. The two vectors are then correlated and the correlation coefficient is then plotted on the map at the associated grid point position. This process is repeated for all corresponding pairs of grid points and the resulting distribution of point values is contoured. Regions of high, low and zero resemblance can then be determined visually. Obviously this method relies upon the validity of the grid values; this is a function of the distribution of the data points and of the complexity of the polynomial surface that is fitted. However, it is superior to either of the other two direct (correlational) methods already discussed since it does attempt to give a spatial picture of the similarity or dissimilarity between the two trend maps.

A question of statistical inference can be raised at this stage. Given that the distribution of the dependent (mapped) variable is continuous, then the observed point pattern on which the trend calculations are based represents a sample drawn from this underlying continuous distribution. There is a finite chance that the pattern of similarity depicted by the correlation surface is due to chance when the two underlying spatial patterns are completely dissimilar. There is, as yet, no statistical theory available to provide a test of this hypothesis; it does mean, however, that a computed correlation surface may show a completely different pattern if it is based on a second set of data points. This problem is not resolved by showing that both trend surfaces separately account for a significant proportion of the sums of squares of the dependent variable, for — even if the assumptions of the $F$ test are satisfied — one surface may, in places over- or underestimate the true trend and the other surface may correspondingly under- (or over-) estimate, leading to a wider or narrower difference between the two sample surfaces than exists between the true (population) surfaces. Other questions concern the distribution of this error, the effect of the orientation of the grid, the number of surrounding points on which each correlation is based, and the reliability of these correlations. The number of points from which each correlation is determined may be as low as eight or ten; quite different values of the correlation coefficient can be expected for different point distributions with such a low value of $n$.

Indirect methods of trend map comparison are based on the vector of trend coefficients rather than on the map itself (though both are expressing the same information). Since the constant term is estimating the height of the trend surface at the origin it is usually omitted, as the investigator is normally interested in differences of shape rather than altitude, especially if the two dependent variables are measured in different, possibly non-comparable units. Merriam and Sneath (1966) propose that the trend coefficients are first standardized to have a mean of zero and unit variance before the degree of similarity or dissimilarity is computed, though other weighting methods can be considered. These authors use two measures of association; the product-moment correlation coefficient and the Euclidean distance (Equation (6.1)). The former is well-known; the latter is described in Chapter 6. It can range from 0 (perfect similarity) to infinity (perfect dissimilarity) though it can be standardized to lie in the range 0–1. Although this method does not rely on point values, it still measures resemblance over the entire map in terms of a single number which does not allow the user to determine the pattern of agreement or disagreement over the map area as a whole. Further difficulties relate to the stability of the trend coefficients themselves. It is known that the matrix of sums of squares and cross-products tends to become badly ill-conditioned as the order of the surface rises. This means that the computed coefficients may change radically if the observed values are only slightly altered. This alteration may not indicate that the shape of the trend surface is different, for different combinations of coefficients can produce a similar surface. However, measures of similarity based on the co-efficients vector could be expected to change significantly. Discussion of this and related points can be found in Mandelbaum (1966), Haggett and Bassett (1970), Macomber (1971) and in Cliff et al. (1975; Chapter 4). A general discussion of the problem of comparing spatial distribution patterns is provided by Miller (1964). Rao (1971) and Rao and Srivastava (1969) discuss an alternative method based upon the sums of squares and cross-products matrix. However, all of the methods mentioned in this section have flaws or disadvantages and none of them can be unreservedly recommended. Further work is required on methods of characterizing continuous surfaces by easily understood and simply compared values before any real progress is made.

### 3.2.3 Examples

Two examples are given in this section. The first is an artificial example which is meant to illustrate the results of trend surface analysis when the form of the population trend is known. The use of the $Z^2$ array and the statistical tests described in the previous section are both demonstrated in this example. The second example is concerned with the use of the spatial autocorrelation statistics of Cliff and Ord (1973) in the context of trend surface analysis. The computations in this second example also serve to outline the method of testing for spatial autocorrelation among the residuals from a multiple regression (Section 2.4(iii)).

In the first example the spatial coordinates of a set of 50 points were randomly selected in the range $0 \leqslant u, v \leqslant 4$ using a pseudo-random number generator (Appendix A). At each point $(u_i, v_i)$ the value of a dependent variable $z_i$ was computed from the equation

$$z_i = 10.0 + 1.5u_i + 2.74v_i + 0.9u_i^2 - 1.3u_iv_i + 0.35v_i^2 + \epsilon_i \qquad (3.14)$$

Equation 3.14 is the population trend equation; the results of the trend surface analysis described below can thus be compared with the known true equation. In Equation 3.14, the term $\epsilon_i$ represents a random normal deviate, generate by the subroutine DEVIAT (Appendix A).

Table 3.11
Data for trend surface analysis

| U | V | Z |
|---|---|---|
| 0.34568 | 1.63375 | 15.50510 |
| 2.12204 | 1.50033 | 15.65517 |
| 3.59207 | 1.74668 | 25.59478 |
| 3.03284 | 3.40023 | 23.90152 |
| 2.50076 | 3.94623 | 22.61969 |
| 2.12755 | 0.05777 | 16.83189 |
| 1.50919 | 2.89395 | 19.03233 |
| 0.99888 | 0.94603 | 14.25926 |
| 0.31310 | 2.88471 | 19.48856 |
| 3.81815 | 2.05421 | 24.32168 |
| 0.39387 | 1.71776 | 15.25599 |
| 2.30015 | 2.19620 | 20.20652 |
| 1.82713 | 2.84910 | 17.24281 |
| 3.44922 | 0.10048 | 25.40588 |
| 2.12258 | 1.57505 | 19.69229 |
| 3.83657 | 2.03112 | 25.52720 |
| 2.84509 | 0.99712 | 22.84070 |
| 0.15892 | 3.69023 | 23.11204 |
| 3.76708 | 3.51807 | 25.16849 |
| 3.82833 | 0.15148 | 27.37079 |
| 2.30205 | 1.61928 | 19.65154 |
| 1.34327 | 3.57494 | 20.82506 |
| 1.96948 | 1.66936 | 18.16886 |
| 0.19098 | 2.45325 | 17.46781 |
| 3.64088 | 0.51394 | 27.29056 |
| 2.03107 | 2.13888 | 18.64183 |
| 3.31717 | 0.59681 | 23.93013 |
| 2.91035 | 0.40277 | 20.94453 |
| 2.18726 | 3.36225 | 19.68072 |
| 2.25893 | 1.77414 | 17.63129 |
| 0.71787 | 2.68860 | 17.94553 |
| 0.41758 | 0.41164 | 11.65821 |
| 2.52468 | 2.10316 | 17.99768 |
| 2.44630 | 3.25563 | 21.85324 |
| 3.77461 | 1.52822 | 25.77245 |
| 3.96416 | 3.21180 | 25.07276 |
| 1.86946 | 1.30846 | 17.56846 |
| 0.77042 | 2.11604 | 17.05352 |
| 3.16073 | 1.26948 | 21.41214 |
| 1.99685 | 1.23050 | 17.21846 |
| 3.44571 | 2.55773 | 23.18665 |
| 0.54134 | 3.02093 | 20.77807 |
| 0.99487 | 2.01151 | 17.18739 |
| 0.53274 | 1.96529 | 16.97701 |
| 3.86038 | 2.81338 | 24.67132 |
| 3.57695 | 0.14361 | 26.24609 |
| 3.21601 | 3.17373 | 22.95910 |
| 1.53424 | 1.35905 | 16.28678 |
| 2.47820 | 2.79563 | 21.61406 |
| 1.57791 | 3.54026 | 21.82708 |

The resulting data matrix is listed in Table 3.11 and the location of the 50 points is shown in Figure 3.13(a).

The first order sample trend (Figure 3.13(b)) is represented by the equation

$$z_i = 13.23 + 2.60u_i + 0.80v_i$$

This equation is not very similar to the population model (Equation

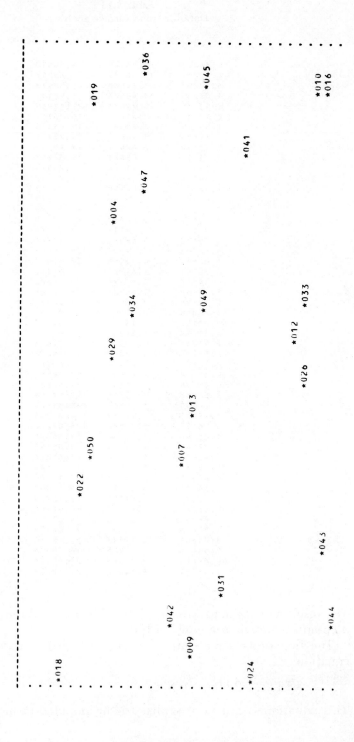

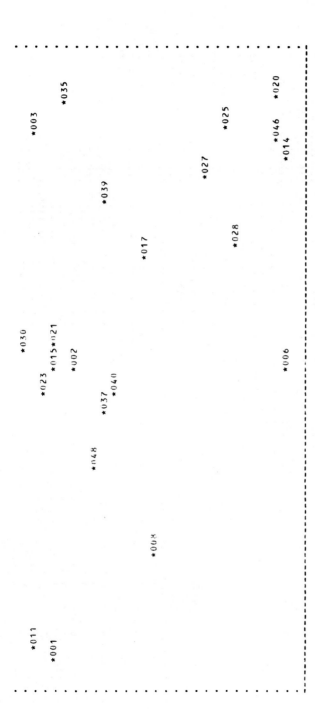

Figure 3.13. (a) Location of data points, using data listed in Table 3.11

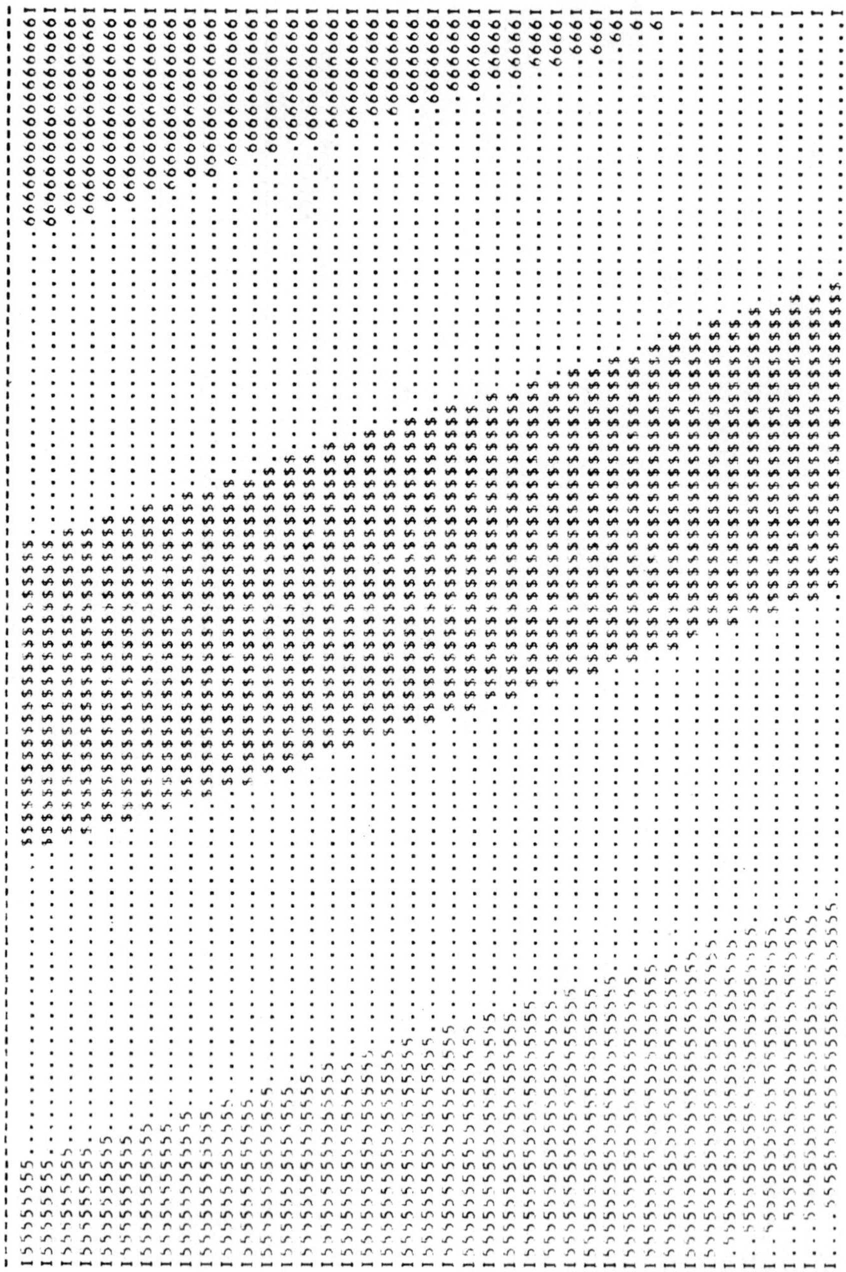

```
REFERENCE CONTOUR  =    20.0000
CONTOUR INTERVAL   =     2.5000

KEY TO CONTOUR VALUES

......     12.5000  TO  15.0000
555555     15.0000  TO  17.5000
......     17.5000  TO  20.0000
$$$$$$     20.0000  TO  22.5000
......     22.5000  TO  25.0000
666666     25.0000  TO  27.5000

COORDINATES OF MAP CORNERS ARE:
    TOP LEFT        0.1589      3.9462
    TOP RIGHT       3.9642      3.9462
    BOTTOM LEFT     0.1589      0.0578
    BOTTOM RIGHT    3.9642      0.0578
```

Figure 3.13. (b) First order trend surface, using data listed in Table 3.11

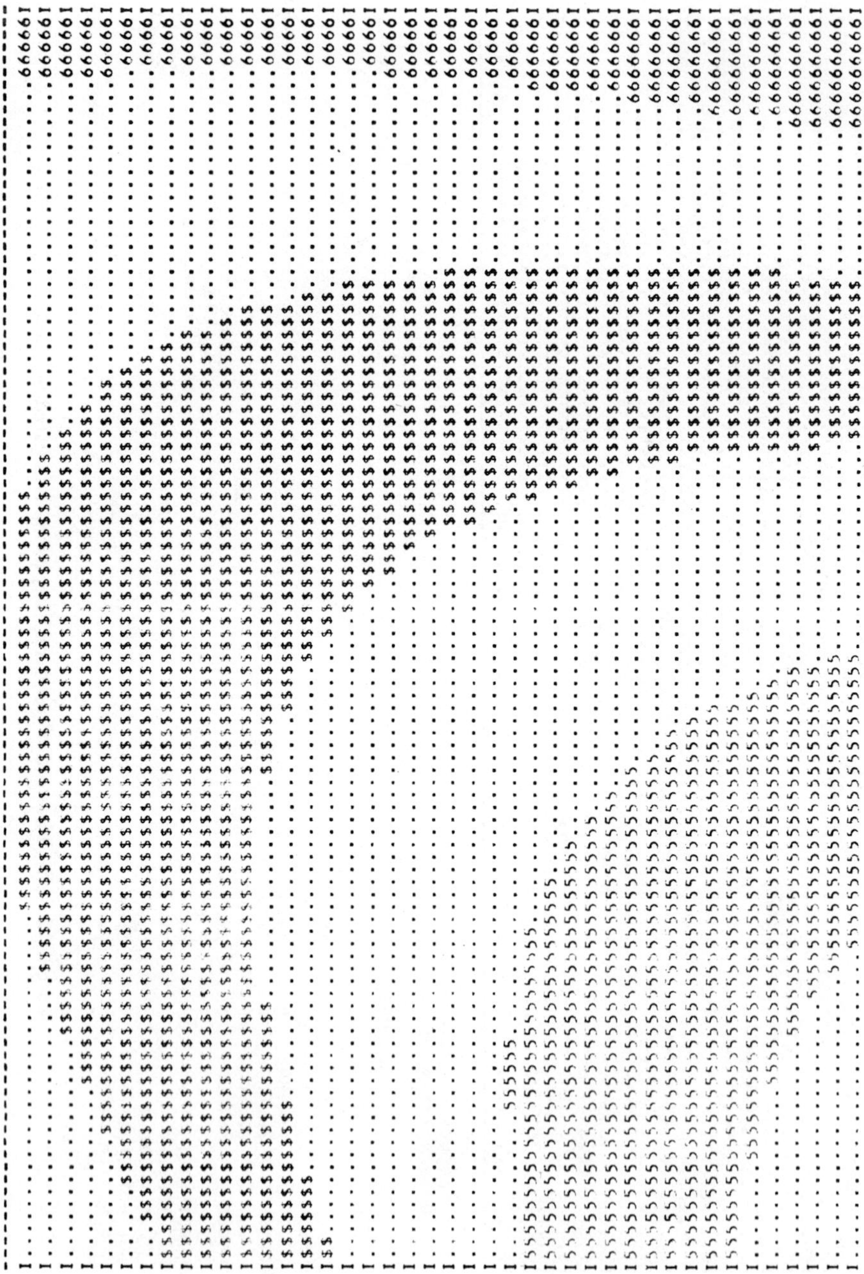

```
I..............................................5555555555555555.................................6666666661
I..............................................5555555555555555555..............................6666666661
I.............................................$5555555555555555555.............................66666666661
I............................................$$55555555555555555555.........................666666666661
I...........................................$$$5555555555555555555555......................66666666666I
I..........................................$$$$$555555555555555555555.....................666666666666I
I.........................................$$$$$$55555555555555555555555..................6666666666666I
I........................................$$$$$$$5555555555555555555555555...............66666666666666I
I4......................................$$$$$$$$5555555555555555555555555..............666666666666666I
I44....................................$$$$$$$$$55555555555555555555555555............6666666666666666I
I444..................................$$$$$$$$$$5555555555555555555555555555..........6666666666666666I
I4444................................$$$$$$$$$$$55555555555555555555555555555........66666666666666666I
I44444..............................$$$$$$$$$$$$5555555555555555555555555555555......666666666666666666I
I444444............................$$$$$$$$$$$$5555555555555555555555555555555555...6666666666666666666I
I4444444..........................$$$$$$$$$$$$$$555555555555555555555555555555555..66666666666666666666I
I44444444........................$$$$$$$$$$$$$$$5555555555555555555555555555555...666666666666666666666I
I444444444......................$$$$$$$$$$$$$$$$$55555555555555555555555555555...6666666666666666666666I
I4444444444...................$$$$$$$$$$$$$$$$$$$5555555555555555555555555555..666666666666666666666666I
I44444444444.................$$$$$$$$$$$$$$$$$$$$555555555555555555555555555..6666666666666666666666666I
I444444444444..............$$$$$$$$$$$$$$$$$$$$$$55555555555555555555555555..66666666666666666666666666I
I4444444444444............$$$$$$$$$$$$$$$$$$$$$$$5555555555555555555555555..666666666666666666666666666I
I44444444444444.........$$$$$$$$$$$$$$$$$$$$$$$$$555555555555555555555555..6666666666666666666666666666I

REFERENCE CONTOUR  =   20.0000
CONTOUR INTERVAL   =    2.5000

KEY TO CONTOUR VALUES

444444         10.0000  TO   12.5000
......         12.5000  TO   15.0000
555555         15.0000  TO   17.5000
......         17.5000  TO   20.0000
$$$$$$         20.0000  TO   22.5000
......         22.5000  TO   25.0000
666666         25.0000  TO   27.5000
......         27.5000  TO   30.0000

COORDINATES OF MAP CORNERS ARE:
    TOP LEFT         0.1589        3.9462
    TOP RIGHT        3.9642        3.9462
    BOTTOM LEFT      0.1589        0.0578
    BOTTOM RIGHT     3.9642        0.0578
```

Figure 3.13. (c) Second order trend surface, using data listed in Table 3.11

3.14) although it accounts for 66.19 per cent of the sum of squares. An $F$ test of the significance of the first order trend gave $F = 46.00$ with 2 and 47 degrees of freedom. Since the value of the $F$ ratio at the 0.05 significance level with 2 and 47 degrees of freedom is 3.23, it is unlikely that the first order sample trend surface is estimating a horizontal population surface. Although a level of explanation of 66 per cent and a significant $F$-ratio may be deemed sufficient evidence to stop fitting, we know that the true surface is second order, and that the first order coefficients are considerably different from those used in Equation 3.14. Consequently a second order surface (Figure 3.13(c)) was fitted, giving the equation

$$z_i = 10.20 + 1.68u_i + 2.50v_i + 0.79u_i^2 - 1.19u_iv_i + 0.32v_i^2 \quad (3.15)$$

This function accounts for 94.31 per cent of the observed sum of squares. The coefficients are very close to those of Equation (3.14). An $F$-test of the null hypothesis that a value of explained sum of squares as high as this could be expected due to sampling fluctuations when the population explained sum of squares is zero was rejected at the 0.05 significance level (the calculated value of $F$ was 145.94 with 5 and 44 degrees of freedom, while the tabled $F$ at the 0.05 per cent level was 2.44). Since the definition of the observed $z_i$ included a random normal error term, this $F$-test is more likely to satisfy the assumptions of the test, especially as the computed values of the coefficients are very close to the actual values (cf. Equations (3.14)) and (3.15)). The $F$-test can also be used to assess the significance of the increase in explained sums of squares from the linear to the linear plus quadratic trend. The null hypothesis is that the increase in the explained sum of squares due to adding the quadratic term is due to chance. The value of $F$ computed according to the procedure of Table 3.10 above is 72.52; with 3 and 44 degrees of freedom it comfortably exceeds the tabled value of $F$ at a significance level of 0.05 ($F_{3,44,0.05} = 2.84$). Hence the null hypothesis is not accepted.

Adding the cubic terms gives the equation

$$z_i = 9.39 + 0.91\ u_i + 4.99\ v_i + 1.34\ u_i^2 - 1.50\ u_iv_i - 0.94\ v_i^2$$
$$-0.076\ u_i^3 - 0.020\ u_i^2 v_i + 0.097\ u_iv_i^2 + 0.17\ v_i^3$$

which accounts for 94.66 per cent of the observed sum of squares, an increase of 0.35 per cent. The $F$-test of the null hypothesis that the increase in explanation due to adding the cubic terms could have arisen by chance gave a value of $F$ equal to 0.64. Since $F_{4,40,0.05} = 2.61$ the null hypothesis is accepted and, on statistical grounds alone, it can be concluded that the true surface is most probably described by a second-order (linear plus quadratic) surface. Since the data is artificially generated, this fact can be checked. In this case the statistical procedure has provided the correct answer. It should be noted that the level of explanation in this example is high, that is, the residuals are small. Consequently, the procedure might be expected to be rather more accurate than in a situation in which the trend is less well defined. Also, the sequential use of the $F$-test does violate a fundamental assumptions of statistical hypothesis testing, because the acceptance of a null hypothesis relating to the $k$th degree trend depends on the rejection of hypotheses concerning the $1, 2, \ldots, (k-1)$th degree trends. The hypotheses are thus not independent. In addition, the test requires that the residuals are normally distributed, independent and have homoscedastic variances.

The $Z^2$ array for the linear-plus-quadratic trend can give additional information about the nature of the trend surface. (See Section 3.2.1 (C) and Table 3.9.) The $Z^2$ array for the second-order trend surface is shown in Table 3.12. The $u^2$ term has a relatively low contribution (6.014) and, if Grant's (1957) example were followed, this term would be excluded from the trend equation. The $u$ term contributes most (426.5) to the explained sum of squares (which is the sum of the elements of the $Z^2$ array, or 657.57 in this case). Turning to the $Z^2$ array for the linear-plus-quadratic-plus-cubic trend surface (Table 3.13) it is clear that the additional cubic terms do not add much to the explained sum of squares, and the terms $u$, $v$, $uv$ and $u^2$ alone account for a high proportion of the explained sum of squares. The study of the $Z^2$ array is an alternative way of discriminating between the trend and the residual variation. In this example, the second-order terms have been selected (as in the $F$-test) but additional information regarding the relative importance of the individual terms is useful.

Table 3.12
$Z^2$ array for second-order surface

| --- | 34.97 | 6.014 |
|---|---|---|
| 426.5 | 118.4 | |
| 71.69 | | |

Table 3.13
$Z^2$ array for third-order surface

| --- | 34.97 | 6.014 | 1.619 |
|---|---|---|---|
| 426.5 | 118.4 | 0.299 | |
| 71.69 | 0.123 | | |
| 0.348 | | | |

The question of distinguishing between systematic and residual variation in trend surface analysis has been approached from yet another angle by Cliff and Ord (1973). If the mapped variable $z$ is spatially autocorrelated, then it can be said to vary systematically over the study area. A first-order trend surface is then fitted and the residuals from this surface are tested for spatial autocorrelation; if there is still evidence of such autocorrelation a second order surface is fitted and the test repeated. When the null hypothesis that the spatial autocorrelation coefficient is zero in the population is accepted, then the residual variation can be considered to be random. The evaluation of the coefficient I was described in Section 2.4(iii), and a significance test — based on the assumption that the set of $n\ z$ values was a random sample drawn from a population in which the disturbances are normally distributed — was outlined. If the data points constitute a fixed pattern, that is, if it is not possible to measure $z$ at all points within the study area, then the evaluation of $E(I)$ and $\text{var}(I)$ should be based on Cliff and Ord's Assumption $R$. This is fully described in Cliff and Ord (1973, p. 15).

As an example of the use of autocorrelation measures in trend surface analysis, polynomial trend surfaces of orders one, two and three were fitted to a set of 41 points in the Nottingham district at which rainfall records were available for the year 1973. The location of these points is shown in Figure 3.14 and the input data is given in Table 3.14. The weighting matrix was chosen to be $w_{ij} = 1$ if points $i$ and $j$ are first nearest neighbours and $w_{ij} = 0$ otherwise. Initially the null hypothesis that there was no spatial autocorrelation among the *observed* $z$ values was tested. This test involves a slightly different procedure from that given in Section 2.4(iii); the spatial autocorrelation statistic I is computed from:

$$I = \frac{n \sum_{\substack{i=1 \\ (i \neq j)}}^{n} \sum_{j=1}^{n} w_{ij} z_i z_j}{W \sum_{i=1}^{n} z_i^2}$$

where, as in Section 2.4(iii), $W$ is the sum of the weights (in this case $n$, since there are $n$ nearest neighbours) and $z_i$ is the observed value at point $i$ expressed as a deviation from the mean of the $z$'s. The $w_{ij}$ are the elements of the weight matrix as described above. For the example data,

$W = 41$,

$\sum_{i=1}^{n} z_i^2 = 65160.2031$,

Table 3.14
Data for autocorrelation test example

| Identifier | Grid Reference | Location | 1973 Total Precipitation (mm) | Altitude (m) |
|---|---|---|---|---|
| 1 | 442 517 | Codnor | 679 | 74 |
| 2 | 583 521 | Papplewick | 691 | 97 |
| 3 | 649 518 | Oxton | 672 | 126 |
| 4 | 696 524 | Brackenhurst | 573 | 70 |
| 5 | 697 542 | Southwell | 562 | 38 |
| 6 | 503 456 | Watnall | 705 | 119 |
| 7 | 538 468 | Bulwell Hall | 649 | 62 |
| 8 | 596 483 | Ramsdale Res. | 632 | 153 |
| 9 | 627 451 | Lambley | 614 | 64 |
| 10 | 654 443 | Burton Joyce | 579 | 20 |
| 11 | 684 446 | Hoveringham | 576 | 18 |
| 12 | 622 414 | Netherfield | 568 | 27 |
| 13 | 635 417 | Stoke Bardolph | 567 | 24 |
| 14 | 667 349 | Cotgrave | 558 | 97 |
| 15 | 565 429 | Basford | 585 | 51 |
| 16 | 428 408 | W. Hallam | 607 | 70 |
| 17 | 468 395 | New Stanton | 653 | 46 |
| 18 | 481 395 | Ilkeston | 615 | 39 |
| 19 | 568 395 | Nottingham Castle | 585 | 58 |
| 20 | 592 393 | Sneinton | 570 | 24 |
| 21 | 596 399 | Colwick Hill | 573 | 94 |
| 22 | 615 393 | Holme Sluices | 583 | 23 |
| 23 | 588 376 | W. Bridgford | 607 | 27 |
| 24 | 538 382 | Lenton | 608 | 49 |
| 25 | 541 365 | Beeston | 591 | 34 |
| 26 | 507 375 | Bramcote | 609 | 101 |
| 27 | 581 351 | Wilford Hill | 563 | 90 |
| 28 | 501 340 | Long Eaton | 640 | 30 |
| 29 | 477 344 | Long Eaton | 608 | 33 |
| 30 | 449 328 | Draycott | 636 | 40 |
| 31 | 507 261 | Sutton Bonington | 621 | 48 |
| 32 | 508 277 | Kingston Hall | 606 | 43 |
| 33 | 517 269 | Moulter Hill | 569 | 52 |
| 34 | 542 270 | Fox Hill | 521 | 73 |
| 35 | 542 249 | White Hill | 584 | 66 |
| 36 | 561 252 | Lings Farm | 637 | 61 |
| 37 | 606 274 | Wysall | 609 | 76 |
| 38 | 567 279 | East Leake | 557 | 91 |
| 39 | 660 225 | Old Dalby | 636 | 125 |
| 40 | 650 253 | Broughton Lodge | 623 | 113 |
| 41 | 684 315 | Kinoulton | 562 | 43 |

The input to the trend surface analysis program was the coordinates of each station after division by 100 to reduce the magnitude of the powers and cross-products, followed by the values of 1973 total precipitation. The altitude data are used in a later example (Figure 3.16).
(data abstracted from: *Hydrometric Yearbook,* 1973, Trent River Authority, Nottingham.)

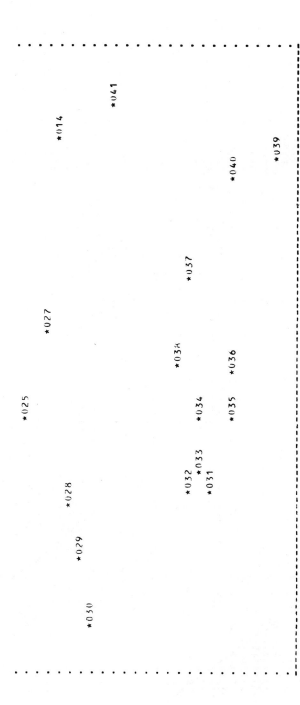

Figure 3.14. Location of data points, spatial autocorrelation example. The data points are the rainfall recording station in the Nottingham area

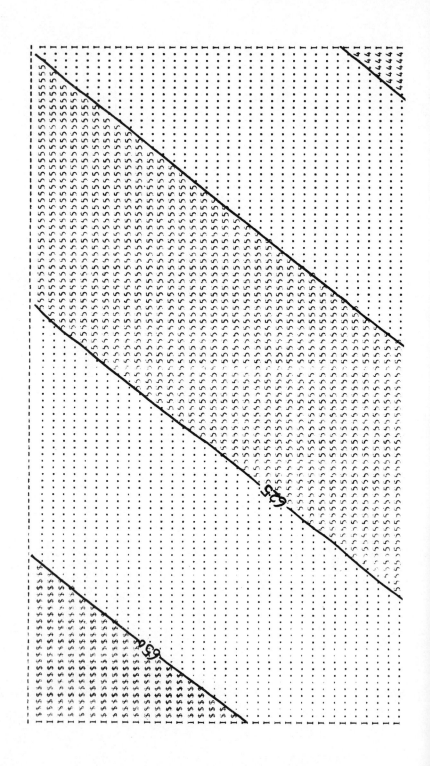

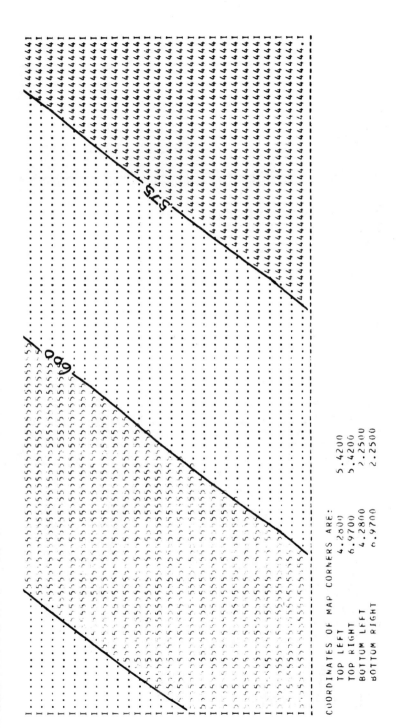

COORDINATES OF MAP CORNERS ARE:
```
    TOP LEFT       4.2800      5.4200
    TOP RIGHT      6.9700      5.4200
    BOTTOM LEFT    4.2800      2.2500
    BOTTOM RIGHT   6.9700      2.2500
```

Figure 3.15. (a) First order trend surface fitted to data listed in Table 3.14

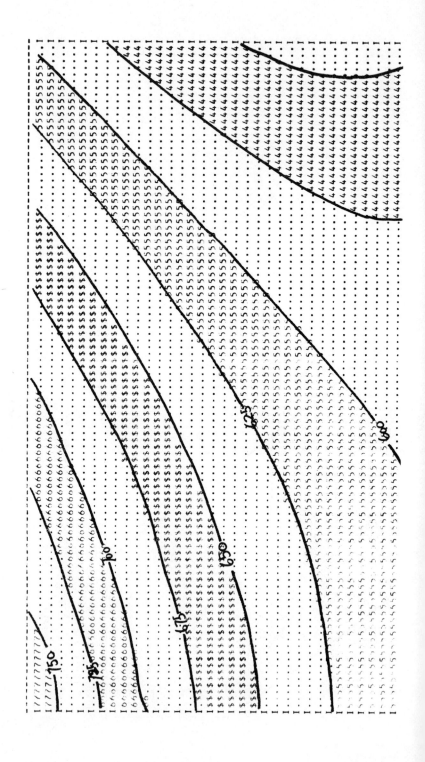

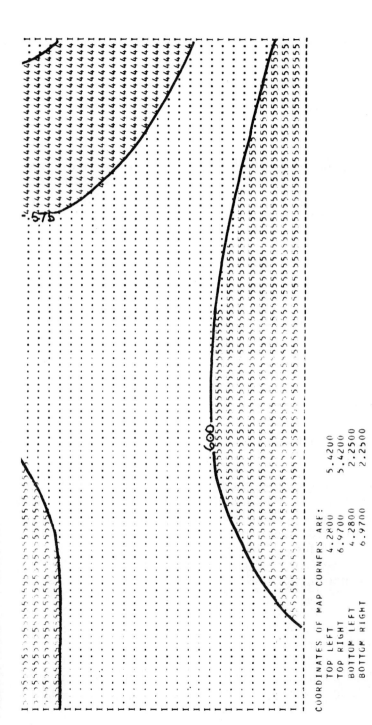

Figure 3.15. (b) Second order trend surface fitted to data listed in Table 3.14

and

$$\sum_{\substack{i=1 \ j=1 \\ (i \neq j)}}^{n} \sum w_{ij} z_i z_j = 32900.273$$

giving $I = 0.5049$.

The expected value of $I$, $E(I)$, is simply:

$$E(I) = -1/(n-1)$$

which, in this example, is $-1/40 = -0.0250$. The standard error of $I$ is the square root of:

$$E(I^2) = \frac{4An^2 - 8n(A+D) + 12A^2}{4A^2(n^2-1)}$$

where

$$A = \tfrac{1}{2} \sum_{i=1}^{n} L_i$$

$$D = \tfrac{1}{2} \sum_{i=1}^{n} L_i(L_i - 1)$$

and $L_i$ is the number of nonzero entries in the $i$th row of the weighting matrix. If binary weights are used, as in the example, $A$ is equal to $W/2$ and $D$ is zero. With $A = 20.5$ and $n = 41$, the standard error of $I$ is 0.2222. For large $n$, a standard normal deviate can be obtained from the formula

$$\text{dev} = (I - E(I))/\text{s.e.}(I)$$
$$= (0.5097 + 0.0250)/0.2222 = 2.3844$$

This value comfortably exceeds the 95 per cent point of the standard normal distribution, which allows the non-acceptance of the null hypothesis. It is therefore probable that the observed $z$ values are spatially autocorrelated, thus it is possible to fit a first order (linear) trend surface to see if this removes the systematic variation.

A first order trend surface fitted to the rainfall data of Table 3.14 above gave the following equation:

$$z_i = 686.58 - 24.73 u_i + 15.61 v_i$$

This surface (Figure 3.15(a)) accounted for 26.125 per cent of the corrected sums of squares. The procedures of Section 2.4(iii) were applied to the residuals from this surface in order to test the null hypothesis that there is no spatial autocorrelation among these *residuals*. The results were: $I = 0.4073$, $I_x = 2.8069$, $E(I) = -0.0739$ and s.e. $(I) = 0.2139$ giving a standard deviate value of 2.2495. This value exceeds the 95 per cent point of the standard normal distribution so the null hypothesis is not accepted. Since the residuals from the first order trend surface are still systematically related, a second order (quadratic) surface was fitted, producing the following equation:

$$z_i = 246.108 + 155.759u_i - 30.125v_i - 8.914u_i^2$$
$$-21.617u_iv_i + 23.086v_i^2$$

The quadratic surface (Figure 3.15(b)) explains 52.538 per cent of the corrected sum of squares. The null hypothesis of zero autocorrelation among the second order residuals was tested as previously, with the following results: $I = 0.0583, I_x = 5.0699, E(I) = -0.1499$ and s.e.$(I) = 0.2583$, giving a standard normal deviate of 0.7865. Since this is less than the 95 per cent point of the standard normal distribution, the null hypothesis of zero autocorrelation can be accepted. This implies that the residuals from the quadratic trend surface are spatially random, given that the weighting matrix is appropriate. Obviously, the choice of weighting matrix is of prime importance — the nearest neighbour criterion used in this example was selected only on the grounds of simplicity. Details of alternative methods of calculating the $w_{ij}$ can be found in Section 2.4(iii).

The approach based on spatial autocorrelation measures is not altogether successful, although it has a certain theoretical appeal. The choice of weights $w_{ij}$ is a difficulty, as discussed earlier in Section 2.4(iii). Another problem, which is likely to be more prominent in trend surface applications, is the fact that, as presented by Cliff and Ord (1973), the inverse of $(X'_{(1)}X_{(1)})$ is required in order to evaluate $E(I)$. This matrix is the sums of squares and cross products of the geographical coordinates and their powers and cross products, not corrected for the means. It is likely to become near singular as the order of the trend surface rises, so that the accuracy of the computed inverse will decline. This loss in accuracy can have a very significant effect on the value of $E(I)$; in some cases — including the example given by Cliff and Ord — the value of $E(I)$ varies considerably even for the first order surface if calculations are performed to 4, 8 and 10 significant figures. Hence, the effects of computational errors can lead to a meaningless value of the test statistic.

To illustrate the methods of trend map comparison described above a second analysis was performed for the same data point distribution, that is, the rainfall recording stations in the Nottingham area. The dependent variable for this second analysis is height above mean sea level. The second-order trend surface for this data (Figure 3.16) is compared to the second order trend surface for the rainfall data used in the autocorrelation example above. A comparison of the two maps should allow some conclusions to be drawn concerning the spatial relationship between altitude and rainfall totals in the Nottingham area.

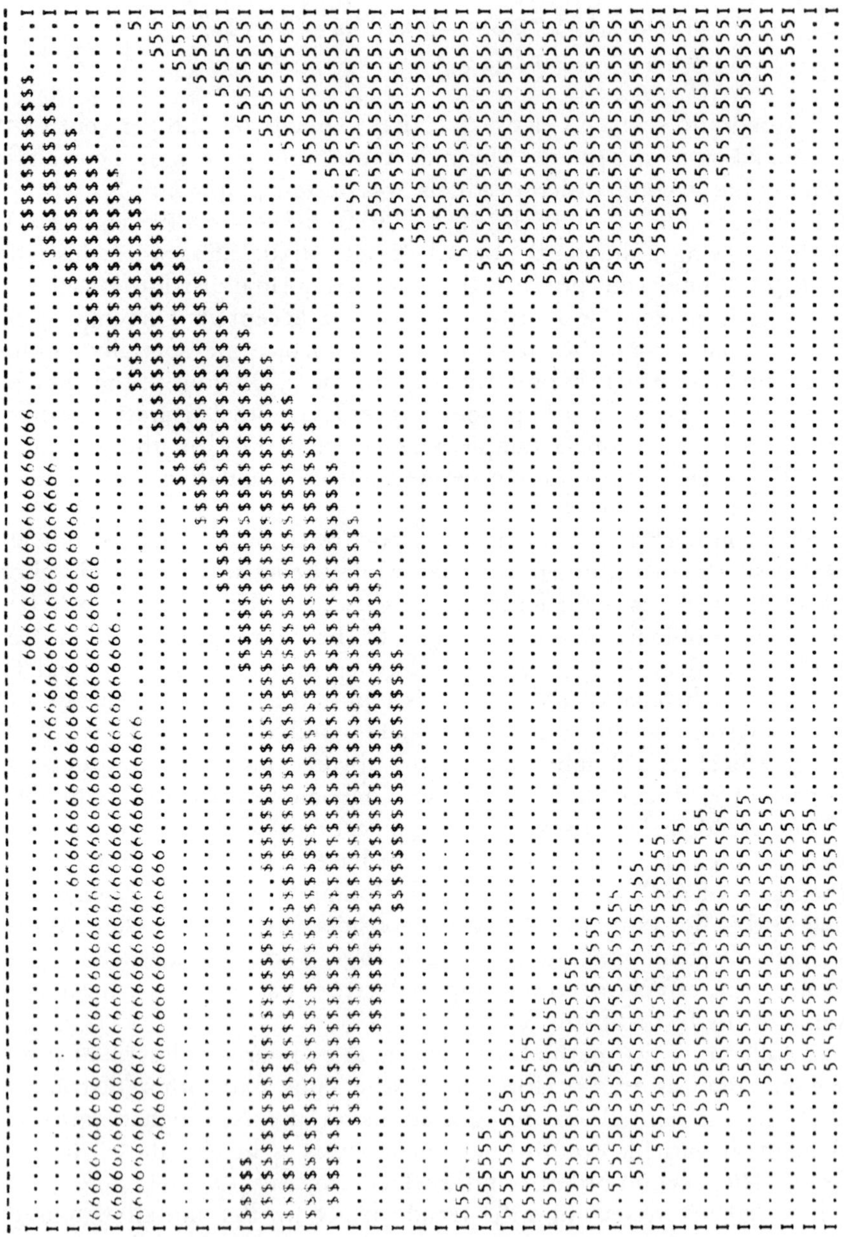

```
I.....555555555555555555.................................I
I....5555555555555555555.................................I
I....55555555555555555555................................I
I....555555555555555555555...............................I
I....5555555555555555555555..............................I
I....555555555555555555555$$.............................I
I....5555555555555555555$$$$.............................I
I....555555555555555555$$$$$$$...........................I
I...555555555555555555$$$$$$$$$$.........................I
I...55555555555555555$$$$$$$$$$$$$.......................I
I...5555555555555555$$$$$$$$$$$$$$$$.....................I
I...555555555555555$$$$$$$$$$$$$$$$$$$...................I
I...55555555555555$$$$$$$$$$$$$$$$$$$$$$.................I
I...5555555555555$$$$$$$$$$$$$$$$$$$$$$$$$...............I
I...555555555555$$$$$$$$$$$$$$$$$$$$$$$$$$$$.............I
I..5555555555555$$$$$$$$$$$$$$$$$$$$$$$$$$$$$$...........I
I..555555555555$$$$$$$$$$$$$$$$$$$$$$$$$$$$$$$$$.........I
I..55555555555$$$$$$$$$$$$$$$$$$$$$$$$$$$$$$$$$$$$.......I
I..5555555555$$$$$$$$$$$$$$$$$$$$$$$$$$$$$$$$$$$$$$.....I
I..555555555$$$$$$$$$$$$$$$$$$$$$$$$$$$$$$$$$$$$$$$$...I
I..55555555$$$$$$$$$$$$$$$$$$$$$$$$$$$$$$$$$$$$$$$$$$.I
I..5555555$$$$$$$$$$$$$$$$$$$$$$$$$$$$$$$$$$$$$$$$$$$$I
I..555555$$$$$$$$$$$$$$$$$$$$$$$$$$$$$$$$$$$$$$$$$$$$$I
I.555555$$$$$$$$$$$$$$$$$$$$$$$$$$$$$$$$$$....666666666I
I.55555$$$$$$$$$$$$$$$$$$$$$$$$$$$$$$$$...6666666666666I
I.5555$$$$$$$$$$$$$$$$$$$$$$$$$$$$$$...666666666666666I
I.555$$$$$$$$$$$$$$$$$$$$$$$$$$$$$..66666666666666666I
I-----------------------------------------------------I

REFERENCE CONTOUR  =    70.0000
CONTOUR INTERVAL   =    20.0000

KEY TO CONTOUR VALUES

:::::         10.0000    TO     30.0000
55555         30.0000    TO     50.0000
:::::         50.0000    TO     70.0000
$$$$$         70.0000    TO     90.0000
:::::         90.0000    TO    110.0000
66666        110.0000    TO    130.0000
:::::        130.0000    TO    150.0000

COORDINATES OF MAP CORNERS ARE:
    TOP LEFT         4.2800         5.4200
    TOP RIGHT        6.9700         5.4200
    BOTTOM LEFT      4.2800         2.2500
    BOTTOM RIGHT     6.9700         2.2500
```

Figure 3.16. Second order trend surface fitted to station altitude data, Table 3.14. This surface is compared to Figure 3.15(b) in the example in the text. The resulting correlation surface is shown in Figure 3.17

The first method of map comparison involved the simple correlation of the predicted (trend) values for altitude and rainfall measured over a grid pattern which ranged from coarse to fine. The correlations for each grid size are 0.7789 (5 x 5), 0.7635 (10 x 10), 0.7576 (15 x 15) and 0.7546 (20 x 20). The grid size is given in brackets. It can be seen that the measured correlation falls as the grid mesh becomes finer, and that the value of the correlation is approaching 0.75. What this number actually means in spatial terms is open to question, although it can be used as evidence of the overall general similarity of the two map patterns. However, if a spatial measure is needed, the correlation at each grid point must be calculated and mapped, as explained above. A 10 x 10 grid was chosen, and the correlation between the predicted values on the two surfaces at eight points surrounding each selected grid intersection point was computed. The eight points were obtained using the following scheme: taking the coordinates of the grid intersection point to be $(u, v)$, the coordinates of eight surrounding points were defined by firstly selecting increments $\delta u$ and $\delta v$ in the $u$ and $v$ directions respectively, and then taking the coordinates $(u + \delta u, v)$, $(u, v + \delta v)$, $(u, v - \delta v)$, $(u - \delta u, v)$, $(u + \delta u, v + \delta v)$, $(u + \delta u, v - \delta v)$, $(u - \delta u, v + \delta v)$ and $(u - \delta u, v - \delta v)$. The values for rainfall and altitude were computed from the trend equations for each of these points using $\delta u$ and $\delta v$ as one half of the distance between grid points in the $u$ and $v$ directions respectively to give two sets of nine estimated values including the value at the grid point itself. The correlation coefficient for these two sets of values was plotted at the point $(u, v)$ and the exercise repeated at each of the grid points. A contour map of the resulting pattern can be termed a correlation surface. The areas of greatest mutual similarity and dissimilarity can then be visually identified. In the present example, a prominent area of negative correlation occurs in the south-central part of the map displaying the correlation surface (Figure 3.17). The correlation between the two trend maps in the remainder of the area is high (above 0.8). The area of negative correlation corresponds to the Trent valley south-west of Nottingham, which is not picked out by the trend surface for altitude, possibly due to a lack of control points in this area. Thus in this instance the correlation surface is perhaps indicating that the trend surface for altitude is not a particularly good fit. Inspection of the $R^2$ value confirms this — the second order surface accounts for only 35 per cent of the sum of squares of the dependent variable.

The comparison of maps is an unduly neglected topic in quantitative geography. It is obvious that further research is necessary. Such research may concentrate upon methods of characterizing the continuous geographical surface other than by sampling the surface at a number of points. The point pattern methods discussed earlier do have some use, but they should not be considered as the ultimate answer.

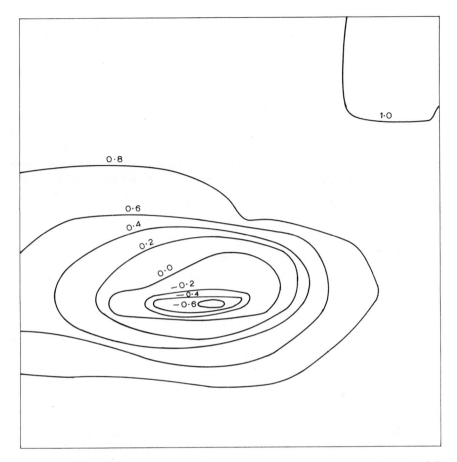

Figure 3.17. Correlation surface for the second-order trend surfaces for rainfall (Figure 3.15(b)) and station altitude (Figure 3.16)

## 3.3 Selection of Explanatory Variables in Multiple Regression
### 3.3.0 Introduction
In Chapter 2 it was assumed that the investigator could specify a set of $k$ explanatory variables which together accounted for the systematic variation in a dependent variable. This is frequently an ideal state of affairs and, as such, is one that does not often occur in practice. In some instances it is possible to say no more than that each member of the set of explanatory variables should exert some influence on the variation in the dependent variable. The degree of influence exerted by some of the explanatory variables may be so low that it is not worth using them either as predictors or as elements in a hypothesis which purports to account for the behaviour exhibited by the dependent variable. A technique of discriminating between significant and non-

significant controls over the variation in the dependent variable would thus appear to be useful. In practice, things are not quite as simple as they initially appear to be. Unless the explanatory variables are completely uncorrelated among themselves, it is difficult to disentangle their relative contributions to the regression sum of squares. This problem was discussed in Chapter 2 under the heading of multicollinearity (Section 2.4(i)). A second difficulty is the definition of the term 'significant'. One often speaks loosely of the 'best' set of explanatory variables or the 'significant' contributors to the regression sum of squares but no objective measure of significance has yet been defined. Most comparisons of regression functions are based on the coefficient of determination $(R^2)$ or on some function of $R^2$. The sampling distribution of $R^2$ is not known except when the population value is zero, so exact statistical tests based on $R^2$ are not feasible. Most selection procedures do use some quasi-statistical measure as a guide, but these should never be given a strict interpretation.

Two approaches to the problem of selecting the 'best' subset from $k$ explanatory variables are in general use. These are:

(i) the selection approach, which involves the addition to or subtraction from the regression equation of explanatory variables according to their contribution to the regression sum of squares. If the method begins with the regression of $y$ on all $k$ explanatory variables and proceeds by removing a single explanatory variable at each cycle until the optimum subset is left, then it is termed backward elimination. Forward selection procedures begin by regressing $y$ on the most significant explanatory variable and continue by adding a single explanatory variable at each cycle until the subset of significant explanatory variables has been identified. A third method alternates between adding and subtracting explanatory variables. This is the stepwise process. All three methods have the aim of finding the optimum subset of explanatory variables without evaluating every possible regression of $y$ on subsets of the $k$ explanatory variables.

(ii) The combinational approach was considered infeasible until relatively recently. It involves the calculation of $R^2$ or some other optimality criterion for every possible combination of the explanatory variables. This involves a considerable amount of computation and requires efficient programs and large, fast computers to be worthwhile. Several algorithms implementing this approach have been published in the last decade; the one discussed in this section will handle up to 40 explanatory variables in a relatively short time.

## 3.3.1 The Selection Approach

The backward elimination and forward selection procedures are described in detail by Draper and Smith (1966, Chapter 6). They conclude that the backward elimination method is 'slightly inferior' to the forward selection method but that the stepwise method is superior to both. Stepwise regression is a variation of the forward selection algorithm. Instead of simply adding an explanatory variable to the regression at each cycle until no 'significant' explanatory variables remain, the stepwise method incorporates a check of the 'significance' of the explanatory variables that are in the regression equation at each cycle. A single cycle in stepwise regression thus involves (a) the examination of the variables currently included in the regression equation to see whether any should be deleted (this step is obviously omitted at the first cycle) and (b) the examination of the variables not yet entered into the regression equation to see if any should be added. The procedure terminates when none of the explanatory variables in the regression are to be deleted and none of the unentered variables are to be added. The test of significance used in this procedure is the $t$-test of the null hypothesis that the value of a partial regression coefficient is zero in the population. Draper and Smith (1966) call this a 'partial $F$-test' because they use the fact that $t^2$ with $m$ degrees of freedom is distributed as $F$ with 1 and $m$ degrees of freedom. Consequently, they use $F$ rather than $t$ as the test statistic. The significance levels are unaltered. The term 'partial $F$-test' is used to distinguish this test from the $F$-test of the significance of the entire regression (Chapter 2).

Draper and Smith's stepwise procedure is based on the method described by Efroymson (1960). Hemmerle (1967) also discusses the method, and Breaux (1968) has published a computer algorithm. The description given here is brief and is meant only to provide an outline of the working of the computer program listed in Table 3.15. Readers who wish to follow through a detailed worked example should refer to one of the publications mentioned above. The method of use of the computer program is described in the listing in the form of comments.

The stepwise regression method proceeds as follows:

1. Calculate **R**, the matrix of correlations among the explanatory variables ($x$'s) and the vector **g** of correlations between the dependent variable ($y$) and the explanatory variables.

2. Select the $x_i$ having the highest correlation with $y$ and enter it in the regression.

3. Select from the remaining $x_i$ the one that has the highest partial correlation with $y$, holding constant the contribution of the included $x$'s.

4. Compute individual $t$-tests of the null hypothesis $\beta_i = 0$ for the included $x_i$ and delete any $x_i$ for which the null hypothesis is accepted. (This $t$-test will be two-tailed for the alternative hypothesis is $\beta_i \neq 0$.)

5. Repeat steps 3 and 4 until no $x$ variable is a candidate for inclusion (step 3) or deletion (step 4).

It is not clear from the literature whether $x$'s that have been deleted at earlier cycles should be allowed to enter at a later stage. Obviously if the $x$ variable deleted at cycle $n$ is selected for entry at cycle $n + 1$ then an endless loop would be set up. Similarly, to allow a deleted $x$ variable to re-enter at a later stage might set up a longer repetitive chain. In the accompanying program a deleted variable is never allowed to re-enter, and a limit of 50 cycles is set to prevent the program from looping indefinitely.

The computations at steps 2, 3 and 4 are now described more fully. Let **A** be the partitioned matrix

$$\mathbf{A} = \begin{bmatrix} \mathbf{R} & \mathbf{g}' & \mathbf{I} \\ \hline \mathbf{g} & s & \mathbf{O} \\ \hline -\mathbf{I} & \mathbf{O} & \mathbf{O} \end{bmatrix}$$

where **R** is the $(k \times k)$ matrix of correlations among the $x$'s, **g** the $(1 \times k)$ column vector of correlations between $y$ and the $x$'s, with **g**' being its transpose, **I** the $(k \times k)$ identity matrix, **O** the $(k \times k)$ null matrix, and $s$ the scalar 1. **A** is thus a $(2k + 1) \times (2k + 1)$ matrix. Firstly, compute the vector **v** where $v_i = (a_{i,k+1} a_{k+1,i})/a_{ii}$ $(i = 1, k)$ and select the largest element. This gives the first variable to be entered. The $F$-ratio to test whether this variable should be entered or not is given by $(\phi \cdot \max[v_i])/(a_{k+1,k+1} - \max[v_i])$. In this equation $\phi$ is the number of degrees of freedom assuming that the particular $x$ has been entered. If $m$ is the number of currently entered $x$'s (including the one just selected) $\phi$ is equal to $(n - m - 1)$ where $n$ is the number of observations. If the resulting $F$ ratio exceeds a preselected critical value the variable is entered, otherwise the procedure terminates.

The next stage is the modification of **A** to allow for the entry of the chosen variable, which will be denoted by $x_j$. This is accomplished in two stages:

(i) $a_{ji}^{(1)} = a_{ji}/a_{jj}$ $(i = 1, 2, \ldots, 2k + 1)$

(ii) $a_{mi}^{(1)} = a_{mi} - (a_{mj} a_{ji})/a_{jj}$ $(m, i = 1, 2, \ldots, 2k + 1; i, m \neq j)$

The elements $a_{ij}^{(1)}$ form the transformed matrix $\mathbf{A}^{(1)}$ which is **A** adjusted for the entry of $x_j$.

This is equivalent to one 'sweep' in the Gaussian elimination procedure mentioned in Chapter 1 in connection with matrix inversion. In fact, the lower right $(k \times k)$ portion of matrix $\mathbf{A}^{(1)}$ contains the elements of $\mathbf{R}^{-1}$ with zeros corresponding to as yet unentered $x$'s. The standardized regression coefficient for $x_j$, i.e. $\hat{\beta}_{(s)j}$, is now held in element $(2k + 1, k + 1 + j)$ of the matrix $\mathbf{A}^{(1)}$. It can be converted to a partial regression coefficient in the usual way (Equation (2.21)).

If two or more variables are included at a given cycle each must be tested for significance, using the $t$-test. It is more convenient to compute $F(=t^2)$ for each entered variable $x_j$. The test ratio is

$$F = (\phi\, a_{j,k+1}^{(1)2})/(a_{k+1,k+1}^{(1)} a_{k+j,k+j}^{(1)}) \tag{3.17}$$

If any variable is found not to contribute significantly to the regression sum of squares it is eliminated by reversing the Gaussian sweep outlined above, i.e.

(i) $\quad a_{ji}^{(2)} = a_{ji}^{(1)}/a_{k+1+j,k+1+j}^{(1)} \qquad (i = 1,2,\ldots, 2k+1)$

(ii) $\quad a_{m\,i}^{(2)} = a_{m\,i}^{(1)} - (a_{m,k+1+j}^{(1)} \cdot a_{ji}^{(1)})/a_{k+1+j,k+1+j}^{(1)}$

$$(m,i = 1,2,\ldots, 2k+1; m,i \neq j)$$
$$\tag{3.18}$$

where $x_j$ is the variable to be eliminated.

The test for entry of the next variable is carried out by computing the vector **v** (see above) for all $x$'s not yet included, selecting the largest element, testing the contribution of the corresponding variable to the regression sum of squares and adjusting matrix $A^{(q)}$ to give matrix $A^{(q+1)}$ where $q$ is the current cycle number (Equation (3.16)). The individual $x$'s that are now entered into the equation are then checked to see whether they should be deleted, using Equation (3.18). This process is repeated until no variable can be added or deleted, bearing in mind the comments made earlier.

Computational accuracy is an important consideration in the stepwise procedure. In theory the forward Gaussian sweep on row $j$ of matrix A (performed when $x_j$ is entered) and a backward sweep on the same row should be self-cancelling, i.e. the initial value of row $j$ should be produced by the backward sweep. In practice this will not be the case, though rounding error can be reduced if double-precision arithmetic is used. Nevertheless, errors may still accumulate if the number of cycles is large.

The stepwise method has been criticized on the grounds that it may not pick up the 'best' subset of explanatory variables. Hamaker (1962) and Oosterhoff (1960) demonstrate this. Also, it is based on the implied proposition that there is a single 'best' equation. There may be several near-optimal combinations of the $x$'s and the choice between them could depend on subjective decisions, for example, ease of measurement or cost of sampling if prediction is the purpose of the analysis.

An important consideration, which is mentioned in Section 3.2 in connection with the test of the significance of the fit of successive trend surfaces, and in Chapter 5 in connection with the sequential $\chi^2$

test of the number of maximum likelihood factors, is that the $t$-tests (or 'partial $F$-tests') which are carried out at each cycle to determine whether or not a variable should be added to or deleted from the regression, are sequential and not independent. Pope and Webster (1972) point out that the use of the $F$ statistic may appear to be more objective or statistically sounder than the use of criteria such as $R^2$, $\bar{R}^2$ or $C_p$ or any of the other criteria mentioned below in connection with combinatorial methods. This is not so, because the distribution of $F$ under conditions of dependence of successive null hypotheses is not the same as the central $F$-ratio distribution which is listed in statistics tables. Consequently, the 'partial $F$-test' cannot be related to a particular significance level. The use of the term 'significance level' is therefore misleading for '... it is not known how to relate the area above this value in the $F$-distribution to the probability of any type of error in the choice of a set of predictors (Pope and Webster, 1972, p. 330). The choice of a cut-off point or 'probability' value for the partial $F$-test is thus as subjective as the use of $R^2$ or any of the other measures of goodness of fit. Furthermore, its use may give a false sense of security.

A Fortran program for stepwise regression based on the algorithm described above is listed in Table 3.15.

### 3.3.2 The Combinatorial Approach

Computational methods for the comparison of all possible regressions of a set of $k$ explanatory variables on a single dependent variable are given by Furnival (1971), Garside (1967, 1971A,B), Gorman and Thomin (1966), Hocking (1972), Hocking and Leslie (1967), Lamotte and Hocking (1970), Morgan and Tatar (1972) and Schatzoff, Feinberg and Tsao (1968). A Fortran computer program for the Furnival method is available from Grosenbaugh (1967). These methods involve procedures similar to the Gaussian elimination algorithm which was described in Section 3.3.1, using a forward sweep to add a variable and a backward sweep to delete it. This is done for all the $2^k - 1$ possible regressions. However, it is difficult to decide which of the $2^k - 1$ equations is optimal. In some cases, the decision may rest partly on nonstatistical considerations, for some explanatory variables may be easier or cheaper to measure than others. Thus, a combination of explanatory variables that is slightly less than optimally efficient from the statistical point of view may be selected on grounds of cost or availability. Even so, a measure of efficiency is still necessary.

An obvious candidate is $R^2$, the coefficient of determination. It was shown in Chapter 2 that $R^2$ is a function of the number of explanatory variables and so it cannot be used for the comparison of regressions having different numbers of explanatory variables. The adjusted $\bar{R}^2$ value reflects the number of $x$'s in the equation, and is therefore a better criterion. Its calculation was discussed in Chapter 2. Other

## Table 3.15
### Computer program for stepwise regression

```
C STEPWISE REGRESSION PROGRAM BY P. M. MATHER, SEPTEMBER 1972.
C SEE CHAPTER 3, SECTION 3 FOR ADDITIONAL DETAILS
C  :::  NOTE: ALL CONTROL CARDS READ FROM CHANNEL 5.
C DATA IN ORDER:
C NB: THE TERM 'CARD' SHOULD BE READ AS 'SET OF CARDS' FOR
C LARGE PROBLEMS.
C CARD 1:    JOB IDENTIFICATION. USE UP TO 80 COLS AS REQUIRED.
C CARD 2:    KORIN   NV   NM
C            KORIN = 1 MEANS THAT THE LOWER TRIANGLE + DIAGONAL OF THE
C            CORRELATION MATRIX ARE TO BE SUPPLIED FOLLOWED BY THE VECTORS
C            OF MEANS AND STANDARD DEVIATIONS
C            KORIN = 0 MEANS THAT THE RAW DATA IS TO BE INPUT.
C            PUNCH VALUE OF KORIN IN COLUMN 1
C            COLS 2-4: NUMBER OF PREDICTOR VARIABLES.
C            COLS 5-7: NUMBER OF OBSERVATIONS PER VARIABLE.
C CARD 3:    NAMES OF PREDICTOR VARIABLES IN SAME ORDER AS ENTERED ON
C            THE DATA CARDS. 8 CHARACTERS PER NAME AND 10 NAMES PER CARD.
C            USE AS MANY CARDS AS NECESSARY.
C CARD 4     PROBABILITY LEVEL FOR USE IN F-TEST. (SEE TEXT) MUST BE
C            IN THE RANGE 0 TO 1. PUNCH IN COLS 1-10, WITH DECIMAL POINT.
C CARD 5:    THE INPUT CHANNEL NUMBER FOR THE DATA MATRIX OR THE
C            CORRELATION MATRIX + MEANS AND STANDARD DEVIATIONS.
C            IF NEGATIVE, INPUT ASSUMED TO BE UNFORMATTED.
C            IF POSITIVE, INPUT ASSUMED TO BE FORMATTED.
C            IF UNFORMATTED, THE DATA MATRIX IS ASSUMED TO FORM ONE
C            LOGICAL RECORD. IF THE CORRELATION MATRIX IS READ, THEN IT
C            IS ASSUMED TO FORM ONE LOGICAL RECORD WHILE THE VECTORS OF
C            MEANS AND STANDARD DEVIATIONS EACH OCCUPY ONE LOGICAL RECORD
C            ON THE SPECIFIED DEVICE.
C            PUNCH VALUE IN COLUMNS 1-3.
C CARD 6:    INPUT FORMAT FOR DATA. LEAVE BLANK IF UNFORMATTED DATA
C            INPUT. PUNCH IN COLS 1-80.
C CARD 7:    INTEGER GIVING NUMERICAL POSITION OF DEPENDENT VARIABLE IN
C            THE DATA, E.G. IF DEPENDENT VARIABLE IS THE 5TH ENTRY ON EACH
C            CARD, PUNCH 5. ********** NOTE: THIS CARD IS USED ONLY
C            IF RAW DATA IS PROVIDED. OMIT IF CORRELATION MATRIX IS INPUT.
C            PUNCH VALUE IN COLS 1-3.
C CARD 8:    IF KORIN = 1 ENTER CORRELATION MATRIX (LOWER TRIANGLE PLUS
C            DIAGONAL) WITH EACH ROW BEGINNING ON A NEW CARD. FOLLOW BY
C            VECTORS OF MEANS AND STANDARD DEVIATIONS, STARTING EACH ON A
C            FRESH CARD. PUNCH IN FORMAT SPECIFIED ON CARD 5.
C            ***$$$*** THE DEPENDENT VARIABLE MUST BE LAST***$$$***
C            IF KORIN = 0 ENTER DATA MATRIX ROWWISE, STARTING EACH ROW ON
C            A FRESH CARD. USE FORMAT SPECIFIED ON CARD 5.
C            REFER TO CARD 5 FOR DETAILS OF UNFORMATTED INPUT.
C CARD 9:    OUTPUT FORMAT SPECIFICATION FOR SUBROUTINE MATPR. SEE
C            APPENDIX A.
C CARD 10:   REPEAT FROM CARD 1 OR PUNCH STOP IN COLS 1-4.
C
      DIMENSION A(2500)
      DATA STOP/4HSTOP/
      NIG=2500
C DON'T FORGET TO CHANGE THE VALUE OF NIG IF DIMENSION OF A IS ALTERED!
    1 READ(5,101) (A(I),I=1,20)
      IF(A(1).EQ.STOP) STOP
  101 FORMAT(20A4)
      WRITE(6,201) (A(I),I=1,20)
  201 FORMAT(24H1TITLE OF THIS JOB        ,20A4//)
      READ(5,9999) KORIN, NV, NM
 9999 FORMAT(I1,2I3)
      IF(NV.EQ.0) GOTO 999
C SET UP BASE ADDRESSES
      NT=NV+1
      L=NT+NV
      LL=L*L+1
```

Table 3.15 *continued*

```
      L1=LL+NT
      MP=NM
      MQ=NT
      IF(KORIN.NE.1) GOTO 3
      L1=1
      MP=1
      MQ=1
   3  L5=L1+NV
      L6=L5+NV
      L7=L6+NV
      L8=L7+NV
      L9=L8+NT
      L10=L9+NT
      L11=L10+L*L
      L12=L11+NV
      L13=L12+NV
      L14=L13+NV
      I=L14+NV+NV
  75  FORMAT('OCORE REQUESTED = ',I4,' WORDS'/)
      WRITE(6,75) I
      IF(I.LE.NIG) GOTO 2
      WRITE(6,202) I
 202  FORMAT('OSORRY, YOUR DATA MATRIX IS TOO BIG - INCREASE DIMENSION O
     *F A IN MASTER PROGRAM TO',I6)
      GOTO 500
   2  WRITE(6,200) NV,NM
 200  FORMAT('OSTEPWISE REGRESSION PROGRAM BY PAUL M. MATHER'//
     *'OYOUR SPECIFICATION IS'//'   NO. OF INDEPENDENT VARIABLES =',
     *I5//'   NO. OF OBSERVATIONS            =',I5//)
      CALL REX(A(1),A(LL),A(L1),A(L5),A(L6),A(L7),A(L8),A(L9),A(L10),
     #A(L11),A(L12),A(L13),A(L14),NV,NM,NT,L,KORIN,MP,MQ)
      GOTO 1
 999  WRITE(6,444)
 444  FORMAT(' NV SET TO 0. JOB TERMINATED')
      GOTO 500
 445  WRITE(6,446)
 446  FORMAT(///'ONORMAL END OF RUN ENCOUNTERED')
 500  STOP
      END




      SUBROUTINE REX(A,DATA,V,BETA,STER,FRAT,AV,S,B,NIN,LABEL,LOUT,NAMES
     #,NV,NM,NT,L,KORIN,MP,MQ)
      DIMENSION  A(L,L),DATA(   MQ),NIN(NV),LABEL(NV),LOUT(NV),
     1BETA(NV),STER(NV),FRAT(NV),AV(NT),S(NT),V(NV),B(L,L),FMT(20),
     # NAMES(NV,2)
      WRITE(6,7771)
7771  FORMAT('OWARNING - DO NOT TAKE THE TERM PROBABILITY TOO SERIOUSLY
     * IN THE CONTEXT OF THIS PRINTOUT!'///)
      Z=1.0/FLOAT(NM)
      NL=NV
      NIND=0
      READ(5,9997) ((NAMES(I,J),J=1,2),I=1,NV)
      READ(5,5566) FCRIT
5566  FORMAT(F10.0)
      WRITE(6,800) FCRIT
 800  FORMAT(' ''PROBABILITY'' LEVEL = ',F8.5)
      IF(FCRIT.LT.1.0.OR.FCRIT.GT.0.0) GOTO 809
      WRITE(6,815)
 815  FORMAT('***PROBABILITY LEVEL MUST BE IN RANGE 0.0-1.0 EXCLUSIVE. TH
     *IS RUN TERMINATED.')
      RETURN
 809  CONTINUE
C
      DO 176 I=1,NV
```

Table 3.15   *continued*

```
  176 NIN(I)=I
      READ(5,1001) ICHAN
      IFORM=0
      IF(ICHAN.LT.0) IFORM=1
      ICHAN=IABS(ICHAN)
      READ(5,9997) FMT
      IF(KORIN.EQ.0) GOTO 7072
      IF(IFORM.EQ.1) GOTO 939
      DO 938 I=1,NT
  938 READ(ICHAN,FMT) (A(I,J),J=1,I)
      READ(ICHAN,FMT) AV
      READ(ICHAN,FMT) S
      GOTO 940
  939 READ(ICHAN) ((A(I,J),J=1,I),I=1,NT)
      READ(ICHAN) AV
      READ(ICHAN) S
  940 WRITE(6,6066)
 6066 FORMAT('1 MEANS AND STANDARD DEVIATIONS'///)
      WRITE(6,7071) (I,AV(I),S(I),I=1,NT)
 7071 FORMAT(' ',I6,2E17.8)
      GOTO 7073
 7072 WRITE(6,201)
  201 FORMAT('1',20X,'DATA MATRIX'///)
      REWIND 3
      READ(5,1001) NDEP
 1001 FORMAT(I3)
      WRITE(6,1002) NDEP
 1002 FORMAT('0DEPENDENT VARIABLE = NO',I3)
      DO 1 I=1,NM
      IF(IFORM.EQ.1) GOTO 941
      READ(IFORM,FMT) DATA
      GOTO 942
  941 READ(ICHAN) DATA
  942 WRITE(6,200) DATA
      IF(NDEP.EQ.MQ) GOTO 1
      NH=NDEP+1
      T=DATA(NDEP)
      DO 188 M=NH,NT
  188 DATA (M-1)=DATA(M)
      DATA(NT)=T
    1 WRITE(3) DATA
      ENDFILE 3
      REWIND 3
      IF(NDEP.NE.MQ) WRITE(6,1199) MQ
 1199 FORMAT('0DEPENDENT VARIABLE MOVED TO COL.',I4,' BUT THE ORDER OF T
     1HE EXPLANATORY VARIABLES IS UNALTERED'///)
  200 FORMAT(' ',10G12.4)
      READ(5,9997) FMT
 9997 FORMAT(20A4)
      IND=2
      II=1
      CALL SSCP(AV,A,S,NT,NM,L,IND,3,6,FMT,II)
      REWIND 3
 7073 CALL SWAP(A,L,NT)
      MM=L*L
      IP=1
      CALL MATPR(64H        CORRELATION MATRIX.DEPENDENT VARIABLE IN LAST R
     1OW/COLUMN    ,8,FMT,6,A,MM,L,NT,NT,IP,2,8)
      DO 349 I=1,NT
  349 S(I)=S(I)*S(I)*NM
      IL=NT+1
      LL=IL+NV-1
      K=0
      DO 9 I=IL,LL
      DO 8 J=1,NV
```

Table 3.15  *continued*

```
      A(J,I)=0.
    8 A(I,J)=0.
      K=K+1
      A(I,K)=-1.
    9 A(K,I)=1.
C
      DO 10 I=IL,LL
      A(I,NT)=0.
   10 A(NT,I)=0.
      DO 11 I=IL,LL
      DO 11 J=IL,LL
   11 A(I,J)=0.
C START MAIN LOOP
      KOT=0
   30 KOT=KOT+1
C THE NUMBER OF ITERATIONS IS LIMITED ARBITRARILY BY THE NEXT
C INSTRUCTION. ALTER IT IF NECESSARY.
      IF(KOT.GT.50) GOTO 414
      CALL VCALC(V,NL,A,L,N,NIN,NV)
C
C   CALCULATES VECTOR V AND PUTS IDENT OF HIGHEST IN N.
C
      IF(IRE.NE.1) GO TO 45
      IF(N.EQ.LOUT(1)) GO TO 46
   45 NIND=NIND+1
      N1=NIND
      N2=NM-NIND-1
      NN=N
      N=NIN(N)
      DF=N2
      LP=N+NT
      LABEL(NIND)=N
      WRITE(6,550) KOT
  550 FORMAT('1SUMMARY OF INFORMATION AT ITERATION  ',I6///)
      WRITE(6,400) N
  400 FORMAT('0VARIABLE',I5,'  HAS BEEN ADDED TO THE REGRESSION EQUATION
     *'//' INDEPENDENT VARIABLES ARE NOW')
      DO 51 I=1,NIND
      J=LABEL(I)
   51 WRITE(6,52)J, (NAMES(J,K),K=1,2)
   52 FORMAT(' ',I4,2X,2A4)
   50 CALL RESET (N,A,L,B)
      F=(DF*A(N,NT)**2)/(A(NT,NT)*A(LP,LP))
      PR=FISHER(1,N2,F)
      IF(PR.GT.FCRIT) GOTO 71
      WRITE(6,335) N
  335 FORMAT('0PARTIAL F TEST ON VARIABLE',I4,',WHICH HAS JUST ENTERED,
     *IS NOT SIGNIFICANT')
      WRITE(6,817) F,PR
  817 FORMAT(' VALUE OF F =',F12.4/' PROBABILITY = ',F12.4)
      NIND=NIND-1
      GOTO 35
   71 IF(N.EQ.NIN(NL)) GOTO 14
      I1=NL-1
      DO 13 I=NN,I1
   13 NIN(I) = NIN(I+1)
   14 NL=NL-1
C     NIN = IDENTS OF VARIABLES NOT YET ENTERED
C
  929 IREM=0
      R=1.0-A(NT,NT)
      SSREG=R*S(NT)
      SSRES =S(NT)-SSREG
      SDRES=SQRT(SSRES/DF)
      SNT=S(NT)
      DO 15 I=1,NIND
      LK=LABEL(I)
```

Table 3.15   *continued*

```
      LP=NT+LK
      BETA(I)=A(LK,NT)*SQRT(SNT/S(LK))
      IF(KORIN.EQ.1) GOTO 831
      GOTO 832
  831 STER(I)=SQRT((SSRES/DF)*A(LP,LP)/(S(LK)*S(LK)))
      GOTO 15
  832 STER(I)=SQRT((SSRES/DF)*A(LP,LP)/S(LK))
   15 CONTINUE
      CON=0.0
      DO 48 I=1,NIND
      LK=LABEL(I)
   48 CON=CON+BETA(I)*AV(LK)
      CON=AV(NT)-CON
      R=R*100.0
      WRITE(6,600)SSREG,SSRES,R,SDRES
  600 FORMAT(////' SUM OF SQUARES EXPLAINED BY REGRESSION =',E20.11/
     1' RESIDUAL SUM OF SQUARES =',E20.11/
     2' PERCENT OF VARIANCE EXPLAINED = ',F12.5/
     3' STANDARD DEVIATION OF RESIDUALS = ',F12.5)
      WRITE(6,614)
  614 FORMAT('0STANDARDIZED REGRESSION COEFFICIENTS'//)
      DO 615 I=1,NIND
      LK=LABEL(I)
  615 WRITE(6,616)   LABEL(I),A(LK,NT)
  616 FORMAT(' ',I6,2X,F12.5)
      WRITE(6,625) CON,(LABEL(I),BETA(I),STER(I),I=1,NIND)
  625 FORMAT(' CONSTANT TERM =',F12.5/
     *' PARTIAL REGRESSION COEFFICIENTS AND THEIR STANDARD ERRORS'//
     * 50(' ',I6,3X,2F12.5/))
      F=(SSREG/FLOAT(N1))/(SSRES/DF)
      PR=FISHER(N1,N2,F)
      WRITE(6,500) F,PR,N1,N2
  500 FORMAT(///' F-RATIO FOR TESTING SIGNIFICANCE OF THE REGRESSION SS
     1IS',  E20.11/
     3' PROBABILITY LEVEL = ',F9.4/
     4' DEGREES OF FREEDOM = ',I4,' AND',I4//)
      IF(NIND.EQ.1) GO TO 229
      WRITE(6,640)
  640 FORMAT('0PARTIAL F-TESTS ON INDEPENDENT VARIABLES'////)
      DO 16 I=1,NIND
      LK=LABEL(I)
      LP=LK+NT
   16 FRAT(I)=(DF*A(LK,NT)**2)/(A(NT,NT)*A(LP,LP))
      IRE=0
      I=0
   17 I=I+1
      IF(I.GT.NIND) GOTO 4141
      PR=FISHER( 1,N2,FRAT(I))
      WRITE(6,660) LABEL(I),FRAT(I),PR
      IF(PR.GT.FCRIT) GOTO 17
      LK=LABEL(I)
      WRITE(6,700)LK
  700 FORMAT('0*****VARIABLE',I4,'  HAS BEEN REMOVED FROM THE REGRESSION
     1***'//)
      CALL REMOVE(A,L,LK,NT,B)
      IRE=IRE+1
      LOUT(IRE)=LK
C LOUT = IDENTS OF VARIABLES REMOVED
      DO 67 K=1,NIND
      IF(LK.NE.LABEL(K)) GOTO 67
      I1=NIND-1
      DO 68 J=K,I1
      FRAT(J)=FRAT(J+1)
   68 LABEL(J)=LABEL(J+1)
      GOTO 69
   67 CONTINUE
   69 IREM=1
```

Table 3.15   *continued*

```
       NL=NL+1
       NIND=NIND-1
       NIN(NL)=LK
       I=I-1
       GOTO 17
 4141 WRITE(6,670) N1,N2
  670 FORMAT('0 DEGREES OF FREEDOM ARE   ',I4,' AND',I4)
  660 FORMAT(' ',I6,F12.3,F12.7)
  229 WRITE(6,661)
  661 FORMAT('0PARTIAL CORRELATIONS  OF UNENTERED VARIABLES WITH Y'///)
       DO 662 I=1,NL
       K=NIN(I)
       PR=A(K,NT)*A(K,NT)/(A(K,K)*A(NT,NT))
       PR=SQRT(PR)
  662 WRITE(6,663) K, PR
  663 FORMAT(' ',I6,F12.5)
       N1=NIND
       N2=NM-NIND-1
       DF=N2
       IF(IREM.EQ.1.AND.NIND.NE.0) GOTO 929
   25 IF(NL)35,35,30
   35 WRITE(6,750)
  750 FORMAT('0END OF STEPWISE PROCEDURE. THE ''BEST'' REGRESSION IS THA
      *T LISTED ON PREVIOUS ITERATION')
       IF(KORIN.EQ.1) RETURN
       CALL RESID (DATA,NM,NT,BETA,NIND,LABEL,CON,NV)
       RETURN
   46 WRITE(6,751) N
  751 FORMAT('0VARIABLE',I4,' IS THE CANDIDATE FOR ADMISSION AT THIS CYC
      1LE'// SINCE IT WAS REMOVED AT THE PREVIOUS CYCLE EXECUTION IS NOW
      2TEMINATED')
       GOTO 35
  414 WRITE(6,415)
  415 FORMAT('0 NO. OF CYCLES BECOMING EXCESSIVE. PLEASE CHECK YOUR DATA
      1.'// EXECUTION TERMINATED.'//)
       RETURN
       END




       SUBROUTINE SWAP(A,L,N)
       DIMENSION A(L,N)
       DO 1 I=1,N
       DO 1 J=1,N
    1 A(I,J)=A(J,I)
       RETURN
       END




       SUBROUTINE RESID(DATA,NM,NT,BETA,NIND,LABEL,CONST,NV)
C COMPUTES RESIDUALS FROM FINAL REGRESSION EQUATION
       DIMENSION DATA(NT), BETA(NIND), LABEL(NV)
  100 FORMAT('1RESIDUALS FROM THIS REGRESSION'///       OBS.  Y     CALC.
      *Y       RESIDUAL   CASE'//)
  200 FORMAT(' ',3F11.4,I6)
       WRITE(6,100)
       DO 1 I=1,NM
       Y=CONST
       READ(3) DATA
       DO 2 J=1,NIND
       LK=LABEL(J)
    2 Y=Y+BETA( J)*DATA(LK)
       RES=DATA(NT)-Y
```

Table 3.15 *continued*

```
    1 WRITE(6,200) DATA(NT), Y, RES, I
      RETURN
      END




      SUBROUTINE REMOVE(A,L,LK,NT,B)
C PERFORMS 1 BACKWARD GAUSSIAN SWEEP ON ROW LK
      DOUBLE PRECISION P,Q,R,S
      DIMENSION A(L,L),B(L,L)
      K=LK+NT
      S=DBLE(A(K,K))
      P=1.0D0/S
      DO 1 I=1,L
      S=DBLE(A(LK,I))
    1 B(LK,I)=SNGL(S*P)
      DO 2 I=1,L
      IF(I.EQ.LK) GO TO 2
      R=DBLE(A(I,K))
      DO 3 J=1,L
      Q=DBLE(A(I,J))
      S=DBLE(A(LK,J))
    3 B(I,J)=SNGL(Q-R*S*P)
    2 CONTINUE
      DO 4 I=1,L
      DO 4 J=1,L
    4 A(I,J)=B(I,J)
      RETURN
      END




      SUBROUTINE RESET (N,A,L,B)
C PERFORMS ONE FORWARD GAUSSIAN SWEEP ON ROW N
      DOUBLE PRECISION P,Q,R,S
      DIMENSION A(L,L),B(L,L)
      S=DBLE(A(N,N))
      P=1.0D0/S
      DO 1 I=1,L
      S=DBLE(A(N,I))
    1 B(N,I)=SNGL(S*P)
      DO 2 I=1,L
      IF(I.EQ.N) GO TO 2
      DO 3 J=1,L
      R=DBLE(A(I,N))
      Q=DBLE(A(I,J))
      S=DBLE(A(N,J))
    3 B(I,J)=SNGL(Q-R*S*P)
    2 CONTINUE
      DO 4 I=1,L
      DO 4 J=1,L
    4 A(I,J)=B(I,J)
      RETURN
      END




      SUBROUTINE VCALC(V,NL,A,L,N,NIN,NV)
C CALCULATES VECTOR V
      DIMENSION V(NL),A(L,L),NIN(NV)
      M=NV+1
      IF(NL.EQ.1) GOTO 2
```

Table 3.15   *continued*

```
      DO 1 I=1,NL
      J=NIN(I)
    1 V(I)=A(J,M)*A(M,J)/A(J,J)
      N=MAX(V,NL)
      RETURN
    2 N=NIN(1)
      RETURN
      END




      FUNCTION MAX(V,NL)
      DIMENSION V(NL)
      X=V(1)
      MAX=1
      DO 1 I=2,NL
      IF(X.GE.V(I)) GO TO 1
      X=V(I)
      MAX=I
    1 CONTINUE
      RETURN
      END




      FUNCTION FISHER(M,N,X)
C   THIS IS A TRANSLATION OF CACM ALGORITHM 322 BY E. DORRER.
C   LATER AMENDMENTS INCORPORATED.
C   M,N ARE DEGREES OF FREEDOM
      INTEGER A,B
      ZM=M
      ZN=N
      A=2*(M/2)-M+2
      B=2*(N/2)-N+2
      W=X*ZM/ZN
      Z=1.0/(1.+W)
      IF(A.NE.1) GOTO 2
      IF(B.NE.1) GOTO 1
      P=SQRT(W)
      Y=.3183098862
      D=Y*Z/P
      P=2.*Y*ATAN(P)
      GOTO 4
    1 P=SQRT(W*Z)
      D=.5*P*Z/W
      GOTO 4
    2 IF(B.NE.1) GOTO 3
      P=SQRT(Z)
      D=.5*Z*P
      P=1.-P
      GOTO 4
    3 D=Z*Z
      P=W*Z
    4 Y=2.*W/Z
      IF(A.NE.1) GOTO 6
      JJ=B+2
      DO 5 J=JJ,N,2
      D=(1.+A/FLOAT(J-2))*D*Z
    5 P=P+D*Y/(FLOAT(J)-1.)
      GOTO 7
    6 ZK=Z**((N-1)/2)
      D=D*ZK*N/FLOAT(B)
      P=P*ZK+W*Z*(ZK-1.)/(Z-1.)
    7 Y=W*Z
```

Table 3.15    *continued*

```
        Z=2./Z
        B=N-2
        JJ=A+2
        DO 8 I=JJ,M,2
        J=I+B
        D=Y*D*J/FLOAT(I-2)
      8 P=P-Z*D/FLOAT(J)
        IF(P.GT.1.) P=1.
        IF(P.LT.0.) P=0.
        FISHER=P
        RETURN
        END
```

measures that have been proposed are $C_p = (n - k)(\text{RSS}/(n - k)/2 - 1) + k$ and $J_p = (n + k)/(n - k)\text{RSS}$ where RSS is the residual sum of squares. Kennard (1971) points out that $\bar{R}^2$ and $C_p$ have a one-to-one correspondence. Grosenbaugh (1967) uses a measure termed 'relative mean-square residual' or RMSR which is the ratio of the mean square of the residuals ($\Sigma_{i=1}^{n} e_i^2 /n$) to the variance of $y$. Each of these criteria is widely used but it is difficult to suggest a means to choose rationally between them.

The use of the selection and combinatorial approaches is demonstrated with reference to a set of hydrological data published by Benson (1962). This data set consists of 164 measurements on six explanatory and one dependent variable. It is listed in Table 3.16. Benson (1962) found that the relationship between the dependent variable and the explanatory variables was log-linear rather than linear, hence the data was transformed by taking the logarithms of all the observations. The proposed regression model is:

$$\log y = f(\log X_1, \log X_2, \ldots, \log X_6) + \log e$$

This shows that it is the sum of squares of the logarithms of the errors that is minimized. Conversion from logarithmic form by taking antilogarithms of the regression estimates does not result in a least-squares fit to the untransformed data. This can only be achieved by a nonlinear model. (See Chapter 2 and Section 3.4.)

The dependent variable in the data selected for this example is $Q_{2.33}$, the mean annual flood, defined as the discharge having a recurrence interval of 2.33 years. The six explanatory variables are: $X_1$, area of the drainage basin, in square miles; $X_2$, average slope of the main channel in feet per mile, measured between the points located 10 per cent and 85 per cent of the total distance from mouth to source; $X_3$, the percentage of the area of the basin covered by ponds and lakes; $X_4$, the 24-hour maximum precipitation intensity having a recurrence interval of 2.33 years; $X_5$, the average number of degrees Fahrenheit below freezing point in January, and $X_6$, an orographic factor.

Table 3.17 shows the results obtained from applying the stepwise

## Table 3.16
Data for stepwise and combinatorial regression examples. (Data from M. A. Benson, *Water Supply Paper 1580-B*, U.S. Geological Survey, Washington, D.C., 1962). The logarithms of the data are listed

|    | X1     | X2     | X3      | X4     | X5     | X6     | Y       |
|----|--------|--------|---------|--------|--------|--------|---------|
| 1  | 7.1309 | 1.4679 | 1.4516  | 0.5306 | 3.1355 | 0.0000 | 9.6486  |
| 2  | 6.7696 | 1.0296 | 1.9006  | 0.7419 | 3.1355 | 0.0000 | 9.0478  |
| 3  | 8.6465 | 1.6715 | 1.0296  | 0.6419 | 3.1355 | 0.0000 | 11.3504 |
| 4  | 9.0204 | 1.5369 | 1.0116  | 0.6419 | 3.1355 | 0.0000 | 11.5617 |
| 5  | 7.3902 | 1.4207 | 0.9561  | 0.7419 | 3.0910 | 0.0000 | 10.1186 |
| 6  | 5.1668 | 2.9897 | 0.7572  | 0.8755 | 2.9444 | 0.0000 | 8.2428  |
| 7  | 4.9972 | 2.1029 | 1.4085  | 1.1632 | 2.7081 | 0.0000 | 7.4731  |
| 8  | 7.2442 | 1.4134 | 1.2119  | 0.9933 | 2.8332 | 0.0000 | 9.7981  |
| 9  | 5.6951 | 3.2351 | 0.9858  | 1.0647 | 2.8904 | 0.0000 | 9.0768  |
| 10 | 5.7746 | 3.4883 | 1.3788  | 1.0647 | 2.8904 | 0.0000 | 9.0180  |
| 11 | 5.1818 | 2.4732 | -0.3711 | 1.0296 | 2.6391 | 0.0000 | 8.1775  |
| 12 | 4.9972 | 2.6735 | 1.3271  | 0.9933 | 2.3979 | 0.0000 | 7.5496  |
| 13 | 6.7708 | 1.9373 | 0.8713  | 0.7419 | 2.9444 | 0.0000 | 9.7111  |
| 14 | 4.5120 | 4.0724 | 0.8461  | 1.0647 | 2.8904 | 0.0000 | 7.9374  |
| 15 | 5.8693 | 3.8286 | 0.3293  | 0.8755 | 2.8332 | 0.0000 | 9.5539  |
| 16 | 6.2422 | 2.6174 | 0.4574  | 0.9555 | 2.7726 | 0.0000 | 9.5610  |
| 17 | 5.0504 | 5.7519 | -0.2251 | 0.8329 | 2.8904 | 0.0000 | 8.5660  |
| 18 | 4.5625 | 4.3907 | -0.5567 | 0.9933 | 2.8332 | 0.0000 | 8.7948  |
| 19 | 5.1417 | 2.0295 | 0.3716  | 0.9933 | 2.5649 | 0.0000 | 8.1315  |
| 20 | 4.3334 | 5.9240 | -0.5447 | 1.0647 | 2.6391 | 0.0000 | 7.8633  |
| 21 | 5.9558 | 5.9180 | -0.1863 | 1.1939 | 2.5649 | 0.0000 | 9.7291  |
| 22 | 5.7941 | 3.3810 | 1.4702  | 1.0296 | 2.5649 | 0.0000 | 8.1687  |
| 23 | 6.1159 | 2.7014 | 1.3888  | 1.0647 | 2.5649 | 0.0000 | 8.4446  |
| 24 | 7.1686 | 2.2039 | 1.1442  | 1.0647 | 2.5649 | 0.0000 | 9.5539  |
| 25 | 5.0814 | 2.4849 | 1.1569  | 1.1314 | 2.3979 | 0.0000 | 7.4384  |
| 26 | 2.4452 | 5.0561 | 0.5988  | 1.0647 | 2.1972 | 0.0000 | 5.7746  |
| 27 | 5.2045 | 2.4032 | 0.7701  | 1.0296 | 2.1972 | 0.0000 | 7.6732  |
| 28 | 4.6444 | 4.6895 | -0.5276 | 0.9933 | 2.7081 | 0.0000 | 8.8593  |
| 29 | 5.2627 | 4.3907 | -0.3147 | 0.9555 | 2.7081 | 0.0000 | 9.5539  |
| 30 | 4.0741 | 5.2311 | -0.5621 | 0.9555 | 2.7081 | 0.0000 | 8.0229  |
| 31 | 4.9628 | 4.6138 | 0.0100  | 0.9555 | 2.7081 | 0.0000 | 8.7948  |
| 32 | 6.4329 | 3.7377 | -0.1278 | 0.9555 | 2.7081 | 0.0000 | 10.0213 |
| 33 | 4.4520 | 5.1179 | 0.3075  | 0.9933 | 2.5649 | 0.0000 | 7.5496  |
| 34 | 4.2210 | 5.5234 | 1.1939  | 1.1632 | 2.3979 | 0.0000 | 7.2371  |
| 35 | 3.8480 | 3.6190 | 1.8542  | 1.1632 | 2.3979 | 0.0000 | 6.6553  |
| 36 | 4.0057 | 3.1946 | 1.4061  | 1.0986 | 2.5649 | 0.0000 | 6.9177  |
| 37 | 4.0146 | 4.3605 | 0.8796  | 1.0986 | 2.5649 | 0.0000 | 7.2152  |
| 38 | 5.9061 | 2.6161 | 1.5110  | 1.0647 | 2.4849 | 0.0000 | 8.5370  |
| 39 | 4.9836 | 3.4595 | 0.9203  | 1.0647 | 2.5649 | 0.0000 | 7.7053  |
| 40 | 4.8598 | 5.1987 | 0.7572  | 0.9933 | 2.6391 | 0.0000 | 7.8240  |
| 41 | 6.6412 | 2.3514 | 0.5596  | 1.0296 | 2.5649 | 0.0000 | 9.2591  |
| 42 | 5.0562 | 2.7442 | 1.0986  | 0.9555 | 2.3026 | 0.0000 | 7.8633  |
| 43 | 4.6444 | 3.4210 | 0.1484  | 1.0986 | 2.3979 | 0.0000 | 7.2497  |
| 44 | 5.3085 | 3.2809 | 0.4055  | 1.0986 | 2.3979 | 0.0000 | 8.2687  |
| 45 | 5.1417 | 3.4404 | -0.1393 | 1.1632 | 2.3026 | 0.0000 | 8.1915  |
| 46 | 4.6728 | 3.7062 | 1.3550  | 1.1632 | 2.1972 | 0.0000 | 7.7874  |
| 47 | 4.7536 | 1.7681 | 0.8198  | 1.0647 | 2.0794 | 0.0000 | 6.9470  |
| 48 | 3.0127 | 1.8294 | 1.1600  | 1.1632 | 1.9459 | 0.0000 | 5.3936  |
| 49 | 5.7705 | 1.6292 | 0.3988  | 1.1632 | 1.7918 | 0.0000 | 5.9269  |
| 50 | 4.8203 | 0.9163 | 0.6206  | 1.0647 | 1.7918 | 0.0000 | 6.9754  |
| 51 | 5.2149 | 1.4974 | 0.7701  | 1.1632 | 1.7918 | 0.0000 | 7.1546  |
| 52 | 3.5610 | 3.1697 | 1.3558  | 1.1939 | 1.6094 | 0.0000 | 5.7203  |
| 53 | 2.1518 | 3.4720 | -0.4780 | 1.2238 | 0.6931 | 0.0000 | 4.9628  |
| 54 | 5.7471 | 2.3135 | 0.1044  | 1.1632 | 1.6094 | 0.0000 | 6.2146  |
| 55 | 4.5358 | 3.1612 | 1.2060  | 1.2528 | 1.7918 | 0.0000 | 7.3460  |
| 56 | 6.0307 | 2.4423 | 2.2556  | 1.2238 | 1.7918 | 0.0000 | 8.6656  |
| 57 | 3.6454 | 3.2921 | 1.6552  | 1.2809 | 1.0986 | 0.0000 | 6.1420  |
| 58 | 4.1558 | 2.6247 | 1.2413  | 1.2238 | 1.3863 | 0.0000 | 6.4922  |
| 59 | 0.8529 | 3.7329 | -0.1165 | 1.2238 | 1.0986 | 0.0000 | 5.7366  |
| 60 | 4.6052 | 1.4839 | 1.1817  | 1.2528 | 1.0986 | 0.0000 | 6.4052  |
| 61 | 4.2822 | 2.7850 | 0.9594  | 1.2528 | 1.3863 | 0.0488 | 6.6554  |

Table 3.16  *continued*

|     | X1     | X2     | X3      | X4     | X5     | X6      | Y       |
|-----|--------|--------|---------|--------|--------|---------|---------|
| 62  | 5.6870 | 0.9969 | 0.9361  | 1.2528 | 1.0986 | 0.0953  | 7.6256  |
| 63  | 4.7958 | 2.7726 | 0.7324  | 1.1632 | 1.7918 | 0.6152  | 7.7187  |
| 64  | 4.3534 | 2.8332 | 0.7655  | 1.1632 | 1.6094 | 0.4700  | 7.6009  |
| 65  | 5.3707 | 4.2863 | 0.5539  | 1.2258 | 1.7918 | 0.6931  | 6.8876  |
| 66  | 5.1628 | 3.1527 | 0.7031  | 1.1632 | 1.6094 | 0.6931  | 8.2161  |
| 67  | 5.9940 | 2.4336 | 0.6881  | 1.1632 | 1.7918 | 0.6152  | 8.8537  |
| 68  | 4.5412 | 4.4976 | 1.0716  | 1.1632 | 1.9459 | 0.5596  | 6.8024  |
| 69  | 5.0362 | 2.6532 | 0.9594  | 1.1632 | 1.9459 | 0.5306  | 7.6109  |
| 70  | 3.3214 | 4.0595 | 1.4907  | 1.1939 | 1.9459 | 0.2624  | 6.0039  |
| 71  | 5.8021 | 2.5649 | 1.2499  | 1.1632 | 1.9459 | 0.4055  | 8.4229  |
| 72  | 4.4248 | 2.7147 | 0.3988  | 1.2258 | 1.6094 | 0.1398  | 7.2862  |
| 73  | 6.5667 | 2.3321 | 1.0453  | 1.1314 | 1.7918 | 0.4055  | 9.0360  |
| 74  | 4.4841 | 3.2229 | 0.9895  | 1.1632 | 1.3863 | 0.6931  | 7.7874  |
| 75  | 5.4467 | 3.3534 | 0.1989  | 0.8329 | 2.7726 | -0.1054 | 8.6827  |
| 76  | 7.3225 | 2.3514 | 0.3565  | 0.8329 | 2.7726 | -0.2231 | 10.1755 |
| 77  | 3.9653 | 4.1026 | 0.1740  | 0.8755 | 2.7726 | -0.4308 | 7.3840  |
| 78  | 4.8363 | 3.6988 | -0.2677 | 0.8329 | 2.7726 | -0.5978 | 8.0520  |
| 79  | -0.8307| 5.0681 | -0.3285 | 0.8329 | 2.7726 | -0.5978 | 8.9359  |
| 80  | 4.4728 | 4.2767 | -0.6733 | 1.1632 | 2.7081 | 0.0000  | 8.4446  |
| 81  | 5.9789 | 3.3569 | -0.3425 | 0.8755 | 2.7726 | -0.1625 | 9.4727  |
| 82  | 4.5890 | 4.4976 | 0.5008  | 0.8329 | 2.7726 | -0.6931 | 7.4085  |
| 83  | 7.9463 | 2.0980 | 0.2624  | 0.8329 | 2.7726 | -0.2877 | 10.4341 |
| 84  | 3.7542 | 4.5612 | -0.6931 | 0.8755 | 2.7726 | -0.7965 | 6.9276  |
| 85  | 4.8675 | 3.9570 | 0.2151  | 0.8755 | 2.7081 | -0.5978 | 7.9010  |
| 86  | 5.4848 | 2.9907 | -0.6931 | 0.8755 | 2.5649 | 0.5008  | 9.3237  |
| 87  | 3.4177 | 4.3870 | -0.6931 | 0.9163 | 2.7081 | -0.3567 | 6.8459  |
| 88  | 6.5367 | 2.5726 | -0.5447 | 0.8755 | 2.7081 | 0.1398  | 9.9233  |
| 89  | 8.3168 | 1.7034 | 0.1398  | 0.8329 | 2.7726 | -0.2231 | 10.8686 |
| 90  | 4.3883 | 3.9160 | 0.8587  | 0.9953 | 2.6391 | 0.0000  | 7.5121  |
| 91  | 5.3982 | 3.2995 | -0.3857 | 0.8755 | 2.6391 | 0.2624  | 8.9872  |
| 92  | 5.5447 | 3.3105 | 1.4279  | 0.9555 | 2.7081 | 0.0000  | 8.5755  |
| 93  | 5.0626 | 3.3286 | 0.1222  | 0.9555 | 2.7081 | 0.4055  | 8.6495  |
| 94  | 4.6547 | 4.0342 | -0.6931 | 0.9933 | 2.6391 | 0.0953  | 8.3494  |
| 95  | 4.2294 | 4.4705 | -0.5798 | 1.0647 | 2.6391 | 0.0000  | 7.8061  |
| 96  | 8.6112 | 1.5261 | 0.2546  | 0.9163 | 2.7081 | -0.1054 | 11.0974 |
| 97  | 4.4152 | 3.8918 | 0.3075  | 0.9933 | 2.6391 | 0.0000  | 7.7007  |
| 98  | 5.7301 | 3.4565 | -0.2485 | 1.0296 | 2.6391 | 0.4383  | 9.4727  |
| 99  | 8.7429 | 1.4110 | 0.2151  | 0.9163 | 2.6391 | -0.0513 | 11.2898 |
| 100 | 4.2641 | 3.6507 | 1.1184  | 0.9953 | 2.5649 | 0.0000  | 7.4442  |
| 101 | 3.7448 | 4.1589 | 0.9282  | 0.9953 | 2.4849 | 0.0000  | 7.1701  |
| 102 | 3.5635 | 4.6052 | 0.6789  | 0.9953 | 2.3979 | 0.0000  | 6.9177  |
| 103 | 2.1163 | 2.9755 | 1.6074  | 1.0647 | 2.3026 | 0.0000  | 7.0648  |
| 104 | 2.9653 | 3.5032 | 0.6098  | 1.0986 | 2.3026 | 0.0000  | 6.0521  |
| 105 | 3.9200 | 3.9493 | 0.8372  | 1.0986 | 2.3026 | 0.0000  | 6.7214  |
| 106 | 2.5096 | 3.8733 | 0.4447  | 1.0986 | 2.3026 | 0.0000  | 5.6099  |
| 107 | 5.9269 | 2.8736 | 1.0403  | 1.0296 | 2.3026 | 0.0000  | 8.4553  |
| 108 | 4.4819 | 4.1856 | -0.4005 | 1.1314 | 2.3026 | 0.0953  | 8.3309  |
| 109 | 8.9690 | 1.3350 | 0.2624  | 0.9555 | 2.1972 | -0.0513 | 11.4721 |
| 110 | 3.9665 | 4.5518 | 0.2700  | 1.1632 | 2.1972 | -0.0513 | 7.7275  |
| 111 | 5.2933 | 2.7408 | 0.5565  | 1.1632 | 2.1972 | 0.0000  | 7.7706  |
| 112 | 3.7773 | 3.6428 | 1.0403  | 1.1632 | 2.1972 | 0.0000  | 6.5367  |
| 113 | 5.2364 | 2.4681 | 0.4700  | 1.0647 | 2.1972 | -0.1054 | 7.5496  |
| 114 | 5.0175 | 2.1541 | 1.1848  | 1.1632 | 2.0794 | 0.1823  | 7.0901  |
| 115 | 6.5338 | 2.6940 | 0.6627  | 1.1632 | 2.0794 | 0.0000  | 8.7641  |
| 116 | 5.0876 | 3.7305 | -0.2614 | 1.1632 | 2.1972 | 0.4700  | 8.8679  |
| 117 | 0.4947 | 4.7449 | -0.0726 | 1.3083 | 2.0794 | 0.5878  | 4.2268  |
| 118 | 3.9027 | 4.3694 | -0.6735 | 1.1632 | 2.0794 | 0.6628  | 8.0709  |
| 119 | 4.5401 | 4.0055 | 0.0198  | 1.1632 | 2.0794 | 0.5008  | 8.5564  |
| 120 | 6.2086 | 3.3604 | 0.1310  | 1.2258 | 1.9459 | 0.5008  | 9.6486  |
| 121 | 9.2064 | 1.2698 | 0.2927  | 1.0296 | 2.4849 | 0.0000  | 11.6527 |
| 122 | 4.5890 | 2.7147 | -0.1625 | 1.1632 | 1.9459 | -0.2231 | 6.9347  |
| 123 | 5.3753 | 3.6610 | 1.0152  | 1.2528 | 1.9459 | 0.5306  | 8.8818  |
| 124 | 1.4159 | 4.4739 | -0.4005 | 1.3083 | 1.7918 | 0.6931  | 5.7858  |
| 125 | 3.8111 | 4.1141 | 0.5539  | 1.2809 | 1.6094 | 0.6931  | 7.3265  |
| 126 | 2.3609 | 3.5056 | 0.6419  | 1.2809 | 1.6094 | 0.6931  | 7.3152  |
| 127 | 3.2308 | 2.6810 | 0.4762  | 1.2809 | 1.7918 | 0.6931  | 7.1309  |

Table 3.16    continued

| | X1 | X2 | X3 | X4 | X5 | X6 | Y |
|---|---|---|---|---|---|---|---|
| 128 | 4.3041 | 3.2958 | 0.5247 | 1.2528 | 1.6094 | 0.6931 | 7.7664 |
| 129 | 4.3108 | 3.3499 | 0.8544 | 1.1939 | 1.7918 | -0.2231 | 6.9847 |
| 130 | 4.6540 | 3.3911 | 0.4121 | 1.1314 | 1.3863 | 0.5008 | 7.9896 |
| 131 | 3.0910 | 3.6376 | 0.0488 | 1.1939 | 1.0986 | 0.6931 | 6.6201 |
| 132 | 2.9652 | 4.0146 | 0.8659 | 1.1939 | 1.0986 | 0.6931 | 6.4770 |
| 133 | 2.4510 | 3.8501 | 0.9085 | 1.1939 | 0.6931 | 0.6931 | 6.2146 |
| 134 | 4.6913 | 3.8912 | 0.6627 | 1.2238 | 1.3863 | 0.6931 | 7.4955 |
| 135 | 4.0448 | 3.8649 | 0.5365 | 1.0986 | 2.1972 | 0.0953 | 7.6505 |
| 136 | 5.6348 | 2.8034 | 0.8920 | 1.0296 | 2.0794 | 0.0000 | 8.4007 |
| 137 | 5.5722 | 2.1158 | 0.7129 | 1.2258 | 2.0794 | 0.0000 | 8.8099 |
| 138 | 5.3181 | 2.7081 | 0.3075 | 1.2238 | 2.0794 | 0.0000 | 8.0130 |
| 139 | 6.9017 | 2.0347 | 0.6098 | 1.1632 | 1.7918 | 0.0000 | 9.4092 |
| 140 | 4.2268 | 2.7279 | 0.9459 | 1.2528 | 1.3863 | 0.6931 | 6.9847 |
| 141 | 4.8903 | 3.5264 | 0.8286 | 1.3083 | 1.7918 | 0.6931 | 8.1490 |
| 142 | 4.3215 | 3.8712 | 0.1598 | 1.3083 | 1.3863 | 0.7419 | 8.0064 |
| 143 | 7.3428 | 2.1114 | 0.8459 | 1.1939 | 1.9459 | 0.3001 | 9.9530 |
| 144 | 4.3707 | 3.5639 | 0.8126 | 1.3083 | 1.7918 | 0.9361 | 8.1915 |
| 145 | 3.1781 | 4.1620 | 0.1579 | 1.3350 | 1.7918 | 1.0296 | 7.5229 |
| 146 | 3.2921 | 3.0301 | 0.8020 | 1.2809 | 1.9459 | 0.9361 | 9.1466 |
| 147 | 5.0259 | 4.1384 | -0.4620 | 1.0647 | 2.4849 | 0.5306 | 8.2161 |
| 148 | 3.8551 | 2.5337 | 0.8671 | 1.0986 | 2.1972 | 0.5716 | 7.1854 |
| 149 | 3.6636 | 4.3490 | -0.3285 | 1.0986 | 2.3026 | 0.6931 | 7.8070 |
| 150 | 4.8828 | 2.9549 | 0.3148 | 1.0647 | 2.1972 | 0.5306 | 8.4118 |
| 151 | 4.7095 | 2.5257 | -0.2614 | 1.0647 | 2.3979 | 0.6152 | 8.1867 |
| 152 | 5.2511 | 3.5410 | 1.2238 | 0.9555 | 2.5649 | 0.2231 | 8.4608 |
| 153 | 5.7268 | 1.9402 | 0.2469 | 0.9555 | 2.5649 | 0.5008 | 8.7323 |
| 154 | 6.4425 | 1.6734 | 0.2511 | 0.9555 | 2.5649 | 0.4383 | 8.4667 |
| 155 | 3.9512 | 4.5788 | -0.6539 | 0.9163 | 2.7081 | -0.1054 | 7.6497 |
| 156 | 4.3520 | 4.1223 | -0.6349 | 0.9163 | 2.7081 | -0.0513 | 8.1942 |
| 157 | 4.9545 | 3.7281 | -0.6931 | 0.8755 | 2.7081 | 0.0000 | 8.7323 |
| 158 | 6.9508 | 2.2555 | 0.1510 | 0.7835 | 2.7726 | -0.0513 | 10.0476 |
| 159 | 2.8904 | 4.0775 | 1.8718 | 0.8329 | 2.7726 | 0.0000 | 6.1992 |
| 160 | 5.7366 | 3.1555 | 0.6152 | 0.7885 | 2.7726 | -0.0513 | 8.9746 |
| 161 | 6.5309 | 2.4510 | -0.1989 | 0.7885 | 2.7726 | 0.0000 | 9.5539 |
| 162 | 4.8752 | 2.9497 | -0.0733 | 0.7419 | 2.8332 | 0.0000 | 8.4118 |
| 163 | 6.1717 | 2.3026 | -0.1744 | 0.7419 | 2.8332 | 0.0000 | 9.2103 |
| 164 | 4.9558 | 2.6532 | 1.6919 | 0.7885 | 2.8332 | -0.0513 | 7.4384 |

procedure to the Benson data expressed in logarithmic form. At iteration 5 variable $X_4$ was the candidate for entry. Its partial $F$-value was 1.24 which, with 5 and 158 d.f., is not significant at the 0.75 level, which was chosen as a cut-off point. The final regression is:

$$\log Y = 3.33 + 0.67 \log X_1 - 0.04 \log X_3$$
$$\quad\quad\quad\quad (0.033) \quad\quad\quad (0.072)$$

$$+ 0.68 \log X_5 + 0.73 \log X_6$$
$$(0.112) \quad\quad\quad (0.163)$$

The standard errors are bracketed beneath the associated coefficient. At this stage the partial correlation coefficients of log $y$ and the logarithms of $X_2$ and $X_4$ are 0.006 and 0.088 respectively, once the effects of $\log X_1$, $\log X_3$, $\log X_5$ and $\log X_6$ are removed. It would appear from this that the linear contributions of $\log X_2$ and $\log X_4$ do not add anything to the explanation of the variation in log $Y$. Any explanation of the variation in log $Y$ should not include these two variables. The

Table 3.17
Results of stepwise regression analysis of data listed in Table 3.16

| Iteration | Regression equation | $R^2$ | $F$-test of entire regresstion | Partial $F$-tests | d.f. |
|---|---|---|---|---|---|
| 1 | $\log Y = 4.52 + 0.72 \log X$, | 0.71 | 397.9** | $X_2$ : 397.9** | 1 and 162 |
| 2 | $\log Y = 4.67 + 0.74 \log X_1$ $-0.50 \log X_3$ | 0.77 | 268.0** | $X_1$ : 517.7** $X_3$ : 40.6** | 2 and 161 |
| 3 | $\log Y = 3.96 + 0.68 \log X_1$ $-0.42 \log X_3 + 0.44 \log X_5$ | 0.79 | 203.7** | $X_1$ : 387.1** $X_3$ : 31.1** $X_5$ : 18.2** | 3 and 160 |
| 4 | $\log Y = 3.33 + 0.67 \log X_1$ $-0.40 \log X_3 + 0.68 \log X_5$ $+0.73 \log X_6$ | 0.82 | 175.8** | $X_1$ : 422.1** $X_3$ : 31.1** $X_5$ : 37.4** $X_6$ : 19.9** | 4 and 159 |
| 5 | At iteration 5 variable $X_4$ was entered. Its partial $F$-value was 1.24 with 5 and 158 d.f. The tabled value of $F$ at $p = 0.05$ is 2.21, so variable $X_4$ was not accepted, and the procedure was terminated. | | | | |

\*\* : significant at 99% level.

stepwise approach would, then, lead to the conclusion that, of the six explanatory variables, only four are important. The four selected variables are: $X_1$, basin area; $X_3$, storage effect (which has a negative relationship with the dependent variable); $X_5$, the temperature effect (which is really estimating the storage influence of winter ice) and $X_6$, the orographic factor. Neither $X_2$ (main channel slope) nor $X_4$ (rainfall intensity factor) appears to be significant when linear (or, rather, log-linear) relationships are considered.

The selection approach proceeds by an iterative process that searches for the optimal subset of explanatory variables but which avoids the evaluation of every possible regression of the explanatory variables on $Y$. The combinatorial approach relies on the speed of the computer to obtain a measure of optimality for all of the $2^k - 1$ possible regressions. The computating time requirements are not intolerable for values of $k$ less than 20. In the present example using Benson's (1962) data (Table 3.16 above), the stepwise procedure had a run-time of 5 seconds on an 1CL 1906A computer. Grosenbaugh's (1967) combinatorial regression program ran for 8 seconds.

A summary of the results from the combinatorial regression program are given in Table 3.18. The values in the table are the RMSQR (root mean-square residual) values for the best of each of the sets of 1, 2, ..., 6 explanatory variables (entries 1 to 6) followed by the six near-optimal subsets (7–12) and the six worst subsets (13–18). The set chosen by the program was $(X_1, X_3, X_4, X_5, X_6)$. Variable $X_4$, the rainfall intensity factor, which was excluded by the stepwise procedure at iteration 5 is included in the combinatorial algorithm, giving the equation

$$\log Y = 4.20 + 0.66 \log X_1 - 0.39 \log X_3 - 0.56 \log X_4$$
$$\quad\quad\quad\quad (0.034) \quad\quad\quad (0.072) \quad\quad\quad (0.504)$$
$$+ 0.57 \log X_5 + 0.78 \log X_6$$
$$(0.148) \quad\quad\quad (0.170)$$

Neither the stepwise nor the combinatorial algorithms include $X_2$ (main channel slope) in the chosen subset. The difference in the RMSQR values between the sets $(X_1, X_3, X_5, X_6)$ and $(X_1, X_3, X_4, X_5, X_6)$ is small (0.003) so one might choose the smaller subset on the grounds of economy, and also because it can be shown that the standard error of $Y$ increases as the number of explanatory variables rises (Walls and Week, 1969).

Both the combinatorial and the stepwise methods have given comparable results in this example. The additional information regarding near-optimal subsets that is produced by the combinatorial approach may be valuable in some circumstances, for example, if one was searching for a set of predictor variables that were both

Table 3.18
Results of combinatorial regression analysis of data listed in Table 3.16

|     | Entry | Min. RMSQR | Explanatory variables |
|-----|-------|------------|-----------------------|
| (a) | 1     | 0.2911     | 1                     |
|     | 2     | 0.2339     | 1,3                   |
|     | 3     | 0.2114     | 1,3,5                 |
|     | 4     | 0.1891     | 1,3,5,6               |
|     | 5     | 0.1888     | 1,3,4,5,6             |
|     | 6     | 0.1899     | 1,2,3,4,5,6           |
| (b) | 7     | 0.1888     | 1,3,4,5,6             |
|     | 8     | 0.1891     | 1,3,5,6               |
|     | 9     | 0.1899     | 1,2,3,4,5,6           |
|     | 10    | 0.1903     | 1,2,3,5,6             |
|     | 11    | 0.2032     | 1,2,3,4,6             |
|     | 12    | 0.2056     | 1,3,4,6               |
| (c) | 13    | 0.9940     | 6                     |
|     | 14    | 0.9798     | 3                     |
|     | 15    | 0.9749     | 3,6                   |
|     | 16    | 0.8757     | 2                     |
|     | 17    | 0.8747     | 2,6                   |
|     | 18    | 0.7853     | 2,3,6                 |

(a) 'Best' subsets of size 1 to 6;
(b) Six 'best' subsets;
(c) Six 'worst' subsets.

near-optimal and relatively easy or cheap to measure. If interest centres on the selection of the best subset then either method could be used. Both suffer from the disadvantage that a single measure of optimality is not available, consequently results may differ depending on the manner in which 'best' is operationally defined.

## 3.4 Use of Binary Explanatory Variables
### 3.4.0 Introduction
In some instances the explanatory variables in a regression equation may take only the values 0 or 1, which may correspond to present or absent, yes or no, or to any other two-state relationship. In other cases, dummy or binary $(1-0)$ variables can be used to indicate class membership or define the conditions that are prevailing when the measurement is made. For example, economists may use binary variables to indicate whether the observation was made during peacetime (0) or wartime (1) conditions (see Johnston, 1972, p. 177). By using more than one binary variable, several classifications can be introduced thus allowing a hydrologist, for instance, to split up his

observations into four groups based on season of the year, or a geomorphologist to divide observations on slopes into several groups based on slope aspect. If this is done, regression analysis can be used to test hypotheses about inter-group differences. This use of regression analysis is identical in concept to the analysis of variance. The regression approach is, however, more flexible (Darlington, 1968, Cohen, 1968).

*3.4.1 Method and Example*

The comparison of regression relationships by means of dummy variables is described by Gujatari (1970). Suppose that $m$ samples of measurements are available, the number of individuals in each sample being $n_i$ ($i = 1, 2, \ldots, m$) with $N = \Sigma_{i=1}^{m} n_i$. The numbers in each of the groups are not necessarily equal. It is interesting to enquire whether the regression coefficients computed for each sample are significantly different. Consider the following example, with $m = 4$ groups each with 10 members, giving $N = 40$. The measurements consist of depth of soil ($Y$), slope angle ($X_1$) and distance from the crest of the slope ($X_2$) at 40 sites which are located on four different rock types (Table 3.19). If it is postulated that the population regression model is:

$$Y_j = \beta_0 + \beta_1 X_{1j} + \beta_2 X_{2j} + \epsilon_j \qquad (3.19)$$

where $\epsilon_j$ is the disturbance term, then an important question is: do the values of the parameters ($\beta_0$, $\beta_1$ and $\beta_2$) differ from one rock type to another? The regression model (Equation 3.19) can be rewritten as:

$$Y_{ij} = \beta_i + \beta_{i1} X_{1j} + \beta_{i2} X_{2j} + \epsilon_{ij} \quad (i = 1,4; j = 1,N) \qquad (3.20)$$

which is a compressed form of the following equations:

$$Y_{1j} = \beta_1 + \beta_{11} X_{1j} + \beta_{12} X_{2j} + \epsilon_{1j} \quad (j = 1,n_1)$$
$$Y_{2j} = \beta_2 + \beta_{21} X_{1j} + \beta_{22} X_{2j} + \epsilon_{2j} \quad (j = 1,n_2)$$
$$Y_{3j} = \beta_3 + \beta_{31} X_{1j} + \beta_{32} X_{2j} + \epsilon_{3j} \quad (j = 1,n_3)$$
$$Y_{4j} = \beta_4 + \beta_{41} X_{1j} + \beta_{42} X_{2j} + \epsilon_{4j} \quad (j = 1,n_4)$$

Three dummy variables are used to indicate group membership; these are

$D_1 = 0$ if the observation comes from rock type 2,
= 1 otherwise.

$D_2 = 0$ if the observation comes from rock type 3,
= 1 otherwise.

$D_3 = 0$ if the observation comes from rock type 4,
= 1 otherwise.

Table 3.19
Data for example of use of binary explanatory variables (Reproduced with permission from D. Gujatari, *The American Statistician*, 24, 18–23 (1970), Table 2)

| $Y_j$ | $D_1$ | $D_2$ | $D_3$ | $X_{1j}$ | $D_1X_{1j}$ | $D_2X_{1j}$ | $D_3X_{1j}$ | $X_{2j}$ | $D_1X_{2j}$ | $D_2X_{2j}$ | $D_3X_{2j}$ |
|---|---|---|---|---|---|---|---|---|---|---|---|
| 1.40 | 0 | 0 | 0 | 78 | 0 | 0 | 0 | 61 | 0 | 0 | 0 |
| 1.79 | 0 | 0 | 0 | 90 | 0 | 0 | 0 | 59 | 0 | 0 | 0 |
| 1.72 | 0 | 0 | 0 | 94 | 0 | 0 | 0 | 76 | 0 | 0 | 0 |
| 1.47 | 0 | 0 | 0 | 71 | 0 | 0 | 0 | 50 | 0 | 0 | 0 |
| 1.26 | 0 | 0 | 0 | 99 | 0 | 0 | 0 | 61 | 0 | 0 | 0 |
| 1.28 | 0 | 0 | 0 | 80 | 0 | 0 | 0 | 54 | 0 | 0 | 0 |
| 1.34 | 0 | 0 | 0 | 83 | 0 | 0 | 0 | 57 | 0 | 0 | 0 |
| 1.55 | 0 | 0 | 0 | 75 | 0 | 0 | 0 | 45 | 0 | 0 | 0 |
| 1.57 | 0 | 0 | 0 | 62 | 0 | 0 | 0 | 41 | 0 | 0 | 0 |
| 1.26 | 0 | 0 | 0 | 67 | 0 | 0 | 0 | 40 | 0 | 0 | 0 |
| 1.61 | 1 | 0 | 0 | 78 | 78 | 0 | 0 | 74 | 74 | 0 | 0 |
| 1.31 | 1 | 0 | 0 | 99 | 99 | 0 | 0 | 75 | 75 | 0 | 0 |
| 1.12 | 1 | 0 | 0 | 80 | 80 | 0 | 0 | 64 | 64 | 0 | 0 |
| 1.35 | 1 | 0 | 0 | 75 | 75 | 0 | 0 | 48 | 48 | 0 | 0 |
| 1.29 | 1 | 0 | 0 | 94 | 94 | 0 | 0 | 62 | 62 | 0 | 0 |
| 1.24 | 1 | 0 | 0 | 91 | 91 | 0 | 0 | 42 | 42 | 0 | 0 |
| 1.29 | 1 | 0 | 0 | 75 | 75 | 0 | 0 | 52 | 52 | 0 | 0 |
| 1.43 | 1 | 0 | 0 | 63 | 63 | 0 | 0 | 43 | 43 | 0 | 0 |
| 1.29 | 1 | 0 | 0 | 62 | 62 | 0 | 0 | 50 | 50 | 0 | 0 |
| 1.26 | 1 | 0 | 0 | 67 | 67 | 0 | 0 | 40 | 40 | 0 | 0 |
| 1.67 | 0 | 1 | 0 | 78 | 0 | 78 | 0 | 80 | 0 | 80 | 0 |
| 1.41 | 0 | 1 | 0 | 83 | 0 | 83 | 0 | 61 | 0 | 61 | 0 |
| 1.73 | 0 | 1 | 0 | 79 | 0 | 79 | 0 | 62 | 0 | 62 | 0 |
| 1.23 | 0 | 1 | 0 | 70 | 0 | 70 | 0 | 47 | 0 | 47 | 0 |
| 1.49 | 0 | 1 | 0 | 85 | 0 | 85 | 0 | 59 | 0 | 59 | 0 |
| 1.22 | 0 | 1 | 0 | 83 | 0 | 83 | 0 | 42 | 0 | 42 | 0 |
| 1.39 | 0 | 1 | 0 | 71 | 0 | 71 | 0 | 47 | 0 | 47 | 0 |
| 1.39 | 0 | 1 | 0 | 66 | 0 | 66 | 0 | 42 | 0 | 42 | 0 |
| 1.56 | 0 | 1 | 0 | 67 | 0 | 67 | 0 | 40 | 0 | 40 | 0 |
| 1.36 | 0 | 1 | 0 | 67 | 0 | 67 | 0 | 40 | 0 | 40 | 0 |
| 1.40 | 0 | 0 | 1 | 77 | 0 | 0 | 77 | 62 | 0 | 0 | 62 |
| 1.47 | 0 | 0 | 1 | 71 | 0 | 0 | 71 | 55 | 0 | 0 | 55 |
| 1.37 | 0 | 0 | 1 | 78 | 0 | 0 | 78 | 62 | 0 | 0 | 62 |
| 1.15 | 0 | 0 | 1 | 70 | 0 | 0 | 70 | 43 | 0 | 0 | 43 |
| 1.22 | 0 | 0 | 1 | 95 | 0 | 0 | 95 | 57 | 0 | 0 | 57 |
| 1.48 | 0 | 0 | 1 | 96 | 0 | 0 | 96 | 51 | 0 | 0 | 51 |
| 1.31 | 0 | 0 | 1 | 71 | 0 | 0 | 71 | 41 | 0 | 0 | 41 |
| 1.27 | 0 | 0 | 1 | 63 | 0 | 0 | 63 | 40 | 0 | 0 | 40 |
| 1.22 | 0 | 0 | 1 | 62 | 0 | 0 | 62 | 45 | 0 | 0 | 45 |
| 1.36 | 0 | 0 | 1 | 67 | 0 | 0 | 67 | 39 | 0 | 0 | 39 |

The dummy variables are incorporated into the regression equation as follows:

$$y_j = \beta_0 + \beta_1 D_1 + \beta_2 D_2 + \beta_3 D_3 + \beta_4 X_{1j} + \beta_5 (D_1 X_{1j})$$
$$+ \beta_6 (D_2 X_{1j}) + \beta_7 (D_3 X_{1j}) + \beta_8 X_{2j} + \beta_9 (D_1 X_{2j})$$
$$+ \beta_{10} (D_2 X_{2j}) + \beta_{11} (D_3 X_{2j}) + \epsilon_j \quad (j = 1, N) \qquad (3.20)$$

where:

$\beta_0$ : intercept for group 1;
$\beta_1$ : differential intercept for group 2;
$\beta_2$ : differential intercept for group 3;
$\beta_4$ : partial regression coefficient of $X_1$ for group 1;
$\beta_5$ : differential partial regression coefficient of $X_1$ for group 2;
$\beta_6$ : differential partial regression coefficient of $X_1$ for group 3;
$\beta_7$ : differential partial regression coefficient of $X_1$ for group 4;
$\beta_8$ : partial regression coefficient of $X_2$ for group 1
$\beta_9$ : differential partial regression coefficient of $X_2$ for group 2
$\beta_{10}$ : differential partial regression coefficient of $X_2$ for group 5
$\beta_{11}$ : differential partial regression coefficient of $X_2$ for group 4

The differential intercepts and differential partial regression coefficients can be used to recover the four original regression equations:

$$Y_1 = \hat{\beta}_0 + \hat{\beta}_4 X_1 + \hat{\beta}_8 X_2 \qquad \text{(Group 1)}$$
$$Y_2 = (\hat{\beta}_0 + \hat{\beta}_1) + (\hat{\beta}_4 + \hat{\beta}_5) X_1 + (\hat{\beta}_8 + \hat{\beta}_9) X_2 \qquad \text{(Group 2)}$$
$$Y_3 = (\hat{\beta}_0 + \hat{\beta}_2) + (\hat{\beta}_4 + \hat{\beta}_6) X_1 + (\hat{\beta}_8 + \hat{\beta}_{10}) X_2 \qquad \text{(Group 3)}$$
$$Y_4 = (\hat{\beta}_0 + \hat{\beta}_3) + (\hat{\beta}_4 + \hat{\beta}_7) X_1 + (\hat{\beta}_8 + \hat{\beta}_{11}) X_3 \qquad \text{(Group 4)}$$

Estimates of the $\beta_i$ can be obtained by using Equation (3.20) alone. The assumptions of the regression model must hold, however, if the estimators of the $\beta_i$ are to have the desirable properties of least squares regression estimates (see Chapter 2). The conventional matrix inversion method can be used to find the $\hat{\beta}_i$, which are the ordinary least squares estimates of the $\hat{\beta}_i$. Subsequently, the $t$-test, outlined in Chapter 2, is used to test the null hypothesis that the individual $\beta_i$ are zero in the population. If none of the differential intercept terms ($\beta_2, \beta_3, \beta_4$) or differential partial regression coefficients ($\beta_5, \beta_6, \beta_7, \beta_9, \beta_{10}, \beta_{11}$) depart significantly from zero then it can be concluded that Equation (3.19) alone can be used, and that the group membership (rock type) has no independent effect on the linear relationship between $Y$ and the $X$'s. Whether $Y$ is significantly related to $X_1$ and $X_2$ is not considered by this technique, for its aim is to isolate group effects.

The data for this example are listed in Table 3.19 above, with the dummy variables included. A conventional regression program (such as that listed in Table 2.1 above) gave the following results:

$$Y_j = 1.357 + 0.044\,D_1 + 0.094\,D_2 - 0.313\,D_3$$
$$\phantom{Y_j =\ }(0.463)\quad\ (0.618)\quad\ (0.481)$$
$$- 0.006\,X_{1j} + 0.001\,(D_1 X_{1j}) - 0.002\,(D_2 X_{1j})$$
$$\phantom{-\ }(0.008)\quad\ \ (0.009)\quad\quad\quad\ (0.011)$$
$$+ 0.007\,(D_3 X_{1j}) + 0.011\,X_{2j} - 0.006\,(D_1 X_{2j})$$
$$\phantom{+\ }(0.009)\quad\quad\quad (0.008)\quad\ (0.001)$$
$$+ 0.000\,(D_2 X_{2j}) - 0.007\,(D_3 X_{3j})$$
$$\phantom{+\ }(0.001)\quad\quad\quad (0.011)$$

The $\beta$'s were tested separately by the $t$-test and all the differential intercept and partial regression coefficients were found not to differ from zero at a significance level of 0.05. There is therefore no statistical justification for believing that the relationship of $Y$ to $X_1$ and $X_2$ differs from group to group.

It is not always necessary to apply the dummy variables to every term in the regression equation; for example, terms $\hat{\beta}_1 D_1$, $\hat{\beta}_2 D_2$ and $\hat{\beta}_3 D_3$ could be omitted if it were thought that the intercept term was constant from group to group.

Notice that, even though there were four groups in the example only three dummy variables were used. There are two reasons for this: (i) the four groups can be completely differentiated by three dummy variables, and (2) if four dummy variables were employed the matrix $(\mathbf{X'X})$ would be singular. (However, as the dummy variables are combined with real variables, and since floating-point computations are known to be prone to error (Chapter 1) the inversion routine would not necessarily find that $(\mathbf{X'X})$ was singular.) Johnston (1972, p. 178) calls this the 'dummy variable trap'. Since four dummy variables are unnecessary in this example anyway, the problem does not arise.

Regression analysis using binary explanatory variables to indicate class membership has been used by Simonett (1967) in an investigation of the relationship between the dependent variables of landslide frequency and volume of earth moved by each landslide, and the explanatory variables of slope angle, length of slope, gradient of streams and distance from the inferred epicentre of the earthquake. A single dummy variable was used to indicate whether the observation came from the Bewani Mountains or the Torricelli Mountains, New Guinea. Ethridge and Davies (1973) used dummy variables in a regression analysis of the relationship between mean quartz grain size ($y$) and quartz content ($x$) as a percentage of the grains counted. The dummy variables were used to distinguish bay-fill, marine-bar and fluvial sands.

The results allowed the authors to conclude that, for the particular rock formation studied, it was possible to distinguish the three environments on the basis of the regression relationship.

## 3.5 Nonlinear Least Squares Methods

### 3.5.0 Introduction

The use of polynomials in a single explanatory variable was described in Section 3.1, where it was stated that the fitting of such a function to the values of a dependent variable could be motivated by a desire either (i) to approximate the form of the curve passing through the scatter of $(x, y)$ points or (ii) to test the hypothesis that the population relationship between a dependent variable and an explanatory variable is of a specified form. In the same way the method of nonlinear least squares can be used (i) to fit a curve to a set of points or (ii) to test a statistical hypothesis. Curve fitting or interpolation requires no assumptions about the distributions of, or relationships among, the variables. In a regression model, however, the assumptions set out in Chapter 2 must hold, if only approximately, if the ordinary least squares estimators are to have the properties of unbiasedness and minimum variance.

### 3.5.1 Methods

Nonlinear relationships between a dependent variable $y$ and a set of explanatory variables $x_1, x_2, \ldots, x_k$ are not defined by the form of the curve (or surface) relating $y$ to the $x$'s. In Section 3.1 curves of the form

$$y = a_0 + a_1 x + a_2 x^2 + \ldots + a_k x^k$$

were fitted. These do not plot as straight lines yet the model is said to be linear in the parameters because it is of the form

$$y = a_0 + a_1 \phi_1 + a_2 \phi_2 + \ldots + a_k \phi_k \tag{3.21}$$

The $\phi$'s can be any function of the explanatory variables (such as squares, reciprocals or logarithms) provided that these functions do not involve the coefficients $a_i$. Nonlinear models include all models not of the form (3.21). For example (noting that $\theta_i$ is used to represent a model parameter)

$$y = \theta_0 x_1^{\theta_1} \tag{3.22}$$

or

$$y = \theta_0 + x_1^{\theta_1} + \theta_2^{x_2} \tag{3.23}$$

are nonlinear models. (See Figure 3.18(a), (b) and (c)). Equation (3.22) can be reduced to linear form by taking logarithms to base e, i.e.

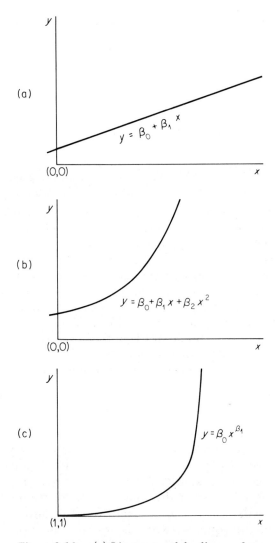

Figure 3.18. (a) Linear model, linear functional relationship;
(b) Linear model, curvilinear relationship;
(c) Nonlinear model, nonlinear relationship

$$\log_e y = \log_e \theta_0 + \theta_1 \log_e x_1$$

and can therefore be treated by conventional linear regression techniques (Chapter 2). Equation (3.23) is not so reducible; it is said to be intrinsically nonlinear, and the parameters $\theta_0$ and $\theta_1$ cannot be found by the methods of Chapter 2. In passing it should be noted that if

(3.22) is written out fully, so as to include the disturbance term, it is possible to distinguish

$$y = \theta_0 x_1^{\theta_1} \epsilon \qquad (3.24)$$

from

$$y = \theta_0 x_1^{\theta_1} + \epsilon \qquad (3.25)$$

In (3.24) the disturbances are multiplicative and so the logarithmic transformation can be made, to give

$$\log_e y = \log_e \theta_0 + \theta_1 \log_e x_1 + \log_e \epsilon \qquad (3.26)$$

whereas (3.25) is intrinsically nonlinear. This assumption concerning the disturbance term is vital, yet it is rarely mentioned in a practical context. Goldfeld and Quandt (1971) give a full discussion of this point. Also notice that the application of linear least squares methods to (3.26) results in minimization of $\Sigma \log_e \epsilon^2$ not $\Sigma \epsilon^2$, that is, the sum of squares of the logarithms of the residuals is minimized rather than the sum of squares of the actual residuals. If (3.26) is back-transformed to the form (3.24) by taking anti-logarithms of the computed estimates $\theta_0$ and $\theta_1$ it is necessary to note that (3.26) is not a true least squares function. In practice this should not lead to difficulties.

Linear least squares problems can be handled analytically; in the absence of complications the vector of estimates $\hat{\boldsymbol{\beta}}$ is obtained from the solution of

$$\hat{\boldsymbol{\beta}} = (\mathbf{X}'\mathbf{X})^{-1}\mathbf{X}'\mathbf{y},$$

as outlined in Chapter 2. No analytical solution is possible for the majority of nonlinear least squares problems, and recourse is made to numerical methods of approximation. Many such methods exist, dating back to Gauss (see, for example, Seal, 1967). In general, each method begins with an initial guess of the values of the $\theta_i$ (termed the starting point) and proceeds by altering the individual $\theta_i$ until a minimum of the residual sum of squares (RSS) is found. The process could be likened to the progress of a blind man making his way from the top of a mountain to the valley below by continually moving downhill. In fact, one of the most widely-used methods (that of steepest descent) can be explained by this analogy. The gradient of the residual sum of squares surface is computed and the $\theta_i$ moved in the direction of steepest descent. This process if repeated until it is impossible to move downslope any further. In theory, a minimum value of RSS will eventually be located by this method, but in practice it may take a long time, and the resulting minimum may be local rather than global. To illustrate the point, consider a nonlinear function which is of the form $y = \theta_0 x^{\theta_1}$. For a given set of $x$'s and $y$'s the values of $\theta_0$ and $\theta_1$ can be set to correspond to the coordinate values of the grid intersections of a

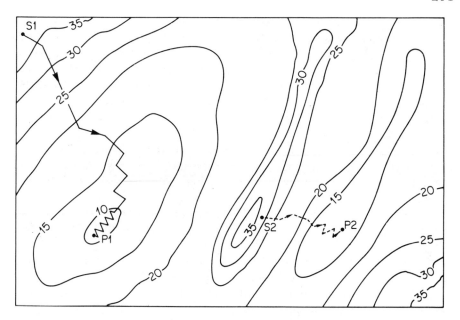

Figure 3.19 Residual sum of squares surface. The lines of steepest descent from points S1 to P1 (————) and from points S2 to P2 (— — — — —) are shown

rectangular coordinate system. At each grid intersection the value of the residual sum of squares is calculated. A contour-type map can then be drawn to represent the shape of the resulting residual sum of squares surface, which may look something like Figure 3.19. A starting-point is chosen (e.g. S1) and the line of steepest descent is followed (shown by ———— on Figure 3.19) until point P1 is reached. No downslope movement from P1 is possible, therefore the algorithm correctly identifies P1 as the point of minimum RSS, giving $\theta_0 \simeq 0.5$ and $\theta_1 \simeq 1.5$. If the steepest-descent iterations were begun from the point S2 rather than S1 then the search would terminate at P2 (where $\theta_0 \simeq 2.2$ and $\theta_1 \simeq 5.5$) which is far from the true (global) minimum. This example serves to show that (i) iterative methods require good starting-points and (ii) they may converge to a local, rather than a global, minimum. The steepest-descent method, outlined above, is not particularly efficient as it tends to require a large number of steps, and it is known to zig-zag towards a minimum rather than proceed directly. Nevertheless, it is widely used (see the sections on nonmetric multidimensional scaling and nonlinear mapping in Chapter 6). An alternative method is based on Newton's algorithm to determine the roots of a polynomial. (See Section 5.3(B) for an application of this algorithm.) This involves approximating the form of the residual sum of squares surface at a particular point by a quadratic surface, finding the minimum on the quadratic surface, moving to this pseudo-minimum, fitting a second quadratic surface at the new point, and so on. This procedure is known as the Newton–Raphson method. It is far more efficient than the method of steepest descent if the starting point is close to the true minimum, because convergence is quadratic (i.e. the number of correct digits in the estimates of the $a_i$ is doubled at each iteration). Bard (1974, p. 89) suggests that for many common cases the Newton–Raphson method is up to 25 000 times more efficient than the method of steepest descent. Unfortunately the Newton–Raphson method is not of great practical use unless modified in some way, for it cannot locate a minimum if the starting point is not close to that minimum. Thus, Joreskog and van Thillo (1971) use steepest-descent iterations to locate the region of the minimum before moving to the Newton–Raphson algorithm to find the minimum itself. (See Chapter 5, maximum likelihood factor analysis.) Secondly, the rate of curvature as well as the gradient of the residual sum of squares surface must be calculated at each point. Expressions to evaluate the rate of curvature tend to be complex and coding them in Fortran in itself can be a considerable problem.

In order to appreciate the points made in the preceding paragraph it is necessary to digress slightly at this stage and introduce the concept of the derivative of a function. In the foregoing pages, the terms 'gradient' and 'curvature at a point' of the residual sum of squares surface were used. In many instances it is necessary to calculate or estimate either the gradient alone or both the gradient and the curvature of the surface. The gradient of the surface parallel to each axis is equivalent to the first derivative of the function defining the surface with respect to the parameter represented by that axis. The first derivatives at a given point are, in fact, the rates of change of the function parallel to each axis at that point. To take a well-known example, the relationship between distance moved ($s$), initial velocity ($u$), elapsed time ($t$) and acceleration is

$$s = ut + \tfrac{1}{2}ft^2$$

Given $t = 3$ seconds, $f = 10$ ft/sec$^2$ and $u = s$ ft/sec the distance travelled from time $t = 0$ is $s = 2.3 + \tfrac{1}{2} \times 10.9 = 6 + 45 = 51$ feet. The rate of change of distance with respect to time, or velocity, at time $t$ is the first derivative of $s$ with respect to $t$, i.e.

$$\frac{ds}{dt} = v = u + ft$$

Therefore, with $u = 2$, $f = 10$ and $t = 3$ the velocity is $2 + 30 = 32$ ft/sec. The rate of change of velocity with respect to time is acceleration, $f$, defined as

$$\frac{dv}{dt} = f = d^2s/dt^2$$

In the example, $f$ is a constant 10 feet per second per second.

In this example the rate of change of $s$ with respect to $t$ is written $ds/dt$. Since $v = ds/dt$ then the first derivative of $v$ with respect to $t$, $dv/dt$, is also the second derivative of $s$ with respect to $t$, or $d^2s/dt^2$.

The example shows how the rate of change over time of a variable, $s$, can be computed analytically by evaluating the first derivative of $s$ with respect to $t$. The rate of change of $ds/dt$ is given by the second partial derivative of $s$ with respect to $t$, or $d^2s/dt^2$. To transfer these concepts to the least-squares problem discussed above (Equation (3.25)), write the residual sum of squares as

$$f^i = \sum_{j=1}^{N} (\theta_0^i x_{j1}^{\theta_1^i} - y)^2$$

where $f^i$ is the sum of squares of the residuals (differences between observed and estimated $y$) at the point on a plot such as Figure 3.19 above given by $(\theta_0^i, \theta_1^i)$. The first partial derivatives of $f^i$ with respect to $\theta_0^i$ and $\theta_1^i$ gives the direction of the greatest increase in the value of $f^i$ parallel to the axes $\theta_0$ and $\theta_1$. The resultant of these orthogonal gradients gives the direction of the maximum rate of increase in the function at the given point $(\theta_0^i, \theta_1^i)$. If we choose to move in the opposite direction then we would follow the line of steepest descent. The second partial derivatives of $f$ with respect to $\theta_0$ and $\theta_1$ evaluated at $(\theta_0^i, \theta_1^i)$ give the rate of curvature of the residual sum of squares at $(\theta_0^i, \theta_1^i)$. A minimum on the residual sum of squares surface has been located when the gradient is zero and when the curvature is concave upwards. All iterative methods are based on these principles.

It is beyond the scope of this book to provide an exhaustive account of differential calculus. Readers lacking a knowledge of this topic are referred to one of the many textbooks available, for example, Open University (1971), Spivak (1967) or Kline (1967). In many cases, it is possible to use numerical techniques to approximate both first and second order derivatives. These are known as difference methods. Two variants are in wide use; one-sided finite differences and central differences. Bard (1974, pp, 117–119) and McCammon (1969, p. 3) discuss the computational aspects. The computer program listed by McCammon (1969) contains a finite-difference algorithm to approximate the first derivatives.

In the following paragraphs the matrix of the first derivatives of the residual sum of squares with respect to each of the parameters in turn (the Jacobean matrix) will be denoted by **J**. The $i$th column of the $(n \times k)$ matrix **J** gives the gradient at each of the $n$ data points with respect to axis (parameter) $\theta_i$. The direction of maximum increase in the residual sum of squares is given by the vector **q** where

$$\mathbf{q} = 2\mathbf{J}'\mathbf{f}$$

where **f** is the vector of residuals. Thus, at point $(\theta_0^i, \theta_1^i)$ the residuals $f_i$ are calculated and the gradient vector **q** computed as above. Let $\mathbf{q} = \{q_0, q_1\}$. Then the direction of maximum increase in the residual sum of squares is towards the point $(\theta_0^i + q_0, \theta_1^i + q_1)$. Since the search is for a minimum rather than a maximum of the residual sum of squares, we move towards the point $(\theta_0^i - q_0, \theta_1^i - q_1)$ by an amount known as the step length. The matrix of second order derivatives, termed the Hessian matrix, will be denoted by **G**. It is now possible to show why the Newton–Raphson method fails when the starting point is not close to a minimum. If the parameter estimates at iteration $i$, namely, $\theta_1^i, \theta_2^i, \ldots, \theta_k^i$, are held in the vector $\boldsymbol{\theta}_i$ then the revised values, held in $\boldsymbol{\theta}_{i+1}$, at iteration $i + 1$ are given by

$$\boldsymbol{\theta}_{i+1} = \boldsymbol{\theta}_i - \mathbf{G}_i^{-1}\mathbf{q}_i \tag{3.27}$$

This depends on the nonsingularity of **G**. If the point represented by the vector $\boldsymbol{\theta}_i$ is far from a minimum, $\mathbf{G}_i$ will be singular and the method fails. Marquardt (1963) proposed a modification which is based on the same principles as those used in ridge regression (Chapter 2) where a similar problem — singularity or near-singularity of the matrix of sums of squares and cross products of the explanatory variables in multiple linear regression — was met. Instead of using Equation (3.27) the following expression is evaluated:

$$\boldsymbol{\theta}_{i+1} = \boldsymbol{\theta}_i - (\mathbf{G}_i + \lambda\mathbf{P})^{-1}\mathbf{q}_i$$

where $\lambda$ is a positive scalar and **P** any positive definite matrix. The matrix $(\mathbf{G}_i + \lambda\mathbf{P})$ will always be nonsingular if $\lambda$ is sufficiently large. A Fortran subroutine for the Marquardt method has been provided by Fletcher (1971).

Many iterative methods other than those of steepest descent, Newton–Raphson and Marquardt for locating a minimum of a nonlinear function are in common use. The Gauss–Newton method is a variant of Newton–Raphson applied to least-squares problems. The Gauss–Newton method involves the use of an estimate of the Hessian matrix **G** by **N** where

$$\mathbf{N} = 2\mathbf{J}'\mathbf{J}$$

or

$$n_{ij} = 2 \sum_{k=1}^{n} (\mathrm{d}f_k/\mathrm{d}\theta_i)(\mathrm{d}f_k/\mathrm{d}\theta_j)$$

where $\mathrm{d}f_k/\mathrm{d}\theta_i$ is the first derivative of the residual sum of squares function with respect to $\theta_i$, the $i$th parameter. N is a good approximation to G if the residuals are small. Hence, a Gauss—Newton iteration consists of finding

$$\theta_{i+1} = \theta_i - N^{-1} q_i$$

where $\theta_i$ is the vector of estimates of the parameters and $q_i$ the gradient vector, as before. N is always nonsingular in the case of nonlinear least squares when a single equation defines $y$. McCammon (1969) gives a Fortran program listing of a Gauss—Newton algorithm. Extensions to this program to deal with weighted nonlinear regression problems are described by McCammon (1973). Bard (1974, pp. 96—106) and L. C. W. Dixon (1972, pp. 41—44) provide further details. Other algorithms involving iteration are described by Fletcher and Powell (1963), Fletcher (1965), Davidon (1968) and Peckham (1970).

The second group of methods are termed direct search techniques. They involve the search for a minimum without the need to compute either first or second order derivatives. The function (e.g. the residual sum of squares) is evaluated at a number of points within the required region (for example, at the intersections of a grid) and the minimum value is selected. The process can then be repeated in the region of the minimum with a finer grid mesh. Because direct search methods do not require derivatives, the effort required of the user is reduced, but they are generally less efficient in terms of computing requirements than the iterative techniques described above. One of the most successful of the direct search algorithms is the Simplex method of Nelder and Meade (1965), which is described by L. C. W. Dixon (1972, pp. 82—85) and programmed in Fortran by O'Neill (1971). In most nonlinear least squares problems, the iterative methods are generally preferable, however, particularly the Gauss—Newton and Marquardt algorithms.

An interesting method which does not require derivatives is that of conjugate gradients (Fletcher, 1972). It is based on the fact that if the surface representing variation in the quantity to be minimized (for example, the residual sum of squares in regression) is quadratic then, if the minima along two parallel subspaces are located, the overall minimum lies between these two local minima. If the true surface is not quadratic, the global minimum will be approached in a series of iterations.

Confidence limits can be calculated for the parameters of a nonlinear model, provided that the necessary assumptions are satisfied (see Chapter 2). Letting J denote the $(n \times k)$ Jacobean matrix of first partial derivatives of the function with respect to the $k$ parameters evaluated at the $n$ data points, then the Hessian matrix G of second partial derivatives can be approximated by $G = 2J'J$, provided that the residuals $f_i$ are small at the minimum. (This is the basis of the Gauss—Newton algorithm, discussed above.) The variance $s_{\hat{\theta}}^2$ of $\hat{\theta}_i$ is given by

$$s_{\hat{\theta}_i}^2 = \text{var}(\hat{\theta}_i) = (SQ/(m-n))[\mathbf{G}^{-1}]_{ii} \tag{3.28}$$

where SQ is the residual sum of squares at the minimum. Confidence intervals around the true parameters are then calculated as

$$\hat{\theta}_i - s_{\hat{\theta}_i} t_{(\beta/2, m-k)} \leq \theta_i \leq \hat{\theta}_i + s_{\hat{\theta}_i} t_{(\beta/2, m-k)} \tag{3.29}$$

The confidence limits indicated by (3.29) are the $100(1-\beta)$ per cent limits. Also, $s_{\hat{\theta}_i}$ is the standard error of $\hat{\theta}_i$ and $t_{(\beta/2, m-k)}$ is the $100(\beta/2)$ percentage point of Student's $t$ distribution with $m-k$ degrees of freedom. The $F$-test of the null hypothesis that the $\theta_i$ are all zero, or alternatively that the explained sum of squares is zero in the population is given, as in Chapter 2, by:

$$F = \frac{(TS - SQ)/(n-k-1)}{TS/(n-1)}$$

which, is distributed as the $F$ ratio with $(n-k-1)$ and $(n-1)$ degrees of freedom.

Wolberg (1967) and Bard (1974) give further details of the computation and use of confidence intervals and significance tests in nonlinear least squares problems.

The choice of a suitable starting point is essential in all the iterative algorithms. A starting point consists of 'guesstimates' of the location of the minimum of the function, and normally it is supplied by the user as the initial values of the parameters to be estimated. If the starting point is far from the position of the minimum then (i) the process may converge to a local rather than a global minimum and (ii) computing time will be increased by a substantial amount. If one has prior knowledge of the likely values of the parameters this can be used to determine starting values. Otherwise, the algorithm should be run using several different starting values. If all lead to the same minimum value of the function then one can be reasonably certain that it is the overall minimum. If differing minima are found, take several new starting points close to the smallest minimum and try again. Sometimes estimates of the location of the minimum can be found by expressing the model in approximate form as a linear equation then using ordinary (linear) least squares methods to find estimates of the parameters of the approximate model. These estimates can be used as the starting point in a nonlinear least squares algorithm. Thus, the model $y = aX^b$ can be approximated by $\log y = \log a + b \log X$, and estimates of $a$ and $b$ obtained. These estimates could subsequently be used as a starting point in the determination of estimates of the true values of $a$ and $b$.

## 3.5.2 Example

To illustrate one use of nonlinear minimization methods a geomorphological example is taken from Bagnold (1954) and Horikawa and Shen (1960). Given the empirical results of earlier workers, it could be postulated that $q$, the volume of sand passing through a vertical plane of unit width, is proportional to the shear velocity of the wind, written $U_*$. This shear velocity is usually defined by an empirical formula such as

$$U_* = 0.053 U_{100}$$

or

$$U_* = 0.527 U_{446.5} - 17.1$$

where $U_{100}$ and $U_{446.5}$ are the wind velocities at heights of 100 and 446.5 cm. above the sand surface (Horikawa and Shen, 1960, p. 6). Bagnold (1954) proposed that the relationship between $q$ and $U_*$ is of the form

$$q = aU_*^b \tag{3.30}$$

where $a$ and $b$ are parameters to be estimated, and $q$ and $U_*$ are measured in the same units (e.g. $q$ in grams per centimetre per second and $U_*$ in centimetres per second). Other workers have suggested similar formulae, for example:

$$q = a(U_* - 10.8)^b \tag{3.31}$$

(see Horikawa and Shen, 1960, p. 32). These equations can be thought of as theoretical models. An experimenter is concerned with fitting such a model to his data and with determining whether the model is or is not acceptable. In this case, suitable data is provided by Horikawa and Shen (1960, pp. A2 and A3) and a part of this data has been selected for use in this example (Table 3.20).

The model to be fitted is

$$\mathbf{q} = a\mathbf{U}_*^b + \epsilon$$

where $\epsilon$ represents a disturbance term which, for the sake of the argument, is assumed to be normally distributed, independent and with homoscedastic variances. The residual (i.e. the sample estimator of the disturbance) for observation $i$ is

$$e_i = \hat{a} U_*^{\hat{b}} - q_i$$

The function to be minimized is $\Sigma_{i=1}^{n} e_i^2$. If Marquardt's method is to be used, the first derivatives of e with respect to $a$ and $b$ are needed. These derivatives can be supplied by the user in the form of a computer routine, or else a difference algorithm can be used, as discussed earlier.

Table 3.20
Data for nonlinear least squares example (Data from K. Horikawa and H. W. Shen, *Tech. Memo No. 119*, U.S. Army, Corps of Engineers, Beach Eroson Board, 1960)

|    | 1      | 2        |
|----|--------|----------|
| 1  | 0.0211 | 37.5000  |
| 2  | 0.0376 | 32.0000  |
| 3  | 0.0398 | 31.1000  |
| 4  | 0.0594 | 37.2000  |
| 5  | 0.0706 | 38.1000  |
| 6  | 0.0836 | 38.1000  |
| 7  | 0.0909 | 38.7000  |
| 8  | 0.0966 | 44.8000  |
| 9  | 0.1200 | 44.8000  |
| 10 | 0.1260 | 44.8000  |
| 11 | 0.1370 | 44.8000  |
| 12 | 0.1390 | 45.7000  |
| 13 | 0.1730 | 57.0000  |
| 14 | 0.1990 | 56.1000  |
| 15 | 0.2080 | 55.8000  |
| 16 | 0.2130 | 54.3000  |
| 17 | 0.2250 | 54.3000  |
| 18 | 0.2320 | 65.8000  |
| 19 | 0.2370 | 57.0000  |
| 20 | 0.2690 | 65.2000  |
| 21 | 0.2880 | 65.8000  |
| 22 | 0.2900 | 64.0000  |
| 23 | 0.3060 | 65.2000  |
| 24 | 0.3150 | 76.2000  |
| 25 | 0.3350 | 66.1000  |
| 26 | 0.3540 | 75.0000  |
| 27 | 0.3800 | 76.2000  |
| 28 | 0.4220 | 74.7000  |
| 29 | 0.4440 | 106.1000 |
| 30 | 0.4530 | 75.3000  |
| 31 | 0.4570 | 89.6000  |
| 32 | 0.4890 | 83.8000  |
| 33 | 0.5070 | 88.4000  |
| 34 | 0.5440 | 69.2000  |
| 35 | 0.5990 | 86.9000  |
| 36 | 0.6000 | 100.3000 |
| 37 | 0.6200 | 87.5000  |
| 38 | 0.6460 | 98.4000  |
| 39 | 0.7130 | 102.7000 |
| 40 | 0.7840 | 100.3000 |
| 41 | 0.8100 | 114.9000 |
| 42 | 0.8380 | 100.3000 |
| 43 | 0.8690 | 110.6000 |
| 44 | 0.9240 | 114.3000 |
| 45 | 1.0300 | 125.6000 |
| 46 | 1.0700 | 111.6000 |
| 47 | 1.1100 | 122.2000 |
| 48 | 1.1200 | 126.2000 |
| 49 | 1.1800 | 126.2000 |
| 50 | 1.1800 | 126.2000 |
| 51 | 1.2200 | 126.2000 |
| 52 | 1.2500 | 126.2000 |
| 53 | 1.2800 | 137.2000 |
| 54 | 1.3200 | 138.4000 |
| 55 | 1.4100 | 125.6000 |
| 56 | 1.5100 | 108.8000 |
| 57 | 1.5200 | 137.2000 |
| 58 | 1.5400 | 137.8000 |

Table 3.20 *continued*

|    | 1      | 2        |
|----|--------|----------|
| 59 | 1.6300 | 136.6000 |
| 60 | 1.7600 | 135.6000 |
| 61 | 1.7900 | 152.1000 |
| 62 | 1.9200 | 123.4000 |
| 63 | 1.9200 | 144.5000 |
| 64 | 1.9900 | 150.9000 |
| 65 | 2.2400 | 151.2000 |
| 66 | 2.2800 | 147.2000 |

Two algorithms from the Numerical Algorithms Group (NAG) library were used. The first of these, subroutine EO4FBF, is based on the method of Powell (1968), which is a compromise between the steepest descent technique and Marquardt's algorithm. Derivatives were estimated numerically by the method of finite differences. The second subroutine, EO4GAF, is a Marquardt algorithm as programmed by Fletcher (1971). Derivatives were supplied as a Fortran subroutine (see below).

Subroutine EO4FBF was initially started with estimates $\hat{a} = 0.1$ and $\hat{b} = 1.5$. At this point the residual sum of squares is $0.7113 \times 10^6$. 98 iterations were required to locate a minimum, which occurred at the point $\hat{a} = 0.000011946$ and $\hat{b} = 2.396851$, where the residual sum of squares was 1.9357. Computing time was 6 seconds on an 1CL 1906A. A second run of EO4FBF was made, with initial estimates $\hat{a} = 0.125 \times 10^{-5}$ and $\hat{b} = 3$. Convergence to a minimum was attained in 71 iterations, and estimates $\hat{a} = 0.000011944$ and $\hat{b} = 2.396889$ were given. The residual sum of squares was 1.9357 — the same as before. A graph of $U_* q = 0.000011944\ U_*^{2.396889}$ is shown in Figure 3.20.

The second subroutine, EO4GAF, required explicit expressions for the first partial derivations of $q$ with respect to $\hat{a}$ and to $\hat{b}$ respectively to be supplied. These expressions are:

$$dq/d\hat{a} = U_*^{\hat{b}}$$

and

$$dq/d\hat{b} = \hat{a} U_*^{\hat{b}} \cdot \log_e U_*$$

The results achieved are shown in Table 3.21.

Table 3.21
Nonlinear least squares example

|   | Starting point | | Terminal point | |
|---|---|---|---|---|
|   | $\hat{a}$ | $\hat{b}$ | $\hat{a}$ | $\hat{b}$ |
| 1 | 0.1194441E−4 | 2.396889 | 0.1194599E−4 | 2.396889 |
| 2 | 0.1E−6 | 3.5 | 0.1194605E−4 | 2.396857 |
| 3 | 0.1E−4 | 2.0 | 0.1194607E−4 | 2.396857 |

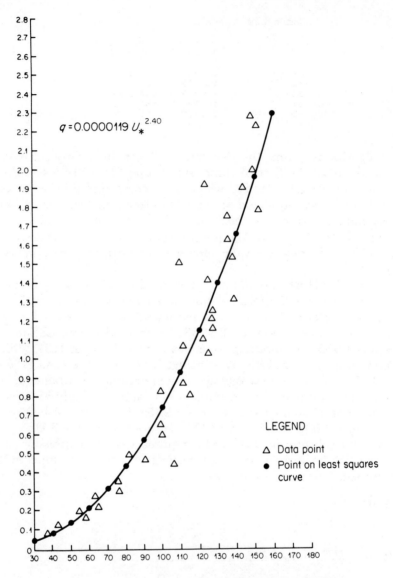

Figure 3.20  Plot of $q = 0.000011944 U_*^{2.396889}$

All three runs identified the minimum residual sum of squares to be 1.935715. Total computing time for all three runs was 6 seconds on an 1CL 1906A. The Hessian matrix $\mathbf{G}$, required in the computation of confidence limits was approximated by $\mathbf{G} = 2\mathbf{J}'\mathbf{J}$ where $\mathbf{J}$ is the Jacobean matrix. For the first parameter, $a$, confidence limits were obtained from Equation (3.28) by firstly determining $\mathbf{G}^{-1}$ to be

$$\mathbf{G}^{-1} = \begin{bmatrix} 0.1192\text{E-}11 & \\ -0.3164\text{E-}13 & 0.3528\text{E-}3 \end{bmatrix}$$

and subsequently finding

$$s_{\hat{a}}^2 = 1.935715/(66\text{-}2) \times 0.1192\text{E-}11 = 0.3605\text{E-}13$$

and

$$s_{\hat{b}}^2 = 1.935715/(66\text{-}2) \times 0.3528\text{E-}3 = 0.1067\text{E-}4$$

The standard errors, $s_{\hat{a}}$ and $s_{\hat{b}}$, are easily found to be 0.1899E-7 and 0.3266E-2 respectively, and the confidence limits are derived using equation (3.29) with $\beta = 0.05$, giving $t_{0.025,64}$ equal to 2.00. The upper and lower 95 per cent confidence intervals are:

$$0.1198405\text{E-}4 \leqslant a \leqslant 0.1190809\text{E-}4$$
$$2.390324 \leqslant b \leqslant 2.40399$$

Lastly, the explained sums of squares can be expressed as a percentage of the total (corrected) sum of squares to give a numerical expression of the closeness of fit of the model. In the example, RSS at the minimum was 1.935715; since TSS is 25.3664 the explained sum of squares (ESS) = 23.4307. The ratio ESS/TSS x 100 is 92.36. Thus, 92.36 per cent of the sums of squares is explained by the model.

The Marquardt algorithm EO4GAF was used to estimate the parameters $a$ and $b$ in a second model, given by Equation (3.31). Three starting points were used: (i) $\hat{a} = 0$, $\hat{b} = 3$; (ii) $\hat{a} = 0.0001$, $\hat{b} = 2$; (iii) $\hat{a} = 0$, $\hat{b} = 2.5$. In each case the algorithm converged to the point $\hat{a} = 0.00000288$, $\hat{b} = 2.645$ with $s_{\hat{a}} = 0.456\text{E-}7$ and $s_{\hat{b}} = 0.370\text{E-}2$. The minimum sum of squares at this point is 1.92291 — a small amount less than the minimum sum of squares using the Bagnold formula. In fact, the least-squares fit of the Bagnold formula gave an explained sum of squares of 92.369 per cent while the alternative formula produced an explained sum of squares of 92.419 per cent — hardly a significant difference.

*Part Three*

# Analysis of Interdependence

Most systems in physical geography are described by a large number of variables. Most of these variables are interrelated, thus information is duplicated and the essential structure of the system is obscured. Principal components analysis provides a means of eliminating redundancies from a set of interrelated variables. The resulting principal components are uncorrelated. Since intercorrelations among the variables are an embarrassment in multiple regression analysis and in many classification techniques, principal components analysis can logically precede such techniques. Factor analysis, on the other hand, is a method of investigating the correlation structure of a multivariate system. In essence, it is an attempt to find groups of variables (factors) that are measuring a single important aspect of the system. Such factors are not necessarily uncorrelated, and so a method of transforming the factors (called rotation) is adopted. This involves a prior hypothesis that the system has a 'simple structure', and the factors are rotated to fit as closely as possible this hypothetical structure. The nature of the factors, as well as the relationships between the factors, can allow the research worker to summarize the basic structures or patterns within the system of study.

Chapter 4

# Principal Components Analysis

## 4.0 Introduction

Most data sets in physical geography consist of measurements on variables which are correlated with each other. For some purposes it is useful to transform this data set into one in which the variables are mutually uncorrelated. The investigator may wish to know the number of linearly independent sources of variation in the system that the chosen variables represent, or he may require linear independence of the variables to overcome problems such as that of multicollinearity in regression analysis. The first requirement could be defined as *economy of description*. The relationships between a number of interconnected variables can generally be summarized in terms of a fewer number of principal components. In the second case, in which principal components analysis is being used as a prelude to multiple linear regression for example, the technique allows the *orthogonalization* of a set of measurements. This is analogous to the use in many national maps of a rectangular coordinate system, such as the Ordnance Survey grid, rather than an oblique coordinate system which — although not altering the position of the points shown on the map — is less efficient and more difficult to use. Rao (1964) summarizes these two points: 'When a large number of measurements are available, it is natural to enquire whether they could be replaced by a fewer number of the measurements or of their functions, *without loss of much information*, for convenience in the analysis and in the interpretation of data' (p. 329).

Principal components analysis thus provides a method of constructing new variables which are pairwise uncorrelated. Each principal component is a linear combination of the observed variables, and these linear functions are chosen to be orthogonal. Thus, the principal components are composites or the original variables. An infinite number of orthogonal bases of a vector space exist, so in order to provide a unique set of coefficients the first principal component is defined as that linear combination of variables which has the maximum variance of all linear functions derivable from the given variables. The second principal component is the linear combination of variables having the maximum variance of all linear functions of the given variables that are orthogonal to the first principal component, and so on. The reasons for this choice are given below. The coefficients of the principal components are termed *principal component loadings* while the measurements of the principal components upon each of the individuals are called *principal component scores*. The loadings matrix and the scores matrix are denoted by **L** and **F** respectively, **L** being a $(p \times p)$ matrix and **L** an $(n \times p)$ matrix, where $n$ is the number of observations and $p$ the number of variables. The set of $n$ measurements on the $p$ variables are contained in the data matrix **X**. **R** holds the correlations among the $x_i$ and **S** the variances and covariances of the $x_i$ (Table 4.1).

Table 4.1
Notation For Chapter 4

| | |
|---|---|
| $n$ | number of individuals measured on each variable |
| $p$ | number of variables |
| **X** | $(n \times p)$ matrix of observed data |
| **R** | $(p \times p)$ matrix of correlations among the variables $(x_i)$ |
| **S** | $(p \times p)$ matrix of variances and covariances of the $x_i$ |
| **A** | $(p \times p)$ matrix of column normalised eigenvectors of either **R** or **S** |
| **L** | $(p \times p)$ matrix of principal component loadings |
| **Z** | $(n \times p)$ matrix of standardized data |
| **D** | $(p \times p)$ diagonal matrix of sample eigenvalues of **R** or **S** |
| $\Delta$ | $(p \times p)$ diagonal matrix of (population) eigenvalues (of $\Sigma$ or **P**) |
| **F** | $(n \times p)$ matrix of principal component scores |
| **F**\* | $(n \times p)$ matrix of standardized principal component scores |
| $\Sigma$ | $(p \times p)$ population variance-covariance matrix |
| **P** | $(p \times p)$ population correlation matrix |

The transformation of $x_i$ into $f_i$ can be explained in geometrical terms. If each column of **X**, that is, each $x_i$, is treated as a geometrical vector in a space of $p$ dimensions then the rows of **X** are the coordinates of the $n$ individuals in terms of these $p$ vectors representing the $x_i$. Without altering the relative positions of these $n$ points (representing the rows of **X**) the vectors given by the $x_i$ can be replaced by a new set of reference vectors which are orthogonal (at right angles to each other). These orthogonal axes are the $f_i$. A two-dimensional example is shown in Figure 4.1. The orthogonal axes are AB and CD; these replace the two oblique axes — representing two observed variables — labelled WX and YZ. Point P1, for example, has coordinates $a1, a2$ with reference to the orthogonal axes AB,CD and coordinates $y1, y2$ on the oblique axes WX,YZ. The projections of the points onto each axis define the elements of an algebraic vector (Section 1.3). Thus, the projections of the points onto axis AB are the elements of a vector, say $f_i$, which is the $i^{th}$ column of **F**. These vector elements or projections are the scores or measurements of the objects or places represented by the points upon the observed variables (oblique axes) or composite variables (orthogonal axes). If all the vectors are scaled to unit length, the cosine of the angle between each pair of vectors is numerically equal to the product-moment correlation between the variables (or composite variables) concerned (Section 1.3). The angle is measured between the pair of positive (or negative) ends of the axes; for example, in Figure 4.1 $\cos(\theta)$ is numerically equal to the product-moment correlation between the variables represented by the axes YZ and WX. Assuming for the moment that the orthogonal axes AB and CD represent principal components, then the projections of the points P1 and P2 upon each axis are equivalent to the 'scores' of these points on the principal components ($f_{ij}$) while the angles between the orthogonal axes and the oblique axes (e.g. $\alpha$ in Figure 4.1) are analogous to principal component loadings. In Figure 4.1, $\cos(\alpha)$ is the loading of the variable represented by the oblique axis WX on the principal component represented by the orthogonal axis CD.

It is sometimes found in practice that an $m$-dimensional space can contain more than $m$ oblique axes — thus, more than two oblique axes

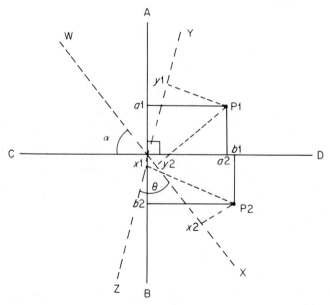

Figure 4.1  Point P1 has coordinates $(a2, a1)$ on axes AB, CD and coordinates $(y2, y1)$ on axes WX, YZ. Point P2 has coordinates $(b1, b2)$ on axes AB, CD and coordinates $(x2, x1)$ on axes WX, YZ

can be drawn in a two-dimensional space. If $p$ oblique axes are used as the reference axes of an $m$-dimensional space, where $p$ is greater than $m$, then those vectors can be replaced by $m$ orthogonal axes. A difficulty arises when some of these $m$ dimensions contain little of the variability in the positions of the points. In Figure 4.2, most of the locational variability of the points is contained in the dimensions represented by axes AB and CD while very little scatter occurs in the direction EF. Straight line distances between the points will be affected only slightly if we disregarded differences in the EF dimension. If we consider the meaning of 'locational variability' in subject-matter terms, we find that it is analogous with the terms 'information' and 'variance'. A set of measurements of some variable exhibiting little, if any, change does not allow much distinction between the objects on which the measurements were made. If the measurements of a variable exhibit considerable variation then we can conclude that the variable concerned represents a significant contributor to (or is a major source of) variability between the individuals. In Figure 4.2, axes AB and CD indicate major 'dimensions of variability' in the locations of the three points shown, but EF does not. A major problem in PCA is the distinction between important and unimportant dimensions of variability.

So far we have avoided any specific justification for the choice of position of the rectangular axes. If we are simply concerned with orthogonality then no position of the axes can logically be preferred to

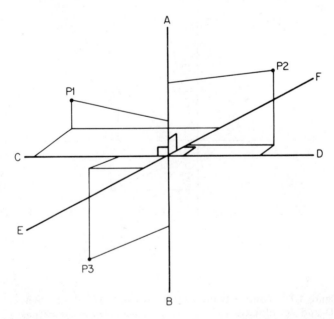

Figure 4.2   Dimensions of variability. See text for explanation

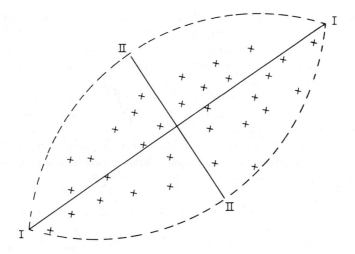

Figure 4.3  Principal axes

any other, as noticed in Chapter 3 in the discussion of orthogonal polynomials. This means, however, that we have no criterion to use in an algorithm designed to find such axes. The discussion in the previous paragraph led towards the conclusion that the axes identified 'dimensions of variability'. It would seem appropriate, then, to choose the axes in such a way that the variation along each axis was the maximum possible, given that each axis was perpendicular to each of the others. Axis I would thus represent the most important dimension of variability, axis II the next most important, and so on. This idea can be represented graphically in Figure 4.3. The points represent, as usual, individuals on which measurements are made. The observed variables (indicated, for example, by oblique axes on Figure 4.1) are not shown; they have been replaced by two orthogonal axes. The first of these, axis I, is drawn in the direction of maximum scatter. This is the first principal axis of the geometrical body enclosing the scatter of points. Axis II is the second principal axis. If the selection of points is a sample from some larger population, the enclosing line can be given a probabilistic interpretation — it defines the region of the space in which $p$ per cent of the population is located. When the whole population is being analysed, no such probability statement can be made. Thus, the use of the 'maximum variance' criterion allows the selection of a unique set of orthogonal axes, which are the principal axes of the geometrical body which encloses the scatter of points.

The shape of this geometrical body depends on the weight given to each of the selected variables. For example, we may consider a situation in which measurements on a variable $x_1$ range from $-10^6$ to $10^6$, while the measurements on a second variable $x_2$ range from 0.1 to $-0.1$. The distribution of the points will approximate to a line, which will be the axis representing variable $x_1$. Variability in $x_2$ will have no significant effect on the shape of the distribution. If, however, the variables are standardized so that each has the same mean and standard deviation (usually 0 and 1 respectively) then variable $x_2$ will increase considerably in importance relative to variable $x_1$. The principal axes of the geometrical bodies enclosing the scatter of points measured on standardized and unstandardized data will thus be quite different; in fact, there will be no simple relationship between them, for the magnification or reduction along each oblique axis will not be constant. This aspect of PCA has received little attention from geographers; most users of the technique implicitly standardize the variables by basing the analysis on the correlation matrix, **R**. Yet an analysis of the unstandardized data, based on the variance—covariance (or dispersion) matrix **S** would, in most cases, produce different results. PCA is therefore dependent on the scales of measurement of the variables. These are often noncomparable, and sometimes arbitrary, so one is usually forced to rely on standardization. The effects of this scaling of the data are reflected in the results.

A geometrical model of PCA is useful in defining the concepts on which the technique is based. Due to the large number of observations, and the fact that one is rarely concerned with the two- or three-dimensional case, resort has to be made to algebraic methods of analysis. Assuming that the data is standardized, the principal axes of the data matrix **X** are found to be the scaled eigenvectors of **R**. The square roots of the corresponding eigenvalues are the scaling factors. The eigenvalues themselves measure the scatter, or variance, along each principal axis. When standardized data are used, the total variance of the system is equal to $p$ units, since **X** has $p$ columns, each containing $n$ measurements standardized so as to have unit variance. The trace of **R** (sum of its diagonal elements) equals $p$, as does the sum of the eigenvalues of **R**. Therefore, each eigenvalue gives the proportion of variance within the associated dimension. If we were working with the variance-covariance matrix **S**, trace (**S**) would give the total variance of the system in the original units. Unless the scales of measurement of the variables were comparable, the 'total variance' would not mean very much, as we would not be adding like to like. The algebraic interpretation of PCA is explained in more detail in the next section. A more detailed account of the geometrical model of PCA is given by Morrison (1967) and Davis (1973).

## 4.1 Mathematical Foundations

**X** is a matrix of dimension $(n \times p)$ holding the $n$ measurements on $p$ correlated variables. The $\mathbf{x}_i$ (columns of **X**) are to be transformed into a new set of variables $\mathbf{f}_i$ which have the properties of mutual orthogonality and maximum variance, as discussed above in Section 4.0. The transformation is achieved by defining $f_i$ as a linear combination of the $x_i$. For example, $f_1$ is given by

$$f_{1i} = a_{11}x_{1i} + a_{12}x_{2i} + \ldots + a_{p1}x_{pi} \quad (i = 1,2,\ldots,n) \quad (4.1a)$$

or, in matrix form,

$$\mathbf{f}_1 = \mathbf{X}\mathbf{a}_1 \quad (4.1b)$$

where $\mathbf{f}_1$ is a vector of $n$ elements, being the first column of a $(n \times p)$ matrix $\mathbf{F}$ of principal component scores, and $\mathbf{a}_1$ is a vector of p elements, the first column of a matrix $\mathbf{A}$ of principal component loadings (Figure 4.4). *eigenvectors*

$$\underset{(n \times 1)}{\mathbf{f}} = \underset{(n \times p)}{\mathbf{X}} \cdot \underset{(p \times 1)}{\mathbf{a}}$$

$$\begin{bmatrix} f_1 \\ f_2 \\ f_3 \\ f_4 \end{bmatrix} = \begin{bmatrix} x_{11} & x_{12} & x_{13} \\ x_{21} & x_{22} & x_{23} \\ x_{31} & x_{32} & x_{33} \\ x_{41} & x_{42} & x_{43} \end{bmatrix} \cdot \begin{bmatrix} a_1 \\ a_2 \\ a_3 \end{bmatrix}$$

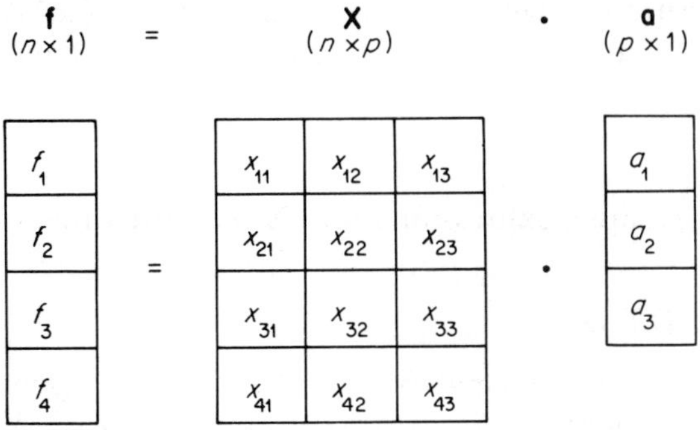

Figure 4.4  Illustrating the derivation of principal components

The sum of squares of the elements of $\mathbf{f}_1$ is

$$\mathbf{f}_1'\mathbf{f}_1 = \mathbf{a}_1'\mathbf{X}'\mathbf{X}\mathbf{a}_1 \quad (4.2)$$

and, in order to achieve the property of maximum variance, $\mathbf{a}_1$ is chosen so as to maximize (4.2). Some constraint must be imposed on the choice of $\mathbf{a}_1$ otherwise the vector $\mathbf{a}_1 = \{\infty, \infty, \infty, \ldots, \infty\}$ could be selected, making (4.2) infinitely large. The constraint is:

$$\mathbf{a}_1'\mathbf{a}_1 = 1 \quad (4.3)$$

Johnston (1972), Morrison (1967) and other authors show that (4.2) is maximized subject to the restriction (4.3) if $\mathbf{a}_1$ is the unit-length eigenvector corresponding to the largest eigenvalue of $\mathbf{X}'\mathbf{X}$. The vectors $\mathbf{a}_2, \mathbf{a}_3, \ldots, \mathbf{a}_p$ can be defined in a similar fashion as the unit-length eigenvectors corresponding to the second, third, ..., pth eigenvalues, in descending order of magnitude, of the matrix $\mathbf{X}'\mathbf{X}$. Since $\mathbf{X}'\mathbf{X}$ will normally be positive-definite, all the eigenvalues will exceed zero. If two or more of the eigenvalues are equal a difficulty arises, for although the elements of the eigenvectors corresponding to equal eigenvalues can be chosen to be orthogonal, there are an infinite number of mutually orthogonal eigenvectors.

The $p$ principal components of $\mathbf{X}$ are therefore the columns of an $(n \times p)$ matrix $\mathbf{F}$:

$$\mathbf{F} = \mathbf{XA} \qquad (4.4)$$

Letting $\mathbf{D}$ be the diagonal matrix of eigenvalues arranged in descending order of magnitude $(d_1 > d_2 > \ldots > d_p)$ and $\mathbf{A}$ the matrix of corresponding unit-length eigenvectors, then the variances of the $\mathbf{f}_i$ can be found from

$$\mathbf{F'F} = \mathbf{A'X'XA} \qquad (4.5)$$

The basic structure of eigenstructure of any square symmetric matrix $\mathbf{P}$ was defined in Chapter 1 in terms of its eigenvalues $\mathbf{D}$ and eigenvectors $\mathbf{A}$ as

$$\mathbf{P} = \mathbf{ADA'} \qquad (4.6)$$

so, substituting this expression for $\mathbf{X'X}$, equation (4.5) can be rewritten:

$$\mathbf{F'F} = \mathbf{A'(ADA')A} \qquad (4.7)$$

However, $\mathbf{A}$ — being the matrix of unit length eigenvectors of a positive-definite matrix — is orthonormal, hence $\mathbf{AA'} = \mathbf{A'A} = \mathbf{I}$. Equation (4.7) therefore simplifies to

$$\mathbf{F'F} = \mathbf{I D I} = \mathbf{D} \qquad (4.8)$$

which shows that the variances of the $\mathbf{f}_i$ are numerically equal to the elements $d_i$ of the diagonal matrix $\mathbf{D}$ of eigenvalues of $\mathbf{X'X}$. The sum of the variances is thus trace $(\mathbf{D})$ or trace $(\mathbf{X'X})$. This fact can be used as a computational check on the accuracy of the calculated eigenvalues. The proportion of the total variance due to each principal component is thus $(d_i/\text{trace}(\mathbf{D}) \times 100)$ per cent.

The elements of $\mathbf{A}$ are sometimes difficult to interpret. They can be converted into correlations between the principal components ($\mathbf{f}_i$) and the original variables ($\mathbf{x}_i$) in the following manner. The cross-products of $\mathbf{X}$ and the first principal component, $\mathbf{f}_1$, are the elements of the vector $\mathbf{X'f}_1$. But

$$\mathbf{X'f}_1 = \mathbf{X'Xa}_1$$
$$= d_1 \mathbf{a}_1$$

since $\mathbf{a}_1$ is defined by

$$(\mathbf{X'X} - d_1 \mathbf{I})\mathbf{a}_1 = 0$$

that is,

$$X'Xa_1 = d_1 Ia_1 = d_1 a_1$$

The correlation between variable $x_i$ and principal component $l$ is now found from

$$l_{i1} = \frac{d_1 a_{i1}}{\sqrt{d_1}\sqrt{\sum_{j=1}^{p} x_{ij}^2}}$$

$$= \frac{\sqrt{d_1} a_{i1}}{\sqrt{\sum_{j=1}^{p} x_{ij}^2}}$$

In general,

$$l_{ij} = \frac{\sqrt{d_j} a_{ij}}{\sqrt{\sum_{k=1}^{p} x_{ik}^2}} \quad (i,j = 1, 2, \ldots, p) \tag{4.9}$$

The elements $l_{ij}$ form the matrix $L$ which is usually termed the principal components loadings matrix. The proportion of the variance of $x_i$ contributed by principal component $j$ is the square of $l_{ij}$. The squares of $l_{ij}$ sum to unity for given $i$, when all $p$ principal components are extracted. It is more common to select $m$ ($m < p$) of the principal components for further analysis. If this is done, then $\Sigma_{j=1}^{m} l_{ij}$ will give the proportion of the variance of variable $x_i$ which is extracted by the $m$ principal components. Some authors refer to this value as the 'communality' of variable $x_i$; this can lead to confusion since the same term has a technical meaning in the context of factor analysis (Chapter 5), and hence its use in principal components analysis should be avoided.

The method of PCA is usually applied not to observed variables $X$ but to their standardized equivalents, $Z$. The matrix $(X'X)$ then becomes the correlation matrix, $R$. This does not affect the above argument, except that the denominator in Equation (4.9) is omitted, as it is by definition unity. The principal component loadings matrix computed from $R$ is thus

$$L = AD^{1/2} \tag{4.10}$$

$A$ is the matrix of unit-length eigenvectors of $R$ and $D$ is the diagonal matrix of eigenvalues.

A second convention is to assume that the columns of $F$ are standardized to have unit variance, thus $F$ is replaced by $F^*$ where

$$F^* = ZLD^{-1} \tag{4.11}$$

Substituting (4.10) into (4.11) gives

$$F^* = ZAD^{-\frac{1}{2}} \tag{4.12}$$

that is, $F^*$ is computed by postmultiplying the matrix $Z$ of standard scores by the matrix $A$ of column-normalized eigenvectors and postmultiplying this product by the diagonal matrix containing the reciprocals of the square roots of the eigenvalues. Since $d_i$ is equivalent to the variance of the $i^{th}$ principal component, then $d_i^{\frac{1}{2}}$ is the corresponding standard deviation, so the above procedure is equivalent to dividing $ZA$ by the standard deviations of the principal components. If the principal components are not to be standardized, use (4.13) instead of (4.12):

$$F = ZL. \tag{4.13}$$

The question of scaling of the columns of the principal components scores matrix is not of academic importance only. Since principal components analysis frequently precedes another form of analysis, such as cluster analysis (Chapter 6) or regression analysis (Chapter 2) the variance given to each $f_i$ is of great significance. In the classification procedures of Chapter 6, the weight (variance) of each character (variable) has a direct bearing on the resulting classification (See the remarks in Section 6.1.B).   see also p 313

So far in this chapter, no reference to statistical inference has been made. In fact, the presentation above is based on the premise that we are dealing with an entire population or that we are solely concerned with description. Lawley and Maxwell (1971, p. 17) and Morrison (1967, p. 247) discuss the case in which the elements of $X$ are a random sample of $n$ measurements on $p$ variables. It is assumed that the $x_i$ follow a multivariate normal distribution, and that $n > p$. Given these assumptions, it is possible to derive the large-sample distributional properties of the sample eigenvalues and principal component loadings. Knowledge of the covariance and variances of the estimates will give an indication of their reliability or stability and furthermore will allow tests of hypotheses and the construction of confidence intervals.

Using $D$ to denote the diagonal matrix of sample eigenvalues of $S$, one may wish to find the $(100 - \alpha)$ per cent confidence level for the corresponding population eigenvalues, $\Delta\{=\delta_i\}$ given the assumptions that $n$ is large and that the $x_i$ are multivariate-normal with covariance matrix $\Sigma$ whose sample estimator is $S$. The quantity

$$\sqrt{\frac{n}{2}} \, (d_i - \delta_i)/\delta_i$$

is then a standard normal deviate and the confidence interval around $\delta_i$

is found from:

$$\frac{d_i}{1 + Z_{\alpha/2}\sqrt{\frac{2}{n}}} \leq \delta_i \leq \frac{d_i}{1 - Z_{\alpha/2}\sqrt{\frac{n}{2}}} \quad (4.14)$$

Here, $Z_{\alpha/2}$ is the upper $\alpha/2$ percentage point of the standard normal distribution. If $(100 - \alpha) = 95$ then $\alpha = 5$ and $Z_{\alpha/2} = 1.96$. For 99 per cent confidence limits $Z_{\alpha/2} = 2.58$. Note that (4.14) applies only to eigenvalues of **S**, the covariance matrix, and not to eigenvalues of **R**.

A second result is also applicable only to the eigenvalues of **S**. This test concerns the equality of the smallest $(p - k)$ eigenvalues of **S**. If these eigenvalues are equal then there is no point in considering more than $k$ principal components, because they represent principal axes of equal length, which define a multi-dimensional sphere and hence have no unique positions. If all $p$ eigenvalues are equal it follows that the $p$ variables have zero covariances. It is necessary to assume that $n$ is large and the $x_i$ are multivariate-normal. The formula for use in the case $k = 0$ (which implies zero covariances in the population) is:

$$(n - \tfrac{1}{6}[2p + 1 + 2/p])(-\log_e[\det(\mathbf{S})] + p \log_e[\operatorname{tr}(\mathbf{S})/p]) \quad (4.15)$$

This quantity has an asymptotic chi-square distribution with $\tfrac{1}{2}(p + 2)(p - 1)$ degrees of freedom. If the quantity computed from (4.15) exceeds $\chi^2$ with the appropriate degrees of freedom and at a preselected level of significance the null hypothesis that the $p$ eigenvalues of $\mathbf{\Sigma}$ are equal can be rejected. In computing (4.15) remember that $\Sigma d_i = \operatorname{tr}(\mathbf{S})$ and $\Pi d_i = \det(\mathbf{S})$.

Where $(p - k) > 0$ the statistic

$$n'[-\log_e\{\det(\mathbf{S})\} + \log_e(d_1 d_2 \ldots d_k) + q \log_e d] \quad (4.16)$$

is computed. In this equation,

$$q = p - k$$
$$d = (\operatorname{tr}(\mathbf{S}) - d_1 - d_2 - \ldots - d_k)/q$$

and

$$n' = n - k - (2q + 1 + 2/q)/6$$

Lawley and Maxwell (1971) suggest that the addition of the quantity

$$d^2 \sum_{i=1}^{k} 1/(d_i - d)^2$$

to (4.16) will give a closer approximation to $\chi^2$. The test statistic is compared to the tabled value of $\chi^2$ with $\tfrac{1}{2}(q + 2)(q - 1)$ degrees of freedom and significance level $\alpha$. The null hypothesis that $\delta_{q+1} =$

$\delta_{q+2} = \ldots = \delta_p$ is rejected if the tabled value of $\chi^2$ with these degrees of freedom is exceeded.

If the sample correlation matrix **R** (estimating **P**) is analysed, then the null hypothesis that all the (population) eigenvalues are equal is equivalent to $H_0$: **P** = **I**, since the sum of the eigenvalues equals the trace of **P**. Hence, if all the eigenvalues are equal, they must be equal to 1. This implies that the $x_i$ are statistically independent. Lawley and Maxwell (1971) provide the test statistic

$$-[n - (2p + 5)/6] \log_e [\det(\mathbf{R})] \tag{4.17}$$

For large $n$, this statistic is distributed approximately as $\chi^2$, with $\frac{1}{2}p(p-1)$ degrees of freedom. The null hypothesis is rejected if the test statistic exceeds the tabled value of $\chi^2$.

There is no good approximation to a $\chi^2$ test corresponding to (4.16) when **R** is used rather than **S**. Thus far, one requirement has been that n is large. Unfortunately, when $n$ becomes large the distribution of criterion (4.17) does not tend towards the $\chi^2$ distribution as for the preceding three test statistics. The test is therefore only a rough and ready guide. The number of degrees of freedom is $(p - k + 2)(p - k - 1)/2$. The null hypothesis that the last $(p - k)$ eigenvalues are equal can be tentatively rejected if the test statistic exceeds the tabled value of $\chi^2$ with the appropriate number of degrees of freedom.

Inferential tests in PCA are seen, then, to have a rather limited value, for despite the assumption of multivariate normality and the requirement that the sample size be large, the more accurate tests are applicable only to the analysis of **S** and not of **R**, while the formulae for computing confidence limits around the eigenvalues of **R** have as yet not been worked out. However, PCA is used mainly as a descriptive tool or as an exploratory technique, so the lack of formal tests is not a great embarrassment.

## 4.2 Computation

In the last section it was noted that the weights given to the observed variables in the calculation of the $i^{th}$ principal component are the elements of the eigenvector associated with the $i^{th}$ eigenvalue of **R** or **S**. An essential requirement for a principal components program is, therefore, a subroutine to compute the basic structure (eigenstructure) of a symmetric matrix. Subroutine HTDQL, described and listed in Appendix A, is used for this purpose. Also in Appendix A are listings of two other subroutines, MATPR and SSCP, which are used in the PCA program. The former produces a neat printout of a matrix and the latter computes either **R** or **S** (correlation or covariance matrix) and the vectors of means and standard deviations of the variables.

Table 4.2 below contains the Fortran listing of a computer program for PCA. In the main routine the base addresses of the necessary arrays are calculated. This is done so that the core requirement for any individual run can be specified by altering only one card, the DIMENSION card. The array Z must have a minimum dimension of $N$ where

$$N = NV^2 + 5NV$$

and $NV$ is the number of variables to be analysed. The arrays used in the program are:

    A: the ($NV$ x $NV$) matrix of correlations or covariances (only the lower triangle plus principal diagonal are in fact computed) which is later over-written by its eigenvectors.
  FMT: contains the output format used in subroutine MATPR.
   XB: vector of means.
    S: vector of standard deviations.
    D: vector of eigenvalues.
    B: workspace.
    F: workspace.
ALPHA: contains text 'Covariance matrix' in column 1 and text 'Correlation matrix' in column 2, both in (6A4) format.

The input channel (assumed to be the card-reader) is given logical number 5, and the output channel (lineprinter) is logical unit 6. Two scratch files (disc or tape) is needed; these are given logical numbers 3 and 4.

The user of the program may provide either raw data or the lower triangle plus principal diagonal of **R** or **S**. In the later case, principal component scores are not calculated. Instructions on the use of the program are given in the form of comments within the listing (Table 4.2).

The output from the program consists of:

   (i) a copy of the data, where raw data is supplied;
  (ii) means and standard deviations (when raw data is input);
 (iii) the lower triangle of the correlation (covariance) matrix;
 (iv) eigenvalues and cumulative proportion of total variance;
  (v) principal components loadings matrix;
 (vi) proportion of the variance of each variable accounted for by $k$ principal components;
(vii) principal components scores (where raw data input).

Various options exist for the calculation of standardized or unstandardized scores, and the input and output of data and results. These are detailed in the comments which introduce the program (Table 4.2).

Table 4.2
Computer program for principal components analysis

```
C  ---------------- USE OF PRINCIPAL COMPONENTS ANALYSIS PROGRAM --------
C  PROGRAM USES I/O UNITS 5 (CARDREADER), 6 (LINEPRINTER), 4 AND 3 (DISC
C  OR MAGTAPE FILES). OPTIONALLY UNIT 7 IS USED IF PRINCIPAL COMPONENT
C  SCORES ARE TO BE OUTPUT TO CARDS OR TO MAGTAPE/DISC FILE IN FORMATTED
C  FORM (SEE BELOW).
C  NOTE THAT THE PRINCIPAL COMPONENTS SCORES MATRIX IS PLACED ON UNIT 3
C  (DISC OR MAGNETIC TAPE FILE) IN UNFORMATTED FORM WITH EACH ROW HELD AS
C  ONE LOGICAL RECORD. IF SCORES ARE REQUIRED FOR LATER USE IT IS MORE
C  ECONOMICAL TO PRESERVE THE FILE ON UNIT 3 THAN TO O/P THE SCORES IN
C  FORMATTED FORM TO UNIT 7.
C  A SECOND OPTION ALLOWS THE DATA OR CORRELATION/COVARIANCE MATRIX
C      (LOWER TRIANGLE + DIAGONAL, READ LINEARLY ROWWISE) TO BE READ IN
C      UNFORMATTED FORM FROM A DISC OR MAGTAPE FILE. THE LOGICAL UNIT
C      NUMBER OF THIS FILE IS SUPPLIED ON CONTROL CARD 3. IF THIS
C      OPTION IS USED, NOTE THAT EACH ROW OF THE DATA MATRIX SHOULD FORM
C      ONE LOGICAL RECORD. IF THE CORRELATION/COVARIANCE MATRIX
C      IS SUPPLIED ON FILE, THEN THE WHOLE (LR. TRIANGLE + DIAGONAL)
C      SHOULD BE WRITTEN ON FILE AS ONE LOGICAL RECORD. EG: IF DATA IS
C      ON FILE, WITH LOGICAL UNIT NO. 9, THEN THE FOLLOWING FORTRAN
C      CODING SHOULD BE USED (ASSUMING MATRIX X HAS BEEN CORRECTLY
C      DIMENSIONED AND THAT NA IS THE NUMBER OF CASES AND NV THE NUMBER
C      OF VARIABLES):
C      DO 1 I=1,NA
C    1 WRITE(9) (X(I,K),K=1,NV)
C      ENDFILE 9
C  IF THE CORRELATION/COVARIANCE MATRIX IS HELD IN ARRAY S THEN USE:
C      WRITE(9) ((S(I,J),J=1,I),I=1,NV)
C      ENDFILE 9
C  CORE REQUIRED IS NV*NV+5*NV WHERE NV IS NUMBER OF VARIABLES, OR NA*NF,
C  WHICHEVER IS LARGER.  (NA IS THE NUMBER OF OBSERVATIONS PER VARIABLE
C  AND NF THE NUMBER OF PRINCIPAL COMPONENTS).  IF DIMENSION OF Z IN THE
C  MAIN PROGRAM IS ALTERED, CHANGE VALUE OF NDIM SO THAT IT IS EQUAL TO
C  THE NEW DIMENSION OF Z.
C
C  DATA IN THE FORM:
C
C  DATA CARD 1: NA, NV
C  NA IS NUMBER OF AREAS. PUNCH IN COLS 1-3.
C  NV IS NUMBER OF VARIABLES. PUNCH IN COLS 4-6.
C
C  DATA CARD 2 - EPSI, KORR, INPUT, CRDOUT, NUM, ISCALE
C  EPSI CUTOFF POINT FOR EIGENVALUES; PUNCH IN COLS 1-10.
C       INCLUDE DECIMAL POINT.
C  KORR INDICATOR OF DATA TYPE  -   IF KORR = 0 RAW DATA IS READ AND THE
C  CORRELATION OR COVARIANCE MATRIX IS COMPUTED# IF KORR=1 THEN THE LOWER
C  TRIANGLE AND DIAGONAL OF THE CORRELATION (COVARIANCE) MATRIX IS READ
C  SEE BELOW FOR DETAILS. PUNCH VALUE OF KORR IN COLUMN 11.
C  INPUT =1: COMPUTE VARIANCE-COVARIANCE MATRIX FROM RAW DATA;
C        =2: COMPUTE CORRELATION MATRIX FROM RAW DATA
C  PUNCH VALUE OF INPUT IN COLUMN 12. IF KORR = 1 THEN SET VALUE OF INPUT
C  TO EITHER 1 OR 2 DEPENDING ON WHETHER THE COVARIANCE(1) OR
C      CORRELATION(2) MATRIX IS TO BE READ.
C  CRDOUT: INTEGER INDICATOR, IF SET TO 1 THE P. C. SCORES WILL BE
C      OUTPUT TO UNIT 7 IN FORMATTED FORM. UNIT 7 COULD BE A CARD PUNCH
C      OR A MAGTAPE OR DISC UNIT. THE O/P FORMAT IS: (I3,7F10.4/(7F10.4))
C      THE FIRST VALUE (IN COLS 1-3 OF THE RECORD) IS THE CASE NUMBER,
C      FOLLOWED BY THE SCORES ON SUCCESSIVE PRINCIPAL COMPONENTS.
C  PUNCH VALUE OF CRDOUT IN COLUMN 13 OF CARD 2.
C  NUM: THE NUMBER OF PRINCIPAL COMPONENTS TO BE RETAINED.
C  PUNCH IN COLS 14-16. MAY BE ZERO.
C  ISCALE =0: SCALE SCORES TO HAVE UNIT VARIANCE;
C         =1: SCALE SCORES TO HAVE VARIANCE OF CORRESPONDING COMPONENT.
C  PUNCH VALUE OF ISCALE IN COL 17.
C  SCORES ON PRINCIPAL COMPONENTS BASED ON [S] RATHER THAN [R] WILL
C      ALWAYS BE UNSTANDARDIZED. SCORES ON PRINCIPAL COMPONENTS BASED
```

Table 4.2  *continued*

```
C      ON [R] WILL BE STANDARDIZED OR UNSTANDARDIZED, DEPENDING ON THE
C      VALUE OF ISCALE.
C
C DATA CARD 3 - IFILE
C IFILE: IF NONZERO INDICATES THAT DATA OR CORRELATION/COVARIANCE MATRIX
C      IS TO BE READ FROM DISC (DEPENDING ON THE VALUE OF KORR ON CARD 2)
C      THE LOGICAL UNIT NUMBER OF THE FILE WILL BE IFILE, WHICH SHOULD,
C      THEREFORE, BE A POSITIVE INTEGER. PUNCH VALUE IN COL. 1.
C ****** IFILE MUST NOT EQUAL 5,6,OR 3 ************
C DATA CARD 4: FMT          ****OMIT IF IFILE (CARD3) IS POSITIVE *****
C COLS 1-80: INPUT FORMAT FOR DATA OR CORRELATION/COVARIANCE MATRIX
C
C DATA CARD 5               ****OMIT IF MATRIX TO BE READ FROM FILE ***
C DATA MATRIX, PUNCHED ROWWISE IN FORMAT SPECIFIED ON CARD 4. START
C      EACH ROW ON A NEW CARD. IF CORREALTION/COVARIANCE MATRIX TO BE
C      READ, PUNCH ROWWISE STARTING EACH ROW ON A NEW CARD AND USING
C      FORMAT GIVEN ON CARD 4.
C
C DATA CARD 6
C FMT: OUTPUT FORMAT FOR SUBROUTINE MATPR. USE ALL OR PART OF 80 COLUMNS
C
C USES SUBROUTINES SSCP, MATPR AND HTDQL FROM APPENDIX A.
C
      DIMENSION Z(10000),FMT(20)
      NDIM=10000
C REMEMBER TO SET NDIM TO CURRENT DIMENSION OF Z.
      READ(5,1) NA, NV
    1 FORMAT(2I3)
C SET UP BASE ADDRESSES FOR CURRENT PROBLEM
      N2=NV*NV+1
      N3=N2+NV
      N4=N3+NV
      N5=N4+NV
      N6=N5+NV
      N7=N6+NV
      IF(N7.GT.NDIM) GOTO 3
C REX IS MAIN ROUTINE
      CALL REX(NA,NV,Z(1),Z(N2),Z(N3),Z(N4),Z(N5),Z(N6),FMT,NC)
C PRINT SCORES, LEFT IN UNFORMATTED FORM ON UNIT 3 AFTER CALL TO SCORE
      IP=1
      NC1=(NC-1)*NA
      DO 5 I=1,NA
      J=NC1+I
    5 READ(3) (Z(M),M=I,J,NA)
      NW=NA*NC
      CALL MATPR(32HPRINCIPAL COMPONENT SCORES         ,4,FMT,6,Z,NW,NA,
     1 NA,NC,IP,2,8)
      GOTO 2
C ERROR EXIT IF DIMENSION OF X IS TOO SMALL.
    3 WRITE(6,4) N7
    4 FORMAT(' DIMENSION OF Z TOO SMALL IN MAIN PROGRAM. SET TO',I10)
    2 STOP
      END



      SUBROUTINE REX(NA,NV,A,XB,S,D,B,F,FMT,NC)
C MAIN ROUTINE FOR PRINCIPAL COMPONENTS ANALYSIS
      INTEGER CRDOUT
      REAL MACHEP
      DIMENSION A(NV,NV),FMT(20),XB(NV),S(NV),D(NV),B(NV),
     *F(NV),ALPHA(6,2)
      DATA ALPHA/'COVA','RIAN','CE  ','MATR','IX  ','    ','CORR',
     *'ELAT','ION ',' MAT','RIX ','    '/
      TRACE = 0.0
      WRITE(6,10)
```

Table 4.2   *continued*

```
      10 FORMAT('0',10X,'PRINCIPAL COMPONENTS ANALYSIS.    P. M. MATHER, 1973')
C CONTROL CARD 2
         READ(5,8) EPSI, KORR, INPUT, CRDOUT, NUM, ISCALE
       8 FORMAT(F10.0,3I1,I3,I1)
C CONTROL CARD 3
         READ(5,751) IFILE
     751 FORMAT(I1)
C BRANCH ACCORDING TO VALUE OF IFILE
         IF(IFILE.GT.0) GOTO 50
C CONTROL CARD 4
         READ(5,752) FMT
     752 FORMAT(20A4)
C BRANCH ACCORDING TO VALUE OF KORR
         IF(KORR.EQ.1) GO TO 20
         REWIND 4
C READ DATA MATRIX ROW BY ROW FROM CARDS AND TRANSFER TO DISC
         DO 80 I=1,NA
         READ(5,FMT) S
      80 WRITE(4) S
         ENDFILE 4
         REWIND 4
         IFILE = 4
         GOTO 52
      50 IF(KORR.EQ.1) GOTO 51
C SUBROUTINE SSCP IS LISTED IN APPENDIX A
C READ DATA FROM DEVICE IFILE AND COMPUT CORRELATION/COVARIANCE MATRIX
      52 CALL SSCP(XB,A,S,NV,NA,NV,INPUT,IFILE,6,FMT,1)
         REWIND IFILE
         GO TO 12
C READ LR. TRIANGLE OF CORRELATION OR COVARIANCE MATRIX IF REQUESTED
C FROM CARDS ........
      20 DO 22 I=1,NV
      22 READ(5,FMT) (A(I,J),J=1,I)
         GOTO 55
C ...... OR FROM UNIT IFILE.
      51 READ(IFILE) ((A(I,J),J=1,I),I=1,NV)
C COPY LR. TRIANGLE TO UPPER.
      55 DO 23 I=1,NV
         S(I)=SQRT(A(I,I))
         DO 23 J=1,NV
      23 A(I,J)=A(J,I)
         WRITE(6,24) (ALPHA(I,INPUT),I=1,6)
      24 FORMAT('0',6A4,'INPUT')
COMPUTE TOTAL VARIANCE
      12 DO 6 I=1,NV
       6 TRACE = TRACE + A(I,I)
C READ O/P FORMAT FOR *MATPR*
         READ(5,101) FMT
         NW=NV*NV
     101 FORMAT(20A4)
         IP=1
C SUBROUTINE MATPR IS LISTED IN APPENDIX A
         CALL MATPR(ALPHA(1,INPUT),6,FMT,6,A,NW,NV,NV,NV,IP,1,4)
C
C THESE TWO PARAMETERS ARE MACHINE-DEPENDENT. SEE WRITEUP OF
C SUBROUTINE HTDQL, APPENDIX A.
         MACHEP=2.0**(-37)
         ETA=2.0**(-255)
         TOL=ETA/MACHEP
C
C SUBROUTINE HTDQL IS LISTED IN APPENDIX A
COMPUTE EIGENVALUES AND EIGENVECTORS OF A
         CALL HTDQL(NV,NV,NV,A,D,B,A,MACHEP,TOL)
         IF(NV.EQ.0) RETURN
         WRITE(6,11)
      11 FORMAT('1 EIGENVALUES AND CUMULATIVE PROPORTION OF TOTAL VARIANCE'
        1//)
```

Table 4.2  *continued*

```
      NC=0
      IF(D(1).GT.EPSI) NC = 1
      B(1)=D(1)
      C=B(1)/TRACE
      I=1
      WRITE(6,2) I,B(1),C
      DO 1 I=2,NV
      IF(D(I).GT.EPSI) NC = NC+1
      B(I)=B(I-1)+D(I)
      C=B(I)/TRACE
    1 WRITE(6,2) I, D(I), C
      WRITE(6,7) TRACE
    7 FORMAT('OTOTAL VARIANCE ANALYSED = ',G15.6//)
    2 FORMAT(' ',I6,2F20.5/)
      IF(NUM.GT.0) NC=MIN0(NC,NUM)
      IF (NC.EQ.0) GO TO 99
      WRITE(6,17) NC
   17 FORMAT('0',I6,' PRINCIPAL COMPONENTS SELECTED'///)
      DO 33 I=1,NC
      P=SQRT(D(I))
C SCALE EIGENVECTORS (DEPENDING ON WHETHER CORRELATION OR COVARIANCE
C MATRIX USED)
      IF(INPUT.EQ.1) GOTO 44
      DO 3 J=1,NV
    3 A(J,I)=A(J,I)*P
      GOTO 33
   44 DO 45 J=1,NV
   45 A(J,I)=A(J,I)*P/S(J)
   33 CONTINUE
    4 IP=1
C SUBROUTINE MATPR IS LISTED IN APPENDIX A
      CALL MATPR(40HPRINCIPAL COMPONENTS LOADINGS MATRIX     ,5,FMT,6,
     * A,NW,NV,NV,NC,IP,2,8)
      WRITE(6,13) NC
   13 FORMAT('0 PROPORTION OF VARIANCE OF EACH VARIABLE ACCOUNTED FOR BY
     *',I4,' PRINCIPAL COMPONENTS'///'    VARIABLE       PROPORTION'///)
      DO 14 I=1,NV
      C=0.0
      DO 15 J=1,NC
   15 C=C+A(I,J)*A(I,J)
   14 WRITE(6,16) I, C
   16 FORMAT(' ',I10,7X,F10.4)
      IF(KORR.EQ.1) STOP
      IND=0
      CALL SCORE(A,NV,B,F,NC,NA,D,XB,S,INPUT,CRDOUT,ISCALE,IFILE)
      RETURN
   99 WRITE(6,100)
  100 FORMAT('0*** CHECK PARAMETER CARD - NUM = 0 AND NO EIGENVALUES ARE
     1 GREATER THAN EPSI!')
      STOP
      END



      SUBROUTINE SCORE(A,N,B,C,K,M,XB,XMN,S,IND,CRDOUT,ISCALE,IFILE)
C CALCULATES PRINCIPAL COMPONENT SCORES
      INTEGER CRDOUT
      DIMENSION A(N,K),B(N),C(K),XB(N),XMN(N),S(N)
      REWIND 3
      IF(ISCALE.EQ.1) WRITE(6,20)
   20 FORMAT('0 NOTE: SCORES ON PRINCIPAL COMPONENTS ARE UNSTANDARDIZED'
     1)
    8 DO 1 I=1,M
      READ(IFILE) B
      DO 5 J=1,N
```

Table 4.2  *continued*

```
    5 B(J)=(B(J)-XMN(J))/S(J)
      DO 2 J=1,K
      C(J)=0.
      IF(IND. EQ. 2. AND. ISCALE. EQ. 0) GOTO 11
      DO 12 L=1,N
   12 C(J)=C(J)+B(L)*A(L,J)
      GO TO 2
   11 DO 6 L=1,N
    6 C(J)=C(J)+B(L)*A(L,J)/XB(J)
    2 CONTINUE
C STORE ROWS OF SCORES MATRIX ON UNIT 3
      WRITE(3) (C(L),L=1,K)
C OUTPUT SCORES IN FORMATTED FORM IF NEEDED
      IF(CRDOUT. NE. 1) GO TO 1
      WRITE(7,7) I,(C(L),L=1,K)
    7 FORMAT(I3,7F10. 4/(7F10. 4))
    1 CONTINUE
      ENDFILE 3
      REWIND 3
      RETURN
      END
```

## 4.3 Example

This example of principal components analysis uses data taken from Melton's (1957) classic study of the relations among elements of climate, surface properties and geomorphology. Nine of the variables measured from maps or in the field have been selected; these reflect the characteristics of 57 drainage basins in the Southwest United States. These are:

$x_1$: total relief of basin, in feet.

$x_2$: height of basin mouth above sea level, in feet.

$x_3$: basin perimeter, in miles.

$x_4$: total channel length, in miles.

$x_5$: total number of channels.

$x_6$: average bifurcation ratio.

$x_7$: maximum valley-side slope angle, degrees.

$x_8$: number of stream sources.

$x_9$: basin area, in square miles.

The data matrix is listed in Table 4.3. Since several different measurement scales are used (feet, miles, square miles, degrees) the variances of individual variables are not directly comparable. Standardization to unit variance was thus necessary. The lower half triangle of the correlation matrix is shown in Table 4.4. Visual analysis of Table 4.4 shows that some strong relationships are exhibited, for example, between $x_8$ and $x_5$ ($r = 0.999$), $x_8$ and $x_4$ ($r = 0.928$), $x_9$ and $x_4$ ($r = 0.937$) and between $x_5$ and $x_4$ ($r = 0.921$). It is also worth noting that the

## Table 4.3
Data for principal components analysis example (Data from M. A. Melton, *Tech. Rept. 11*, Project NR 389–042, Dept. of Geology, Columbia University, New York, 1957)

|    | 1 | 2 | 3 | 4 | 5 | 6 | 7 | 8 | 9 |
|----|---|---|---|---|---|---|---|---|---|
| 1  | 760.0000 | 5490.0000 | 1.7040 | 2.4810 | 30.0000 | 2.7860 | 31.8000 | 20.0000 | 0.1430 |
| 2  | 1891.0000 | 4450.0000 | 2.7650 | 4.3940 | 30.0000 | 5.8330 | 37.0000 | 26.0000 | 0.3120 |
| 3  | 325.0000 | 5525.0000 | 1.5000 | 2.6600 | 36.0000 | 3.0420 | 21.1000 | 25.0000 | 0.1620 |
| 4  | 515.0000 | 4760.0000 | 2.7500 | 5.3200 | 117.0000 | 4.8440 | 30.1000 | 98.0000 | 0.2210 |
| 5  | 513.0000 | 6090.0000 | 1.1420 | 2.0800 | 32.0000 | 5.1000 | 25.7000 | 26.0000 | 0.1010 |
| 6  | 1570.0000 | 8640.0000 | 6.1300 | 10.2100 | 76.0000 | 4.2900 | 24.9000 | 61.0000 | 1.3600 |
| 7  | 2210.0000 | 8415.0000 | 8.7600 | 15.0000 | 66.0000 | 4.5000 | 26.6000 | 56.0000 | 2.9900 |
| 8  | 515.0000 | 7040.0000 | 1.3000 | 1.2600 | 13.0000 | 3.5000 | 22.2000 | 10.0000 | 0.0890 |
| 9  | 1192.0000 | 6258.0000 | 8.4470 | 30.6060 | 286.0000 | 6.5000 | 29.1000 | 225.0000 | 2.0570 |
| 10 | 1540.0000 | 6280.0000 | 5.1740 | 11.3830 | 82.0000 | 4.0700 | 23.3000 | 63.0000 | 0.7633 |
| 11 | 950.0000 | 8520.0000 | 2.8800 | 6.8700 | 62.0000 | 3.6500 | 27.2000 | 47.0000 | 0.4760 |
| 12 | 850.0000 | 9460.0000 | 7.4800 | 7.7900 | 30.0000 | 4.9000 | 11.6000 | 24.0000 | 1.7500 |
| 13 | 1237.0000 | 5937.0000 | 2.0460 | 2.9930 | 28.0000 | 2.7200 | 29.6000 | 19.0000 | 0.2520 |
| 14 | 553.0000 | 7480.0000 | 4.1200 | 22.8000 | 407.0000 | 4.3100 | 21.0000 | 305.0000 | 0.7400 |
| 15 | 281.0000 | 7050.0000 | 3.3600 | 8.2400 | 83.0000 | 4.1900 | 8.2000 | 67.0000 | 0.4810 |
| 16 | 1242.0000 | 6525.0000 | 3.5200 | 7.4900 | 51.0000 | 3.7900 | 29.2000 | 41.0000 | 0.7230 |
| 17 | 889.0000 | 7836.0000 | 3.2950 | 8.6550 | 65.0000 | 3.7400 | 32.4000 | 50.0000 | 0.6270 |
| 18 | 1342.0000 | 5340.0000 | 3.1200 | 7.8100 | 69.0000 | 8.3400 | 33.0000 | 56.0000 | 0.4570 |
| 19 | 4523.0000 | 4879.0000 | 10.3700 | 78.5100 | 507.0000 | 4.4900 | 39.3000 | 398.0000 | 5.4600 |
| 20 | 3275.0000 | 6050.0000 | 5.0500 | 11.5300 | 50.0000 | 3.5700 | 30.4000 | 38.0000 | 1.1530 |
| 21 | 1510.0000 | 5490.0000 | 4.0900 | 12.9600 | 116.0000 | 4.8880 | 30.0000 | 98.0000 | 0.6650 |
| 22 | 1655.0000 | 5245.0000 | 2.5800 | 4.4200 | 30.0000 | 2.8330 | 31.9000 | 21.0000 | 0.2905 |
| 23 | 1655.0000 | 5245.0000 | 2.5600 | 5.4600 | 45.0000 | 3.4200 | 33.7000 | 34.0000 | 0.3125 |
| 24 | 1475.0000 | 4450.0000 | 1.8370 | 2.0640 | 18.0000 | 4.7500 | 37.0000 | 15.0000 | 0.1496 |
| 25 | 2144.0000 | 4197.0000 | 4.1480 | 9.9420 | 71.0000 | 4.2270 | 35.0000 | 57.0000 | 0.6280 |
| 26 | 515.0000 | 6650.0000 | 1.0500 | 1.2600 | 17.0000 | 5.1000 | 27.4000 | 14.0000 | 0.0550 |
| 27 | 834.0000 | 6450.0000 | 5.9090 | 16.0990 | 160.0000 | 6.4400 | 31.1000 | 134.0000 | 1.0680 |
| 28 | 834.0000 | 6450.0000 | 5.3790 | 10.7580 | 110.0000 | 4.6300 | 31.1000 | 90.0000 | 0.6640 |
| 29 | 1010.0000 | 6745.0000 | 4.2420 | 13.6940 | 109.0000 | 4.4300 | 24.6000 | 86.0000 | 0.9250 |
| 30 | 543.0000 | 6745.0000 | 1.8560 | 2.8980 | 18.0000 | 2.4200 | 24.6000 | 13.0000 | 0.1800 |
| 31 | 621.0000 | 7099.0000 | 2.2730 | 3.8630 | 27.0000 | 4.6000 | 24.6000 | 21.0000 | 0.2780 |
| 32 | 290.0000 | 6745.0000 | 4.9240 | 12.9930 | 85.0000 | 4.2500 | 27.8000 | 69.0000 | 0.9470 |
| 33 | 955.0000 | 7080.0000 | 2.0830 | 2.3870 | 20.0000 | 2.7800 | 27.8000 | 16.0000 | 0.1930 |
| 34 | 885.0000 | 7150.0000 | 1.5530 | 1.5540 | 10.0000 | 2.7500 | 27.8000 | 7.0000 | 0.1290 |
| 35 | 847.0000 | 7188.0000 | 1.5910 | 1.6100 | 14.0000 | 3.1700 | 31.3000 | 10.0000 | 0.0940 |
| 36 | 798.0000 | 7188.0000 | 1.0980 | 1.0230 | 11.0000 | 3.0000 | 31.3000 | 8.0000 | 0.0645 |
| 37 | 1039.0000 | 5961.0000 | 2.7270 | 3.2950 | 28.0000 | 5.5000 | 29.6000 | 24.0000 | 0.2520 |
| 38 | 1213.0000 | 5961.0000 | 3.0300 | 6.8940 | 49.0000 | 6.4300 | 29.6000 | 41.0000 | 0.4580 |
| 39 | 1074.0000 | 5813.0000 | 2.5000 | 2.9540 | 30.0000 | 5.3300 | 29.6000 | 26.0000 | 0.3200 |
| 40 | 370.0000 | 8295.0000 | 1.7400 | 2.0000 | 21.0000 | 4.3300 | 17.8000 | 17.0000 | 0.1560 |
| 41 | 430.0000 | 8240.0000 | 2.1300 | 2.3100 | 14.0000 | 3.7500 | 18.9000 | 11.0000 | 0.1820 |
| 42 | 690.0000 | 8410.0000 | 1.6300 | 1.6800 | 12.0000 | 3.2500 | 18.9000 | 9.0000 | 0.1080 |
| 43 | 773.0000 | 8410.0000 | 2.0700 | 2.4100 | 18.0000 | 3.8300 | 18.9000 | 17.0000 | 0.1980 |
| 44 | 100.0000 | 6790.0000 | 0.8300 | 1.4000 | 25.0000 | 4.4000 | 11.4000 | 19.0000 | 0.0429 |
| 45 | 80.0000 | 6790.0000 | 0.5500 | 0.4700 | 10.0000 | 2.7500 | 11.4000 | 7.0000 | 0.0136 |
| 46 | 96.0000 | 6765.0000 | 0.6500 | 0.7300 | 15.0000 | 4.0000 | 11.4000 | 12.0000 | 0.0215 |
| 47 | 2490.0000 | 6535.0000 | 11.9700 | 59.4500 | 363.0000 | 2.8700 | 28.0000 | 293.0000 | 4.9300 |
| 48 | 1765.0000 | 6575.0000 | 7.3500 | 21.7600 | 140.0000 | 3.4600 | 26.7000 | 114.0000 | 1.9400 |
| 49 | 1158.0000 | 6862.0000 | 2.6890 | 4.7170 | 34.0000 | 3.2300 | 32.8000 | 26.0000 | 0.3580 |
| 50 | 1070.0000 | 7055.0000 | 2.1780 | 3.4480 | 26.0000 | 2.7000 | 32.8000 | 18.0000 | 0.2730 |
| 51 | 1495.0000 | 7055.0000 | 2.9170 | 3.3930 | 27.0000 | 2.6700 | 32.8000 | 18.0000 | 0.2995 |
| 52 | 1601.0000 | 6949.0000 | 2.8030 | 4.2050 | 28.0000 | 3.0800 | 32.8000 | 21.0000 | 0.3200 |
| 53 | 1215.0000 | 5135.0000 | 7.7600 | 23.1500 | 160.0000 | 3.8600 | 29.5000 | 131.0000 | 1.1920 |
| 54 | 1587.0000 | 5095.0000 | 6.1600 | 17.0200 | 119.0000 | 4.7100 | 29.9000 | 98.0000 | 1.3900 |
| 55 | 1230.0000 | 5120.0000 | 4.7400 | 8.4600 | 54.0000 | 3.7900 | 23.4000 | 43.0000 | 0.8110 |
| 56 | 1290.0000 | 4960.0000 | 2.0400 | 2.8000 | 24.0000 | 6.2500 | 37.0000 | 21.0000 | 0.1910 |
| 57 | 2400.0000 | 4920.0000 | 2.2600 | 3.2900 | 27.0000 | 5.1600 | 36.2000 | 23.0000 | 0.2580 |

correlations between $x_2$ and $x_3$, $x_1$ and $x_6$, $x_4$ and $x_6$, and $x_9$ and $x_2$ are near zero. The correlations of $x_2$ and the other variables are negative without exception.

The eigenvalues of this correlation matrix, together with individual and cumulative percentages of total variance, are contained in Table 4.5. Nearly 76 per cent of the variability among the 57 sample points lies in only two dimensions, and 86.5 per cent lies in a three-dimensional space. Dimensions 5, 6, 7, 8 and 9 contain only a negligible

Table 4.4
Correlations among the nine morphometric variables detailed in the text

|       | $x_1$  | $x_2$  | $x_3$  | $x_4$  | $x_5$  | $x_6$  | $x_7$  | $x_8$  | $x_9$  |
|-------|--------|--------|--------|--------|--------|--------|--------|--------|--------|
| $x_1$ | 1.000  |        |        |        |        |        |        |        |        |
| $x_2$ | −0.370 | 1.000  |        |        |        |        |        |        |        |
| $x_3$ | 0.619  | −0.017 | 1.000  |        |        |        |        |        |        |
| $x_4$ | 0.657  | −0.157 | 0.841  | 1.000  |        |        |        |        |        |
| $x_5$ | 0.474  | −0.150 | 0.737  | 0.921  | 1.000  |        |        |        |        |
| $x_6$ | 0.074  | −0.274 | 0.167  | 0.094  | 0.165  | 1.000  |        |        |        |
| $x_7$ | 0.607  | −0.566 | 0.162  | 0.217  | 0.158  | 0.170  | 1.000  |        |        |
| $x_8$ | 0.481  | −0.158 | 0.753  | 0.928  | 0.999  | 0.181  | 0.164  | 1.000  |        |
| $x_9$ | 0.689  | −0.016 | 0.910  | 0.937  | 0.788  | 0.071  | 0.158  | 0.799  | 1.000  |

Table 4.5
Eigenvalues and percent explanation for morphometric data of Table 4.3

| Principal component | Eigen value | Individual % variance | Cumulative % variance |
|---|---|---|---|
| 1 | 5.043  | 56.029 | 56.029  |
| 2 | 1.746  | 19.399 | 75.428  |
| 3 | 0.997  | 11.076 | 86.504  |
| 4 | 0.610  | 6.781  | 93.285  |
| 5 | 0.339  | 3.776  | 97.061  |
| 6 | 0.172  | 1.907  | 98.967  |
| 7 | 0.079  | 0.8727 | 99.840  |
| 8 | 0.014  | 0.1556 | 99.996  |
| 9 | 0.0004 | 0.0042 | 100.000 |

proportion of the total variability. Thus, $m$ (the number of principal components chosen to represent the data) lies somewhere in the range 2 to 4. Several considerations might affect the final choice of a value for $m$; if the aim of the analysis is to re-express as much as possible of the variation in the data, $m = 4$ would be appropriate. On the other hand, 'interpretability' of the principal components may be a requirement, in which case $m$ will be set to the number of principal components to which a meaning can be attached. Thirdly, it might be argued that a plot of the eigenvalues $d_i$ against $i$ should show a break of slope at the point of discrimination between the $m$ useful principal components and the $(p - m)$ trivial ones; Cattell (1966B, p. 206) calls this the 'scree test'. It is essentially a graphical method of estimating whether the last $(p - m)$ eigenvalues are approximately equal. A different procedure has been advocated by Kaiser, who argues that, since each $x_i$ has a variance of 1 when expressed in standard form, any principal component with an eigenvalue less than 1 is not worth consideration. This 'eigenvalue-one' criterion seems to have gained almost universal, if uncritical, acceptance. Difficulties arise in cases such as the present example:

Table 4.6
Principal component loadings matrix and percentage of variance accounted for by three principal components

| Variable | Principal component | | | Percent variance explained |
|---|---|---|---|---|
| | I | II | III | |
| $x_1$ | 0.75 | −0.38 | −0.36 | 83.05 |
| $x_2$ | −0.25 | 0.82 | −0.08 | 73.20 |
| $x_3$ | 0.89 | 0.19 | 0.00 | 82.19 |
| $x_4$ | 0.97 | 0.14 | −0.03 | 96.63 |
| $x_5$ | 0.91 | 0.18 | 0.16 | 88.26 |
| $x_6$ | 0.20 | −0.36 | 0.86 | 89.97 |
| $x_7$ | 0.35 | −0.80 | −0.25 | 83.19 |
| $x_8$ | 0.92 | 0.17 | 0.16 | 89.90 |
| $x_9$ | 0.93 | 0.22 | −0.10 | 92.16 |

should $d_3$ (= 0.9969) be counted or not? Furthermore, the sizes of the eigenvalues are related to the order of **R**. Finally, the statistical tests described in Section 4.2 could be used if the assumptions were satisfied.

In this example, $m$ was set to 3 on the grounds that the first 3 principal components accounted for over 86 per cent of total variation, that the 3 principal components were interpretable, and that all had eigenvalues which were almost 1 or well in excess of 1. The (9 x 3) principal component loadings matrix is shown in Table 4.6. The loadings, which represent correlations between the principal components and the original variables, fall into three classes: (i) close to zero, (ii) approaching 1, and (iii) intermediate. The members of class (ii) are 0.75 and above, for a well-defined gap exists between the lowest member of class (ii) and the highest member of class (iii). The high loadings of Principal Component I, which identify the variables which are most closely related to P.C.I., are those of variables $x_1$, $x_3$, $x_4$, $x_5$, $x_8$ and $x_9$. All are positive, which means that they all act in the same direction. The variables concerned are measures of total relief, perimeter, total channel length, total number of channels, number of stream sources, and basin area. They can be thought of as measures of basin magnitude, for all could be expected to increase as basin size increased. In general, the probability of $x_1$ being high is greater in a large basin than in a small one. However there seem to be little connection between basin magnitude and $x_2$, elevation of basin mouth, $x_6$, the average bifurcation ratio, and $x_7$, the maximum valley-side slope angle.

Principal Component II has two high loadings, on variables $x_2$ and $x_7$ These loadings are of opposite sign, and indicate that the maximum valley-side slope angles are greater in basins with mouths at relative low elevations. This is borne out by inspection of the data. The average value of $x_2$ is 6,488 feet, and the standard deviation is 1,190 feet.

The lowest slope angle is 8.2°, in a basin in which $x_2$ = 7,050 feet. Several basins have $x_7$ of 11—12° and $x_2$ ranging from 6,790 feet to 9,460 feet, all above average, while the highest slope angle (37°) occurs in a basin with a mouth elevation of 4,450 feet. The second principal component is therefore a measure of altitude and slope steepness.

The third principal component has only one high loading, on $x_6$ (average bifurcation ratio) which is a measure of the channel network shape. The 57 basins can thus be adequately described in terms of 3 concepts — basin magnitude, altitude and slope steepness and channel network shape. These concepts can be quantified or measured; this is done by calculating the scores on the three principal components, using Equation (4.12). The mean of each principal component score vector is 0 and the standard deviation is 1.0. The matrix **F** can be searched for values in excess of ±1.0 — these are the individuals which most markedly display the effects of the principal components.

Table 4.7
Basins with highest scores on the first principal component

|          |       | Basin number |       |       |       |           |
|----------|-------|------|-------|-------|-------|-----------|
|          |       | 19   | 47    | 9     | 14    | Mean of   |
| Score on PC1 |   | 4.79 | 3.39  | 1.80  | 1.228 | variable  |
| Variable | $x_1$ | 523  | 2490  | 1192  | 553   | 1173      |
|          | $x_3$ | 10.37| 11.97 | 8.44  | 4.12  | 3.55      |
|          | $x_4$ | 78.51| 59.45 | 30.61 | 22.8  | 9.25      |
|          | $x_5$ | 507  | 363   | 286   | 407   | 73.77     |
|          | $x_8$ | 398  | 293   | 225   | 305   | 58.54     |
|          | $x_9$ | 5.46 | 4.93  | 2.05  | 0.74  | 0.71      |

The scores on principal component I range from 4.79 to —1.14. The four basins with the highest positive scores on this principal component are shown in Table 4.7. The scores of each basin on the variables which are highly correlated with PCI are all above average, the only exception being the total relief of basin 14. Numbers and lengths of channels are far above the mean value. For comparison the four lowest scoring basins are given in Table 4.8. The comparison of Tables 4.7 and 4.8 speaks for itself. Numbers of channels, total channel length and area are considerably less for basins 45, 46, 44 and 42 than for the basins shown in Table 4.6.

A similar comparison of the highest and lowest scores could be attempted for the other two principal components, with the aims of establishing more precisely the meaning of the principal components and of identifying the extremes or end-members of the set of individuals.

Table 4.8
Basins with lowest scores on the first principal component

| | | Basin number | | | | |
|---|---|---|---|---|---|---|
| | | 45 | 46 | 44 | 42 | Mean of variable |
| Score on PC1 | | −1.135 | −1.056 | −0.981 | −0.873 | |
| Variable | $x_1$ | 80 | 96 | 100 | 690 | 1173 |
| | $x_3$ | 0.55 | 0.65 | 0.83 | 1.63 | 3.55 |
| | $x_4$ | 0.47 | 0.73 | 1.40 | 1.68 | 9.25 |
| | $x_5$ | 10 | 15 | 25 | 12 | 73.77 |
| | $x_8$ | 0.7 | 12 | 19 | 9 | 58.54 |
| | $x_9$ | 0.0136 | 0.0215 | 0.0429 | 0.1080 | 0.71 |

Some users of the technique plot the principal component scores pairwise, or map their distribution, usually with the aim of identifying groups of mutually similar individuals. This classification process is accomplished more efficiently by the methods described in Chapter 6. However, mapping of the distribution of the principal component scores can be useful in suggesting causes of the variation so exhibited. Principal component scores can also be input to regression and trend surface analysis programs, as described in Chapters 2 and 3.

One consideration which should be taken into account is the fact that each principal component score is the sum of the contributions of all variables loading on that principal component, even those not counted as being significant. It would be possible to exclude nonsignificant variables by replacing all low or near-zero loadings by zero so that the resulting scores represent only the influences of the variables which are thought to be important. This procedure may be useful in cases when the user is sufficiently confident that he is able to specify 'high' and 'low' loadings. In the example given this would be possible, but in many other applications the results may be less clear-cut. Furthermore, the resulting vectors of scores would no longer be orthogonal.

Other examples of the use of principal components analysis are widespread in the literature, covering fields such as geomorphology (Mather and Doornkamp, 1970), sedimentology (McCammon, 1966; Davis, 1970; Griffiths and Ondrick, 1969), hydrology (Eiselstein, 1967; Yevyevitch, 1972; Stammers, 1967), meteorology and climatology (Craddock, 1969, 1973; Christiensen and Bryson, 1966) and biogeography (Barkham and Norris, 1970). In the remainder of this section two areas of application, meteorology and biogeography, will be briefly reviewed.

Craddock (1973) points out that one view of the problem of long-range weather forecasting is to treat it as a vast multivariate problem. The number of variables, which can be as high as 50 000, can be reduced

(a) by discarding all those variables which on the basis of reasoning from known theory are likely to be unimportant, and
(b) by considering only large-scale features which are dependent upon the general atmospheric circulation.

Even so, the number of variables is large and some of the information is redundant in the sense that it is common to two or more variables. The use of principal components analysis is proposed as a means both of providing an economical description of the data set and of allowing the identification of the principal components in terms of recognizable physical processes. A practical problem is that of deciding upon the number of principal components to retain. Craddock uses a method which he calls the LEV (logarithm of eigenvalue) graph, which is a plot of the logarithm (to the base e) of the eigenvalues against the ordinal number of the eigenvalue. This method is similar to that described by Cattell (1966B). The problems which Craddock concludes are most pressing are firstly that the results of principal components analysis are not stable from one data set to another. Although the first few principal components generally agree, the smaller — though still significant — members of the set vary greatly from one experiment to another. Secondly, the number of principal components, which may be as high as 50, is still large. The time taken to interpret the spatial pattern is appreciable. Thirdly, the principal components may not be readily interpretable, perhaps because the underlying physical processes are correlated. The best line of approach may, in fact, involve the use of factor analysis (Chapter 5) in which the orthogonality constraint may be lifted.

Barkham and Norris (1970) discuss the results of applying principal components analysis in biogeography. The particular problem considered was the relationship between vegetation and soil systems of two beechwoods located in the Cotswold Hills of southern England. Hypotheses about the variation within each system were generated by principal components analysis and these hypotheses were subsequently tested using simple and canonical correlation. The principal components of the vegetation system, representing 40.5 per cent of the total variance, were interpreted as (i) an environmental component, mainly reflecting moisture and slope conditions, with mature woodland at the positive end of the scale and steep, exposed situations at the negative end, and (ii) a shade/soil component which is reflected in the distribution of the species *Hedera helix*. Principal components 1 and 2 of the soil system represent almost 35 per cent of the total variance, and are interpreted as: (i) moisture status of the soil (compounded of moisture content and drainage rate) and (ii) stoniness of the soil.

Subsequent analysis revealed relationships between the first principal components of the vegatation and soil systems, However, a large part of the variation remained unaccounted for.

Other uses of principal components analysis in biogeography are concerned with its use in a classification, a topic to which we will return in Chapter 6.

## 4.4 Conclusions

Principal components analysis is a robust technique. This does not provide an excuse for misusing it. Despite the fact that it has been widely used over the last 10 or 20 years, several areas of difficulty remain. These relate, in general, more to the nature of the data than to computational problems which are relatively insignificant given the power of modern computers and the availability of programs. The *choice of variables* is the crux of the issue. It is easy to select a large number of variables simply because they happen to be available. The variables should, as far as possible, cover the 'domain' of interest, and should not be concentrated upon one or two easily-measured facets. Next, the *spatial distribution of the sampling points* must be carefully considered. If the aim of the experiment is to identify large-scale processes then the sample points must cover a large area. As Craddock (1973) points out, data taken from a small area such as the Thames Valley will not allow the identification of the rain-making process, which operates on a scale of several hundreds of miles. Lastly, the *scaling* of the observed variables and of the principal component scores is of paramount importance. Most published examples of the use of principal components analysis fail to indicate (i) why **R** (rather than **S**) was chosen, or (ii) how the columns of **F** are scaled, and on what basis the choice of scaling factor was made. Without serious consideration of these fundamental issues, the results obtained even from the most accurate computer program are quite worthless.

*Chapter 5*

# *Factor Analysis*

## 5.0 Introduction

Widely varying ideas have been expressed regarding the relevance of factor analysis to contemporary research problems. The view that factor analysis is a key technique in multivariate studies is widely held, especially in psychology (Cattell, 1965, 1966B, Horst, 1965) and the social sciences (Rummell, 1970). The opposite view is taken by some physical scientists, such as Matalas and Reiher (1965) and Blackith and Reyment (1971), all of whom believe that principal components analysis (Chapter 4) is both more useful and better defined than factor analysis. It is true to say that, until recently, the statistical and mathematical foundations of factor analysis have been poorly expressed, and the aims of the technique have not been widely understood. A variety of estimation procedures have been suggested, and many applications of factor analysis have been characterized mainly by their lack of correspondence with reality, or by the triviality of their findings. In recent years, considerable theoretical and computational developments have taken place, leading to a new approach that has a more solid basis in theory. The principles of this new approach were put forward over thirty years ago by D. N. Lawley but it has only been possible to put his ideas into practice with the help of modern computers. In addition, the older approach has been rationalized by Harman (1967), Horst (1965), Kaiser (1970), Kaiser and Caffrey (1965) and others, so that it is not now reasonable to dismiss the technique as being unrealistic. Nevertheless, one should approach with caution. In this chapter an attempt is made to summarize recent opinion on the aims and purpose of factor analysis, on the form and assumptions of the underlying model, and on the properties of the alternative estimation procedures and factoring strategies. Subsequently, consideration is given to the question of the meaning of factors. This involves the topic of transformation of the initial solution. Finally, the computation of factor scores is discussed. Fortran subroutines to perform some of the operations described here are listed in the text. Reference is made to other sources of programs to deal with the more complex calculations associated with, for example, maximum likelihood methods, details of which are beyond the scope of this book. The basic ideas are straightforward, though, and an efficient computer program is readily available.

## 5.1 Definition

Most authors have their own definition of the purpose of factor analysis. Some even confuse the technique of principal components and factor analysis (see Chapter 4 for a definition of principal components analysis). To understand the aim of the method, its origins in the field of educational psychology must be considered. In this discipline, interest was focused upon the measurement and explanation of variations in intellectual functions among children. These functions were measured by a variety of tests. Researchers then posed the question: how many basic intellectual functions are there, what are they, and how are they interrelated? Factor analysis resulted from attempts to answer these questions. Spearman, writing in 1904, thought that a child's performance in a series of tests of mental ability could be explained by reference to a 'general intelligence factor', which underlay performance in all tests, and to a series of particular or specific factors relating to performance in individual tests. A child scoring highly on one test would, under this hypothesis, tend to score highly on all of the other tests, with variations between test scores being accounted for by the specific factors. The correlations between vectors of scores on the tests would thus be entirely due to the general factor, the level of correlation between two tests reflecting the degree to which the general intelligence factor influenced the outcome of each of the particular tests, assuming that the tests require a varying degree of intelligence.

Later workers, such as Thurstone (1947), extended this idea of the common (in this case, general) factor and specific factors associated with individual tests. Several common factors, which may or may not be related, were hypothesized under this multiple common factor model to account for correlations between test scores. These common factors were given names such as 'verbal reasoning' or 'numerical ability' as well as 'general intelligence'. Each common factors described some, not necessarily all, the tests. The fact that the tests were correlated was again thought to be related to the level to which performance in the tests was influenced by the common factors.

When the technique of factor analysis was applied in the context of geography, intelligence 'tests' became 'variables' and the persons to whom the tests were given were replaced by areas or objects on which the variables were measured. The basic premise — that the correlations among the variables was due to the influence of the common factors — remained unaltered. The common factors are viewed as *descriptive concepts*, summarizing the behaviour of a set of variables. Notice that the variables could be said to form the interacting components of a unified system. It would not make sense, for example, for the educational psychologist to include as test scores the times achieved by the subjects in a 100 metre race, or their shoe size; the concept of a common factor would be untenable if he were to do so. Relationships *between* systems are best sought out be regression or canonical correlation methods (Hotelling, 1957). Factor analysis is, therefore, the determination of a set of descriptive concepts which summarize the relationships among the components of a system of interacting variables. Any attempt to interpret factors as 'causes' or 'dynamic forces' is meaningless and futile. Any such interpretation must rise out of further investigation, or *a priori* knowledge. It is true to say that the results of a factor analysis are nothing more than a re-expression of the information content of the data. (see Mukherjee, 1973, for an example). This is not intended as a criticism, for the inspection of a given data set from a new viewpoint may reveal previously unsuspected relationships. However, the user must have some knowledge of the system under study, otherwise meaningless results could be accepted uncritically. Armstrong (1967) gives an amusing example from the field of metallurgy to illustrate his conclusion that 'factor analysis, by itself, may be misleading as far as the development of theory is concerned'. He suggests that, prior to factor analyzing a set of data, the research should evaluate such things as:

(a) the types of relationship that exist among the variables. In Section 5.2 it will be pointed out that factor analysis is concerned with *linear* relationships. Digman (1966) and Carroll (1961) discuss this point;
(b) the number of factors to be expected, although it is possible in exploratory studies to rely on empirical or statistical rules;
(c) the nature of factors to be expected. This implies some prior knowledge of the system;
(d) the variables used. Do they adequately represent the system, and are they consistent with (c)?
(e) the inter-factor relationships. The factors may or may not be intercorrelated.

Point (a) in the above list is worth elaborating upon. There are no assumptions necessary for the calculation of product-moment correlation coefficient, but the actual range of possible values will not be given by $-1 \leqslant r \leqslant +1$ unless the variables have a bivariate normal distribution. If the distribution of the variables is non-normal, it is probably wise to use a nonparametric measure, such as Spearman's rho or

Kendall's tau, otherwise the use of Pearson's $r$ could well mask, rather than reveal, relationships. Rummel (1970) gives a good account of alternative correlation measures.

Another consideration concerns the presence in the data-set of a variable which exerts an influence over many, if not all, of the remaining variables. All the correlations then reflect the influence of the 'universal' variable, rather than simple bivariate relationships. The use of partial correlation coefficients seems to be justified in these circumstances. Their use is equivalent to treating the variable to be partialled out as having the status of a factor.

From the preceding discussion it should be apparent that factor analysis is useful in those situations in which the relationships among the elements of a multivariate system are being scrutinized. Prior knowledge of the system is necessary, however, to provide criteria against which the outcome of the analysis can be evaluated. Also, as with other statistical techniques, there must be an adequate experimental design. A thorough discussion of this topic in the context of factor analysis is given by Cattell (1966A: Chapter 2).

## 5.2 Statistical Model

The terminology and notation of the present chapter presents a problem, since no agreed system has as yet been reached. Some authors use the same symbols as those employed in this chapter to indicate different concepts. Psychologists (e.g. Harman, 1967; Mulaik, 1972) use X and F, for example, to indicate the *transpose* of the matrices X and F used in this chapter. However, the notation of this chapter has been kept as close as possible to that used in the remainder of the book, and to standard mathematical and statistical usage. Thus, X represent an $n$ (cases) by $p$ (variables) matrix, and not its transpose. The factor model, in the notation described below, is written

$$X = F\Lambda' + E \tag{5.1}$$

This appears to be similar to the linear regression model (2.4),

$$y = X\beta + E \tag{2.4}(bis)$$

but the major difference is that in (2.4) the elements of matrix X are observable (being the measurements on the explanatory variables) whereas in (5.1) both $\Lambda$ and F must be estimated. $\Lambda$ is the 'factor loadings matrix' and, in geometrical terms, gives the coordinates of the points representing variables relative to the axes of a $k$-dimensional space. $\Lambda$ is thus a $(p \times k)$ matrix where $p$ is the number of variables and $k$ the number of common factors. The columns of $\Lambda$ need not be orthogonal, as is the case with the principal component loadings matrix. F is an $(n \times k)$ factor score matrix, which gives the coordinates of the observations in the $k$-dimensional space defined by $\Lambda$, and the influence of the factors on the individual cases is expressed by the elements of **F**. As usual, **X** is the $(n \times p)$ matrix of observations ($n$ individuals, $p$ variables) while **E** is the residual matrix, which expresses the effects of the specific factors affecting the variables together with measurement error. The individual observations $x_{ij}$ are thus seen to be the product of two matrices, $\Lambda$ and **F**, plus the associated disturbance or residual term:

$$x_{ij} = \sum_{r=1}^{k} \lambda_{ir} f_{jr} + e_{ij} \tag{5.2}$$

The problem of estimating the $\lambda_{ij}$ and the $f_{ij}$ is considered in Sections 5.3 and 5.5 respectively.

A statistical model is not completely defined without assumptions. In the case of the factor model, it is necessary to specify that:

I: $E(\mathbf{f}_i) = E(\mathbf{e}_i) = 0$;
II: The $\mathbf{f}_i$ and the $\mathbf{e}_i$ are independent;
III: The elements $e_{ij}$ are independent of one another;
IV: The variance–covariance matrix $\Psi$ of the $e$'s is diagonal and nonsingular (this follows partly from III);
V: $\Sigma$, the variance–covariance matrix of the $x$'s, has rank $k$;
VI: $k < p$;
VII: Each $x_i$ is correlated with at least one other of the $x$'s.

In addition, one should note that (5.1) implies that the $\mathbf{f}_i$ are linear combinations of the $\mathbf{x}_i$. The use of nonlinear models in factor analysis is relatively new and little experience of such models has been gained. McDonald (1962, 1967) and Bartlett (1953) discuss nonlinearity in detail.

A basic difficulty in the use of the factor model is the fact that neither $\Lambda$ or **F** is uniquely defined for a given set of data **X**. If $T$ is any nonsingular $(k \times k)$ matrix, then

$$\Lambda^* = \Lambda T^{-1}.$$

Furthermore,

$$\mathbf{F}^* = T\mathbf{F}$$

thus

$$\mathbf{F}^* \Lambda^{*\prime} = \mathbf{F}\Lambda',$$

which leads to

$$\mathbf{X} = \mathbf{F}^* \Lambda^{*\prime} + \mathbf{E}. \tag{5.3}$$

In other words, (5.1) and (5.3) cannot be distinguished simply by observing the $x$'s. In the words of Kendall (1954), '... no valid scientific discrimination between rival interpretations is possible for which the observational consequences are the same'. An infinite number of factor solutions is therefore possible. In geometrical terms, the transformation of $\Lambda$ to $\Lambda^*$ can be visualized as a rotation of the axes of the $k$-dimensional factor space. Without reference to external criteria it is therefore impossible to fix the position of these axes. This is the so-called problem of rotation. The indeterminacy can be reduced if $\Phi$, the inter-factor correlation matrix, is required to be an identity matrix, that is, if the factor axes are required to be orthogonal. However, this still does not determine a unique solution; the number of orthogonal transformations is still infinite.

Joreskog (1963) and others (e.g. Anderson and Rubin, 1956) have shown that, if Equation (5.1) correctly represents the structure of the observations X, then $\Sigma$, the variance–covariance matrix derived from X, must have a particular form, namely

$$\Sigma = \Lambda \Phi \Lambda' + \Psi^2 \tag{5.4}$$

Harman (1967) refers to this equation as 'the fundamental postulate of factor analysis'. It implies that the off-diagonal elements of $\Sigma$ can be reproduced from a knowledge of the factor loadings $\Lambda$ and the correlations between the factors, $\Phi$, since $\Psi^2$ is a diagonal matrix. The elements of the diagonal of $\Sigma$ are the sum of two variances; the variance derived from, or attributable to the common factors [i.e. $(\Lambda \Phi \Lambda)_{ii}$] and the variance of the residuals $((\Psi^2)_{ii})$. The first part of the variance of a variable is the 'communality' of the particular variable, the second is its 'uniqueness'. An intuitive explanation of these quantities was given in Section 5.2. It is usually the case that the sample correlation matrix is being analyzed; Equation (5.4) then becomes

$$R = \Lambda \Phi \Lambda' + \Psi^2 \tag{5.5}$$

and $R^* = R - \Psi^2$ is known as the matrix of 'reproduced correlations'. If $k$, the number of common factors, has been chosen correctly, and if the model (5.1) holds, then the 'residual correlations' given by the off-diagonal elements of $R^+ = R - R^*$ should be randomly distributed about a mean of zero. This implies that $R^*$ (or $\Sigma^*$) is of rank $k$, that is, it has exactly $k$ nonzero eigenvalues.

Many other facets of the factor model are open to discussion, and readers are referred to the work of Lawley and Maxwell (1971), Joreskog (1963) and Anderson and Rubin (1956) for detailed expositions. The last-named authors give a very comprehensive account of the common factor model.

## 5.3 Estimation

The computational procedures in this section will be illustrated by reference to a simple example, which is based on the work of Gustafson (1973). This author uses the techniques of orthophotography to extract several measures of drainage-basin morphology, the basins measured being located either in the Wies/Schönau basins in Central Europe or in the Far East. For the purpose of this chapter, 8 of Gustafson's morphological variables were selected namely: basin area (km$^2$), perimeter (km), total stream length (km), total number of streams, length of primary drainage channel (km), basin slope (degrees; defined as the arc tangent of the ratio of the elevation of the headwaters of the primary drainage channel to the length of the primary drainage channel), basin shape (defined as basin length divided by basin width), and relative relief of the basin (km). The first four variables (area, perimeter, total stream length and stream numbers) were expected to correlate highly, because, in a region of relatively homogeneous climate, there should be a close relationship between basin 'size' characteristics. The slope and relative relief variables were also expected to relate strongly, leaving a third, shape dimension of relatively minor importance. (This interpretation is, in part, a function of the variables chosen. However, as stated above, we are not interested in finding 'basic underlying causes' in factor analysis, but solely in identifying common sources of covariation in a given set of measurements.)

The data used were the measurements on these 8 variables for the first 41 of the Central European basins listed by Gustafson (1973, p. 135). These were arranged in order of increasing area. Figure 5.1 shows selected bivariate plots of the data, which are listed in Table 5.1.

Figure 5.1(a) shows a strong linear association ($r = 0.9348$) between basin area and total stream length, which indicates that the largest basins have the longest stream network length. Figure 5.1(b) shows a very weak linear relationship ($r = -0.0949$) between the variables of perimeter length and basin shape. Thirdly, Figure 5.1(c) shows a negative linear relationship ($r = -0.7303$) for variables 4 and 6 (perimeter length and basin slope), the largest basins (in areal extent) having the lowest slopes.

The variance—covariance and correlation matrices derived from this data are given in Table 5.2. These matrices were computed using subroutine SSCP and printed by subroutine MATPR (Appendix A).

At this point it is necessary to distinguish two approaches to the problem of estimation. The first of these is the conventional statistical approach, in which it is assumed that the elements of the ($n \times p$) matrix X make up a random sample of $n$ measurements on a fixed set of $p$ variables and that the structure of the population relationships among these variables is to be inferred from the sample. In the second approach it is assumed that the matrix X contains a set of $n$ measurements on a selection of $p$ variables, the $n$ individuals comprising the complete population of individuals. In other words, there is a 'universe of variables' which is made up of all variables which could conceivably be measured. The $p$ observed variables are a subset drawn from this universe, and the purpose of the factor analysis is to infer the nature of the relationships in the universe of variables from the selection at our disposal. Hotelling (1933) set out this approach succinctly:

> 'Instead of regarding the analysis of a particular set of tests as our ultimate goal, we may look on these merely as a sample of a hypothetical larger aggregate of tests. Our aim is then to learn something of the situation portrayed by the larger aggregate. We are thus brought to a type of sampling theory quite distinct from that which we have heretofore considered. Instead of dealing with the degree of instability of functions of the correlations of the observed tests arising from the smallness of the number of persons, we are now concerned with the degree of instability resulting from the limited number of tests whose correlations now enter our analysis.'

Spearman's uni-factor theory of mental ability, referred to in Section 5.2, could be quoted as an example of the drawing of inferences about a universe of variables (performance in all possible tests of mental ability) from a selection of such tests. In the remainder of this chapter we shall use the term *selection* to mean a set of variables drawn from, or representative of, a universe of variables, restricting the use of the term *sample* to its usual statistical context, i.e. denoting a subset of measurements on a number of individuals drawn from a larger set of such measurements. If the factor analysis is concerned with estimating the parameters of the model from a sample of individuals we will refer to it as 'statistical factor analysis' in distinction to 'nonstatistical factor analysis' in which the direction of inference is from a selection to a universe of variables.

## 5.3.0 Nonstatistical Factor Analysis

In this subsection *alpha factor analysis*, developed by Kaiser and Caffrey (1965), is described. The aim of the alpha method is to estimate common factors from the selection of variables which have maximum correlation with the universe common factors. It is assumed that the individuals measured upon the selection of variables make up the population of interest. Alpha factor analysis takes its name from the alpha coefficient of reliability or generalizability proposed by Cronbach (1951). In the context of this section, the alpha coefficient can be treated as the squared correlation between a common factor in the selection of variables and the corresponding universe common factor. Kaiser and Caffrey (1965, p. 6) show that these alpha coefficients can be computed from

$$\alpha_s = (p/p - 1) \cdot (1 - 1/d_s)$$

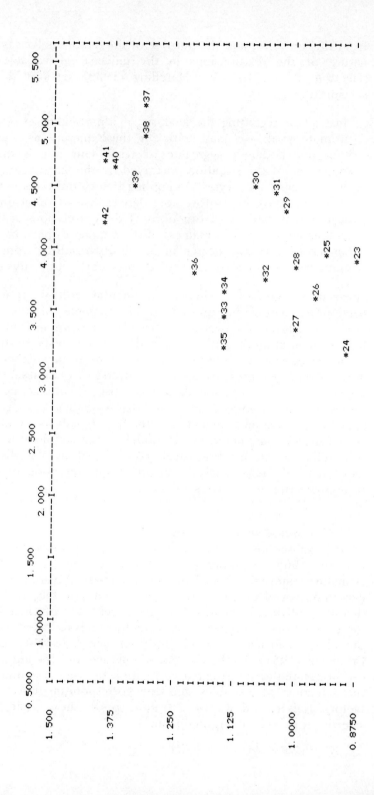

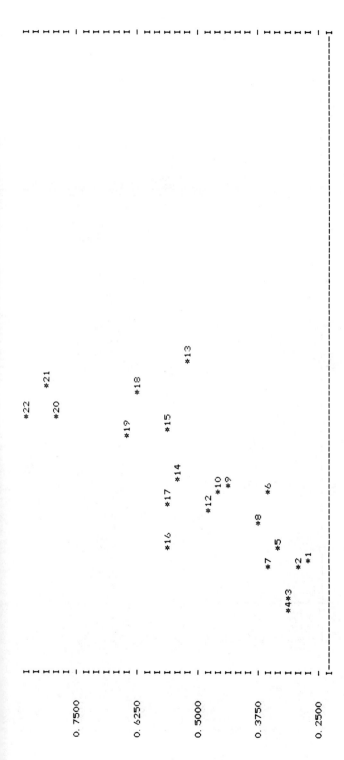

Figure 5.1 Selected bivariate scatter plots of variables listed in Table 5.1

(a) total stream length (horizontal axis) versus basin area

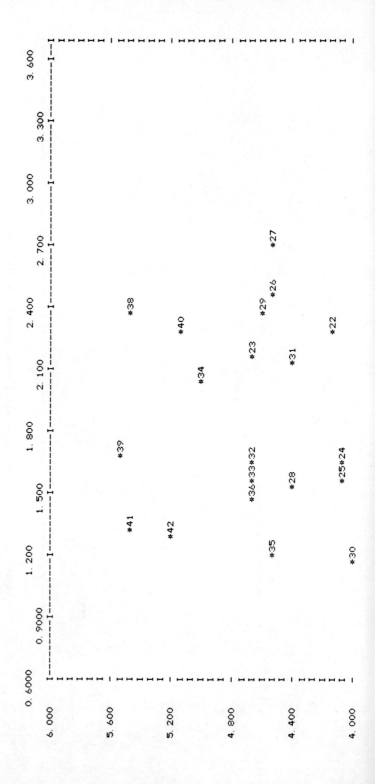

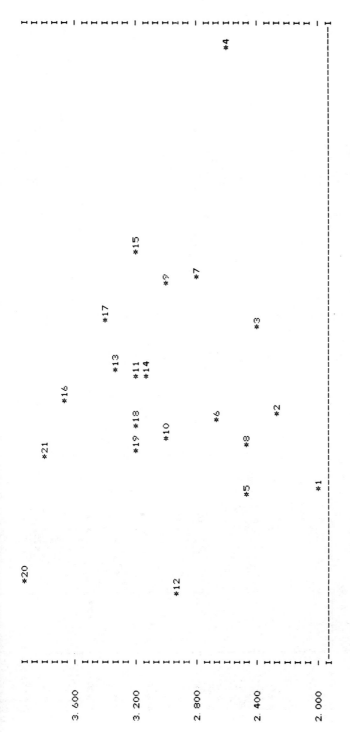

Figure 5.1  *continued*

(b) perimeter length versus basin shape (horizontal axis)

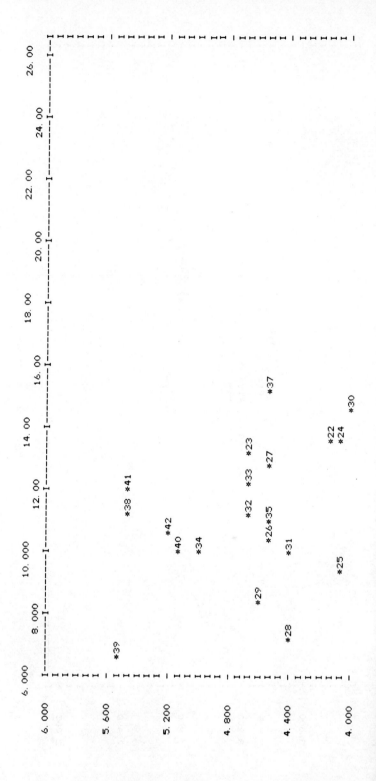

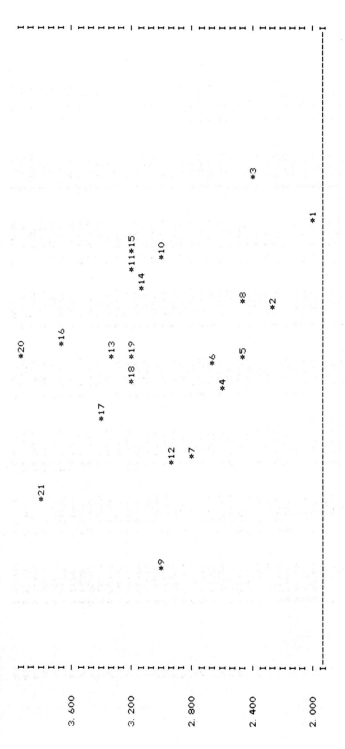

Figure 5.1 *continued*

(c) perimeter length versus basin slope (horizontal axis)

Table 5.1

Data for factor analysis example (Reproduced with permission from G. C. Gustafson, *Münchener Geographical Abhandlungen*, 11, Geographische Institut, München, 1973)

| | 1 | 2 | 3 | 4 | 5 | 6 | 7 | 8 |
|---|---|---|---|---|---|---|---|---|
| 1 | 0.26880 | 1.37500 | 5.00000 | 2.00000 | 0.82500 | 20.35680 | 1.43270 | 0.31200 |
| 2 | 0.27190 | 1.35000 | 5.00000 | 2.22500 | 0.91250 | 17.53360 | 1.80540 | 0.30500 |
| 3 | 0.29350 | 1.07500 | 3.00000 | 2.35000 | 0.95000 | 21.89860 | 2.22860 | 0.41100 |
| 4 | 0.30410 | 0.97500 | 3.00000 | 2.60000 | 1.22500 | 14.94960 | 3.58400 | 0.32800 |
| 5 | 0.32450 | 1.48750 | 4.00000 | 2.42500 | 0.93750 | 16.07980 | 1.41250 | 0.29100 |
| 6 | 0.33680 | 1.93750 | 8.00000 | 2.67500 | 1.20000 | 15.86510 | 1.75760 | 0.34200 |
| 7 | 0.34920 | 1.33750 | 3.00000 | 2.78750 | 1.18750 | 12.86680 | 2.46700 | 0.29200 |
| 8 | 0.36150 | 1.67500 | 4.00000 | 2.41250 | 0.97500 | 17.75050 | 1.65330 | 0.35800 |
| 9 | 0.42950 | 1.98750 | 9.00000 | 2.96250 | 1.28750 | 9.19580 | 2.42000 | 0.21800 |
| 10 | 0.44680 | 1.95000 | 7.00000 | 2.95000 | 1.21250 | 19.18080 | 1.67190 | 0.42300 |
| 11 | 0.45920 | 1.95000 | 6.00000 | 3.17500 | 1.28750 | 18.89680 | 1.97200 | 0.46200 |
| 12 | 0.46660 | 1.81250 | 7.00000 | 2.88750 | 0.97500 | 12.51400 | 0.91710 | 0.26200 |
| 13 | 0.51480 | 2.98750 | 12.00000 | 3.31250 | 1.45000 | 15.96530 | 2.00380 | 0.41600 |
| 14 | 0.53770 | 2.02500 | 5.00000 | 3.08750 | 1.20000 | 18.17120 | 1.99100 | 0.43000 |
| 15 | 0.55000 | 2.43750 | 4.00000 | 3.20000 | 1.50000 | 19.47580 | 2.58850 | 0.57200 |
| 16 | 0.55620 | 1.50000 | 4.00000 | 3.67500 | 1.26250 | 16.49880 | 1.86520 | 0.50500 |
| 17 | 0.56550 | 1.82500 | 3.00000 | 3.37500 | 1.58750 | 13.96910 | 2.24900 | 0.39600 |
| 18 | 0.61490 | 2.75000 | 9.00000 | 3.20000 | 1.23750 | 15.14990 | 1.72630 | 0.35100 |
| 19 | 0.63840 | 2.38750 | 5.00000 | 3.20000 | 1.21250 | 15.92770 | 1.62590 | 0.39500 |
| 20 | 0.79720 | 2.55000 | 9.00000 | 3.91250 | 1.07500 | 15.94500 | 0.97490 | 0.46300 |
| 21 | 0.81700 | 2.81250 | 6.00000 | 3.77500 | 1.53750 | 11.44070 | 1.58020 | 0.36200 |
| 22 | 0.85470 | 2.53750 | 4.00000 | 4.15000 | 1.75000 | 13.49060 | 2.28520 | 0.44000 |
| 23 | 0.87140 | 3.91250 | 12.00000 | 4.67500 | 2.01250 | 13.11020 | 2.17550 | 0.47000 |
| 24 | 0.90040 | 3.17500 | 6.00000 | 4.05000 | 1.67500 | 13.62090 | 1.65400 | 0.45200 |
| 25 | 0.94240 | 3.97500 | 11.00000 | 4.05000 | 1.63750 | 9.40590 | 1.55090 | 0.31200 |
| 26 | 0.96720 | 3.58750 | 9.00000 | 4.54250 | 1.92500 | 10.36350 | 2.45450 | 0.40000 |
| 27 | 0.99620 | 3.37500 | 7.00000 | 4.56250 | 2.01250 | 12.67860 | 2.69890 | 0.47400 |
| 28 | 1.01040 | 3.83750 | 9.00000 | 4.41250 | 1.52500 | 7.15670 | 1.53170 | 0.28200 |
| 29 | 1.01970 | 4.30000 | 13.00000 | 4.63750 | 1.95000 | 8.40770 | 2.35860 | 0.36900 |
| 30 | 1.09700 | 4.52500 | 13.00000 | 3.97500 | 1.50000 | 14.53530 | 1.15780 | 0.53800 |
| 31 | 1.04440 | 4.46250 | 8.00000 | 4.40000 | 2.05000 | 9.96060 | 1.13850 | 0.37100 |
| 32 | 1.06910 | 3.77500 | 10.00000 | 4.68750 | 1.80000 | 11.12700 | 1.65900 | 0.41000 |
| 33 | 1.14330 | 3.46250 | 8.00000 | 4.70000 | 1.95000 | 12.12250 | 1.54420 | 0.45000 |
| 34 | 1.15880 | 3.65000 | 6.00000 | 2.22500 | 2.05000 | 9.87950 | 2.04550 | 0.47300 |
| 35 | 1.16000 | 3.21250 | 5.00000 | 4.56250 | 1.56250 | 10.87480 | 1.20120 | 0.42100 |
| 36 | 1.21750 | 3.81250 | 10.00000 | 4.70000 | 1.78750 | 11.96970 | 1.46980 | 0.47500 |
| 37 | 1.31200 | 5.18250 | 14.00000 | 4.57500 | 1.82500 | 15.28060 | 1.14490 | 0.51000 |
| 38 | 1.31450 | 4.91250 | 15.00000 | 5.52500 | 2.48750 | 11.15820 | 2.38150 | 0.49200 |
| 39 | 1.34420 | 4.50000 | 13.00000 | 5.55000 | 2.32500 | 6.58210 | 1.67500 | 0.31100 |
| 40 | 1.38430 | 4.67500 | 14.00000 | 5.15000 | 2.10000 | 9.96570 | 2.28410 | 0.39000 |
| 41 | 1.40780 | 4.72500 | 12.00000 | 5.46250 | 2.22500 | 11.87120 | 1.30300 | 0.46900 |
| 42 | 1.41340 | 4.22500 | 7.00000 | 5.21250 | 1.87500 | 10.60420 | 1.29370 | 0.41200 |

Table 5.2
(a) Variance–covariance matrix and (b) correlation matrix for the data of Table 5.1.

(a)

|   | 1 | 2 | 3 | 4 | 5 | 6 | 7 | 8 |
|---|---|---|---|---|---|---|---|---|
| 1 | 0.1346 | | | | | | | |
| 2 | 0.4117 | 1.4402 | | | | | | |
| 3 | 0.8450 | 3.4486 | 12.1207 | | | | | |
| 4 | 0.3516 | 1.0690 | 2.2416 | 0.9543 | | | | |
| 5 | 0.1366 | 0.4396 | 0.9263 | 0.3940 | 0.1807 | | | |
| 6 | -0.9230 | -2.8868 | -6.3526 | -2.6366 | -1.0718 | 13.2286 | | |
| 7 | -0.0441 | -0.1209 | -0.3635 | -0.0495 | 0.0350 | -0.0048 | 0.2765 | |
| 8 | 0.0116 | 0.0340 | 0.0457 | 0.0319 | 0.0129 | 0.0690 | -0.0007 | 0.0062 |

(b)

|   | 1 | 2 | 3 | 4 | 5 | 6 | 7 | 8 |
|---|---|---|---|---|---|---|---|---|
| 1 | 1.00000 | | | | | | | |
| 2 | 0.93481 | 1.00000 | | | | | | |
| 3 | 0.66149 | 0.82541 | 1.00000 | | | | | |
| 4 | 0.96584 | 0.89860 | 0.64898 | 1.00000 | | | | |
| 5 | 0.87573 | 0.86182 | 0.62592 | 0.93433 | 1.00000 | | | |
| 6 | -0.69160 | -0.66136 | -0.50169 | -0.73068 | -0.69328 | 1.00000 | | |
| 7 | -0.22843 | -0.20431 | -0.19857 | -0.09494 | 0.15666 | -0.00249 | 1.00000 | |
| 8 | 0.39981 | 0.35933 | 0.16640 | 0.40605 | 0.38558 | 0.24036 | -0.01751 | 1.00000 |

where **D** is the diagonal matrix of eigenvalues of **B** and

$$B = H^{-1}(R - I)H^{-1} + I$$

In this equation **H** is the diagonal matrix with the square roots of communality values in its principal diagonal. (The inverse of a diagonal matrix is found simply by replacing diagonal elements by their reciprocals.) The elements of the diagonal of $H^2$ are unknown, and are estimated by an iterative procedure. At each iteration a new estimate $\hat{H}^2$ is calculated, a new matrix **B** is then found and its eigenvalues **D** are summed. When tr(**D**) differs from $p$ by a specified amount, then the procedure has converged. The alpha factor loadings matrix $\Lambda_\alpha$ is then determined from

$$\Lambda_\alpha = HQD^{1/2} \tag{5.6}$$

where **Q** is the matrix of normalized eigenvectors of **B** and **D** the diagonal matrix of eigenvalues of **B**.

Alpha factors have the property that they each maximize the associated alpha coefficients $\alpha_s$. If the associated $\alpha_s$ is nonpositive then the corresponding universe factor is not considered to be meaningful. In fact, the number of meaningful factors (i.e. those which have positive $\alpha_s$ coefficients) is the same as the number of eigenvalues of **R** that exceed 1.0.

A second property of alpha factors is that the elements of $\Lambda_\alpha$ are independent of the scales used to measure the variables. This is an important feature whenever these scales are arbitrary. The topic of scaling was discussed in Chapter 4 in connection with principal components analysis. On the debit side, the diagonal elements of $H^2$ generated by the iterative procedure outlined above can exceed 1.0, which implies that the common variance of a variable exceeds its total variance. This is generally thought to be unacceptable, though Kaiser and Caffrey (1965, p. 9) do not share this view. Situations involving $h_i^2$ values in excess of 1.0 are termed *Heywood cases*. They appear frequently in analyses of observational data; so much so that the other two factor estimation methods that are described in this section embody precautions to ensure that $h_i^2 \leqslant 1.0$.

The computations involved in alpha factor are straightforward, and can be summarized as follows.

(1) Beginning with $\hat{H}_1^2 = I$ or $\hat{H}_1^2 = I - (\text{diag}(R^{-1}))^{-1}$, the diagonal matrix containing the squared multiple correlation of each variable with the remaining $p - 1$ variables, determine $B = \hat{H}_1^{-1}(R - I)\hat{H}_1^{-1} + I$ and find the $k$ largest eigenvalues and the corresponding unit-length eigenvectors of $B$. The value of $k$ is set to the number of eigenvalues of $R$ that exceed 1.0. Let $D_k$ and $Q_k$ indicate the $(k \times k)$ diagonal matrix holding the first $k$ eigenvalues of $B$ and the $(p \times k)$ matrix holding the corresponding unit-length eigenvectors of $B$, respectively.

(2) Find the first approximation to the factor loading matrix for $B$ from the equation $\hat{C} = Q_k D_k^{1/2}$.

(3) The revised estimate $\hat{H}_2^2$ is now found from $\hat{H}_2^2 = \hat{H}_1^2 \, \text{diag}(\hat{C}\hat{C}')$

(4) Repeat steps (2) and (3) until $\sum_{i=1}^{k} d_i = p \pm \epsilon$ where $\epsilon$ is a predefined tolerance level.

(5) The matrix of alpha factor loadings is now computed from $\Lambda_\alpha = \hat{H}\hat{C}$ where $\hat{H}$ and $\hat{C}$ are the final approximations to $H$ and $C$.

A subroutine to carry out these operations is listed in Table 5.3.

The results achieved by applying the alpha factor analysis method to the Gustafson data are shown in Table 5.6. The initial communality estimates were squared multiple correlations (SMC's) computed from

$$\text{SMC}_j = 1.0 - 1.0/r^{jj}$$

where $r^{jj}$ represents the $j^{\text{th}}$ diagonal element of $R^{-1}$. Some elements of $R^{-1}$ (Table 5.4) are relatively large, indicating either that the $p$ variables are not a good selection from the universe of variables or that the factor analysis model does not hold even as $p$ approaches infinity. (Kaiser, 1963).

Using a convergence criterion of 0.001 the following results were obtained:

| Iteration | Sum of eigenvalues | Sum of squares of rejected eigenvalues |
|---|---|---|
| 1 | 7.691330 | 0.155340 |
| 2 | 7.914813 | 0.094060 |
| 3 | 7.978682 | 0.086658 |
| 4 | 7.994470 | 0.085843 |
| 5 | 7.998002 | 0.085644 |
| 6 | 7.998690 | 0.085433 |
| 7 | 7.998803 | 0.085186 |
| 8 | 7.998841 | 0.084925 |

The sum of squares of the 'rejected' eigenvalues has converged to within $\epsilon = 0.001$ by iteration 8. The rejected eigenvalues are those with negative coefficients of generalizability, as discussed above. In this case, there were 5 rejected eigenvalues, giving 3 generalizable alpha factors.

Variable 5 has a communality greater than 1.0; hence it is a Heywood case. These occur frequently in those approaches to factor analysis that involve iteration. The reponse of the user to this situation can range from indifference to over-concern. Heywood cases can mean that the factor analysis model, for $k$ factors, does not fit, or that the sample size is too small, or that too few variables have been selected. Some authors consider that variables for which computed communalities equal or exceed 1.0 should be treated as factors and that the matrix of partial correlation coefficients (with the offending variable partialled out) should be factored rather than the original correlation matrix. In this example, the length of primary drainage channel could hardly be regarded as a 'factor'. Other authors have viewed the problem of iterating to minimize a given quantity (the sum of squares of rejected eigenvalues) as one of *constrained* minimization subject to the condition that no communality exceeds 1.0. In the next part of this section procedures for ensuring that the minimum lies within the acceptable region defined by $h_i^2 \leqslant 1.0$ will be described for the minres method. No such procedure is available for alpha factor analysis. In this respect, the technique may be regarded as unsatisfactory.

Table 5.6(a) shows that the important loadings on Factor 1 are those of variables 1–5 (positive), 6 (negative) and, to a much lesser extent, 8 (positive). All these variables are related to basin magnitude, with area, perimeter, total stream length, total number of streams and length of the primary drainage channel being positively related (i.e. they increase with basin size) while basin slope declines as the basin magnitude increases. Relative relief shows a tendency to increase as the basin size increases, but this is by no means a strong relationship.

Only two variables relate to Factor 2. If all the signs in column 2 of

## Table 5.3
## Subroutine ALPHA: alpha factor analysis

```
      SUBROUTINE ALPHA(R,N1,RR,N2,H,N3,COM,HOLD,DUM,N4,NV,MAXIT,EPSI,
     + NUF,FMT)
C
C SUBROUTINE FOR ALPHA FACTOR ANALYSIS. SEE CHAPTER 5 FOR DETAILS.
C ARGUMENTS:
C R        REAL ARRAY, MINIMUM DIMENSION NV*NV. HOLDS CORRELATION MATRIX
C          ON INPUT. UNALTERED BY THE ROUTINE.
C N1       INTEGER CONSTANT. FIRST DIMENSION OF R IN CALLING PROGRAM.
C RR       REAL MATRIX,MINIMUM DIMENSION NV*NV,USED AS WORKSPACE.
C N2       INTEGER CONSTANT. FIRST DIMENSION OF RR IN CALLING PROGRAM.
C H        REAL ARRAY,MINIMUM DIMENSION NV*NV. USED AS WORKSPACE.
C N3       INTEGER CONSTANT. FIRST DIMENSION OF H IN CALLING PROGRAM.
C COM      REAL ARRAY,MINIMUM DIMENSION NV USED AS WORKSPACE.
C HOLD     REAL ARRAY,MINIMUM DIMENSION NV,USED AS WORKSPACE.
C DUM      REAL ARRAY,MINIMUM DIMENSION NV*NV. USED AS WORKSPACE.
C N4       INTEGER CONSTANT. FIRST DIMENSION OF DUM IN CALLING PROGRAM.
C NV       INTEGER CONSTANT. NUMBER OF VARIABLES.
C MAXIT    INTEGER. MAXIMUM NUMBER OF ITERATIONS. IF ZERO OR LESS
C          OR GREATER THAN 100 ON ENTRY, MAXIT IS RESET TO 40.
C EPSI     REAL CONSTANT. CONVERGENCE CRITERION. IF ZERO OR LESS THAN
C          0.00000001 ON ENTRY, EPSI IS RESET TO 0.00001
C NUF      INTEGER CONSTANT. NUMBER OF FACTORS. IF ZERO ON ENTRY THE
C          NUMBER OF FACTORS IS SET BY THE ROUTINE.
C FMT      REAL ARRAY. OUTPUT FORMAT FOR USE IN MATPR. SEE APPENDIX A.
C
C OTHER ROUTINES NEEDED:
C MATPR,GJDEF1,COPY,HTDQL
C
      REAL MACHEP
      DIMENSION R(N1,NV),RR(N2,NV),H(N3,NV),COM(NV),HOLD(NV),DUM(N4,NV),
     1 FMT(20)
C *** NOTE: THE FOLLOWING TWO PARAMETERS ARE MACHINE DEPENDENT
C SEE APPENDIX A.
      MACHEP=2.0**(-37)
      EPS=2.0**(-255)
      TOL=EPS/MACHEP
  100 FORMAT(22H1ALPHA FACTOR ANALYSIS/68H PROGRAM BY PAUL M. MATHER, GEO
     *GRAPHY DEPT., UNIVERSITY OF NOTTINGHAM///)
      WRITE(6,100)
      IF(MAXIT.LE.0.OR.MAXIT.GT.100) MAXIT=40
      IF (EPSI.LE.0.0.OR.EPSI.LT.1.0E-7) EPSI=1.0E-4
      WRITE(6,130) NV,MAXIT,EPSI
  130 FORMAT('0INPUT SPECIFICATIONS'/
     *' NUMBER OF VARIABLES (TESTS) ',I6/
     *' MAXIMUM NUMBER OF ITERATIONS',I6/
     *' CONVERGENCE CRITERION (EPSI)',G12.7)
      IP=1
      NN=N1*NV
      CALL MATPR(24HCORRELATION MATRIX      ,6,FMT,6,R,NN,N1,NV,NV,
     1 IP,1,4)
      CALL COPY(RR,N2,R,N1,NV,NV)
      CALL GJDEF1(NV,RR,COM,N2)
      IF(NV) 612,613,612
  613 WRITE(6,614)
  614 FORMAT('0CORRELATION MATRIX IS SINGULAR')
      STOP
  612 IP=1
      NN=N2*NV
      CALL MATPR(32HINVERSE OF CORRELATION MATRIX   ,8,FMT,6,RR,NN,N2,
     1 NV,NV,IP,1,4)
      DO 3 I=1,NV
    3 H(I,I)=1.0-1.0/RR(I,I)
  103 FORMAT('0SMC COEFFICIENTS'// ',5(10(12F8.4)))
      SS=0.
      DO 98 I=1,NV
```

```
      HOLD(I)=H(I,I)
   98 H(I,I)=SQRT(H(I,I))
      WRITE(6,103) (HOLD(I),I=1,NV)
      WRITE(6,108)
      DO 51 KOUNT=1,MAXIT
      CALL COPY(RR,N2,R,N1,NV,NV)
      DO 7 I=1,NV
      S=H(I,I)
      DO 717 J=1,NV
  717 RR(I,J)=RR(I,J)/H(J,J)/S
    7 RR(I,I)=1.0
      CALL HTDQL(NV,N2,N3,RR,COM,DUM,H,MACHEP,TOL)
      IF(KOUNT.GT.1) GOTO 10
      DO 9 I=1,NV
      IF(COM(I).GE.1.0) GOTO 9
      NUF=I-1
      GOTO 10
    9 CONTINUE
   10 S=0.
      DO 12 I=1,NUF
      S=S+COM(I)
      D=SQRT(COM(I))
      DO 12 J=1,NV
   12 H(J,I)=H(J,I)*D
  108 FORMAT(1H1,10X,18HSUM OF EIGENVALUES,5X,26HSS OF REJECTED EIGENVAL
     *UES///)
  109 FORMAT(' ',I6,8X,F10.6,14X,F10.6)
      Y=0.
      K=NUF+1
      DO 21 I=K,NV
   21 Y=Y+COM(I)*COM(I)
      WRITE(6,109) KOUNT, S,Y
      DO 13 I=1,NV
      DO 13 J=1,NV
      RR(I,J)=0.
      DO 13 K=1,NUF
   13 RR(I,J)=RR(I,J)+H(J,K)*H(J,K)
      DO 14 I=1,NV
   14 HOLD(I)=HOLD(I)*RR(I,I)
      IF(ABS(S-SS).LE.EPSI) GOTO 17
      DO 16 I=1,NV
      DO 217 J=1,NV
  217 H(I,J)=0.
   16 H(I,I)=SQRT(HOLD(I))
   51 SS=S
   17 DO 18 I=1,NV
      S=SQRT(HOLD(I))
      DO 18 J=1,NUF
   18 H(I,J)=H(I,J)*S
      IP=1
      NN=N3*NV
      CALL MATPR(24HALPHA FACTOR LOADINGS    ,6,FMT,6,H,NN,N3,NV,NUF,
     1 IP,2,4)
 1010 FORMAT(////' CALCULATED COMMUNALITIES'////)
      WRITE(6,1010)
      DO 67 I=1,NV
      HOLD(I)=0.
      DO 67 J=1,NUF
   67 HOLD(I)=HOLD(I)+H(I,J)*H(I,J)
      WRITE(6,1011) (I,HOLD(I),I=1,NV)
 1011 FORMAT(1H ,10(I4,')',F8.4))
      S=0.
      DO 47 I=1,NV
   47 S=S+HOLD(I)
      SS=(S/FLOAT(NV))*100.0
      WRITE(6,1012) S,SS
 1012 FORMAT('0TOTAL COMMUNALITY =',F8.3,//' EXPRESSED AS A PERCENTAGE OF
     1 TOTAL VARIANCE =',F8.3,'%')
      RETURN
      END
```

Table 5.4
Inverse of correlation matrix shown in Table 5.2(b)

|   | 1 | 2 | 3 | 4 | 5 | 6 | 7 | 8 |
|---|---|---|---|---|---|---|---|---|
| 1 | 44.3580 | | | | | | | |
| 2 | -26.9821 | 31.4690 | | | | | | |
| 3 | 8.9946 | -11.0493 | 5.8785 | | | | | |
| 4 | -37.9741 | 23.9993 | -8.2247 | 58.9546 | | | | |
| 5 | 9.5351 | -13.4247 | 3.1960 | -24.1215 | 22.3509 | | | |
| 6 | -4.6081 | 5.9712 | -2.4373 | 14.3241 | -3.4072 | 10.3007 | | |
| 7 | 1.3554 | 2.3863 | -0.2863 | 3.8312 | -5.6877 | 1.4822 | 2.9508 | |
| 8 | 3.4136 | -4.6914 | 2.1005 | -10.2158 | 2.4237 | -6.8767 | -1.0260 | 5.8408 |

Table 5.5
SMC's for Gustafson data (Table 5.1)

| Variable: | 1 | 2 | 3 | 4 | 5 | 6 | 7 | 8 |
|---|---|---|---|---|---|---|---|---|
| SMC: | 0.9775 | 0.9682 | 0.8299 | 0.9830 | 0.9553 | 0.9029 | 0.6645 | 0.8288 |

Table 5.6
(a) Alpha factor loadings; (b) communalities from alpha factor analysis

(a)

|   | 1 | 2 | 3 |
|---|---|---|---|
| 1 | 0.9649 | -0.0847 | 0.0670 |
| 2 | 0.9909 | -0.0602 | 0.1063 |
| 3 | 0.7177 | 0.0194 | 0.1477 |
| 4 | 0.9659 | -0.0317 | -0.0914 |
| 5 | 0.9307 | 0.0450 | -0.3770 |
| 6 | -0.7293 | -0.5783 | -0.0076 |
| 7 | -0.1442 | 0.2094 | -0.7371 |
| 8 | 0.3451 | -0.8591 | -0.2856 |

Table 5.6 (b)

| Variable: | 1 | 2 | 3 | 4 | 5 | 6 | 7 | 8 |
|---|---|---|---|---|---|---|---|---|
| Communality | 0.9427 | 0.9968 | 0.5373 | 0.9423 | 1.0103 | 0.8663 | 0.6079 | 0.9047 |

Table 5.6(a) are reversed, Factor 2 is seen to be a relief factor, picking out basin slope and relative relief. Basin shape may be slightly related to this factor in a negative sense. Factor 3 is identified by variable 7 (basin shape) with variable 5 (length of primary drainage channel) having weak links with this factor. Relative relief (but not basin shape) may also have an affinity with this factor.

The problem of bringing the factors into sharper focus is discussed in a later section under the heading of factor rotation. We can note in passing, however, that several variables (notably 6 and 8) show linkages across two or even all three of the factors, indicating that these factors are perhaps not independent in subject-matter (rather than in statistical) terms.

## 5.3.1 Statistical Factor Analysis

The factor loadings $\lambda_{ij}$ can be estimated statistically by (i) least squares or (ii) maximum likelihood methods, much in the same way as were the parameters of the linear regression model (Chapter 2). The least-squares method does not require the assumption of multivariate normality of X, whereas the maximum likelihood method does. Thus, statistical tests are, in theory, only applicable when the maximum likelihood method is used.

Harman and Jones (1966) first proposed a feasible algorithm permitting the use of least-squares estimators in factor analysis. They termed their method *minres* which stands for 'minimum residuals'. In Section 5.2 we saw that the off-diagonal correlations could be reproduced from the factor loadings matrix $\Lambda$ (Equation 5.4), $R^*$ being the matrix of reproduced correlations. $R^+$, the residual matrix, is defined as $R^+ = R - R^*$. The aim of the minres approach is to minimize the sums of squares of the off-diagonal elements of $R^+$ or, in other words, find $\Lambda$ subject to the constraint that

$$f(\Lambda) = \sum_{i=j+1}^{p} \sum_{j=1}^{p-1} \left( r_{ij} - \sum_{m=1}^{p} \lambda_{jm} \lambda_{im} \right)^2 = \text{minimum} \tag{5.7}$$

This principle is so well-known that it is not surprising that it had been stated by earlier workers. Harman and Jones (1966) give the first practical method of estimating all the elements of $\Lambda$ simultaneously. The stepwise approach of Comrey (1962) does not produce results that are identical to the simultaneous approach of Harman and Jones.

The computational method adopted by Harman and Jones is described by Harman (1967) as a variant of the Gauss–Seidel procedure which involves successive changes in the elements of an initial approximation to $\Lambda$, $\hat{\Lambda}_0$, to give a new estimate, $\hat{\Lambda}_1$. The process is repeated to produce a series of estimates $\hat{\Lambda}_2, \hat{\Lambda}_3, \ldots \hat{\Lambda}_m$ which eventually converges to the required solution. Joreskog and van Thillo (1971) use a modified form of the Newton–Raphson algorithm (Section 3.5.1) in their computer program for ordinary least-squares factor analysis (minres) and claim that it is more efficient than the Gauss–Seidel method. For moderate values of $p$ the difference is not startling.

The mathematical derivation of the following computing formulae is not given here; it can be found in Harman (1967). Computations begin with the calculation of an initial approximation to $\Lambda$. Usually $\hat{\Lambda}_0$ is taken to be the first k columns of the principal components loadings matrix based on **R** (Chapter 4). Difficulties can be encountered if the PCA solution is used as an initial approximation, and better results are achieved if the principal components of $(\mathbf{R} - \mathbf{I})$ are used, or alternatively the principal components of $(\mathbf{R} - \mathbf{I} + \mathbf{H}^2)$. (H. H. Harman, personal communication, 1974).* The procedure continues as follows:

(1) $m = 0$
(2) $m = m + 1$
(3) Define $\hat{\mathbf{\Lambda}}_{j(}^{m}$ as the current approximation to the factor loadings matrix with row $j$ set to zero and $_0\mathbf{r}_j^m$ as the row vector of residual correlations for variable $j$ with the residual self-correlation set to zero. The displacements $\boldsymbol{\epsilon}_j^m$ for row $j$ of $\hat{\mathbf{\Lambda}}^m$ are given by:

$$\boldsymbol{\epsilon}_j^m = {}_0\mathbf{r}_j^m \hat{\mathbf{\Lambda}}^m \left( \hat{\mathbf{\Lambda}}_{j(}^{m\prime} \hat{\mathbf{\Lambda}}_{j(}^{m} \right)^{-1}$$

(4) Determine the new approximation to row $j$ of $\hat{\mathbf{\Lambda}}^{m+1}$ from

$$\hat{\boldsymbol{\lambda}}_j^{m+1} = \hat{\boldsymbol{\lambda}}_j^m + \boldsymbol{\epsilon}_j^m$$

(5) Repeat steps 3 and 4 for each row of $\hat{\mathbf{\Lambda}}^m$ in turn. $(j = 1, 2, \ldots, p)$.
(6) Test for convergence by comparing $|\epsilon_{ij}^m|_{\max}$, the absolute largest increment at iteration $m$, with a preset tolerance (0.001 is suitable). If convergence is not yet attained, return to step 2.

During each minor cycle, that is, as each new approximation to the factor loadings matrix is computed, a check is made to see whether the communality of the row currently being operated upon exceeds 1.0. The communality of row $j$ of the current estimate of the factor loadings matrix $\hat{\mathbf{\Lambda}}^m$ is:

$$h_j^2 = \sum_{i=1}^{k} (\hat{\lambda}_{ji}^m)^2 \tag{5.8}$$

Unlike alpha factor analysis, the minres method incorporates a procedure which corrects the entries in a particular row if the communality of that row exceeds 1.0. Algorithms to achieve this are described by Harman (1967) and by Harman and Fukuda (1966) but the method of Browne (1967) as detailed by Mulaik (1972) and Cramer (1974) is preferred due to its computational simplicity. Browne's method is summarized as follows:† let $\hat{\boldsymbol{\lambda}}_j$ be the current row of $\hat{\mathbf{\Lambda}}$, that is, row $j$ of $\hat{\mathbf{\Lambda}}$ is being modified by the Gauss–Seidel process described above. If the $j^{\text{th}}$ row of the correlation matrix **R** is denoted by $\mathbf{r}_j$ where

---

*$\mathbf{H}^2$ is here the diagonal matrix containing the squared multiple correlations of each variable with the remaining $p - 1$ variables.

†For clarity, the superscript $m$'s, indicating the iteration number, are omitted in the following paragraphs.

$r_{jj} = 0$ and if $\Lambda_{)j(}$ is the current factor loadings matrix with the entries in row $j$ set to zero, then $\lambda_j^*$, the modified $j^{th}$ row of $\hat{\Lambda}$, is calculated from

$$\lambda_j^* = Pw$$

where $P$ is the matrix of eigenvectors of the matrix $\hat{\Lambda}_{)j(}'\hat{\Lambda}_{)j(}$. The vector $w$ is defined as

$$w_i = u_i/(d_i - \theta)$$

which requires the eigenvalues $d_i$ of $\hat{\Lambda}'_{)j(}\hat{\Lambda}_{)j(}$ and the vector $u$. This vector is obtained from

$$u = P'\Lambda'_{)j(}r_j$$

The parameter $\theta$ is estimated by an iterative process based on Newton's method for the solution of polynomial equations. If $\theta$ is the required root, and if $\theta_j$ is the $j^{th}$ approximation, then an improved approximation is given by

$$\theta_{j+1} = \theta_j - s_j$$

where

$$s_j = \left[ \frac{h(\theta)}{h'(\theta)} \right]_{\theta = \theta_j}$$

The values $h(\theta)$ and $h'(\theta)$ are:

$$h(\theta) = \sum_{i=1}^{k} \frac{u_i^2}{(d_i - \theta)^2} - 1.0$$

$$h'(\theta) = 2 \sum_{i=1}^{k} \frac{u_i^2}{(d_i - \theta)^3}$$

The iterations for $\theta$ continue until the absolute difference between successive approximations is less than a specified small quantity ($0.0^61$ is used in the accompanying program). The initial approximation to $\theta$, $\theta_{0_2}$ is taken to be one half of the smallest eigenvalue, $d_k$, of $\hat{\Lambda}'_{)j(}\hat{\Lambda}_{)j(}$. Successive approximations $\theta_1, \theta_2, \ldots, \theta_q$ are computed by the method given above. If at the $(j+1)^{th}$ iteration the value of $\theta_{j+1}$ exceeds $d_k$ then the approximation to $\theta_{j+1}$ must be $\frac{1}{2}(d_k + \theta_j)$ rather than $\theta_j - s_j$. The new row of $\hat{\Lambda}$ found by this method should have a communality of 1.0 exactly.

The above procedure has been programmed in Fortran as subroutine ADJUST and is listed with the main minres routine in Table 5.7. It should be noted that the same subroutine can be used to carry out a restricted oblique transformation to fit a reference structure matrix, as described by Mulaik (1972, pp. 304–308).

## Table 5.7
### Subroutines for minres factor analysis

```
      SUBROUTINE MINRES(R,N1,RR,N2,V,N3,EIG,ST,   TEMP,N5,DATA,STORE,NV,
     1 NF,MAXIT,EPSI,FMT1,STOR2)
C
C MINRES FACTOR ANALYSIS
C
C ARGUMENTS:
C R       REAL ARRAY,MINIMUM DIMENSION NV*NV. HOLDS FULL CORRELATION
C         MATRIX (UPPER + LOWER TRIANGLES). UNALTERED BY ROUTINE.
C N1      INTEGER. FIRST DIMENSION OF R IN CALLING PROGRAM.
C RR      REAL ARRAY, MINIMUM DIMENSION NV*NV. UNDEFINED ON ENTRY.
C         HOLDS FACTOR LOADINGS ON EXIT.
C N2      INTEGER. FIRST DIMENSION OF RR IN CALLING PROGRAM.
C V       REAL ARRAY,MINIMUM DIMENSION NV*NV. WORKSPACE.
C N3      INTEGER. FIRST DIMENSION OF V IN CALLING PROGRAM.
C EIG     REAL ARRAY, MINIMUM DIMENSION NV. WORKSPACE.
C ST      REAL ARRAY, MINIMUM DIMENSION NV. WORKSPACE.
C TEMP    REAL ARRAY, MINIMUM DIMENSION NV*NV. WORKSPACE.
C N5      INTEGER. FIRST DIMENSION OF TEMP IN CALLING PROGRAM.
C DATA    REAL ARRAY, MINIMUM DIMENSION NV. WORKSPACE.
C STORE   REAL ARRAY, MINIMUM DIMENSION NV, USED AS WORKSPACE.
C NV      INTEGER. NUMBER OF VARIABLES.
C NF      INTEGER. NUMBER OF FACTORS TO BE EXTRACTED.
C MAXIT   INTEGER. MAXIMUM NUMBER OF ITERATIONS IN GAUSS-SEIDEL PROCESS
C         (SEE CHAPTER 5 FOR DETAILS).
C EPSI    CONVERGENCE CRITERION. MUST BE SMALL POSITIVE REAL VALUE, E G.
C         0.0001. DO NOT SET TOO LOW.
C FMT1    REAL ARRAY, DIMENSION 20, HOLDING FORMAT SPECIFICATION FOR USE
C         IN SUBROUTINE MATPR (SEE APPENDIX A).
C STOR2   REAL ARRAY, MINIMUM DIMENSION NV. USED AS WORKSPACE.
C
C       SUBROUTINES NEEDED:
C HTDQL,MATPR,TRMULT,MTMULT,GJDEF1 FROM APPENDIX A.
C ALSO ADJUST,FVAL,CANON AND IMPORT, WHICH ARE LISTED BELOW.
C
      REAL MACHEP
      DIMENSION R(N1,NV),RR(N2,NV),V(N3,NV),EIG(NV),TEMP(N5,NV),ST(NV),
     1 DATA(NV),STORE(NV),STOR2(NV),FMT1(20)
      WRITE(6,401)
  401 FORMAT('1MINRES FACTOR ANALYSIS'//' PROGRAM BY P. M. MATHER'/' DEPART
     *MENT OF GEOGRAPHY'/' UNIVERSITY OF NOTTINGHAM'/' JUNE, 1972'//)
      NK=NV-1
C THE FOLLOWING TWO PARAMETERS ARE MACHINE DEPENDENT AND MUST BE RESET
C FOR COMPUTERS OTHER THAN ICL 1900 SERIES MACHINES. SEE APPENDIX A.
      MACHEP=2.0**(-37)
      ACC=2.0**(-255)
      TOL=ACC/MACHEP
C SET UP INITIAL SOLUTION USING 1'S IN DIAGONAL OF R OR USER-
C SUPPLIED COMMUNALITIES.
      CALL HTDQL(NV,N1,N3,R,EIG,TEMP,V,MACHEP,TOL)
      WRITE(6,400) (EIG(I),I=1,NV)
  400 FORMAT('OEIGENVALUES'/(' ',12F9.3))
      DO 2 I=1,NF
      S=SQRT(EIG(I))
      DO 2 J=1,NV
    2 V(J,I)=V(J,I)*S
      IP=1
      MM=N3*NV
      CALL MATPR(16HINITIAL SOLUTION,2,FMT1,6,V,MM,N3,NV,NF,IP,2,8)
      WRITE(6,202)
  202 FORMAT(///' ',100('-')///'   ITER    MAX INC'///)
C BEGIN GAUSS-SEIDEL ITERATIONS
      DO 605 IT=1,MAXIT
      SV=0.
      DO 6 J=1,NV
      DO 11 K=1,NV
```

Table 5.7   *continued*

```
        IF(K.EQ.J) GOTO 11
        S=0.
        DO 10 I=1,NF
     10 S=S+V(K,I)*V(J,I)
C EIG ( ) ARE THE RESIDUAL CORRELATIONS FOR ROW J
        EIG(K)=R(J,K)-S
     11 CONTINUE
        EIG(J)=0.
C SET ROW J OF FACTOR LOADINGS MATRIX TO ZERO
        DO 13 JJ=1,NF
        ST(JJ)=V(J,JJ)
     13 V(J,JJ)=0.0
C SOLVE LINEAR EQUATIONS
        CALL TRMULT(TEMP,V,V,N5,N3,N3,NF,NF,NV)
        CALL GJDEF1(NF,TEMP,DATA,N5)
        IF(NF) 20,21,20
     21 RETURN
     20 DO 15 I=1,NF
        DO 15 L=1,I
     15 TEMP(L,I)=TEMP(I,L)
        DO 14 JJ=1,NF
     14 V(J,JJ)=ST(JJ)
        CALL MTMULT(RR,V,TEMP,N2,N3,N5,NV,NF,NF)
C COMPUTE INCREMENTAL CHANGES FOR ROW J
        DO 16 JK=1,NF
        DATA(JK)=0.
        DO 16 JL=1,NV
     16 DATA(JK)=DATA(JK)+RR(JL,JK)*EIG(JL)
C    DATA = INCREMENTAL CHANGES
C COMPUTE NEW ROW J OF FACTOR LOADINGS MATRIX. STORE THE MAXIMUM CHANGE
C IN SV
        DO 31 JK=1,NF
        DT=ABS(DATA(JK))
        IF(DT.GT.SV) SV=DT
     31 V(J,JK)=V(J,JK)+DATA(JK)
C STORE CONTAINS THE MAXIMUM INCREMENTAL CHANGES FOR EACH ROW OF THE
C FACTOR LOADINGS MATRIX
        STORE(J)=SV
C CHECK THAT ROW J HAS A COMMUNALITY LESS THAN OR EQUAL TO 1.0 AND
C CALL ADJUSTING ROUTINE IF NECESSARY.
        COMM=0.0
        DO 37 JK=1,NF
     37 COMM=COMM+V(J,JK)*V(J,JK)
        IF(COMM.LE.1.0) GOTO 6
        CALL ADJUST(V,R(1,J),EIG,DATA,TEMP,STOR2,NF,NV,N3,N5,J,TOL,MACHEP)
        SV=0.0
        DO 694 JK=1,NF
        DT=ABS(V(J,JK)-ST(JK))
    694 IF(DT.GT.SV) SV=DT
        STORE(J)=SV
      6 CONTINUE
        L=IT
        SV=0.
C FIND MAXIMUM INCREMENTAL CHANGE AT ITERATION CYCLE I.
        DO 33 J=1,NV
     33 IF(STORE(J).GT.SV) SV=STORE(J)
        WRITE(6,331) L, SV
    331 FORMAT(' ',I6,F12.6)
C EXIT FROM LOOP IF CONVERGENCE IS ATTAINED
        IF(SV.LE.EPSI) GOTO 25
    605 CONTINUE
        WRITE(6,26) MAXIT,EPSI
     26 FORMAT('0 NO CONVERGENCE AFTER',I4,' ITERATIONS WITH EPSI =',
       *F12.5/' THE FOLLOWING RESULTS ARE THEREFORE PROVISIONAL'///)
        GOTO 27
     25 WRITE(6,115) L
    115 FORMAT('0',20X,I6,'   ITERATIONS ACCOMPLISHED'///)
```

Table 5.7 *continued*

```
   27 IP=1
C COMPUTE FUNCTION VALUE AT THE MINIMUM
      F=FVAL(NK,NV,NF,R,N1,V,N3)
      WRITE(6,332) F
  332 FORMAT('OF VALUE AT MINIMUM:',E17.7)
C TRANSFORM TO CANONICAL FORM
      CALL CANON(V,N3,TEMP,N5,RR,N2,EIG,STORE,NV,NF,MACHEP,TOL)
C COMPUTE VARIANCE OF FACTORS
      CALL IMPORT(V,N3,RR,N2,NV,NF)
      WRITE(6,333)
  333 FORMAT('OCOMMUNALITIES'/)
      DO 334 I=1,NV
      COMM=0.0
      DO 335 J=1,NF
  335 COMM=COMM+RR(I,J)*RR(I,J)
  334 WRITE(6,336) I, COMM
  336 FORMAT(' ',I6,2X,F8.4)
      RETURN
      END



      SUBROUTINE CANON(A,N1,T,N2,B,N3,TEST,WORK,NV,NF,MACHEP,TOL)
C SUBROUTINE TO TRANSFORM TO CANONICAL FORM
      REAL MACHEP
      DIMENSION A(N1,NF),B(N3,NF),T(N2,NF),TEST(NV),WORK(NV),FMT(6)
      DATA FMT(1),FMT(2),FMT(3),FMT(4)/4H(I8,,4H4X,8,4HF12.,4H4)  /
      CALL TRMULT(B,A,A,N3,N1,N1,NF,NF,NV)
      CALL HTDQL(NF,N3,N2,B,TEST,WORK,T,MACHEP,TOL)
      CALL MTMULT(B,A,T,N3,N1,N2,NV,NF,NF)
      IP=1
      MM=N3*NF
      CALL MATPR(40HMINRES SOLUTION IN CANONICAL FORM        ,5,FMT,6,B,
     *MM,N3,NV,NF,IP,2,8)
      RETURN
      END



      SUBROUTINE IMPORT(A,N1,B,N2,N,M)
C SUBROUTINE TO COMPUTE THE VARIANCE OF THE FACTORS
      DIMENSION A(N1,3),B(N2,M)
  100 FORMAT(1H0,20X,'CONTRIBUTION OF EACH FACTOR TO TOTAL COMMUNALITY'/
     *' ',10X,'  PROPORTION     PERCENTAGE     CUMUL. PERCENT'///
     *50(' ',10X,3(F8.3,7X)/))
      DO 1 J=1,M
      A(J,1)=0.0
      DO 1 I=1,N
    1 A(J,1)=A(J,1)+B(I,J)*B(I,J)
      S=0.
      DO 2 I=1,M
    2 S=S+A(I,1)
      DO 3 I=1,M
    3 A(I,2)=(A(I,1)*100.0)/S
      A(1,3)=A(1,2)
      DO 4 I=2,M
    4 A(I,3)=A(I-1,3)+A(I,2)
      WRITE(6,100) ((A(I,J),J=1,3),I=1,M)
      RETURN
      END
```

Table 5.7   *continued*

```
      FUNCTION FVAL(NK,NV,NF,R,N1,V,N2)
C FVAL EVALUATES THE FUNCTION VALUE
      DIMENSION R(N1,NV),V(N2,NV)
      FVAL=0.
      DO 40 J=1,NK
      JP=J+1
      DO 40 K=JP,NV
      FF=0.
      DO 41 I=1,NF
   41 FF=FF+V(K,I)*V(J,I)
      FF=(R(J,K)-FF)**2
   40 FVAL=FVAL+FF
      RETURN
      END




      SUBROUTINE ADJUST(A,B,T,D,P,U,NF,NV,N1,N2,I,TOL,MACHEP)
C SUBROUTINE TO ADJUST FOR HEYWOOD CASES USING BROWNE'S(1967) PROCEDURE
      REAL MACHEP
      DIMENSION A(N1,NF),B(NV),T(NF),D(NF),P(N2,NF),U(NF)
      DO 1 J=1,NF
    1 A(I,J)=0.0
      B(I)=0.0
      CALL TRMULT(P,A,A,N2,N1,N1,NF,NF,NV)
      CALL HTDQL(NF,N2,N2,P,D,T,P,MACHEP,TOL)
   77 THETA=D(NF)*0.5
      NIT=0
      DO 2 J=1,NF
      T(J)=0.0
      DO 2 K=1,NV
      S=0.0
      DO 3 L=1,NF
    3 S=S+P(L,J)*A(K,L)
    2 T(J)=T(J)+S*B(K)
    4 DIFF1=0.0
      NIT=NIT+1
      IF(NIT.GT.500) GOTO 10
      DIFF2=0.0
      DO 5 J=1,NF
      C=D(J)-THETA
      U2=T(J)*T(J)
      DIFF1=DIFF1+U2/C**3
    5 DIFF2=DIFF2+U2/C**2
      S=(DIFF2-1.0)/(DIFF1+DIFF1)
      IF(ABS(S).LE.0.0000001) GOTO 6
      C=THETA-S
      IF(C.LE.D(NF)) GOTO 7
      THETA=(D(NF)+THETA)*0.5
      GOTO 4
    7 THETA=C
      GOTO 4
    6 DO 8 J=1,NF
    8 U(J)=T(J)/(D(J)-THETA)
      DO 9 J=1,NF
      DO 9 K=1,NF
    9 A(I,J)=A(I,J)+P(J,K)*U(K)
      RETURN
   10 WRITE(6,12)
   12 FORMAT(''NOT CONVERGED AFTER 500 ITERATIONS')
      RETURN
      END
```

The minres method gives the results shown in Table 5.8 when applied to the Gustafson data. For the sake of comparison the solutions for $k = 2,3$ and 4 factors are shown. The $f$- value at the minimum is 0.14860 for $k = 2$, dropping to 0.0245 for $k = 3$ and to 0.00051 for $k = 4$. The identity of factors 1 and 2 remain stable across all three solutions, with factor 1 accounting for a large proportion of the variance analysed. Factor 1 represents a general size factor, as shown by the high loadings on all variables except $X_7$ (shape). The loading on variable $X_8$ (relative relief) is moderate. Factor 2 picks out $X_6$ (slope) and $X_8$ (relative relief) and is therefore identifying relief variations. Factor 3 is characterized by $X_7$ (shape) and to a lesser extent $X_5$ (perimeter length) while factor 4 has only one loading greater than 0.2. This is on variable $X_3$ (number of streams) and may be identifying variations in network geometry. This factor accounts for only 4.22 per cent of the total variance analyzed. The minres solution for $k = 3$ factors is substantially the same as the image factor analysis solution (Table 5.11 below) though $X_8$ has a much higher loading on factor 2 in the minres solution (0.927) than in the image factor analysis solution (0.775). Image factor 3 picks out $X_3$ (0.390) which has a loading of only 0.180 on minres factor 3. Variable $X_3$ (0.390) which has a loading of only 0.180 on minres factor 3. Variable $X_3$ (number of streams) could relate to perimeter length and basin shape, though it may appear more logical to consider it separately.

The correspondence between the minres and alpha solutions (Table 5.6(a) above) is closer than the image-minres relationship. The high loading of $X_3$ on factor 3 does not appear in the alpha solution, though the loading of $X_8$ on factor 2 is still lower on the alpha solution (0.839) than on the minres solution (0.927). The maximum likelihood solution (Table 5.9) is more similar to the image solution than to either the alpha or minres results. Overall, however, the four solutions are very closely related. The rotation procedures described in Section 5.4 might be expected to bring the factors derived by the four methods into sharper focus and make them more closely comparable.

The *maximum likelihood* (ML) method of estimation is favoured by statisticians because of the fact that the ML estimator has, generally, such desirable properties of sufficiency, efficiency and consistency. Usually ML estimators are normally distributed; this allows the derivation of confidence limits and the testing of hypotheses. (Chapter 2 contains further discussion of the problem of estimation. See also Kmenta (1971), Harman (1967, p. 211—213) and Yamane (1964)). D. N. Lawley, writing in 1940, was the first to put forward a scheme for computing ML estimates of factor loadings, but his method was rarely used because of the enormity of the calculations in all but the simplest cases. The method was iterative and it proved difficult to determine whether convergence had been attained. Harman (1967,

Table 5.8
Minres factor loadings for $k = 2, 3$ and 4 factors

| Variable | Factor | | | | | | | | |
|---|---|---|---|---|---|---|---|---|---|
| | 1 | 2 | 1 | 2 | 3 | 1 | 2 | 3 | 4 |
| 1 | 0.970 | −0.041 | 0.968 | −0.043 | 0.118 | 0.967 | −0.061 | −0.085 | −0.152 |
| 2 | 0.969 | −0.010 | 0.971 | −0.013 | 0.157 | 0.974 | −0.031 | −0.110 | 0.194 |
| 3 | 0.713 | 0.072 | 0.715 | 0.071 | 0.180 | 0.745 | 0.064 | −0.174 | 0.419 |
| 4 | 0.979 | −0.030 | 0.974 | −0.030 | −0.055 | 0.976 | −0.037 | 0.064 | −0.175 |
| 5 | 0.914 | −0.022 | 0.938 | −0.023 | −0.345 | 0.926 | −0.011 | 0.303 | 0.029 |
| 6 | −0.726 | −0.543 | −0.726 | −0.552 | 0.109 | −0.733 | −0.582 | −0.064 | 0.165 |
| 7 | −0.106 | 0.012 | −0.118 | 0.025 | −0.732 | −0.129 | 0.099 | 0.927 | 0.129 |
| 8 | 0.371 | −0.929 | 0.369 | −0.927 | −0.068 | 0.363 | −0.900 | 0.135 | −0.052 |
| Percentage Communality | 80.65 | 19.35 | 71.87 | 17.20 | 10.93 | 66.30 | 15.69 | 13.79 | 4.22 |
| $f$-value | 0.1486 | | 0.0245 | | | 0.00051 | | | |

Ch.10) gives details of the Lawley method of ML estimation but, as other authors have shown (e.g. Jennrich and Robinson, 1969) the results achieved by this method — and by other iterative methods such as that developed by Hemmerle (1967) — do not in general give true ML solutions because of the difficulty in determining whether the solution has converged.

Developments in numerical analysis — particularly in nonlinear optimization techniques and in the size and speed of available computers — led to a renewal of interest in the ML factor loading estimators. Joreskog (1967) applied the Fletcher–Powell algorithm to locate a minimum of the likelihood ratio. These results are discussed by Joreskog and Lawley (1968). Later, Clarke (1970) developed exact formulae for the second partial derivatives of the likelihood ratio with respect to the parameters. This allowed the use of the Newton–Raphson minimization algorithm, leading to a more efficient computing method.

The computational details of ML estimation using the Newton–Raphson algorithm are beyond the scope of this chapter. Mulaik (1972) covers the mathematical and computational details, though he restricts himself to the use of the Fletcher–Powell algorithm. A more terse account is provided by Lawley and Maxwell (1971). Joreskog and van Thillo (1971) have made available a Fortran-IV program which implements the Newton–Raphson algorithm.

ML estimators of the factor loadings matrix are based on the assumption that the distribution of the variables is multivariate-normal. However, Howe (1955) and Fuller and Hemmerle (1967) show that the estimates are not adversely affected even by radical departures from normality. The validity of the statistical tests of significance which are associated with the ML method is likely to be questionable, however, if the assumption of normality is not approximately satisfied.

Like alpha factor loadings — but unlike those produced by minres — ML factor loadings are independent of the scale of measurement of the variables. In other words, the estimated factor loadings for a particular variable are proportional to the standard deviation of the variable. Consequently, factors derived from **R** and **S** will be related by scale factors. This was a source of difficulty in PCA, as shown in Chapter 4.

The hypothesis that $k$ is the true number of common factors can be tested statistically when the ML method of estimation is used. Details of the derivation of the test statistic, which is based upon the value of the likelihood ratio computed during the estimation procedure, are given by Lawley and Maxwell (1971, p. 35). It is distributed as $\chi^2$ when the hypothesis is true, but is reliable only when $(n - p)$ exceeds 50. The test statistic is, in fact, a measure of the degree to which the residual correlations differ from zero. If the difference is statistically significant then it is probable that the value of $k$ is too low, and the ML estimation is repeated with $k_{new} = k_{old} + 1$ factors. As with the F-ratio test used in stepwise regression or in trend-surface analysis (Chapter 3), the significance level for the test statistic does not take into account the fact that a sequence of dependent null hypotheses is being tested. Acceptance of the hypothesis that $(k + 1)$ factors are sufficient to account for the correlations is dependent upon the rejection of a similar hypothesis for $k$-factors. The error resulting from this deviation from the theoretical requirements is likely to be small; Lawley and Maxwell (1971, p. 37–38) indicate that, if the true number of factors is $k^*$, then the probability that $k$, the number of factors selected by the test procedure, will exceed $k$ is always less than the probability level used in the test. That is, if the test statistic is correct to $\chi^2$ at $P = 0.05$ then the probability that $k > k^*$ will be less than 0.05.

It is worth noting that the value of the test statistic is dependent on $n$ and, for the same fixed set of $p$ variables, different values of $k$ may result if $n$ is varied. Some users of factor analysis have taken the view that it is reasonable to expect that the number of common factors should be a function of the $p$ variables, not of the $n$ observations. Increasing $n$ will, however, increase the amount of information available and may allow the identification of common factors which would otherwise have been missed. Tucker and Lewis (1970) outline what they term a 'reliability coefficient' which is an attempt to resolve this dilemma; it measures how well the factor model with $k$ common factors represents the covariances or correlations among the variables for the population of individuals. The reliability coefficient, like the chi-square test, can be used only when a ML solution has been obtained. No significance level is given by Tucker and Lewis, but it appears that values above 0.95 are acceptable. A second point is that factors which may be statistically significant can, nevertheless, lack practical meaning. Digman (1966) considers that nonlinearities in the factor-variable relationships may produce uninterpretable factors if the *linear* factor model is used. On the other hand, Harman (1967, p. 229) notes that '... one tends to underestimate the number of factors that are statistically significant'. In the example he uses to illustrate this point three factors are shown to be statistically significant, but he dismisses the third as having little 'practical significance'. Rather than reject such uninterpretable factors out of hand one should consider whether the assumptions underlying the test have been satisfied, or whether the linear common factor model is in fact appropriate. In addition, the possibility that an 'uninterpretable' factor is a spurious artefact of the technique of factoring used in its extraction, should be considered.

The ML solution for the Gustafson data with $k = 3$ factors is shown in Table 5.9. The pattern of high and low loadings is similar to that produced by the alpha factor analysis procedure (Table 5.6(a) above). Factor 2 is more clearly defined, but factor 3 slightly less so. However,

Table 5.9
Maximum likelihood factor loadings

| Variable | Factor | | |
|---|---|---|---|
| | I | II | III |
| 1 | 0.9736 | 0.0472 | 0.0693 |
| 2 | 0.9495 | 0.0250 | 0.1803 |
| 3 | 0.7198 | −0.0600 | 0.3520 |
| 4 | 0.9791 | 0.0174 | −0.0815 |
| 5 | 0.9318 | −0.0017 | −0.2527 |
| 6 | −0.7338 | 0.5871 | 0.0679 |
| 7 | −0.1202 | −0.0974 | −0.7403 |
| 8 | 0.3790 | 0.8164 | −0.1314 |

a major difference is that the communalities are constrained by the program to be less than 1.0 (in other words, the maximum of the likelihood function is sought in the region in which all $h_i^2 < 1.0$). These communalities are:

| Variable: | 1 | 2 | 3 | 4 | 5 | 6 | 7 | 8 |
|---|---|---|---|---|---|---|---|---|
| Communality: | 0.950 | 0.936 | 0.673 | 0.960 | 0.917 | 0.851 | 0.623 | 0.772 |

Reference to Table 5.6(b) shows that the communalities for variables 1, 4, 6 and 7 are approximately the same as those produced by alpha factor analysis, but that there are differences between the communalities of the remaining variables, especially variable 5 which is no longer a Heywood variable. Variable 8 has a much lower communality in the ML solution (0.772 as against 0.905 for the alpha solution). The chi-square test for $k = 3$ factors gives $\chi^2 = 89.4$ with 7 degrees of freedom. This exceeds the tabled value even at the 0.001 significance level ($\chi^2_{7,0.001} = 24.3$) so the hypothesis of $k = 3$ factors should be rejected. This implies that, to reduce the matrix of residual correlations to zero further factors should be extracted.* The reliability coefficient for $k = 3$ factors is very low (0.24). The number of factors produced by the alpha, minres and image procedures all give $k = 3$; the ML solution is the odd one out in this respect.

### 5.3.2 Image Analysis
Image Analysis represents a break from the traditional common factor approach in which attention if often focused on the problem of determining a valid set of communalities for the variables rather than

---

*This proved impossible with the Joreskog–van Thillo (1971) program, apparently because of the fact that two of the eight variables become Heywood variables at some stage in the estimation process (K. G. Joreskog, *Personal Communication*, 1974)

on the relationships between the variables. Guttman (1953) made an extensive study of the common-factor approach and introduced the method of image analysis which, while retaining many of the features of common-factor analysis, provides an explicit definition of communality. In his paper, Guttman suggested that the main difference between the common-factor approach and the image analysis approach is that the former implies that common factors are extracted sequentially until the residual correlations are negligible. (See, for example, the discussion of the $\chi^2$ test in Section 5.3.1.) The common-factor method could be considered to involve the extraction of factors one at a time and the replacement of the correlation matrix with the matrix of partial correlations of the variables with the effects of these factors removed. The procedure terminates when the remaining partial correlations are negligible. In image analysis the multiple correlations between the variables and the factors are considered, rather than the separate partial correlations. Later work by Harris (1962) related Guttman's image theory and Lawley's ML approach in image factor analysis (IFA).

Guttman defined the common part of a variable as: '... that part which is predictable by linear multiple correlation from all the other variables in the universe of variables'. This part he termed the 'image' of the variable. It is rarely possible to measure all the members of a universe of variables, so the image is essentially a theoretical concept. An approximation, based on the selection of variables available, can by made, and as $p$ rises to 15 or 20 the values of the squared multiple correlations (SMC's) approach the 'universe' values. Guttman went further than to give an explicit definition of common variance; he proposed that the correlations between or covariances of the *common parts* of the variables should be analyzed, rather than the correlation or covariance matrix of the observed variables with communality values along the principal diagonal. These suggestions mean that (i) Heywood cases (which cause trouble in some of the estimation methods discussed in Sections 5.3.A and 5.3.B) will never occur and (ii) 'imaginary' factors, i.e. factors with negative eigenvalues, will not occur in image analysis. Such imaginary factors are produced if (squared) multiple correlations are used in the principal axis method of factor analysis, which — apart from computational simplicity — has nothing to recommend it.

The method of large analysis can be summarized as follows (Kaiser, 1963):

(a) the basic postulate of image analysis is —

$x_j = \hat{x}_j + e_j$

where $x_j$ is the vector of measurements on the $j^{th}$ variable, $\hat{x}_j$ is the part of $x_j$ that is predictable by multiple correlation methods, and $e_j$ is the nonpredictable random part of $x_j$.

(b) $\hat{x}_j$ is derivable by regression methods from the formula

$$\hat{x}_j = \sum_{k=1}^{p} w_{jk} x_j$$

(see Chapter 2) and $e_j$ is then calculated from $e_j = x_j - \hat{x}_j$.

In practice $\hat{x}_j$ is not found by this method (though it is useful to see the relationship between image theory and regression analysis). The diagonal matrix of unique parts, $\mathbf{J}^2$, is found from

$$\mathbf{J}^2 = \text{diag}(\mathbf{R}^{-1})^{-1}$$

The values in the diagonal of $\mathbf{J}^2$ are the reciprocals of the diagonal elements of the inverse of the correlation matrix. The method is thus dependent on the nonsingularity of $\mathbf{R}$.

(c) The correlations $g_{ij}$ among the $\hat{x}_j$ (or 'images') are now recovered from $\mathbf{R}$ by the formula

$$\mathbf{G} = \mathbf{R} + \mathbf{J}^2 \mathbf{R}^{-1} \mathbf{J}^2 - 2\mathbf{J}^2$$

The corresponding matrix, $\mathbf{Q}$ of correlations among the $e_j$ (or 'anti-images')

$$\mathbf{Q} = \mathbf{J}^2 \mathbf{R}^{-1} \mathbf{J}^2$$

Thus

$$\mathbf{R} = \mathbf{G} - \mathbf{Q} + 2\mathbf{J}^2$$

and

$$\mathbf{J}^2 = \text{diag}(\mathbf{Q})$$

Kaiser (1963) notes that, if the factor model adequately represents the structure of the data, the off-diagonal elements of $\mathbf{Q}$ should be near-zero. Kaiser (1970) and Kaiser and Rice (1974) go on to present a measure of the 'diagonality' of $\mathbf{Q}$ which is termed the 'measure of sampling adequacy' or MSA. If the MSA is low it follows either that the number of observations is insufficient or that the factor analysis model is not appropriate.

An important contribution to image analysis was made by Harris (1962) who showed that the method could be made scale-free in the sense defined above. The matrix

$$\mathbf{G}^* = \mathbf{J}^{-1} \mathbf{G} \mathbf{J}^{-1}$$

is used rather than $\mathbf{G}$. However, the same factors are derivable from $\mathbf{R}^* = \mathbf{J}^{-1} \mathbf{R} \mathbf{J}^{-1}$ in the sense that the correlation between the corresponding factors of $\mathbf{R}^*$ and $\mathbf{G}^*$ is unity. It is more convenient to factor $\mathbf{R}^*$ by taking its eigenvalues $\mathbf{C}$ and eigenvectors, $\mathbf{V}$. Then, the $j^{th}$ column eigenvector is multiplied by $[(c_j - 1)^2/c_j]^{1/2}$, where $c_j$ is the $j^{th}$ eigenvalue, to give $\mathbf{V}^*$ which is then scaled to give the (Harris) Image factor loadings matrix $\Lambda_I$:

$$\Lambda_I = JV^*$$

The question of the number of factors is not answered by image analysis. Empirical rules are generally resorted to, the most favoured one being the setting of $k$, the number of factors, to the number of eigenvalues of **R** that exceed 1.0.

A subroutine to carry out image factor analysis is listed in Table 5.10. Details of use can be found in the comments included in the listing. A comprehensive image analysis program, which includes the rotation of image factors, is described by Kaiser and Rice (1974). This program could be readily implemented using subroutines from Appendix A.

The image factor analysis program (Table 5.10) produces the results given in Table 5.11 when applied to the Gustafson data used in previous examples. The number of factors was set to 3. If Kaiser's (1963) criterion — number of eigenvalues of $(J^{-1}RJ^{-1})$ that exceed 1.0 — was used, five factors would have resulted. This criterion has subsequently been abandoned by Kaiser (e.g. Kaiser and Rice, 1974) who now prefers to rely on the 'eigenvalue $-1$' criterion. The reader can verify from Table 5.11 that no Heywood cases occur. The pattern of factor loadings does not differ substantially from that produced by alpha factor analysis, except that all communalities are now less than 1.0. As noted earlier, the loading of $X_3$ on factor 3 is higher in the image factor solution than in either the alpha or minres solutions. The image factor loadings matrix is more closely related to the maximum likelihood solution in this respect.

## 5.4 Rotation of Factors

In Section 5.2 it was concluded that the factor model was basically indeterminate in that, given a matrix $\Lambda$ of factor loadings estimated by any of the methods given in Section 5.3, any transformation $\Lambda^* = \Lambda T$ (where **T** is any nonsingular $k \times k$ matrix) is mathematically equivalent. By observing **X** alone it is not possible to distinguish between $\Lambda$ and $\Lambda^*$. There are, of course, a multiplicity of possible $\Lambda^*$'s. Not all $\Lambda^*$'s are interpretable in subject-matter terms, and the aim of factor rotations or transformations is to find that $\Lambda^*$ which is most easily interpreted.

Thurstone (1947) laid down a set of rules which he termed 'simple structure' criteria, which were designed to select the most easily interpretable $\Lambda^*$. The basis of the simple structure concept is the principle of simplicity, or Occam's Razor. Thurstone's criteria were:

(i) each row of $\Lambda^*$ should have at least one zero entry;
(ii) for $k$ factors, each column of $\Lambda^*$ should have at least $k$ zero entries;

## Table 5.10
## Subroutine IMAGE: image factor analysis

```
      SUBROUTINE IMAGE(S,RR,N3,EIG,V,NF,NV,FMTO,TOL,MACHEP)
C
C IMAGE FACTOR ANALYSIS BASED ON KAISER(1963) AFTER HARRIS(1962)
C PROGRAM BY P.M.MATHER
C SEE CHAPTER 5 FOR FURTHER DETAILS
C ARGUMENTS TO THE SUBROUTINE:
C  S      REAL ARRAY, MINIMUM DIMENSION NV, USED AS WORKSPACE.
C  RR     REAL ARRAY,MINIMUM DIMENSION (NV,NV),HOLDING CORRELATION MATRIX
C         ON ENTRY.  HOLDS IMAGE FACTOR LOADINGS ON EXIT.
C  N3     INTEGER; FIRST DIMENSION OF RR AS DECLARED IN CALLING PROGRAM
C  EIG    REAL ARRAY,MINIMUM DIMENSION NV. UNDEFINED ON ENTRY.  CONTAINS
C         EIGENVALUES OF J(INV)RJ(INV) ON EXIT.
C  V      REAL ARRAY, MINIMUM DIMENSION NV, USED AS WORKSPACE.
C  NF     INTEGER; NUMBER OF FACTORS. IF ZERO ON ENTRY, THE NUMBER
C         OF FACTORS WILL BE DEFINED AS THE NUMBER OF EIGENVALUES OF
C         J(INV).R.J(INV) THAT EXCEED 1.0;  SEE TEXT.
C  NV     INTEGER; THE NUMBER OF VARIABLES. IF THE CORRELATION MATRIX
C         IS SINGULAR, A MESSAGE IS PRINTED, NV IS SET TO ZERO, AND
C         CONTROL RETURNS TO THE CALLING PROGRAM.
C  FMTO   REAL ARRAY, DIMENSION 20, HOLDING OUTPUT FORMAT SPECIFICATION
C         FOR USE IN SUBROUTINE MATPR (SEE APPENDIX A)
C  TOL    REAL QUANTITY EQUAL TO MACHEP/ETA WHERE MACHEP IS THE LEAST
C         POSITIVE REAL NUMBER FOR WHICH 1.0+MACHEP>1.0 AND ETA IS THE
C         LEAST POSITIVE FLOATING POINT NUMBER THAT EXCEEDS ZERO.
C  MACHEP REAL CONSTANT, DEFINED ABOVE (SEE TOL).
C
C NEEDS SUBROUTINES HTDQL, GJDEF1 AND MATPR FROM APPENDIX A
C
      REAL MACHEP
      DIMENSION RR(N3,NV),S(NV),EIG(NV),FMTO(20),V(NV)
      WRITE(6,603)
  603 FORMAT('1HARRIS(1962) IMAGE FACTOR ANALYSIS'//' ',34('*'))
C INVERT CORRELATION MATRIX
      CALL GJDEF1(NV,RR,EIG,N3)
C IF SINGULAR, OUTPUT MESSAGE, SET NV=0 AND RETURN
      IF(NV) 2,1,2
    1 WRITE(6,56)
   56 FORMAT('0CORRELATION MATRIX WILL NOT INVERT')
      RETURN
C DETERMINE ELEMENTS OF DIAGONAL MATRIX J AND STORE IN S
    2 DO 6 I=1,NV
    6 S(I)=SQRT(RR(I,I))
C SCALE CORRELATION MATRIX -- RR HOLDS J(INV)RJ(INV) --- SEE TEXT
      DO 23 I=2,NV
      SI=S(I)
      I1=I-1
      DO 23 J=1,I1
   23 RR(I,J)=RR(J,I)*SI*S(J)
C FIND EIGENVALUES (EIG) AND EIGENVECTORS (RR) OF MATRIX RR
      CALL HTDQL(NV,N3,N3,RR,EIG,V,RR,MACHEP,TOL)
      WRITE(6,108) (I, EIG(I), I=1,NV)
  108 FORMAT('0      EIGENVALUES'//(' ',I6,2X,F12.5))
C IF NF=0 FIND NUMBER OF EIGENVALUES > 1.0
      IF(NF.GT.0) GOTO 55
      NF=0
      DO 30 I=1,NV
   30 IF(EIG(I).GT.1.0) NF=NF+1
      WRITE(6,32) NF
   32 FORMAT('0NUMBER OF FACTORS = ',I3)
C CONVERT EIGENVALUES OF J(INV)RJ(INV) TO EIGENVALUES OF J(INV)GJ(INV)
C --- SEE TEXT
   55 DO 31 I=1,NF
      SS=SQRT(((EIG(I)-1.0)**2)/EIG(I))
      DO 31 J=1,NV
C AND SCALE EIGENVECTORS APPROPRIATELY.
```

Table 5.10 continued

```
   31 RR(J,I)=RR(J,I)*SS
C NOW RESCALE THE RESULTING EIGENVECTORS TO GIVE IMAGE FACTOR LOADINGS
      DO 41 I=1,NV
      SI=1.0/S(I)
      DO 41 J=1,NF
   41 RR(I,J)=RR(I,J)*SI
C AND OUTPUT THE MATRIX OF LOADINGS
      N=N3*NV
      IP=1
C NOTE: THIS CALL TO MATPR ASSUMES THAT EACH STORE LOCATION CAN HOLD
C 4 ALPHANUMERIC CHARACTERS.
      CALL MATPR(24HIMAGE FACTOR LOADINGS    ,6,FMTO,6,RR,N,N3,NV,NF,IP,
     1 2,4)
      RETURN
      END
```

Table 5.11
Image factor loadings

| Variable | Factor | | |
|---|---|---|---|
| | I | II | III |
| 1 | 0.9761 | −0.0458 | 0.0596 |
| 2 | 0.9463 | −0.0312 | 0.2086 |
| 3 | 0.7102 | 0.0401 | 0.3898 |
| 4 | 0.9815 | −0.0049 | −0.0932 |
| 5 | 0.9311 | 0.0220 | −0.2215 |
| 6 | −0.7307 | −0.5798 | 0.0347 |
| 7 | −0.1229 | 0.1344 | −0.6216 |
| 8 | 0.3848 | −0.7753 | −0.1705 |

(iii) for every pair of columns of $\Lambda^*$ there should be some variables which have a zero entry in one of the columns but not the other;
(iv) if $k$ is 4 or more a large proportion of the corresponding elements of every pair of columns should be zero;
(v) for every pair of columns only a small number of variables should have nonzero entries in each column.

$\Lambda^*$ may refer either to correlated or uncorrelated factors even though the original solution, $\Lambda$, was an orthogonal matrix. If the factors given by $\Lambda^*$ remain uncorrelated the rotation is said to be orthogonal, otherwise it is an oblique rotation. Note that the factor structure and the factor pattern are no longer identical for oblique rotations. The *factor structure matrix* gives the correlations between variables and factors while the *factor pattern matrix* contains the estimates of the parameters in the factor model. The factor structure coefficients are used in the calculation of factor scores, while the factor pattern matrix is the basis of any interpretation of the meaning of the factors. The two

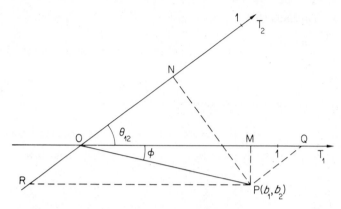

Figure 5.2  Distinction between coordinate and correlation in an oblique reference system (Reprinted from *Modern Factor Analysis, Second Edition*, Figure 13.1, by H. H. Harman by permission of The University of Chicago Press. Copyright © 1967 The University of Chicago Press)

are identical in orthogonal solutions and are collectively termed the factor loadings matrix or the factor matrix. A geometrical explanation of the difference between 'structure' and 'pattern' is given in Figure 5.2. The obliquely transformed factors are represented by the axes $OT_1$ and $OT_2$, the angle $\theta_{12}$ measuring the degree of resemblance of the factors. A standardized variable $z_j$ can be represented by a point P whose coordinates $b_1$ and $b_2$ are equal to the line lengths OQ and OR respectively. These lengths, representing factor pattern coefficients, can thus be greater than unity for oblique factors. The angle between OP and $OT_1$ is $\phi$ and the projections of P on $T_1$ and $T_2$ are given by M and N. The cosine of $\phi$ is equal to the correlation between $z_j$ and $T_1$ while $\cos(\phi + \theta)$ is equal to the correlation of $z_j$ and $T_2$. (See Section 1.3 for a discussion of this point.) The correlations between $z_j$ and the factors are the factor structure coefficients. It is easily verified that, if $\theta$ is a right-angle, the projections OQ and OR are equal to the cosines of $\phi$ and $(\theta + \phi)$ respectively, showing that in the orthogonal case the structure and the pattern are mathematically equal.

It should be noted that there are two kinds of factor pattern and factor structure matrices. The correlation/coordinate description given above refers to what are termed *primary factors*, hence the matrices are termed the primary factor structure and the primary factor pattern matrices, respectively. Some factor analysis texts and computer programs use what are termed *reference axes* rather than the primary factor axes. Reference axes are perpendicular to factor axes, and it can be shown (e.g. Rummell (1970) p. 406) that the primary pattern coefficients are proportional to the reference structure coefficients, and that primary structure coefficients are proportional to the reference pattern coefficients. The reference structure matrix is denoted by **V** and the reference pattern by **W**. $\Lambda_P$ and **V** are equivalent in that the same factors are identifiable. However, the correlations among primary factors do not, in general, equal the correlations among reference factors. Confusion can arise if it is not clear whether the reference structure or the primary pattern is being used. This is important when factor scores are being calculated. In the following pages, the symbols $\Lambda_P$ and $\Lambda_S$ refer to the pattern and structure matrices associated with primary factors and **W** and **V** to the pattern and structure of the reference factors. Rummell (1970, Chapter 17) gives a full account of these matrices. It should be recalled that the pattern and structure of orthogonal factors are numerically equivalent, and that the primary/reference distinction is made only in the context of oblique factor transformations.

Transformations, either orthogonal or oblique, can be 'blind' in that an analytical procedure is applied which will attempt to convert $\Lambda$ to a $\Lambda^*$ which approximates Thurstone's simple structure criteria more closely. Alternatively the matrix $\Lambda$ can be transformed to match a specified structure; for example, it could be hypothesized that the factor pattern matrix has a particular form, and then attempt to fit $\Lambda$ to this specified target matrix. The target matrix represents a hypothesis, and a measure of the goodness of fit of $\Lambda^*$ to the target matrix may help in deciding whether the hypothesis is tenable. Such confirmatory studies have not been widespread; factor analysis has been used in an exploratory fashion in the past. This has led to controversy, much in the same vein as that surrounding trend surface analysis (Chapter 3). Methods such as the 'exploratory' analysis of one half of the sample followed by 'confirmatory' analysis of the other half can be considered in situations in which sampling is impossible.

The earliest attempts to define an analytical rotation, as distinct from a visual or graphical rotation, began in the 1950's and are described by Harman (1967). The most successful of these procedures are termed quartimax, varimax and equamax. Crawford and Ferguson (1970) have shown that all three are in fact special cases of a single function. One of these orthogonal rotation methods — varimax — has wide usage, and it has been claimed by Warburton (1963) that it is the most satisfactory orthogonal rotation. Most factor analysis programs incorporate a varimax routine [see Davis (1973) and Horst (1965)]. Varimax was first proposed in 1958 by Kaiser, who attempted to simplify the rows and columns of $\Lambda$ to give a simple structure solution $\Lambda^*$. His first criterion was termed 'raw varimax' and involved the maximization of

$$V_R = \sum_{j=1}^{k} \left\{ \left( p \sum_{i=1}^{p} (\lambda_{ij}^2)^2 - (\sum_{i=1}^{p} \lambda_{ij}^2)^2 \right) \bigg/ p^2 \right\} \tag{5.9}$$

It was later noted that if the variables were given equal weight (or normalized) by dividing the $\lambda_{ij}$'s in the above formula by $h_i$, the square root of the communality of variable $i$, the results were improved. This amendment produces the normal varimax criterion, $V_N$.

The mechanics of the original varimax procedure are described by Harman (1967). In this method, all possible pairs of factors are rotated separately until the $V_N$ is maximized. Horst (1965) terms this the successive varimax method, and it is the usual algorithm (e.g. Davis, 1973; Cooley and Lohnes, 1971). However, the simultaneous varimax method, in which all the factors are rotated together, was thought by Horst to be more accurate and more rapid, at least for large values of $p$, the number of variables.

Factor transformations are of the form $\Lambda^* = \Lambda T$ and the rotation process in factor analysis involves determining that $(k \times k)$ matrix $T$ which will transform $\Lambda$ to simple structure. In the simultaneous varimax method $T$ is found as follows: define $E$ as a diagonal matrix with elements

$$e_k = \sum_{i=1}^{p} (\lambda_{ik}^*)^2 = (\lambda_i' t_k)^2 \tag{5.10}$$

and a matrix $C$ with elements

$$c_{ij} = (\lambda_i' t_j)^3 \tag{5.11}$$

where $\lambda_i$ is the $i$th column of $\Lambda$, $\lambda_i^*$ the $i$th column of $\Lambda^* = \Lambda T$ and $t_j$ the $j$th column of $T$. Initial approximations to $E$ and $C$, denoted by $E_0$ and $C_0$, can be found by setting $T_0 = I$. The matrix $T_1$ is calculated as follows:

$$T_1 = B_0 G_0^{-1} \tag{5.12}$$

where

$$B_0 = (C_0 - \Lambda E_0/p), \tag{5.13}$$

and

$$G_0 = QD^{\frac{1}{2}}Q'. \tag{5.14}$$

$D$ is the diagonal matrix of eigenvalues of $(B_0' B_0)$ and $Q$ the matrix of unit length eigenvectors. Once $T_1$ has been found, $\Lambda_1^*$ is computed from the relationship $\Lambda_1^* = \Lambda T_1$. Successive approximations $\Lambda_2^*$, $\Lambda_3^*, \ldots, \Lambda_q^*$ are then derived. When the difference in tr($G$), the varimax criterion, is within a predefined tolerance from one approximation to the next the process is halted.

It is usual to maximize the normal varimax criterion $V_N$ rather than the raw varimax criterion, $V_R$. To do this, divide the elements of row $i$ of $\Lambda$ by the square root of the sum of squares of row $i$ (i.e.

$h_i = (\Sigma_{j=1}^{k} \lambda_{ij}^2)^{\frac{1}{2}}$). The final varimax-rotated matrix $\Lambda_q^*$ is then rescaled by multiplying the elements of its row $i$ by $h_i$. In practice, the convergence of tr(G) is quite rapid and there seems little point in using a convergence criterion smaller than $10^{-4}$ (Table 5.12).

The raw and normal varimax factor loadings matrices for the Gustafson data are shown in Table 5.13(a) and 5.13(b). The initial solution in both cases was the minres solution for three factors. In this particular example there is very little to choose between the two solutions. Both have 25 per cent of their elements less than 0.1 in absolute value. The overall picture is perhaps less clear than in the varimax-rotated solution than in the initial minres solution. This is especially true of factor 2, which has two clearly significant loadings (on variables 6 and 8) in the unrotated solution, but in the rotated solution other variables begin to come in. The loading on variable 8 (relative relief) improves from 0.9268 (minres) to 0.9808 (normal varimax), but the loading on variable 6 drops from 0.5516 to 0.4037. This example shows that it is not always true to say that the varimax rotation necessarily improves the interpretability of the resulting factors. In most cases there will be an improvement but it is always as well to look at both the initial and the rotated solutions.

The factor analysis model [Equation (5.4)] specifically allows for the possibility of correlated factors. Until the late 1950's few successful methods of transforming an orthogonal solution into an oblique one had been developed, and manual (graphical) methods were widely used. The oblimax, oblimin, direct oblimin, maxplane and Harris–Kaiser methods were developed during the 1960's, together with several less successful methods. These are described by Mulaik (1972) and Hakstian (1971). One of the most straightforward and widely-used oblique transformations is Promax (Hendrickson and White, 1964). It differs somewhat from the other methods mentioned earlier in that it is based on an attempt to find the best (least-squares) fit between the oblique factor-pattern or factor-structure matrix and a target matrix, which is thought to represent a simple structure solution. For this reason, Mulaik (1972, Chapter 12) calls Promax a procrustean transformation after the legendary Greek highwayman Procrustes who trimmed or stretched unsuspecting passers-by until they fitted his bed. The target matrix in the Promax transformation is normally based on an orthogonal approximation to simple structure produced, for instance, by the varimax rotation, though solutions derived from other orthogonal rotations could equally well be used.

The target matrix G can be obtained from $\Lambda_V$, the varimax factor loadings matrix, in a variety of ways. Some users firstly divide each element of row $i$ of $\Lambda_V$ by the square root of the communality of variable $i$, then divide the elements of column $j$ of the resulting matrix by the square root of the sums of squares of column $j$. Mulaik (1972,

## Table 5.12
## Subroutine VARMAX: simultaneous varimax rotation

```
      SUBROUTINE VARMAX(LAM,Q,M,C,D,A,H,P,K,LEN,MAXIT,EPSI,N1,N2,N3,N4,N
     *5)
C
C SIMULTANEOUS VARIMAX ROTATION AFTER HORST(1965)
C BASED ON 'COMPUTER APPLICATIONS' NO. 13(B), 1972, BY P. M. MATHER
C  *** ARGUMENTS ***
C    LAM  REAL ARRAY OF ORDER P * K HOLDING UNROTATED FACTOR LOADINGS.
C         UNALTERED BY THE ROUTINE.
C    Q    REAL ARRAY OF ORDER P * K. UNDEFINED ON ENTRY. HOLDS VARIMAX
C         ROTATED LOADINGS ON EXIT.
C    M    REAL ARRAY OF DIMENSION K * K.  UNDEFINED ON ENTRY. HOLDS THE
C         TRANSFORMATION MATRIX ON EXIT.
C    C    REAL ARRAY OF ORDER K*K. WORKSPACE.
C    D    REAL VECTOR, MINIMUM DIMENSION K, USED AS WORKSPACE.
C    A    REAL ARRAY, ORDER K * K, USED AS WORKSPACE.
C    H    REAL VECTOR, MINIMUM DIMENSION P. UNDEFINED ON ENTRY.
C         CONTAINS THE SQUARE ROOTS OF THE COMMUNALITIES ON EXIT IF LEN
C         IS SET TO 1 (SEE BELOW).
C    P    INTEGER. NUMBER OF VARIABLES.
C    K    INTEGER. NUMBER OF FACTORS.
C    LEN  SET ON ENTRY TO 0 FOR RAW VARIMAX OR TO 1 FOR NORMAL VARIMAX.
C    MAXIT INTEGER. MAXIMUM NUMBER OF ITERATIONS. THIS SHOULD BE CHECKED
C         ON RETURN FOR NO ERROR MESSAGE IS ISSUED IF MAXIT CYCLES ARE
C         PERFORMED.
C    EPSI REAL CONSTANT. CONVERGENCE CRITERION IN ITERATIVE SCHEME. A
C         VALUE OF 1.E-3 TO 1.E-5 IS APPROPRIATE.
C    N1   FIRST DIMENSION OF LAM AS DECLARED IN THE CALLING PROGRAM.
C    N2   FIRST DIMENSION OF Q.
C    N3   FIRST DIMENSION OF M.
C    N4   FIRST DIMENSION OF C.
C    N5   FIRST DIMENSION OF A.
C      NOTE : N1 TO N5 ARE INTEGER CONSTANTS.
C  *** SUBROUTINES NEEDED ***
C    MTMULT, TRMULT AND HTDQL
C  *** ERROR MESSAGES ***
C    NONE.
C
      INTEGER P
      REAL M,LAM,MACHEP
      DIMENSION LAM(N1,K),Q(N2,K),M(N3,K),C(N4,K),D(K),A(N5,K),H(P)
C THE FOLLOWING TWO PARAMETERS ARE MACHINE-DEPENDENT
      MACHEP=2.0**(-37)
      ETA=2.0**(-255)
      TOL=ETA/MACHEP
      NO=0
      WRITE(6,101)
  101 FORMAT('0 SIMULTANEOUS VARIMAX ROTATION. PROGRAM BY P. M. MATHER'/
     +4X,' TRACE OF A'///)
      BT=0.0
      IF(LEN) 20,6,20
   20 DO 3 I=1,P
      H(I)=0.0
      DO 4 J=1,K
    4 H(I)=H(I)+LAM(I,J)*LAM(I,J)
      H(I)=SQRT(H(I))
      DO 3 J=1,K
    3 LAM(I,J)=LAM(I,J)/H(I)
    6 CALL COPY(Q,N2,LAM,N1,P,K)
   16 DO 191 KOT=1,MAXIT
      DO 5 I=1,K
      D(I)=0.0
      DO 7 J=1,P
    7 D(I)=D(I)+Q(J,I)*Q(J,I)
    5 D(I)=D(I)/FLOAT(P)
      DO 8 I=1,P
```

Table 5.12    *continued*

```
      DO 8 J=1,K
    8 Q(I,J)=Q(I,J)**3-Q(I,J)*D(J)
      CALL TRMULT(A,LAM,Q,N5,N1,N2,K,K,P)
      CALL TRMULT(M,A,A,N3,N5,N5,K,K,K)
      CALL HTDQL(K,N3,N4,M,D,Q,C,MACHEP,TOL)
      S=0
      DO 11 I=1,K
      S=S+D(I)
   11 D(I)=1.0/SQRT(D(I))
      WRITE(6,100)S
  100 FORMAT(' ',E16.7)
      DO 12 I=1,K
      DO 12 J=1,K
      M(I,J)=0.0
      DO 12 L=1,K
   12 M(I,J)=M(I,J)+C(I,L)*D(L)*C(J,L)
      CALL MTMULT(C,A,M,N4,N5,N3,K,K,K)
      CALL MTMULT(Q,LAM,C,N2,N1,N4,P,K,K)
      AF=BT
      BT=S
      IF((BT-AF)-EPSI) 15,15,17
   15 NO=NO+1
      IF (NO-4) 191,25,25
   17 IF(NO.GT.0) NO=0
  191 CONTINUE
   25 IF(LEN) 60,70,60
   60 DO 22 I=1,P
      S=H(I)
      DO 22 J=1,K
   22 Q(I,J)=Q(I,J)*S
   70 MAXIT=KOT
      RETURN
      END
```

p. 301) divides the elements of column $j$ by the highest absolute value in column $j$ at this second stage, so that the result is a matrix which has a maximum absolute value of 1 in each column. This approach is used in the accompanying subroutine. The target matrix **G** is then found by raising the elements of the scaled varimax factor loadings matrix to the power m, where m is a positive integer (usually, 2, 3 or 4). The signs of the original elements of $\Lambda_V$ are retained.

Other approaches to the calculation of **G** omit the first stage (division of each row of $\Lambda_V$ by the corresponding $h_i$) or simply use the first $k$ normalized eigenvectors of **R**. In each case, the matrix is modified by raising each element to the power $m$ and retaining the original sign. A different value of $m$ could be used for each column, but there are no objective criteria on which to base the choice of an $m$ value.

The effect of powering is to make the small elements of the orthogonal approximation matrix ($\Lambda_V$ in the case discussed above) approach zero more rapidly than the larger elements. No firm guide to the choice of $m$ can be given; Hendrickson and White (1964), Mulaik (1972) and Lawley and Maxwell (1971) agree that $m = 4$ is the most usual value. In the accompanying subroutine the transformation can be

Table 5.13
(a) Raw varimax solution; (b) normal varimax solution. Both solutions based on the minres factor loadings matrix shown in Table 5.8.

(a)

|  | FACTOR | | |
|---|---|---|---|
| VARIABLE | I | II | III |
| 1 | 0.9363 | 0.1954 | 0.1907 |
| 2 | 0.9411 | 0.1636 | 0.2349 |
| 3 | 0.7020 | 0.0380 | 0.2337 |
| 4 | 0.9559 | 0.1927 | 0.0249 |
| 5 | 0.9436 | 0.1956 | -0.2673 |
| 6 | -0.8142 | 0.4182 | 0.0730 |
| 7 | -0.0587 | -0.0059 | -0.7391 |
| 8 | 0.2125 | 0.9772 | 0.0006 |

(b)

|  | FACTOR | | |
|---|---|---|---|
| VARIABLE | I | II | III |
| 1 | 0.9350 | 0.2117 | 0.1859 |
| 2 | 0.9408 | 0.1800 | 0.2240 |
| 3 | 0.7040 | 0.0502 | 0.2255 |
| 4 | 0.9525 | 0.2095 | 0.0138 |
| 5 | 0.9367 | 0.2124 | -0.2782 |
| 6 | -0.8206 | 0.4037 | 0.0851 |
| 7 | -0.0673 | -0.0045 | -0.7364 |
| 8 | 0.1952 | 0.9808 | -0.0008 |

TRACE OF A

0.2356014E 01
0.2538480E 01
0.2540158E 01
0.2539897E 01
0.2539909E 01
0.2539904E 01
0.2539904E 01
0.2539904E 01
0.2539904E 01
0.2539904E 01
0.2539904E 01

accomplished for several different integral values in the range $m_{min} \leq m \leq m_{max}$.

As in the varimax method the procedure involves the determination of a transformation matrix T such that $\Lambda_P = \Lambda_V T$ where $\Lambda_P$ is the oblique primary factor-pattern matrix. The distinction between $\Lambda_P$ and $\Lambda_S$, the oblique primary factor-structure matrix, was made earlier. The transformation matrix is obtained in two stages; first, obtain a matrix T* from

$$T^* = (\Lambda_V' \Lambda_V)^{-1} \Lambda_V' G \tag{5.15}$$

which, in matrix form, can be seen to be analogous to the least-squares equations $\hat{\beta} = (X'X)^{-1} X'y$ of Chapter 2. The matrix T* does not produce unit-length oblique factors (that is, the factors are not standardized, so $\Phi$ is a covariance and not a correlation matrix). Consequently, since there is usually no reason to assume that factors are not unit length, $T^*$ is modified to ensure that $\Phi$ is a correlation matrix. If $D$ is a diagonal matrix given by

$$D^2 = \text{diag}(T^{*'}T^*)^{-1} \tag{5.16}$$

then the transformation matrix $T$ that will give unit-length oblique factors is calculated from

$$T = T^*D \tag{5.17}$$

The primary factor correlation matrix $\Phi$ is obtained from

$$\Phi = (T'T)^{-1} \tag{5.18}$$

and $\Lambda_S$, the oblique primary factor structure matrix from

$$\Lambda_S = \Lambda_P \Phi \tag{5.19}$$

The results of applying the Promax rotation to the Gustafson data are shown in Table 5.14. Table 5.14(a) shows the target matrix, which is scaled so that the highest absolute value in each column is 1.0. This target matrix was obtained by the method described earlier, using an exponent of 2. The basic matrix was the normal varimax factor loadings matrix of Table 5.13(b). The results for exponent 2 and exponent 4 are shown in Table 5.14(b) and 5.14(c) respectively. The effect of the oblique rotation is to make the identity of factor 1 far clearer; variables 7 and 8 have primary pattern coefficients which are very close to zero on this factor, while the other variables retain their high coefficients. Factor 2 is also 'clearer' in that all variables except 6 and 8 are increased slightly. A similar increase in previously high coefficients and a reduction in previously low coefficients is seen in oblique factor 3 vis-à-vis the corresponding orthogonal varimax factor. This effect is more clearly seen in the exponent-4 solution; apart from this the two Promax solutions are very similar.

The primary-factor correlation matrix indicates that factor 3 is not linearly related to either factor 1 or factor 2. The correlation of 0.21 between factors 1 and 2 is worthy of consideration. A correlation of 0.21 is equivalent, in geometrical terms, to an angle of 78° between the factor axes. This is a quite considerable departure from the original 90° angle.

A listing of subroutine PROMAX, together with instructions for its use, is given in Table 5.15. Subroutines TRMULT, GJDEF1, MTMULT and MATPR are listed in Appendix A. An auxiliary subroutine, SWAP, is also listed in Table 5.15. It copies the lower triangle of a square matrix into the upper triangle.

Of course, Varimax and Promax are not the only rotation procedures. Varimax is certainly the most widely-used and generally-

## Table 5.14

(a) Target matrix, (b) primary factor structure, primary factor pattern and factor intercorrelation matrices for Promax rotation with exponent 2; (c) target matrix, (d) primary factor structure, primary factor pattern and factor intercorrelation matrices for Promax rotation with exponent 4. Both solutions based on normal varimax solution, Table 5.13(b).

(a)

FACTOR

| VARIABLE | 1 | 2 | 3 |
|---|---|---|---|
| 1 | 0.96129 | 0.04886 | 0.03655 |
| 2 | 0.95913 | 0.03480 | 0.05227 |
| 3 | 0.94664 | 0.00476 | 0.09340 |
| 4 | 1.00000 | 0.04795 | 0.00020 |
| 5 | 0.92009 | 0.04691 | -0.07805 |
| 6 | -0.83732 | 0.20093 | 0.00826 |
| 7 | -0.00865 | -0.00003 | -1.00000 |
| 8 | 0.03997 | 1.00000 | -0.00000 |

TARGET MATRIX

(b)

FACTOR

| VARIABLE | 1 | 2 | 3 |
|---|---|---|---|
| 1 | 0.92595 | 0.12499 | 0.15435 |
| 2 | 0.93678 | 0.09095 | 0.19329 |
| 3 | 0.71307 | -0.01992 | 0.20526 |
| 4 | 0.94062 | 0.12609 | -0.01828 |
| 5 | 0.91828 | 0.13900 | -0.31026 |
| 6 | -0.88417 | 0.48523 | 0.09240 |
| 7 | -0.08274 | 0.02318 | -0.73709 |
| 8 | 0.07102 | 0.98377 | -0.03700 |

PRIMARY FACTOR PATTERN MATRIX

FACTOR

| VARIABLE | 1 | 2 | 3 |
|---|---|---|---|
| 1 | 0.95553 | 0.33409 | 0.17611 |
| 2 | 0.95942 | 0.30469 | 0.21321 |
| 3 | 0.71195 | 0.14621 | 0.21515 |
| 4 | 0.96757 | 0.32815 | 0.00377 |
| 5 | 0.94348 | 0.31896 | -0.28779 |
| 6 | -0.77795 | 0.29975 | 0.10737 |
| 7 | -0.08917 | -0.03827 | -0.73701 |
| 8 | 0.28290 | 0.99692 | 0.02228 |

PRIMARY FACTOR STRUCTURE MATRIX

FACTOR

| FACTOR | 1 | 2 | 3 |
|---|---|---|---|
| 1 | 1.00000 | | |
| 2 | 0.21596 | 1.00000 | |
| 3 | 0.01552 | 0.05913 | 1.00000 |

PRIMARY FACTOR CORRELATION MATRIX

Table 5.14  *continued*

(c)

TARGET MATRIX

| VARIABLE | FACTOR 1 | 2 | 3 |
|---|---|---|---|
| 1 | 0.92408 | 0.00239 | 0.00134 |
| 2 | 0.91992 | 0.00121 | 0.00273 |
| 3 | 0.89612 | 0.00002 | 0.00872 |
| 4 | 1.00000 | 0.00230 | 0.00000 |
| 5 | 0.84657 | 0.00220 | -0.00609 |
| 6 | -0.70111 | 0.04037 | 0.00007 |
| 7 | -0.00007 | -0.00000 | -1.00000 |
| 8 | 0.00160 | 1.00000 | -0.00000 |

(d)

PRIMARY FACTOR PATTERN MATRIX

| VARIABLE | FACTOR 1 | 2 | 3 |
|---|---|---|---|
| 1 | 0.93158 | 0.11779 | 0.14339 |
| 2 | 0.94283 | 0.08277 | 0.18231 |
| 3 | 0.71780 | -0.02748 | 0.19713 |
| 4 | 0.94248 | 0.12348 | -0.02968 |
| 5 | 0.91387 | 0.14426 | -0.32185 |
| 6 | -0.87643 | 0.48407 | 0.10230 |
| 7 | -0.09870 | 0.04316 | -0.73725 |
| 8 | 0.08303 | 0.98195 | -0.03940 |

PRIMARY FACTOR STRUCTURE MATRIX

| VARIABLE | FACTOR 1 | 2 | 3 |
|---|---|---|---|
| 1 | 0.95759 | 0.32275 | 0.16348 |
| 2 | 0.96202 | 0.29328 | 0.19964 |
| 3 | 0.71429 | 0.13761 | 0.20284 |
| 4 | 0.96775 | 0.31640 | -0.00899 |
| 5 | 0.94020 | 0.30711 | -0.29976 |
| 6 | -0.77494 | 0.31084 | 0.13255 |
| 7 | -0.09795 | -0.03822 | -0.73478 |
| 8 | 0.28615 | 0.99590 | 0.04267 |

PRIMARY FACTOR CORRELATION MATRIX

| FACTOR | 1 | 2 | 3 |
|---|---|---|---|
| 1 | 1.00000 | | |
| 2 | 0.20730 | 1.00000 | |
| 3 | 0.01113 | 0.08264 | 1.00000 |

acceptable orthogonal rotation method, but more controversy surrounds the choice of an oblique transformation. Promax is one of the simpler methods computationally, and it appears to work well (Lawley and Maxwell, 1971, p. 77; Hakstian, 1971) though difficulty is often found in deciding on the value of the parameter $m$ used in the derivation of **G**, the target matrix. Values of 2, 3 or 4 are generally adequate — it is certainly unnecessary to use values in excess of 4 or 5. Hakstian (1971) suggests that '... the more factorially complex the preliminary orthogonal solution and the less clearly defined the

## Table 5.15
### Subroutine PROMAX: Promax oblique rotation

```
      SUBROUTINE PROMAX(A,N1,B,N2,C,N3,R,N4,H,FMT,WORK,NV,NF,K1,K2,INC)
C
C PROMAX ROTATION BY HENDRICKSON AND WHITE(1964)
C PROGRAM BY P.M.MATHER,1969, MODIFIED 1974.
C ARGUMENTS
C A       MATRIX TO BE ROTATED. USUALLY VARIMAX OR OTHER ORTHOGONAL
C           SOLUTION OR ANY TARGET MATRIX ( SEE TEXT.
C         MINIMUM DIMENSION (NV,NF)
C N1      FIRST DIMENSION OF A AS DECLARED IN CALLING PROGRAM
C B       REAL MATRIX. UNDEFINED ON ENTRY, CONTAINS OBLIQUE FACTOR
C         STRUCTURE ON EXIT, DIMENSION AT LEAST (NV,NF)
C N2      FIRST DIMENSION OF B AS DECLARED IN CALLING PROGRAM.
C C       REAL MATRIX,MINIMUM DIMENSION (NV,NF). HOLDS FACTOR PATTERN
C         MATRIX ON EXIT.
C N3      FIRST DIMENSION OF C AS DECLARED IN CALLING PROGRAM.
C R       REAL ARRAY, MINIMUM DIMENSION (NF,NF), CONTAINS INTER-FACTOR
C         CORRELATION MATRIX ON EXIT.
C N4      FIRST DIMENSION OF R AS DECLARED IN CALLING PROGRAM.
C H       REAL ARRAY, MINIMUM DIMENSION NV HOLDING THE SQUARE ROOTS
C         OF THE COMMUNALITIES ( AS OUTPUT BY VARIMAX)
C         UNALTERED BY THE ROUTINE.
C         *MATPR*
C WORK    REAL ARRAY,MINIMUM DIMENSION (NV), WORKSPACE.
C NV      NUMBER OF VARIABLES. INTEGER.
C NF      NUMBER OF FACTORS. INTEGER.
C K1      STARTING EXPONENT FOR PROMAX. INTEGER GREATER THAN 1.
C K2      TERMINAL EXPONENT FOR PROMAX. INTEGER NOT LESS THAN K1.
C INC     K1 IS INCREASED BY INC AT END OF EACH CYCLE. SUBROUTINE
C           EXITS WHEN K1 EXCEEDS K2. INC MUST NOT BE SET TO ZERO
C           OR TO A NEGATIVE VALUE.
C           VALUE.
C
C SUBROUTINES CALLED - TRMULT,ACCINV,MATPR -- LISTED IN APPENDIX A.
C
      DIMENSION A(N1,NF),B(N2,NF),C(N3,NV),R(N4,NF),H(NV),FMT(20),
     ? WORK(NV)
  101 FORMAT(45H1       PROMAX ROTATION WITH EXPONENT EQUAL TO   ,I3//)
  105 FORMAT(1H0,30X,44HPROMAX OBLIQUE ROTATION - PAUL M MATHER,1969///)
      IF(NF.EQ.1) GOTO 888
      NF1=NF+1
      IF(K1.EQ.0.OR.K2.EQ.0.OR.INC.EQ.0) GOTO 75
      KZ=K1
C EPS CONTAINS THE SMALLEST POSITIVE FLOATING POINT VALUE THAT IS
C REPRESENTIBLE IN THE MACHINE.
C ALTER FOR MACHINES OTHER THAN ICL 1900 SERIES.
      EPS=2.0**(-37)
      DO 777 K1=KZ,K2,INC
      WRITE(6,101) K1
C TRMULT AND ACCINV ARE LISTED IN APPENDIX A.
      CALL TRMULT(R,A,A,N4,N1,N1,NF,NF,NV)
      CALL ACCINV(R,C,WORK,NF1,NF,L,N4,N3,EPS)
C TEST FOR ERROR IN ACCINV.
      IF(NF.EQ.0.OR.NF1.EQ.0) GOTO 890
      CALL SWAP(R,N4,NF)
C
C     R HOLDS (A(T)*A) INVERTED
C
      DO 3 I=1,NV
C NOTE ARRAY H HOLDS SQUARE ROOTS OF THE COMMUNALITIES.
      AA=H(I)
      DO 3 J=1,NF
    3 B(I,J)=A(I,J)/AA
      DO 5 I=1,NF
      AA=-100.0
      DO 6 J=1,NV
```

Table 5.15  *continued*

```
      ABSB=ABS(B(J,I))
    6 IF(ABSB.GT.AA) AA=ABSB
      DO 5 J=1,NV
    5 B(J,I)=B(J,I)/AA
      K=K1+1
      DO 7 I=1,NV
      DO 7 J=1,NF
      IF(B(I,J).EQ.0.0) GOTO 7
      B(I,J)=(ABS(B(I,J))**K)/B(I,J)
    7 CONTINUE
C
C     B HOLDS ROW/COL NORMALIZED VARIMAX MATRIX RAISED TO POWER OF THE
C     SPECIFIED EXPONENT
C
      NN=N2*NF
      IP=0
C MATPR IS LISTED IN APPENDIX A.
      CALL MATPR(16HTARGET MATRIX    ,2,FMT,6,B,NN,N2,NV,NF,IP,2,8)
      CALL TRMULT(C,A,B,N3,N1,N2,NF,NF,NV)
C C# A TRANSPOSE POSTMULTIPLIED BY B.
      CALL MTMULT(B,R,C,N2,N4,N3,NF,NF,NF)
C B = (A-TRANSPOSE * A) (INVERSE) * (A-TRANSPOSE * B)
      CALL TRMULT(C,B,B,N3,N2,N2,NF,NF,NF)
      CALL ACCINV(C,R,WORK,NF1,NF,L,N3,N4,EPS)
      IF(NF1.EQ.0.OR.NF.EQ.0) GOTO 890
C SUBROUTINE SWAP IS LISTED IN THIS TABLE.
      CALL SWAP(C,N3,NF)
      DO 9 I=1,NF
      AA=SQRT(C(I,I))
      DO 9 J=1,NF
    9 B(J,I)=B(J,I)*AA
      CALL MTMULT(C,A,B,N3,N1,N2,NV,NF,NF)
      NN=N3*NF
      IP=1
   77 CALL MATPR(32HPRIMARY FACTOR PATTERN MATRIX    ,4,FMT,6,C,NN,N3,NV,
     1 NF,IP,2,8)
C C IS FACTOR PATTERN MATRIX
      CALL TRMULT(R,B,B,N4,N2,N2,NF,NF,NF)
      CALL ACCINV(R,B,WORK,NF1,NF,L,N4,N2,EPS)
      IF(NF1.EQ.0.OR.NF.EQ.0) GOTO 890
      CALL SWAP(R,N4,NF)
      IP=1
      NN=N4*NV
      CALL MATPR(40HPRIMARY FACTOR CORRELATION MATRIX        ,5,FMT,6,R,
     1NN,N4,NF,NF,IP,1,8)
C R IS FACTOR CORRELATION MATRIX
      CALL MTMULT(B,C,R,N2,N3,N4,NV,NF,NF)
C B IS FACTOR STRUCTURE MATRIX
      NN=N2*NF
      IP=0
      CALL MATPR(32HPRIMARY FACTOR STRUCTURE MATRIX  ,4,FMT,6,B,NN,N2,
     1NV,NF,IP,2,8)
  777 CONTINUE
      WRITE(6,105)
      GOTO 875
   75 WRITE(6,213)
  213 FORMAT(46H K1,K2 OR INC SET TO ZERO - CHECK CONTROL CARD  )
      GOTO 875
  888 WRITE(6,889)
  889 FORMAT(47H ONLY 1 FACTOR TO BE ROTATED - EXIT FROM PROMAX)
  875 RETURN
  890 WRITE(6,891)
  891 FORMAT(28H FAIL IN ACCINV - ERROR EXIT)
      STOP
      END
```

Table 5.15   *continued*

```
      SUBROUTINE SWAP(R,M,N)
      DIMENSION R(M,N)
      NN=N+1
      DO 700 I=2,NN
      II=I-1
      DO 700 J=1,II
      R(II,J)=R(I,J)
  700 R(J,II)=R(II,J)
      RETURN
      END
```

separation between large and near-zero entries the lower should be the value' of $m$. One useful guide is the number of near-zero entries in $\Lambda_P$, sometimes called the 'hyperplane count'. Near-zero is taken to mean less than 0.1 in absolute value. However, nothing can replace a thorough understanding of the system of variables under investigation; as Lawley and Maxwell (1971, p. 79) point out, '. . . the interpretation of factors depends ultimately on prior knowledge of the variates being analysed'.

This leads to the possibility that the investigator is able to specify in advance the distribution of high and near-zero loadings in the factor pattern matrix. The resulting matrix, with 1's representing 'high' loadings and 0's the 'near-zero' loadings can be thought of as a hypothesis matrix. This matrix could, for example, be derived from a study of oblique factor patterns determined by previous investigations, or by the splitting of the available sample into two parts. One half of the sample could then be subjected to an 'exploratory' factor analysis, culminating in a blind rotation to simple structure using, for example, Promax. The position of the high values in the Promax oblique primary factor pattern matrix indicates which factors are contributing strongly to the variance of each variable. This leads to a hypothesis matrix of the form

$$\mathbf{B}' = \begin{bmatrix} 1 & 1 & 0 & 1 & 0 & 0 & 1 & 0 \\ 0 & 0 & 1 & 1 & 1 & 0 & 0 & 0 \\ 0 & 0 & 0 & 0 & 0 & 1 & 0 & 1 \end{bmatrix}$$

where the ones and zeros represent high and low coefficients respectively. Mulaik (1972, pp. 314–318) describes a technique which he attributes to Joreskog (1966) that allows the transformation of a given reference *structure* matrix $\mathbf{V}$ to fit the zero elements of $\mathbf{B}$. In other words, the problem is to find a transformation matrix $\mathbf{T}$ such that $\mathbf{V}_T = \mathbf{VT}$. $\mathbf{V}_T$ is the transformed reference structure matrix. The sum of squares of the elements of $\mathbf{V}_T$ corresponding to zero elements of $\mathbf{B}$ is to be minimized, while nonzero elements of $\mathbf{V}_T$ are free to take on any value. To give the factors unit length, a second criterion is adopted, that is, diag $(\mathbf{T}'\mathbf{T}) - \mathbf{I} = 0$. $\mathbf{T}'\mathbf{T}$ is the matrix of correlations among the *reference* factors.

If $\mathbf{t}_j$ is the $j$th column of $\mathbf{T}$ and $\mathbf{b}_j$ the corresponding column of $\mathbf{B}$ then define a matrix $\mathbf{V}_j$ which consists of the rows of $\mathbf{V}$ corresponding to zero elements in $\mathbf{b}_j$. This implies that each column of $\mathbf{B}$ must have at least one zero element. Mulaik (1972, p. 316) shows that the elements of $\mathbf{t}_j$ which ensure the satisfaction of the two criteria set out above are given by the unit length eigenvector corresponding to the smallest eigenvalue of $\mathbf{V}'_j\mathbf{V}_j$. If $\mathbf{V}'_j\mathbf{V}_j$ has multiple zero eigenvalues, any of the corresponding eigenvectors can be chosen. These computations are performed for each of the $k$ columns of $\mathbf{T}$ and $\mathbf{B}$ in turn, leading to the determination of the elements of $\mathbf{T}$. The following sequence of operations will then provide the required matrices:

$\mathbf{V}_T = \mathbf{V}\mathbf{T}$ (transformed reference structure)

$\mathbf{C} = \mathbf{T}'\mathbf{T}$ (correlations among reference factors)

$\Lambda_P = \mathbf{V}_T \mathbf{D}$ (transformed primary pattern)

where

$\mathbf{D}^2 = \text{diag}\,(\mathbf{C}^{-1})$

$\Phi = \mathbf{D}^{-1}\mathbf{C}^{-1}\mathbf{D}^{-1}$ (correlations among primary factors)

$\Lambda_S = \Lambda_P \Phi$ (primary structure)

The sum of the $k$ smallest eigenvalues provides an estimate of the error of fit of $\mathbf{V}_T$ to the zero elements of $\mathbf{B}$. If this sum is small relative to the number of zero elements in $\mathbf{B}$ then the data fit the simple structure hypothesis. As yet, there is no test of significance available, and the decision as to what is 'small' is entirely subjective. A Fortran routine implementing the above algorithm is described and listed in Table 5.16.

Other approaches to this problem are described by Browne (1967), Gruvaeus (1970), Joreskog (1966, 1969), Hakstian (1972) and Mulaik (1972). Lawley and Maxwell (1971) discuss a general approach to confirmatory factor analysis involving maximum likelihood estimation. The use of these methods appears to be justified whenever a body of experience has been built up through the use of exploratory analyses. However, no application of these ideas has yet been published in the geographical literature.

The use of subroutine TARGET is illustrated using the minres results for 3 factors for the Gustafson data (Table 5.6). The transpose of the target matrix $\mathbf{B}$ is:

$$\mathbf{B}' = \begin{bmatrix} 1 & 1 & 1 & 1 & 1 & 1 & 0 & 0 \\ 0 & 0 & 0 & 0 & 0 & 1 & 0 & 1 \\ 0 & 0 & 0 & 0 & 1 & 0 & 1 & 0 \end{bmatrix}$$

## Table 5.16
## Subroutine TARGET: rotation to specified target

```
      SUBROUTINE TARGET(A,B,C,D,IND,T,P,K,SUM,ITOT,N1,N2,N3,N4,FMT)
C     OBLIQUE ROTATION TO SPECIFIED SIMPLE STRUCTURE
C     MULAIK  1972  P.314  AFTER JORESKOG, PSYCHOMETRIKA,1966.
C ARGUMENTS ARE:
C A       REFERENCE STRUCTURE MATRIX TO BE TRANSFORMED.
C         MINIMUM DIMENSION (P*K).
C B       TARGET MATRIX. MINIMUM DIMENSION (P,K)
C C       UNDEFINED ON ENTRY. DIMENSION AT LEAST (K,K)
C D       REAL VECTOR, DIMENSION AT LEAST K. WORKSPACE
C IND     INTEGER VECTOR, DIMENSION AT LEAST P. WORKSPACE.
C T       REAL ARRAY, DIMENSION AT LEAST (K,K). WORKSPACE.
C P       INTEGER. NUMBER OF VARIABLES (TESTS)
C K       INTEGER. NUMBER OF FACTORS.
C SUM     REAL VARIABLE. UNDEFINED ON ENTRY.
C ITOT    INTEGER VARIABLE. UNDEFINED ON ENTRY.
C N1      FIRST DIMENSION OF A AS DECLARED IN CALLING PROGRAM.
C N2      FIRST DIMENSION OF B AS DECLARED IN CALLING PROGRAM.
C N3      FIRST DIMENSION OF C AS DECLARED IN CALLING PROGRAM.
C N4      FIRST DIMENSION OF T AS DECLARED IN CALLING PROGRAM.
C FMT     REAL ARRAY, MINIMUM DIMENSION 20, CONTAINING THE OUTPUT FORMAT
C         FOR SUBROUTINE MATPR, INCLUDING OPENING AND CLOSING BRACKETS.
C         SEE APPENDIX A.
C
C OUTPUT FROM THE ROUTINE:
C B       PRIMARY FACTOR PATTERN MATRIX
C C       CORRELATIONS AMONG PRIMARY FACTORS
C SUM     SUM OF EIGENVALUES (SEE TEXT)
C ITOT    NUMBER OF ZEROS (SEE TEXT)
C
C ERROR INDICATIONS
C P IS SET TO ZERO AND SUBROUTINE IS EXITED IF ANY COLUMN OF THE TARGET
C MATRIX CONTAINS NO ZERO ENTRIES.
C
C SUBROUTINES REQUIRED:
C GJDEF1, TRMULT,  MTMULT,  MATPR AND HTDQL (APPENDIX A)
C
      REAL MACHEP
      INTEGER P
      DIMENSION A(N1,K),B(N2,K),C(N3,K),T(N4,K),D(K),IND(P)
      DIMENSION FMT(20)
      SUM=0.0
      ITOT=0
C THE FOLLOWING TWO PARAMETERS ARE MACHINE-DEPENDENT ----SEE APPENDIX A
      ETA=2.0**(-124)
      MACHEP=2.0**(-24)
      TOL=ETA/MACHEP
      DO 1 I=1,K
      MS=0
      DO 2 J=1,P
      IF(B(J,I).GT.ETA) GOTO 2
      MS=MS+1
      IND(MS)=J
    2 CONTINUE
      IF(MS.EQ.0) GO TO 6
      ITOT=ITOT+MS
      DO 3 J=1,K
      DO 3 L=1,K
      C(J,L)=0.0
      DO 3 M=1,MS
      MM=IND(M)
    3 C(J,L)=C(J,L)+A(MM,L)*A(MM,J)
      CALL HTDQL(K,N3,N3,C,D,B,C,MACHEP,TOL)
      DO 4 J=1,K
    4 T(J,I)=C(J,K)
C T IS THE TRANSFORMATION MATRIX
```

Table 5.16 *continued*

```
    1 SUM=SUM+D(K)
C FORM REFERENCE STRUCTURE MATRIX
      CALL MTMULT(B,A,T,N2,N1,N4,P,K,K)
      IPR=1
      MM=N2*K
      CALL MATPR(40HOBLIQUE REFERENCE STRUCTURE MATRIX      ,10,FMT,6,
     1 B,MM,N2,P,K,IPR,2,4)
      CALL TRMULT(C,T,T,N3,N4,N4,K,K,K)
      IPR=1
      MM=N3*K
      CALL MATPR(40HCORRELATIONS AMONG REFERENCE FACTORS    ,10,FMT,6,
     1 C,MM,N3,K,K,IPR,2,4)
      CALL GJDEF1(K,C,D,N3)
      DO 60 I=1,K
      DO 60 J=I,K
   60 C(I,J)=C(J,I)
      DO 57 I=1,K
   57 D(I)=SQRT(C(I,I))
      DO 58 J=1,K
      DJ=D(J)
      DO 58 I=1,P
   58 B(I,J)=B(I,J)*DJ
      IPR=0
      MM=N2*K
      CALL MATPR(32HPRIMARY FACTOR PATTERN MATRIX     ,8,FMT,6,B,MM,N2,
     1 P,K,IPR,2,4)
      DO 62 I=1,K
      DI=D(I)
      DO 62 J=1,K
   62 C(I,J)=C(I,J)/D(J)/DI
      IPR=0
      MM=N3*K
      CALL MATPR(40HCORRELATIONS AMONG PRIMARY FACTORS      ,10,FMT,6,
     1 C,MM,N3,K,K,IPR,2,4)
      RETURN
    6 WRITE(6,7) I
    7 FORMAT('0COL.',I4,' OF TARGET HAS NO ZEROS')
      P=0
      RETURN
      END
```

The positions of the zero elements in **B** was determined partly by reference to the relative sizes of the entries in the unrotated minres factor loadings matrix and to expectations based on an appraisal of other studies of drainage-basin morphometry, as set out in Section 5.3. The transformed reference structure matrix **V**, the primary factor pattern matrix $\Lambda_P$ and the correlations among the reference and primary factors are shown in Table 5.17. These matrices can be compared to the Promax results shown in Tables 5.14(b) and (c). The first noteworthy point is that the matrices in Table 5.17 are more clearly structured and are easier to interpret than the corresponding Promax matrices in Table 5.14. Secondly, the Promax primary factors are less strongly correlated, especially factors 1 and 3. The sum of the eigenvalues gives an indication of the goodness of fit of the zero elements of **B** to the corresponding elements of **V**, as explained above. In the example, the sum is 0.0844 and the ratio of the sum of the eigenvalues to the number of zero elements in **B** is 0.00603. Both values

Table 5.17
Reference structure, primary pattern and factor intercorrelation matrices produced by target rotation of minres solution shown in Table 5.8

OBLIQUE REFERENCE STRUCTURE MATRIX

|   | 1 | 2 | 3 |
|---|---|---|---|
| 1 | -0.8557 | -0.0297 | 0.0597 |
| 2 | -0.8645 | 0.0005 | 0.0982 |
| 3 | -0.6581 | 0.0810 | 0.1360 |
| 4 | -0.8895 | -0.0170 | -0.1134 |
| 5 | -0.8982 | -0.0111 | -0.4008 |
| 6 | 0.8876 | -0.5615 | 0.1569 |
| 7 | -0.0000 | 0.0217 | -0.7237 |
| 8 | 0.0000 | -0.9221 | -0.0829 |

CORRELATIONS AMONG REFERENCE FACTORS

|   | 1 | 2 | 3 |
|---|---|---|---|
| 1 | 1.0000 | -0.3870 | 0.1931 |
| 2 | -0.3870 | 1.0000 | -0.0064 |
| 3 | 0.1931 | -0.0064 | 1.0000 |

PRIMARY FACTOR PATTERN MATRIX

|   | 1 | 2 | 3 |
|---|---|---|---|
| 1 | -0.9485 | -0.0323 | 0.0610 |
| 2 | -0.9582 | 0.0005 | 0.1004 |
| 3 | -0.7294 | 0.0881 | 0.1390 |
| 4 | -0.9859 | -0.0185 | -0.1159 |
| 5 | -0.9957 | -0.0121 | -0.4096 |
| 6 | 0.9838 | -0.6107 | 0.1604 |
| 7 | -0.0000 | 0.0236 | -0.7397 |
| 8 | 0.0000 | -1.0029 | -0.0847 |

CORRELATIONS AMONG PRIMARY FACTORS

|   | 1 | 2 | 3 |
|---|---|---|---|
| 1 | 1.0000 | 0.3932 | -0.2067 |
| 2 | 0.3932 | 1.0000 | -0.0755 |
| 3 | -0.2067 | -0.0755 | 1.0000 |

SUM OF EIGENVALUES =    0.8439612E-01
PROPORTION OF TOTAL NUMBER OF ZEROS =    0.6028294E-02

are small, hence it can be concluded that the reference structure matrix in Table 5.17 fits the simple structure hypothesis quite well. The lack of a precise statistical test of the hypothesis is a drawback, however.

## 5.5 Factor Scores

The problem of estimating the elements of the factor pattern matrix was discussed in Section 5.3, where it was shown [Equations (5.1) and (5.2)] that the factor pattern coefficients are equivalent to weights which are used to express the variables in terms of the factors. In the present section, the converse of this procedure is discussed, namely, the expression (or measurement) of the factors in terms of the variables. These *factor scores*, as they are termed, measure the influence of each factor on each case, area or individual, and are thus of importance in studies in which the identification of patterns among the observed cases is a primary aim. Thus, factor scores have been widely used in human and economic geography, for example in studies of urban social structure (Giggs, 1973; Rees, 1971). Other uses of factor scores include the correlation of variables not included in the original analysis with the factors and the construction of indices or scales of the type mentioned below in Section 6.2.B in connection with multidimensional scaling methods.

Unfortunately common factor scores are not directly computable, as is the case with principal component scores, though scores on image factors can be found exactly. The reason for this is that the common-factor model specifies $k$ common factors plus $p$ unique or specific factors, not all of which are identifiable from $p$ variables alone. In mathematical terms, this fact means that the matrix of loadings on the $(p + k)$ factors is not square and therefore does not have a unique inverse. A second consideration is that factor scores cannot be 'estimated' in the statistical sense of the term, since they are the values of unobservable factors and not population parameters. The term 'estimated factor scores' will, however, be used with the proviso that the scores are not estimated statistically but calculated according to some approximation.

Several such formulae have, in fact, been put forward. The most widely-used is the regression-type least-squares solution of Thurstone. Bartlett's method uses a slightly different approach, in that the sum of squares of the scores on the specific factors is minimized rather than the sum of squares of the deviations of the estimated from the true common-factor scores. A third method, that of ideal variables, is described by Harman (1967).

The *least-squares* or regression method of estimating the matrix $F$ of true factor scores is very similar to the procedure used in Chapter 2 to find the vector $\hat{\beta}$ of estimated partial regression coefficients. The estimator is:

$$\hat{F} = ZR^{-1} \Lambda_S \qquad (5.20)$$

where $\hat{F}$ is the $(n \times k)$ matrix of estimated factor scores, $Z$ is the $(n \times p)$ matrix of standardized variables (that is, each column of $X$ is standardized to have unit variance and zero mean), $R$ is the $(p \times p)$ matrix of correlations among the variables and $\Lambda_S$ is the $(n \times k)$ matrix of primary factor structure coefficients. The matrix $W = (R^{-1}\Lambda_S)$ can be thought of as containing $k$ columns of regression weights. Coefficients of determination $(R^2)$ for each of these $k$ regressions can be computed from

$$R_q^2 = \sum_{j=1}^{p} w_{jq}^2 + 2 \sum_{j=1}^{p-1} \sum_{m=j+1}^{p} w_{jq} w_{mq} r_{jk} \qquad (q = 1,k) \qquad (5.21)$$

In this equation, $w_{ij}$ is an element of $\mathbf{W}(=\mathbf{R}^{-1}\mathbf{\Lambda}_S)$ and $r_{jk}$ an element of $\mathbf{R}$, the matrix of correlations among the variables. The term $w_{1q}^2$ gives the direct contribution of variable $x_1$ to $R_q^2$ while the term $w_{1q}w_{mq}r_{1q}$ measure the indirect joint contribution of variables $x_1$ and $x_m$. This joint contribution is equally distributed between the two variables concerned. In common factor analysis, $R_q^2$ will usually be less than 1.0 in contrast with principal components analysis, in which the coefficient of determination for any principal component score vector will be exactly one. The total (direct plus indirect) contributions of the variables to $R_q^2$ can be used to assess their relative importance in contributing to the 'meaning' of factor $q$. The use of the $R^2$ value in assessing the determinancy of the factor scores is discussed below. In passing we can note that factor scores estimated by this method are sometimes standardized by the formula:

$$\hat{f}_q^* = 10\hat{f}_q/R_q$$

This has the effect of eliminating negative scores and, according to Harman (1967, p. 358), of improving their 'interpretability'. A Fortran subroutine which carries out the estimation procedure is listed in Table 5.18.

*Bartlett's method* of minimizing the sum of squares of the standardized residuals (or scores on the specific factors) produces estimates $\hat{\mathbf{F}}_B$ that are unbiased, unlike the least squares estimates $\hat{\mathbf{F}}$, but which have a greater variability. The basic equation for this method is:

$$\hat{\mathbf{F}}_B = \mathbf{Z}\mathbf{\Psi}^{-2}\mathbf{\Lambda}_P(\mathbf{\Lambda}_P\mathbf{\Psi}^{-2}\mathbf{\Lambda}_P)^{-1} \tag{5.22}$$

in which $\mathbf{\Lambda}_P$ is a primary factor pattern matrix (for orthogonal or oblique factors) and $\mathbf{\Psi}^2$ is the diagonal matrix of unique variances (i.e. the differences between the total variance and the common variance of each variable). Harman (1967, Tables 16.10A and 16.10B) gives a guide to the computation of $\hat{\mathbf{F}}_B$ from Equation (5.22). He notes that an measure equivalent to the coefficient of determination can be worked out for each column of $\hat{\mathbf{F}}_B$ as follows:

$$R_q^2 = 1.0 - \det(\mathbf{J}_{qq})/\det(\mathbf{J})$$

where $\mathbf{J} = \mathbf{\Lambda}_P\mathbf{\Psi}^{-2}\mathbf{\Lambda}_P$,
  $\det(\mathbf{J})$ is the determinant of $\mathbf{J}$, and
  $\det(\mathbf{J}_{qq})$ is the determinant of the matrix formed from $\mathbf{J}$ by deleting the $q$th row and column.

A subroutine for Bartlett's method is given in Table 5.19.

The *ideal variables* method is based on an approximation to the common parts of the variables. The estimator is:

$$\hat{\mathbf{F}}_I = \mathbf{Z}\mathbf{\Lambda}_P(\mathbf{\Lambda}_P'\mathbf{\Lambda}_P)^{-1} \tag{5.23}$$

## Table 5.18
### Subroutine FSCORE: factor score estimation by regression

```
      SUBROUTINE FSCORE(S,R,Z,BETA,RF,N1,N2,N3,N4,NA,NV,NF,IND)
C FACTOR SCORES BY REGRESSION METHOD (SEE CHAPTER 5 FOR DETAILS)
C ARGUMENTS:
C S         REAL ARRAY, DIMENSION (NV+1)*NF; FACTOR STRUCTURE MATRIX.
C           S IS OVERWRITTEN BY THE ROUTINE.
C R         REAL ARRAY, DIMENSION (NV+1)*NV. CONTAINS THE UPPER
C           TRIANGLE + DIAGONAL OF THE CORRELATION MATRIX IF IND=0 OR
C           THE FULL INVERSE IF IND = 1.
C           CONTAINS LOWER TRIANGLE OF INVERSE OF COFRELATION MATRIX
C           ON EXIT.
C Z         REAL ARRAY. MINIMUM DIMENSION NA*NV. MATRIX OF STANDARD SCORES.
C           CONTAINS REGRESSION ESTIMATES OF FACTOR SCORES ON EXIT.
C BETA      REAL ARRAY. MINIMUM DIMENSION NV*NV. UNDERFINED ON ENTRY.
C           USED TO HOLD FACTOR WEIGHT MATRIX.
C RF        REAL ARRAY. MINIMUM DIMENSION NV. UNDERFINED ON ENTRY.
C           HOLDS THE CORRELATIONS BETWEEN TRUE AND ESTIMATED FACTORS
C           ON EXIT.
C N1,N2,N3,N4  INTEGER CONSTANTS, THE FIRST DIMENSIONS OF S,R,Z,BETA AS
C           DECLARED IN THE CALLING PROGRAM.
C NA        INTEGER CONSTANT. THE NUMBER OF CASES (INDIVIDUALS, AREAS,
C           SUBJECTS).
C NV        INTEGER CONSTANT. THE NUMBER OF VARIABLES(TESTS)
C NF        INTEGER CONSTANT. THE NUMBER OF FACTORS.
C IND       =0: INDICATES THAT R HOLDS THE CORRELATION MATRIX.
C IND       =1: INDICATES THAT R HOLDS THE INVERSE OF THE CORRELATION
C           MATRIX.
C
C SUBROUTINES REQUIRED:
C ACCINV, TRMULT -- LISTED IN APPENDIX A.
C
C ******NOTE: THE UPPER TRIANGLE + PRINCIPAL DIAGONAL OF THE CORRELATION
C MATRIX SHOULD BE HELD ON A DISC OR MAGTAPE FILE DEFINED AS
C CHANNEL 3 IN THE PROGRAM DESCRIPTION. THE MATRIX SHOULD BE WRITTEN TO
C THE FILE BY A STATEMENT OF THE FOLLOWING FORM.
C         WRITE(3) ((R(I,J),J=I,NV),I=1,NV)
C         ENDFILE 3
C
      DIMENSION S(N1,NF),R(N2,NV),Z(N3,NV),BETA(N4,NV),RF(NV)
      WRITE(6,90)
   90 FORMAT('1 FACTOR SCORE ESTIMATION BY REGRESSION METHOD'///)
      IF(IND) 40,41,40
C INVERT THE CORRELATION MATRIX (NOTE: THIS MAY ALREADY BE AVAILABLE
C FROM THE INITIAL FACTORING ROUTINE. E.G. ALPHA METHOD).
   41 EPS=2.0**(-37)
      NV1=NV+1
      CALL ACCINV(R,BETA,RF,NV1,NV,L,N2,N4,EPS)
      IF(NV1.EQ.0) GOTO 700
      IF(NV.EQ.0) GOTO 702
      DO 43 I=2,NV1
      II=I-1
      DO 43 J=1,II
      R(II,J)=R(I,J)
   43 R(J,II)=R(II,J)
C FORM THE MATRIX OF REGRESSION COEFFICIENTS (FACTOR WEIGHTS) FROM
C B=S/R WHERE B=REGRESSION COEFFICIENTS, S=FACTOR STRUCTURE MATRIX
C (OBLIQUE OR ORTHOGONAL) AND R IS THE INVERSE OF THE CORRELATION
C MATRIX.
   40 CALL TRMULT(BETA,S,R,N4,N1,N2,NF,NV,NV)
      WRITE(6,2)
    2 FORMAT('0 FACTOR WEIGHT MATRIX (TRANSPOSE)'///)
      DO 3 I=1,NF
      WRITE(6,20) I
   20 FORMAT('0FACTOR NO. ',I4)
    3 WRITE(6,4) (J, BETA(I,J),J=1,NV)
    4 FORMAT(6(' ',I4,'...',F10.4))
```

Table 5.18   *continued*

```
            NV1=NV-1
            REWIND 3
            READ(3) ((R(I,J),J=1,NV),I=1,NV)
C CALCULATE MULTIPLE CORRELATIONS AND CONTRIBUTIONS (INDIRECT,DIRECT AND
C TOTAL) OF THE VARIABLES.
            DO 5 IP=1,NF
C INITIALIZE A,B AND S.  A WILL BE USED TO ACCUMULATE THE SUM OF THE
C DIRECT CONTRIBUTIONS, B TO ACCUMULATE SUM OF INDIRECT CONTRIBUTIONS.
C COL. 1 OF S WILL BE USED TO STORE THE INDIVIDUAL (DIRECT + INDIRECT)
C CONTRIBUTIONS OF EACH VARIABLE TO FACTOR IP.
C THE MULTIPLE CORRELATION FOR FACTOR IP WILL BE THE SQUARE ROOT OF A+B.
            DO 21 I=1,NV
         21 S(I,1)=0.0
            A=0.0
            B=0.0
          8 FORMAT('0',40('*'),'FACTOR',I4,60('*')//)
            WRITE(6,8) IP
            WRITE(6,88)
C LOOP 6 COMPUTES DIRECT CONTIBUTIONS OF VARIABLES TO FACTOR IP.
            DO 6 J=1,NV
            C=BETA(IP,J)**2
            S(J,1)=S(J,1)+C
            S(J,2)=C
          6 A=A+C
         88 FORMAT(' DIRECT CONTRIBUTION OF VARIABLES TO FACTOR'//)
          9 FORMAT(6(' ',I4,'...',F10.4))
            WRITE(6,9) (J, S(J,2),J=1,NV)
            WRITE(6,10)
C LOOPS 30 AND 7 CONTROL CALCULATION OF INDIRECT CONTRIBUTIONS OF ALL
C PAIRS OF VARIABLES.
            DO 30 J=1,NV1
            J1=J+1
            DO 7 K=J1,NV
            C=BETA(IP,J)*BETA(IP,K)*R(J,K)*2.0
            S(J,1)=S(J,1)+C/2.0
            S(K,1)=S(K,1)+C/2.0
            S(K,2)=C
         10 FORMAT('0INDIRECT CONTRIBUTIONS OF VARIABLES TO FACTOR'//)
          7 B=B+C
            WRITE(6,11) (J,K,S(K,2),K=J1,NV)
         30 CONTINUE
         11 FORMAT(5(' ',I3,',',I3,'...',F11.5))
C COMPUTE MULTIPLE CORRELATION FOR FACTOR IP.
            RF(IP)=SQRT(A+B)
            WRITE(6,12) RF(IP)
         12 FORMAT('0MULTIPLE CORRELATION FOR THIS FACTOR = ',F10.5)
            WRITE(6,22)
            WRITE(6,9) (J, S(J,1),J=1,NV)
         22 FORMAT('0 TOTAL CONTRIBUTION OF EACH VARIABLE TO THE VARIANCE OF T
           1HIS FACTOR'//)
          5 CONTINUE
C CALCULATE FACTOR SCORES
            WRITE(6,23)
            DO 13 K=1,NA
            DO 14 I=1,NF
            SD=0.0
            DO 15 J=1,NV
         15 SD=SD+BETA(I,J)*Z(K,J)
C SCORES FOR CASE (AREA) K ON THE NF FACTORS ARE HELD PRIOR TO TRANSFORM
C ATION IN THE FIRST COLUMN OF ARRAY S.
         14 S(I,1)=SD
C THE TRANSFORMED SCORES ARE NOW OVERWRITTEN ON ARRAY Z. "HEY REPLACE
C THE STANDARD SCORES ON THE VARIABLES, WHICH WERE INITIALLY HELD IN Z.
            DO 16 J=1,NF
         16 Z(K,J)=10.0*S(J,1)/RF(J)+50.0
         13 WRITE(6,24) K,(J,Z(K,J),J=1,NF)
C IF NECESSARY, INSERT STATEMENTS AT THIS POINT TO WRITE FACTOR SCORE
```

Table 5.18  *continued*

```
C MATRIX Z TO DISC, OR TO OUTPUT THE MATRIX TO CARD-PUNCH.
C E. G. : WRITE(8) ((Z(I,J),J=1,NF),I=1,NA) WHERE DEVICE 8 IS A MAGTAPE OR
C DISC FILE THAT IS PRESERVED AT THE END OF THE JOB.
   24 FORMAT('OCASE ',I4/(6(' ',I4,'...',F8.3)))
   23 FORMAT('1         FACTOR SCORES MATRIX (REGRESSION ESTIMATES)'///)
      WRITE(6,25)
   25 FORMAT('O  ***  THE FACTOR SCORES HAVE BEEN SCALED SO THAT EACH FA
     1CTOR HAS A '//'         MEAN OF 50.0 AND VARIANCE EQUAL TO ITS MULTIP
     2LE CORRELATION')
      RETURN
C ERROR CONDITIONS, RELATING TO SINGULARITY OF CORRELATION MATRIX
C AND TO THE CONVERGENCE OF THE ITERATIVE PROCESS IN *ACCINV*.
C SEE CHAPTER 1 AND APPENDIX A.
  700 WRITE(6,701)
  701 FORMAT(' NOT CONVERGING IN ACCINV - SEE APPENDIX A')
      GOTO 703
  702 WRITE(6,704)
  704 FORMAT(' MATRIX SINGULAR. STOP IN SUBROUTINE FSCORE')
  703 STOP
      END
```

where Z is, as previously, the matrix of standardized variables and $\Lambda_P$ is the primary factor pattern matrix. This method should be used only in cases in which the communalities are high. A Fortran subroutine is listed in Table 5.20.

Computation of *scores on image factors* is not complicated by the postulate of unique and common factors, hence image factor scores can be computed directly rather than estimated. The formulae given below are derived by Hakstian (1973). The image factor score matrix $\hat{F}$ is

$$\hat{F} = ZH\Lambda_S \quad (5.24)$$

where, as usual, Z is the $(n \times p)$ matrix of standard scores, $\Lambda_S$ is the $(p \times k)$ image factor structure matrix and H is computed from the expression

$$H = J^{-1}V(D - I)^{-1}V'J^{-1} \quad (5.25)$$

in which $J_i^{-1} = \{\text{diag}(R^{-1})\}^{\frac{1}{2}}$ and D and V are the first k eigenvalues and associated column eigenvectors of the rescaled correlation matrix, defined in Section 5.3.2 as $J^{-1}RJ^{-1}$.

Unless image analysis has been used, it is not clear which of the three methods of estimating common factor scores is most appropriate, though in cases when the data are not well-structured differences between the methods can be expected to occur. Harris (1967) compares several methods of estimating true factor scores (after remarking that: 'It is well known that 'true' factor scores are not uniquely computable and thus in one practical sense are of no use at all.') He points out that the regression method yields scores that are intercorrelated even if the true factors are orthogonal. The strength of the correlation between the true and estimated factors is shown to be a function of the number of variables loading on that factor, thus poorer results can be expected

Table 5.19

**Subroutine BART: factor score estimation by Bartlett's method (Subroutine GJDEF3 is also listed.)**

```
      SUBROUTINE BART(D2,A,GAMMA,Z,XJ,DT,N1,N2,N3,N4,NA,NV,NF)
C
C FACTOR SCORES BY BARTLETT'S METHOD
C  D2      SQUARE ROOTS OF UNIQUENESSES, I.E. SQUARE ROOTS OF (1.0-H(I))
C          FOR I=1 TO NV. H(I) IS THE COMMUNALITY OF THE ITH VARIABLE.
C  A       FACTOR STRUCTURE MATRIX. DIMENSION AT LEAST (NV*NF)
C  GAMMA   REAL ARRAY, MINIMUM DIMENSION NF*NV. UNDEFINED ON ENTRY.
C          HOLDS FACTOR SCORE COEFFICIENTS (FACTOR WEIGHTS) ON EXIT.
C  Z       MATRIX OF STANDARD SCORES. DIMENSION AT LEAST(NA*NV)
C  XJ      WORKSPACE. DIMENSION AT LEAST(NF*NF)
C  DT      REAL ARRAY, MINIMUM DIMENSION NF, USED AS WORKSPACE.
C  N1,N2,N3,N4 INTEGERS. THE FIRST DIMENSIONS OF ARRAYS A,GAMMA,Z AND
C          XJ RESPECTIVELY AS DECLARED IN CALLING PROGRAM.
C  NA      NUMBER OF OBSERVATIONS PER VARIABLE
C  NV      NUMBER OF VARIABLES
C  NF      NUMBER OF FACTORS
C
C OTHER SUBROUTINES REQUIRED:
C  GJDEF3
      DIMENSION D2(NV),A(N1,NF),GAMMA(N2,NV),Z(N3,NV),XJ(N4,NF),DT(NF)
      DO 1 J=1,NV
      DO 1 I=1,NF
    1 A(J,I)=A(J,I)/D2(J)
      DO 2 I=1,NF
      DO 2 J=1,NF
      XJ(I,J)=0.0
      DO 2 K=1,NV
    2 XJ(I,J)=XJ(I,J)+A(K,I)*A(K,J)
      DO 3 J=1,NV
      DO 3 I=1,NF
    3 A(J,I)=A(J,I)/D2(J)
      NF1=NF-1
      DO 20 I=1,NF
      M=0
      DO 21 J=1,NF
      IF(I-J) 25,21,25
   25 M=M+1
      MM=0
      DO 22 K=1,NF
      IF(K-I) 30,22,30
   30 MM=MM+1
      GAMMA(M,MM)=XJ(J,K)
   22 CONTINUE
   21 CONTINUE
   20 CALL GJDEF3(NF1,GAMMA,D2,N2,DT(I))
      CALL GJDEF3(NF,XJ,D2,N4,DET)
      DO 12 I=1,NF
      DO 12 J=I,NF
   12 XJ(I,J)=XJ(J,I)
      DO 4 I=1,NF
      DO 4 J=1,NV
      GAMMA(J,I)=0.0
      DO 4 K=1,NF
    4 GAMMA(J,I)=GAMMA(J,I)+XJ(I,K)*A(J,K)
      WRITE(6,5)
    5 FORMAT('1 MATRIX OF FACTOR WEIGHTS'///)
      DO 6 I=1,NF
      WRITE(6,13) I
   13 FORMAT('0FACTOR',I5)
    6 WRITE(6,7) (GAMMA(J,I),J=1,NV)
    7 FORMAT(' ',10F10.4)
      WRITE(6,11)
   11 FORMAT('0    FACTOR SCORES (BARTLETT''S METHOD)'/)
      WRITE(6,17) (J,J=1,NF)
   17 FORMAT(' ',I5,9I10)
```

Table 5.19  *continued*

```
      DO 15 K=1,NA
      DO 10 I=1,NF
      SD=0.0
      DO 8 J=1,NV
    8 SD=SD+GAMMA(J,I)*Z(K,J)
   10 D2(I)=SD
   15 WRITE(6,16)  K,(D2(J),J=1,NF)
   16 FORMAT(' ',I4/(' ',10F10.4))
      WRITE(6,24)
      DO 23 I=1,NF
      R=1.0-DT(I)/DET
   23 WRITE(6,26) I, R
   24 FORMAT('0MULTIPLE CORRELATIONS BETWEEN TRUE AND ESTIMATED FACTORS'
     1///' FACTOR     RSQUARED'/)
   26 FORMAT( ' ',I4,F14.4)
      RETURN
      END




      SUBROUTINE GJDEF3(N,A,H,IROW,D)
C SUBROUTINE TO FIND INVERSE OF A SYMMETRIC MATRIX AND DETERMINANTS
C OF SELECTED MINORS.
      DIMENSION A(IROW,N),H(N)
      IF(N.EQ.1) GOTO 3
C SET ZERO TO MACHINE ZERO
      ZERO=2.0**(-124)
      D=1.0
      L=N+1
      DO 1 K=1,N
      L=L-1
      P=A(1,1)
      D=D*P
      IF( ABS(P).LE.ZERO) GOTO 10
      DO 2 I=2,N
      Q=A(I,1)
      Z=Q
      IF(I.LE.L) Z=-Q
      H(I)=Z/P
      DO 2 J=2,I
    2 A(I-1,J-1)=A(I,J)+Q*H(J)
      A(N,N)=1.D0/P
      DO 1 I=2,N
    1 A(N,I-1)=H(I)
      D=ABS(D)
      GOTO 900
   10 WRITE(6,11)
   11 FORMAT('0***FAIL IN GJDEF3 - MATRIX SINGULAR'///)
      N=0
      GOTO 900
    3 D=A(1,1)
      A(1,1)=1.0/D
  900 RETURN
      END
```

Table 5.20
Subroutine IKLZD: factor score estimation by ideal variables

```
      SUBROUTINE IDLZD(P,S,Z,WT,N1,N2,N3,N4,NA,NV,NF)
C
C COMPUTATION OF FACTOR SCORES BY IDEALIZED VARIABLES. SEE CHAPTER 5
C P. M. MATHER, 1974
C ARGUMENTS:
C P       REAL ARRAY, MINIMUM DIMENSION (NV*NF). FACTOR PATTERN MATRIX.
C S       REAL ARRAY, MINIMUM DIMENSION (NV*NF)F*NF). WORKSPACE.
C Z       REAL ARRAY, MINIMUM DIMENSION NA*NV. HOLDS MATRIX OF
C         STANDARD SCORES.
C WT      REAL ARRAY, MINIMUM DIMENSION NF*NV. UNDEFINED ON ENTRY.
C         HOLDS FACTOR WEIGHT MATRIX ON EXIT.
C N1,N2,N3,N4 INTEGER CONSTANTS - THE FIRST DIMENSIONS OF ARRAYS
C         P,S,Z AND WT AS DECLARED IN THE CALLING PROGRAM.
C NA      INTEGER. NUMBER OF INDIVIDUALS (AREAS,SUBJECTS)
C NV      INTEGER. NUMBER OF VARIABLES(TESTS)
C NF      NUMBER OF FACTORS. INTEGER.
C
C SUBROUTINES REQUIRED:
C TRMULT;  GJDEF1
C
      DIMENSION P(N1,NF),S(N2,NF),Z(N3,NV),WT(N4,NV)
      CALL TRMULT(S,P,P,N2,N1,N1,NF,NF,NV)
      CALL GJDEF1(NF,S,WT,N2)
      DO 1 I=1,NF
      DO 1 J=1,NF
    1 S(I,J)=S(J,I)
      DO 2 I=1,NF
      DO 2 J=1,NV
      WT(I,J)=0.0
      DO 2 K=1,NF
    2 WT(I,J)=WT(I,J)+S(I,K)*P(J,K)
      WRITE(6,3)
    3 FORMAT('1 FACTOR  WEIGHTS')
      DO 4 I=1,NF
      WRITE(6,5) I
    5 FORMAT('0FACTOR',I4)
    4 WRITE(6,6) (WT(I,J),J=1,NV)
    6 FORMAT(' ',10F10.4)
      WRITE(6,7)
    7 FORMAT('0   FACTOR   SCORES  (IDEAL   VARIABLES')
      DO 8 I=1,NA
      DO 9 J=1,NF
      SD=0.0
      DO 10 K=1,NV
   10 SD=SD+WT(J,K)*Z(I,K)
    9 S(J,1)=SD
    8 WRITE(6,6) (S(J,1),J=1,NF)
      RETURN
      END
```

when the factor is only tenuously identified. Results of a similar study are reported by McDonald and Burr (1967), and further discussion can be found in Tucker (1971).

Two problems are met within the estimation of common factor scores. The first of these was discussed by Guttman (1955) but his conclusions do not appear to have had a great deal of influence. Guttman found that, for a given correlation matrix, common factor loading matrix and data matrix, distinctly different common factor score matrices can be postulated. Scores on common factors have, therefore, a degree of indeterminacy. This is a serious matter if scores are to be used as indices in subsequent work. It should be noted in passing that scores on image factors are completely determinate, as are principal component scores.

The degree of determinacy of a vector of estimated scores on a single common factor can be inferred from the multiple correlation, $R$, of the estimated factor scores and the true factor scores. The calculation of $R^2$ was outlined above. Assuming that the estimated scores form the columns of a ($p \times k$) matrix, $\hat{F}$ then, if $R$ is less than 1.0 for any column of $\hat{F}$, then other common factor score matrices can be shown to exist and the factor score vector is thus not unique. If the smallest correlations between the columns of $\hat{F}$ and the columns of the most different alternative factor score matrix are $R_1^{min}$, $R_2^{min}$, ..., $R_k^{min}$, an equation for measuring the uniqueness of any column of $\hat{F}$ is

$$R_i^{min} = 2R_i^2 - 1$$

Guttman (1955, p. 74) and Rummell (1970, p. 445) give the following table of relationships between $R_i^{min}$ and $R_i$:

| $R_i$ | 0.0 | 0.20 | 0.40 | 0.60 | 0.80 | 0.90 | 0.95 | 0.98 | 0.99 | 1.00 |
|---|---|---|---|---|---|---|---|---|---|---|
| $R_i^{min}$ | −1.0 | −0.92 | −0.68 | −0.28 | 0.28 | 0.62 | 0.81 | 0.92 | 0.96 | 1.00 |

Thus, if $R_i = 1.0$ then the corresponding column of $\hat{F}$ contains a unique set of common factor scores. Seemingly high multiple correlations can hide the fact that alternative, and quite distinct, alternative vectors of factor scores can be postulated. For example, if $R = 0.90$ then other sets of factor scores can be shown to exist. The correlation between the obtained set of factor scores and the maximally-different set is only 0.62. For $R = 0.8$, this correlation falls to 0.28. This highlights a problem in the geographical use of factor scores that hitherto has received little attention. Mulaik (1972, pp. 327–331) gives a comprehensive account of the determinacy of factor scores, and reference should also be made to McDonald (1974).

The second problem is concerned with a recurring topic in this book — computational accuracy. The correlation matrices used in factor analysis are sometimes near-singular, since many closely-related variables are frequently included. The formula for deriving regression estimates of common-factor scores involves inversion of $R$ (Equation 5.20). This can involve considerable error, which is compounded by the fact that the factor structure matrix $\Lambda_S$ (which also appears in Equation 5.20)) is derived from $R$ by an eigenvector transformation which itself may contain inaccuracies. Thus, computational errors can build up and lead to unacceptable results.

## 5.6 Summary

Factor-analytical techniques can be divided into two types: (i) nonstatistical, such as alpha factor analysis, in which the direction of inference is from the selection of variables to the universe of variables, and (ii) statistical, such as minres and maximum-likelihood methods, where the usual convention of drawing inferences about a larger population of individuals from a random sample drawn from that population is followed. Image analysis is a departure from the classical common-factor approach. It is nonstatistical in the sense defined above. Unlike the common-factor techniques, image factor analysis has the property that scores are uniquely computable. Some methods (alpha, maximum-likelihood and image factor analysis) are scale-free, in that the same factors are determined independently of whether $\mathbf{R}$ or $\mathbf{S}$ is analysed. These methods are therefore preferable in situations in which the measurement scales are arbitrary or are not strictly comparable. Thus, for nonstatistical analyses, image factor analysis appears to have advantages not possessed by alpha factor analysis, such as computable (rather than estimated) scores, and avoidance of Heywood cases. For statistical work the maximum-likelihood method seems to have most to recommend it (scale-free, computational mechanisms to avoid Heywood cases, large sample $\chi^2$ test for number of factors). (Refer to Browne, 1968, 1969, and Harris, 1964.)

Once an initial factor pattern matrix has been computed it is necessary to interpret it. Orthogonal and oblique rotation techniques have been developed to aid this process. Oblique solutions are preferable because they are more natural. The Promax rotation is one of the simplest and most reliable of the oblique transformations. The distinction between pattern and structure and between reference and primary factors in an oblique solution should be noted. Finally, the difficulty of estimating common factor scores, and the possibility of variation between the results derived from different estimators, need to be borne in mind. If regression estimates of factor scores are obtained, the comments made in Chapter 2 regarding computational accuracy in multiple regression should be noted.

On neglected aspect of factor analysis is its use in a confirmatory rather than an exploratory way. Recent papers by Joreskog (1973) and Mukherjee (1973) provide the basic theory.

Several points not specifically covered earlier will now be briefly dealt with.

(i) *Number of factors*. This is a basic difficulty in exploratory factor analysis. In principle, enough factors should be extracted so that the matrix of residual correlations $(\mathbf{R} - \mathbf{\Lambda}'\mathbf{\Lambda})$ is not significantly different from the identity matrix. The distribution theory required for this approach to be completely successful is still in the process of being developed, though a large-sample $\chi^2$ test is available in maximum likelihood factor analysis. Such tests are only applicable when a 'statistical' approach is followed. Critics of this test point to its dependence on $n$, the number of observations, and also to the fact that in some instances reliance on a $\chi^2$ test may lead to over-factoring in that the last few factors are uninterpretable. Linn (1968) covers a variety of nonstatistical procedures for determining $k$, the number of factors. He concludes, naturally enough, that the experimental design should be good (i.e. large samples, well-chosen variables) and that the factor model should fit the data. Ths use of the eigenvalue-one criterion ($k$ = number of eigenvalues of **R** that exceed 1.0) may lead to too many factors if the sample size is small. Plotting of the eigenvalues, or the first differences of the eigenvalues, against their rank order is somewhat better, though there may be several break-points in the curve. Post-hoc rationalization of the results of a factor analysis is usually discouraged at a formal level, though in an exploratory study there may be little else that can be done. The only sensible solution is replication, either by repeating the sampling process or by randomly splitting a given sample. The problems of exploratory data analysis are not restricted to factor analysis; see Tukey (1965, 1969) and Digman (1966) for further comments.

(ii) *The assumption of normality* of the observed variables is not explicitly required except in the case of maximum-likelihood methods. However, the interpretation of the factor pattern may be affected by non-normality since the range of the correlation coefficient may be restricted, for example due to the effects of skewed distributions. The size of the correlation coefficients has a direct bearing on the size of the factor pattern coefficients. Carroll (1961) provides a lucid discussion of this and related topics, and the paper by Kowalski (1972) is also relevant.

(iii) Lastly, the question of what is or is not a *significant pattern coefficient* must be faced. Although some recent work has gone into the methods of computing standard errors of the factor loadings (Lawley and Maxwell, 1971; Jennrich and Thayer, 1973; Archer and Jennrich, 1973, Jennrich, 1974) the results apply only to maximum-likelihood factor analysis. Intuition and, sometimes, a Nelson-like approach have been used, neither with any great success. If one adopts a nonstatistical approach (in the sense used above) then any factor loading that is nonzero is, in the strict sense, significant. There is no way at the present time of testing the null hypothesis that a given factor loading is zero in the population when statistical factoring methods are adopted. Monte Carlo studies (Mather, 1969; Armstrong and Soelberg, 1968) show up the dangers of blind 'interpretation' of factor pattern matrices. For example, given a 50 x 10 factor pattern matrix at least 25 entries could be expected to be accepted as significantly different from zero at the 5 per cent level even if all the population factor pattern coefficients were really zero. Again, the answer seems to lie in the fact that one is often 'interrogating' one's data; some wrong answers are to be expected. This should lead to replication of the study (to check that regularities observed in one sample are present in others) and to a readiness to admit that, in an area of inadequate theory, mistakes are possible.

*Part Four*

# Classification

Classification is essentially the identification of groups of similar objects within the set of objects under study. It is necessary so that generalizations concerning within and between group similarities and differences can be made. Such generalizations can be purely descriptive, or they can form the basis of a hypothesis which can then be tested by the use of other techinques, such as the analysis of variance. Two approaches to classification are possible — the identification of groupings, termed classification proper, and the allocation of individuals to existing groups, termed discrimination. Classification proper can be further subdivided into clustering methods and ordination methods. Clustering methods provide for the extraction of discrete groups, which may be either hierarchical (so that each group is a part of another group at the next level of the hierarchy) or nucleated, so that the groups are discrete. Ordination methods involve the display in graphical form of the inter-object similarities in a low-dimensional space. Points, representing objects, are close together if their mutual similarity is high or they are far apart if their mutual similarity is low. A visual check of the two-dimensional scatter diagrams should indicate whether groups (defined as areas of relatively high point density) are present. Cluster analysis emphasises discontinuities, whereas ordination methods display the continuity of the data.

Both cluster analysis and ordination methods rely upon a satisfactory definition of the similarity between the objects to be classified. Many similarity and dissimilarity coefficients have been proposed, ranging from the correlation coefficient to subjective estimates. The first part of Chapter 6 is a survey of methods of measuring similarity. The choice of an appropriate similarity coefficient is important if the results of the classification are to be meaningful. Hierarchical methods of cluster analysis are discussed next, with emphasis being placed on the properties of the various methods. A computer program for the major hierarchical methods is included. The program also allows for the calculation of a goodness of fit coefficient, the cophenetic correlation coefficient, which can be thought of as a measure of the distortion produced by the use of the particular hierarchical method chosen.

Nonhierarchical cluster analysis is considered next. Although hierarchical methods have been widely used in physical geography, there may be no logical or theoretical reason to expect a hierarchical structure in reality. If this is so, then the methods producing nucleated clusters should be applied. A great advantage of these methods is that large numbers of individuals can be classified without inordinate demands being made on the computer.

The third part of Chapter 6 is a consideration of ordination methods. These methods have been widely used in the last few years in ecology and psychology, but have not as yet found favour among physical geographers. The fact that the methods emphasise continuity and not discontinuity in the data is an advantage in situations in which tight groupings are not to be expected.

Discrimination is the topic of Chapter 7. As well as the two-group and multiple-group methods, recently developed pattern recognition techniques for presence-absence variables are discussed. In addition, some techniques for testing for inter-group differences are described.

*Chapter 6*

# *Classification*

## 6.0 Introduction

The word 'classification' has two meanings. It can connote the process of identifying the number, nature and composition of relatively homogeneous groups which together make up the data set under scrutiny. Alternatively it can be used as a noun in the sense of 'arrangement' or 'scheme'. It is usually clear from the context whether the word is used as a noun or as a verb. The purpose of classifying is, usually, to allow the consideration of the characteristics of a small number of types rather than those of a large number of individuals. Classifications can, of course, have other uses — for example, in information storage and retrieval schemes, or in the development of terminology. Each classificatory scheme must have a purpose, and it is this which will primarily determine the type of classification strategy which is employed.

Ordination is distinguished from clustering on the grounds that ordination procedures are concerned primarily with reducing the dimensionality of the space in which the objects are considered to lie, whereas a clustering process has the aim of ordering the objects into discrete sets. The reduction of dimensionality is one of the aims of principal components analysis, which is described in Chapter 4. In that chapter PCA was considered mainly as a method of deriving orthogonal linear combinations of a given set of variables, whereas an ordination technique is concerned with expressing the relative coordinate positions of the objects with respect to a small number of mutually orthogonal axes. However, pairwise plots of the first few principal component scores would constitute part of an ordination process. Visual inspection of such plots may reveal structures which permit the intuitive development of a classificatory scheme. Thus, ordination may accompany clustering or it may be independent of it.

Discrimination is the assessment of the degree of separateness of groups of individuals which are postulated either by a classification or by *a priori* information. Discriminant analysis, in the strict sense, is concerned with the assignment of previously unallocated individuals to their most likely class, given an existing classification. Techniques of discriminant analysis and related problems of measuring multivariate differences between groups, which is usually referred to as multivariate analysis of variance (MANOVA), are considered in Chapter 7.

The techniques of cluster analysis and ordination are described in the following sections. The philosophical basis of classification is not considered in any detail, since many excellent discussions are available (Harvey, 1969; Sneath and Sokal, 1973; Cormack, 1971; L. A. S. Johnson, 1968). The purposes for which the classification is required are nevertheless of vital importance, for there is no overall optimum classificatory method which is appropriate in all circumstances. Each classification should therefore be devised with a particular aim in mind.

It would be manifestly untrue to state that no meaningful or useful classifications were developed before methods of numerical taxonomy were introduced. Nevertheless, if one's aim is to consider simultaneously many characteristics of a large number of objects then some form of automatic selection procedure becomes essential. Unfortunately, the logical processes underlying classificatory processes have proved difficult to identify objectively; this has led to the development of many competing methods both of cluster analysis and ordination. All have some disadvantages or failings, and only empirical use and comparative studies will establish their suitability in particular situations. It is for this reason that the present chapter adopts a catholic approach to the procedures of numerical taxonomy. Once it is established that automatic methods do not always, indeed rarely, produce identical results then attention can be focussed upon (a) the characteristics of each technique and (b) the requirements of the user, with the aim of matching the two. Studies of the merits of different procedures have shown that some methods are of more general applicability than others, but this may be of little help in a particular situation.

## 6.1 Cluster Analysis
### 6.1.0 Types of Cluster Analysis
Many methods of cluster analysis exist. They can be differentiated according to whether they are (a) divisive or agglomerative, and (b) hierarchical or nucleated. In agglomerative procedures one begins with a set of individuals and then builds up groups, by a process of accumulation whereas the reverse procedure is followed in divisive methods. In general, agglomerative methods allow the simultaneous consideration of all characteristics of the object that are considered to be of interest, whereas in divisive methods the whole is split into parts in stages, a different criterion (usually a single variable) being used at each stage. The classification of clustering methods that opens this paragraph

is divisive. Agglomerative methods are also to be preferred on grounds of computational speed, for divisive methods tend to be slow when the number of individuals is increased to 100 or so; refer, for example, to Edwards and Cavalli–Sfortza (1965). McCammon (1968A) indicates that there are 42, 335, 950 different ways of partitioning 15 objects into 4 groups. (The general formula for $n$ individuals and $m$ groups is $1/m! \Sigma_{j=0}^{m} (-1)^j (m/j)(m-j)^n$.) To determine the optimum split would obviously be very time-consuming. Although more efficient ways of finding an optimum split than the head-on approach have been suggested (see Williams and Lambert, 1959; Gower, 1967B) agglomerative methods need much less computer time, although this may still be considerable for large values of $n$.

Agglomerative methods are either hierarchical or nucleated. Hierarchical agglomerative methods are based on the idea that each group at level $i$ is part of a larger group at level $(i+1)$, all groups being subsumed into a universal cluster at level $(n-1)$. This type of structure could be expected in evolutionary taxonomy for evolution is commonly regarded as being a divergent process that can, to some extent, be represented hierarchically (Rohlf, 1970). However, there may be no prior grounds for suspecting hierarchical structures in all situations, and in fact such a representation may lead to considerable distortion.

Rather than regard groups at a particular level being part of larger groups at a higher level one can attempt to represent the structure in the data set in terms of discrete, non-overlapping clusters. If the n points representing the $n$ objects in a $p$-dimensional space could be viewed certain parts of the space might have a higher density of points than others. Clusters could be defined as those portions of space with relatively high point densities, just as villages could be defined as parts of the earth's surface having a relatively high density of buildings. If one has no prior reason to suspect a hierarchical structure, the nucleated model may prove to be more realistic and more valuable than the hierarchical representation.

*6.1.1 Measurement of Similarity*
Methods of numerical classification are based not upon the $(n \times p)$ matrix of observations but on the $(n \times n)$ matrix of measurements of the similarity or dissimilarity of the $n (n-1)/2$ pairs of objects over the $p$ variables. Strictly speaking, measures of similarity and dissimilarity should be distinguished. However, the two are obviously related so the distinction will not be made unless it is necessary in the interests of clarity to do so. The first choice to be made is that of the similarity measure to be used. There is a basic division between measures which can be used with interval or ratio scale data and those which apply to presence–absence measurements. The most commonly-used similarity measure applying to interval or ratio-scale data is the

*Euclidian distance coefficient* $d_{ij}$. Given two individuals $i$ and $j$ the coefficient $d_{ij}$ is defined as:

$$d_{ij} = \left( \sum_{k=1}^{p} (x_{ik} - x_{jk})^2 \right)^{1/2} \Big/ p \qquad (6.1)$$

assuming that the $p$ variables are orthogonal. The situation with $p = 2$ is shown graphically in Figure 6.1, which depicts two objects, labelled $i$ and $j$, and located at points B and C respectively. The distance BC is the Euclidean distance (straight-line distance) of $i$ from $j$, and it can be found from Pythagoras's theorem. Since variables (axes) $x_1$ and $x_2$ are orthogonal, ABC is a right triangle, hence

$$BC = (AC^2 + BA^2)^{1/2}$$

But

$$AC^2 = (x_{i1} - x_{j1})^2$$

and

$$BA^2 = (x_{i2} - x_{j2})^2$$

so

$$BC = ((x_{i2} - x_{j2})^2 + (x_{i1} - x_{j1})^2)^{1/2} = d_{ij}$$

If $p$ is greater than 2 it can be seen easily that the higher-dimensional analogue of BC is found by adding $(x_{i3} - x_{j3})^2, \ldots, (x_{ip} - x_{jp})^2$ to the right hand side. This gives Equation (6.1), except that the divisor $p$ needs some explanation. If values of $d_{ij}$ are compared for differing $p$'s then apparent differences will be at least partially due to the fact that $d_{ij}$ is bound to increase as $p$ rises. Hence $d_{ij}$ is normalized by division by $p$. The value of $d_{ij}$ ranges from 0.0 (perfect similarity) to infinity (complete dissimilarity). The value of $d_{ij}$ can be transformed into a correlation measure if necessary. (Harbaugh and Merriam, 1968).

A second problem involves the scale of the axes $x_1$ and $x_2$ in Figure 6.1. The scale will, quite obviously, influence the distance BC. Since scales of measurement are often quite arbitrary some standardization process is adopted. The convention most commonly used is to give each variable equal weight by transforming observed values so that each variable has a mean of zero and unit variance:

$$z_{ik} = (x_{ik} - \bar{x}_k)/s_k \qquad (6.2)$$

Here, $z_{ik}$ is the standard score, equivalent of $x_{ik}$, the observed score of individual $i$ on variable $k$, $\bar{x}_k$ is the mean value of the observations on variable $k$ and $s_k$ is the standard deviation of variable $k$.

If the $x_{ik}$ are principal component scores they should *not* be automatically standardized. The original variables which were subjected

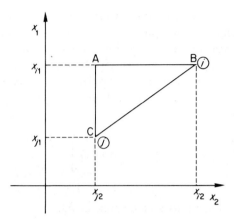

Figure 6.1 Calculation of Euclidean distance coefficient (See text for explanation)

to a principal components transformation may have been standardized (i.e. the PCA based upon **R** rather than **S** — see Chapter 4) in which case each principal component will have a variance which is already expressed in terms of standardized units. To standardize units which are already standardized seems, at best, unnecessary. Hence, standardized principal component scores derived from **R** should not be re-standardized. However, standardization is in many cases carried out automatically before principal component scores are printed, though this is not true of the program given in Chapter 4. To 'unstandardize' the scores on principal component $k$, simply multiply each element of the kth column of **F** (the scores matrix) by $\lambda_k^{1/2}$ or, better still, modify the program to allow output of unstandardized scores. The standardization of principal component scores implies a weighting of the variables which should not be carried out without a reason.

The third difficulty encountered in the use of the Euclidean distance measure is the lack of orthogonality of observed variables (Rohlf, 1967). If variables $i$ and $j$ are inter-correlated then $d_{ij}$ will be over- or under-estimated by an amount depending upon the sign and magnitude of $r_{ij}$. To remove the effects of such lack of orthogonality the variables are frequently subjected to a principal components transformation and the scores on the principal components are used in place of the observed scores. This is usually acceptable provided that the points made earlier regarding standardization are observed. If all principal components are retained, then the inter-point distances are unaffected. The use of Equation (6.1) assumes a rectangular or orthogonal coordinate system, however. However, the last few (nonzero) principal components are often disregarded on the grounds that they are unimportant.

In terms of contribution to total variance this may well be true, but a relatively insignificant principal component may occasionally provide important information concerning the mutual similarities of a particular small group of individuals. If principal components analysis is considered to take the problem one step further away from reality then formula (6.1) can be modified to take into account the correlation between $x_k$ and $x_l$.

$$d_{ij} = \left[ \sum_{k=1}^{p} \sum_{l=1}^{p} (x_{ik} - x_{jk})(x_{il} - x_{jl}) r_{kl} \right]^{1/2} \qquad (6.3)$$

Formula (6.3) can also be used if the $x_{ij}$ are the scores on obliquely-transformed factors (see Chapter 5).

Other similarity coefficients which have been applied to ratio or interval-scale data are the correlation coefficient and the $\cos \theta$ coefficient. The use of the former is not recommended (Imbrie, 1963; Eades, 1965) — it is not meaningful to correlate two individuals, for the 'standard deviation' of an individual is not a meaningful concept. The $\cos \theta_{ij}$ coefficient was introduced by Imbrie (1963). It has the advantage of lying in the range $0 \leqslant \cos \theta_{ij} \leqslant 1$, unlike $d_{ij}$ which is bounded at its lower end but not at its upper end. Gower (1967A) suggests that in some situations the Euclidean representation will produce a set of points lying on a sphere. In this case the distance between any pair of points measured along the great circle may be thought of as a more 'natural' measure than the straightforward Euclidean distance. The $\cos \theta_{ij}$ coefficient is proportional to the great circle distance $r\theta$, and is given by (6.4)

$$\cos \theta_{ij} = \sum_{k=1}^{p} x_{ik} x_{jk} \left[ \sum_{k=1}^{p} x_{ik}^2 \sum_{k=1}^{p} x_{jk}^2 \right]^{1/2} \qquad (6.4)$$

Notice that, if $x_i$ and $x_j$ are standardized to zero mean and unit variance, $\cos \theta_{ij} = r_{ij}$.

Similarity coefficients for use with 'presence–absence' type data are numerous. Several in common use are compared in a study by Hazel (1970). These include the coefficients of Jaccard, $J = C/(N_1 + N_2 - C)$; Dice, $D = 2C/(N_1 + N_2)$; Simpson, $S = C/N_1$; Fager, $F = C/N_1 N_2 - 1/2N_2$ and Otsuka, $O = C/N_1 N_2$ where, in each case, $N_1$ and $N_2$ are the numbers of presence–absence characteristics observed for the two individuals being compared, with $N_1 < N_2$, and $C$ is the number of characteristics possessed by both individuals. Of these coefficients, $J$ is the most widely used but tends to emphasize difference and does not take mismatches into account. $D$ is shown by Hazel to give matches twice the weight of mismatches. $O$ is the binary equivalent of the $\cos \theta$ coefficient discussed earlier, and $F$ is simply $O$ with the subtraction of a quantity proportional to the number of characteristics in the larger of the two samples. The Simpson coefficient $S$ has also been widely used, and is conservative in the sense that it emphasizes similarity rather than difference. The study of Hazel (1970) showed that there was very little difference in the outcome, whichever of these coefficients was used. Another useful survey of similarity coefficients for binary data is Cheetham and Hazel (1969).

A problem as yet unsolved is the calculation of a similarity measure when the data consists of mixed 'quantitative' and 'qualitative' variables. Parks (1969) has suggested an ingenious answer, which involves the scaling of the quantitative variables onto the range 0—1 by the formula

$$z_j = \frac{(x_j - x_{\min})}{(\text{range of } x)}$$

and the mapping of the qualitative variables onto the range 0—1. A two-state variable could be coded 0 (absent) and 1 (present) or a three-state variable 0 (absent), ½ (sparse) and 1 (abundant). Qualitative measure cannot be ranked, so the choice of 1 or 0 for the 'absent' end of the range is arbitrary. This may lead to differences in results. The choice, however, should be consistent within one analysis. Once the variables have been scaled and coded, Parks uses the ordinary Euclidean distance coefficient to measure similarity. Note that the scaling used by Parks does not ensure equality of *scale* for each axis but equality of axis *length*. This may lead to changes in the relative positions of points before and after scaling.

Recent work by Gower (1971) has produced what he terms a general similarity coefficient, $S_G$, which is applicable to binary, categorical and quantitative scales. It is defined as

$$S_G = \frac{\sum_{i=1}^{n} w_{ijk} s_{ijk}}{\sum_{i=1}^{n} w_{ijk}} \tag{6.5}$$

where individuals $j$ and $k$ are being compared over the $i$th variable. The value $s_{ijk}$ lies between 0 and 1 inclusive and $w_{ijk}$ is the weight assigned to variable $i$. The weight is 1 when comparison between objects $j$ and $k$ for the $i$th variable is possible, and 0 when variable $i$ is not measured or is unknown for either or both objects $j$ and $k$. If binary (presence/ absence) data is being used $s_{ijk}$ is 1 for a match and 0 for a mismatch. When the match is negative, i.e. both $j$ and $k$ do not possess the particular attribute, $w_{ijk}$ can be set to 0 if such negative matches are not to be considered. This is in line with the thinking behind the Jaccard and Dice coefficients. If negative matches are to be counted,

$w_{ijk}$ is set to 1. For multistate data $s_{ijk}$ is set to 1 for a match and 0 for a mismatch. The value of $s_{ijk}$ for quantitative data is

$$s_{ijk} = 1 - |x_{ij} - x_{ik}|/R_i \tag{6.6}$$

Here, $x_{ij}$ and $x_{ik}$ are observations of the $i$th variable on objects $j$ and $k$ and $R_i$ is the range of $x$. Sneath and Sokal (1973) refer to the fact that Gower's coefficient in its different forms is similar to one or other of the multifarious similarity coefficients presently in use. For example, when used with multistate data it resembles the simple matching coefficient. With quantitative data, it is the converse of the mean character difference. The general similarity coefficient will be used in Section 6.1.3 in connection with the technique of Principal Coordinates Analysis. However, the methods of hierarchical cluster analysis of Section 6.1.2 are not restricted to a particular similarity or dissimilarity coefficient so, though the Euclidean distance coefficient is used in the program and the examples, any other of the coefficients mentioned here could be used providing that it is appropriate.

Given an $(n \times n)$ matrix of similarities, which we will be denoted by **D** irrespective of the coefficient used, we now have to decide upon the particular agglomerative method of cluster analysis to be used. Firstly, the choice between hierarchical and non-hierarchical methods must be made. This might involve an *a priori* model such as the evolutionary model in biology. Some phenomena (e.g. drainage basins, sedimentary environments) fall naturally into hierarchies but others do not. Ultimately it is the purpose of the classification which determines the approach to be used.

*6.1.2 Hierarchical Clustering Methods*

The number of hierarchical clustering methods is large and to attempt a description of each method would be excessively time-consuming and probably confusing. Several of the more generally-used techniques are described and these are incorporated into a single computer program. It should be made clear at the outset that no single method is superior to the others in all circumstances. A knowledge of the characteristics of each method will perhaps allow the choice of an algorithm appropriate to the particular problem.

For purposes of exposition it will be assumed that the similarity coefficient used in each case is the Euclidean distance, $d_{ij}$. This is, in fact, a measure of *dissimilarity* so that obvious changes must be made to the algorithms described below if a true *similarity* measure is substituted.

The general strategy underlying each of the clustering methods is similar and can be represented as follows: the closest pair of points $i$ and $j$ (which may represent single objects or groups of objects) in the

$p$-dimensional Euclidean space are combined into a single group. The distances of all other points to this group replace their distances to $i$ and $j$. The process is repeated until all points have been incorporated into a single group. This process can be summarized as follows:

(1) Find the smallest element $d_{ij}$ remaining in **D**;
(2) Fuse $i$ and $j$ into a single group, $k$.
(3) Compute new distances $d_{km}$ (where $m$ represents each of the remaining points). These distances replace $d_{im}$ and $d_{jm}$ in **D**.
(4) Repeat from step (1) for $(n-1)$ cycles.

All procedures of this type are limited by the need to store the lower triangle of **D**, and by the need to search for the smallest $d_{ij}$ at each of $(n-1)$ cycles. McCammon (1968A) illustrates the exponential growth in memory requirements and in the number of comparisons as $n$ increases (Table 6.1).

Table 6.1
Number of operations and number of store locations required for hierarchical clustering algorithms

| Number of individuals ($n$) | Number of comparisons needed to find min $(d_{ij})$ at each cycle | Memory for lower triangle of **D** (words) |
|---|---|---|
| 10 | 165 | 45 |
| 50 | 20 825 | 1 225 |
| 100 | 166 650 | 4 950 |
| 200 | 1 333 300 | 19 900 |
| 500 | 20 444 250 | 124 750 |
| 1000 | 166 666 500 | 499 500 |

The figures in column 2 of Table 6.1 can be reduced by searching for mutually independent minimum $d_{ij}$'s. These are $d_{ij}$ values which are the smallest in a particular pair of rows and columns. Thus, several minimum $d_{ij}$'s can be picked out at one cycle, which reduces the total number of cycles from $(n-1)$. However, additional coding is required to check their mutual independence. For n greater than 50 or so it is a worthwhile investment. Memory requirements are a greater drawback. Even with modern computers $n = 200$ is a practical maximum unless **D** is held on backing store (disc, tape or drum). This inevitably results in a considerable increase in execution time, for backing store transfers are slow in comparison with internal transfers. This is a serious limitation which can only be overcome by sophisticated programming.

Lance and Williams (1967) point out three general properties of hierarchical strategies:
(a) combinatorial or non-combinatorial. Assume two groups or

individuals ($i$) and ($j$) with $n_i$ and $n_j$ members respectively and at a distance apart of $d_{ij}$. If $d_{ij}$ is the smallest element remaining in **D** then ($i$) and ($j$) fuse to form new group ($k$) with $n_k = n_i + n_j$ members at this level. (Step 2 in the outline above.) Suppose that the distance $d_{km}$ is required where $m$ is one of the remaining points in the Euclidean space. The values $d_{im}$, $d_{jm}$, $d_{ij}$, $n_i$ and $n_j$ are all known. If $d_{km}$ can be calculated from these five values then the strategy is *combinatorial*. The relationship used in computing $d_{km}$ in a combinatorial strategy is of the form:

$$d_{km} = \alpha_i d_{im} + \alpha_j d_{jm} + \beta d_{ij} + \gamma \mid d_{im} - d_{jm} \mid \qquad (6.7)$$

Different combinatorial clustering algorithms are distinguished by different values of $\alpha_i$, $\alpha_j$, $\beta$ and $\gamma$. When $\gamma = 0$ then the string of minimum $d_{ij}$ values determined at each of the $(n-1)$ cycles will be monotonic (i.e. no backward links will occur in the linkage tree — see below) providing that

$$\alpha_i + \alpha_j + \beta \geqslant 1. \qquad (6.8)$$

Combinatorial strategies do not require the original measurements once **D** has been calculated. They are therefore to be preferred to noncombinatorial strategies on grounds of computational efficiency.

(b) *Compatible or incompatible.* A compatible strategy is one in which measures calculated at successive cycles (e.g. $d_{km}$ in paragraph (a)) are of exactly the same kind as those used initially. This makes interpretation more straightforward.

(c) *Space-conserving or space-distorting.* The initial matrix of similarities **D** defines a space containing all the points, representing the individuals. As groups form, the updated **D** matrix may not define a space with the original properties. If this occurs the strategy is space-distorting. Sometimes the effect is to make the space in the immediate vicinity of a newly-formed group to appear contracted, i.e. the group will seem to move closer to some or all the remaining points. Thus, the chance that an object will join an existing group rather than act as the nucleus of a new group is increased. Such a situation may result in 'chaining'. This may or may not be desirable, depending on the purpose of the classification. Other strategies may produce space-dilating effects, when groups appear to recede on formation. This type of strategy will tend to produce small, compact, apparently well-separated groups. Lance and Williams (1967) call these 'non-conformist' groups. These may exist because their members are rather unlike any other of the objects, including each other.

The hierarchical clustering methods which are now reviewed are all of the combinatorial, compatible type. They do, however, have different space-conserving properties. Nomenclature is a problem; the various algorithms are given different names by different authors. Synonyms will therefore be given.

(i) Nearest neighbour. This is also known as the single linkage and the minimum method. It is one of the oldest of the hierarchical clustering techniques. The distance between two groups $(i)$ and $(j)$ is defined as the distance between their two closest members. Connections between objects, or between objects and groups or between groups are those established by single links between pairs of points. The method is combinatorial, the coefficients of equation (6.7) being $\alpha_i = \alpha_j = 0.5, \beta = 0, \gamma = -0.5$:

$$d_{km} = 0.5 d_{im} + 0.5 d_{jm} - 0.5 \mid d_{im} - d_{jm} \mid \qquad (6.9)$$

It is also compatible. Its space-contracting properties are well-known, and it very often results in a hierarchical structure in which individuals are added sequentially to a single group. Nearest neighbour clustering can also be performed in a divisive manner by finding the minimum spanning tree (Gower and Ross, 1969). This is said to be the most efficient computing method (Rohlf, 1973).

(ii) Furthest neighbour (complete linkage, maximum method) is the converse of (i) for the distance between pairs of groups is defined as the distance between their furthest members. This will normally result in a hierarchical structure that consists of well-separated compact groups. The method is combinatorial, and $d_{km}$ can be found from:

$$d_{km} = 0.5 d_{im} + 0.5 d_{jm} + 0.5 \mid d_{im} - d_{jm} \mid \qquad (6.10)$$

Since a group will tend to recede from some points (but move closer to none) it is markedly space-dilating in contrast to (1).

Because of the marked space-distorting properties of nearest-neighbour and furthest-neighbour methods, various modifications have been proposed. These are described by Sneath and Sokal (1973). In addition, alternative methods — which are based on an average similarity or dissimilarity between-group measure — have been proposed. Some of these methods are:

(iii) Centroid (unweighted pair-group centroid). In this method, inter-group distance is defined as the distance between the centroids (centres of gravity, multivariate means) of the two groups. The combinatorial formula (6.7) is valid only if $d_{ij}^2$ is used. In this formula the coefficients are: $\alpha_i = n_i/n_k$, $\alpha_j = n_j/n_k$, $\beta = -\alpha_i \alpha_j$ and $\gamma = 0$ where $n_i$ is the number of individuals in group $i$, $n_j$ the number in group $j$, and $n_k = n_i + n_j$. The method is compatible and space-conserving in the sense defined above. Because it is simple to understand it has found wide use. However, condition (6.8) is not met, and consequently the set of minimum $d_{ij}$'s is not monotonic, i.e. backward links can occur in the dendrogram representation (Section 6.1.3). These are difficult to interpret in practice.

(iv) Median (weighted pair-group centroid). If $n_i$ and $n_j$ are widely different then the centroid of the group resulting from the fusion of groups $(i)$ and $(j)$ will lie close (and possibly within) the larger group, and the characteristics of the smaller group are lost. For this reason, a modification has been introduced by Gower (1967B) which would give $(i)$ and $(j)$ equal weight. This results in the centroid of the new group lying at the mid-point of the shortest side of the triangle joining the centroids of $(i)$, $(j)$ and any other group or point $(m)$. Again, the combinatorial formula is only valid for squared Euclidean distances. The coefficients in this formula are: $\alpha_i = \alpha_j = 0.5$, $\beta = -0.25$, $\gamma = 0$. Also, backward links can occur.

(v) Group average (unweighted pair-group method using arithmetic averages). This method is probably the most widely used of all hierarchical clustering algorithms. The similarity, or dissimilarity, of two groups is defined as the arithmetic average of the similarities between pairs of members of $(i)$ and $(j)$, i.e. as $1/n_i n_j \, \Sigma_i \Sigma_j d_{ij}$. It is combinatorial with coefficients $\alpha_i = n_i/n_k$, $\alpha_j = n_j/n_k$ and $\beta = \gamma = 0$. Since $\alpha_i + \alpha_j + \beta = 1.0$, condition (6.8) is satisfied and the dendrogram is necessarily monotonic. If measures other than $d_{ij}$ are used, the concept of average similarity should be given some thought. For example, the 'average correlation coefficient' is not easily interpreted. Lance and Williams (1967) suggest the transform

$$d_{ij} = \cos\left(\frac{1}{n_i n_j} \sum_{i,j} \cos^{-1} r_{ij}\right) \qquad (6.11)$$

The group-average method is space-conserving, but less rigorously so than the centroid method.

(vi) Simple average (weighted pair-group method using arithmetic averages) has a similar relationship to the group average method as median has to the centroid strategy. The two groups to be joined are given equal weight irrespective of the values of $n_i$ and $n_j$. This may help preserve the characteristics of small groups, but on the other hand subsequent relationships will be biased in favour of the most recently amalgamated group, which may tend to distort the space to make groups appear to be further apart than they do when the Group Average scheme is used. The method is combinatorial, with $\alpha_i = \alpha_j = 0.5$ and $\beta = \gamma = 0$.

(vii) Minimum variance. An intuitive belief that the groups produced by a clustering procedure should have maximum internal homogeneity lies behind many approaches to classification (Wishart, 1969). Ward (1963) proposed a technique which was specifically developed to minimize the pooled within-group sums of squares (defined as the sum of the squared distances from each point to its cluster centre) at each level. In other words, the two groups to be combined at any given level are those whose fusion produces the least increase in the within-group sum of squares (or 'error sum of squares objective function'). The method can be adapted to fit the conbinatorial scheme (6.7) with coefficients $\alpha_i = (n_m + n_i)/(n_m + n_k)$, $\alpha_j = (n_m + n_j)/(n_m + n_k)$, $\beta = -n_m/(n_m + n_k)$ and $\gamma = 0$. (Cunningham and Ogilvie, 1972). Since condition (6.8) is satisfied, the minimum $d_{ij}$ can be expected to increase monotonically with the hierarchical level, thus ensuring that no backward links will occur in the linkage tree. Little practical experience of Ward's method has been reported in the literature. However, Wishart (1969) thinks that it will favour the fusion of small close clusters.

(viii) *Flexible Strategy.* This method is suggested by Lance and Williams (1967) following from their combinatorial relationship (6.7). The user is allowed to select values of $\alpha_i$, $\alpha_j$ and $\beta$ provided that:

$$\left.\begin{array}{c} \alpha_i + \alpha_j + \beta = 1 \\ \alpha_i = \alpha_j \\ |\beta| \leq 1 \\ \gamma = 0 \end{array}\right\} \quad (6.12)$$

Given a value of $\beta$ in the range $-1 \leq \beta \leq 1$ the values of $\alpha_i$ and $\beta_j$ follow automatically. Thus, by choosing values of $\beta$ in the range $-1$ to $+1$ the method can be made to range from space-dilating ($\beta = -1$) to space-contracting ($\beta = 1$). Lance and Williams state that they have been unable to define the value of $\beta$ for which the system is space-conserving, this being dependent upon the values of the elements of **D**. A small negative value is suggested by comparison with other methods. The flexible strategy approach cannot be used if $d_{ij}$ is a correlation measure. Not all authors agree that the flexible strategy approach is worthwhile — it seems to Sneath and Sokal (1973) too much like cooking the results to make them conform to a desired end-product. Better the devil one knows...! An example of dendrograms produced by the use of this method is given in Figure 6.2. (A description of the dendrogram can be found in Section 6.1.3.) The values of $\alpha_i$, $\alpha_j$, $\beta$ and $\gamma$ in Equation (6.7) for the hierarchical methods discussed above are summarized in Table 6.2.

In this table, as in the text above, $(i)$ and $(j)$ are the groups or individuals currently being joined; $(k)$ is the group resulting from their fusion and $(m)$ is any other point or group, while $n_i$, $n_j$, $n_k$ and $n_m$ are the numbers of individuals in each respective group.

*6.1.3 Display of Hierarchical Relationships*
The output from the algorithms described in the preceding section is both tabular and graphical. The fusions at successive hierarchical levels together with the associated minimum $d_{ij}$ value are printed in a tabular

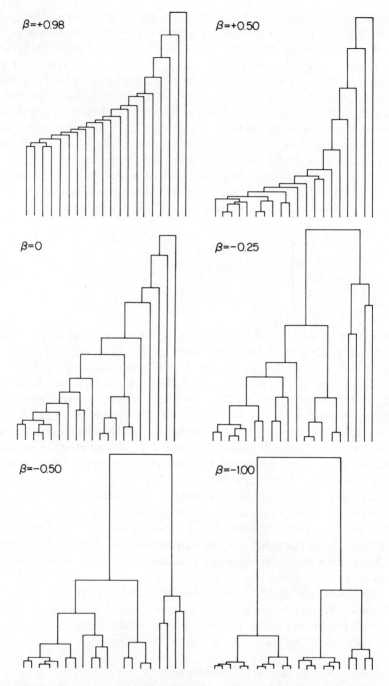

Figure 6.2  Flexible strategy — effects of varying $\beta$ (Reproduced with permission from G. N. Lance and W. T. Williams, *Computer Journ.*, **9**, 373–380 (1967), Figure 1)

Table 6.2
Coefficients in combinatorial equation for the hierarchical clustering methods mentioned in the text

| Method | $\alpha_i$ | $\alpha_j$ | $\beta$ | $\gamma$ |
|---|---|---|---|---|
| NN | 0.5 | 0.5 | 0 | −0.5 |
| FN | 0.5 | 0.5 | 0 | 0.5 |
| GA | $n_i/n_m$ | $n_j/n_m$ | 0 | 0 |
| C  | $n_i/n_m$ | $n_j/n_m$ | $-n_i n_j/n_m^2$ | 0 |
| M  | 0.5 | 0.5 | −0.25 | 0 |
| SA | 0.5 | 0.5 | 0 | 0 |
| WE | $\dfrac{n_m + n_i}{n_m + n_k}$ | $\dfrac{n_m + n_j}{n_m + n_k}$ | $\dfrac{n_m}{n_m + n_k}$ | 0 |

form known as the linkage order. The same information can be represented diagrammatically in the form of a linkage tree or dendrogram (sometimes termed a phenogram). An example of a linkage order might be:

Table 6.3
Example of linkage order

| Item joins | item at | similarity level |
|---|---|---|
| 2 | 4 | 0.65 |
| 3 | 6 | 0.75 |
| 2 | 3 | 1.01 |
| 1 | 5 | 1.52 |
| 1 | 2 | 2.65 |

A graphical representation of the same information is shown in Figure 6.3(a). The identifiers of the individuals (items) are arranged in such a way that the stems of the linkage tree do not cross. There are several permutations of the identifiers that satisfy this requirement. One of these is used in the dendrogram (Figure 6.3(a)) derived from the linkage order shown in Table 6.3. An alternative form of dendrogram is preferred by some authors (e.g. Rohlf, 1974) but this seems less desirable (Figure 6.3(b)) as the tree-like form is lost. However, the computer program required to produce Figure 6.3(b) is less complex than that which produces Figure 6.3(a). McCammon (1968B) has introduced a form of representation which he calls a dendrograph; in this, the shape of the dendrogram of the form shown in Figure 6.3(a) is made as nearly triangular as possible, while the spacing between the stems is not even but is proportional to the distance from the tip of the stem to the cross-bar. This is said to give a clearer visual impression of

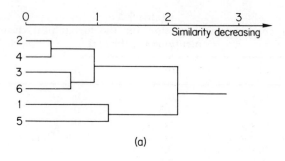

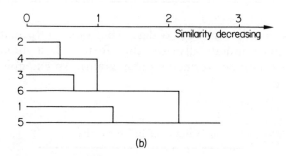

Figure 6.3 (a) Dendrogram derived from linkage order given in Table 6.3; (B) alternative form of dendrogram

the within-group heterogeneity. A Fortran program to produce dendorgraphs on a graph-plotter is listed by McCammon and Wenniger (1970).

*6.1.4 Comparison of Hierarchical Clustering Strategies*

Considering the number of apparently different hierarchical grouping techniques and the voluminous literature on the subject, the number of comparative studies is small and the results inconclusive. On mathematical grounds Jardine and Sibson (1971) prefer the nearest neighbour method, but empirical comparisons have not proved to be so unequivocal. Part of the difficulty lies in the objective comparison of the results of different strategies, which necessitates some measure of the 'goodness of fit' of the hierarchical structure generated by the procedure to the original data. One approach involves the comparison of the coefficients of similarity $d_{ij}^*$ derived from the hierarchical structure and the similarity coefficients $d_{ij}$ measured on the original data. Clearly, if $\mathbf{D}^*$ $(= d_{ij}^*)$ and $\mathbf{D}$ $(= d_{ij})$ closely resemble one another then the structure present in the data can be closely modelled by the

Table 6.4
D* matrix derived from the dendogram shown in Figure 6.3 (a)

|   | 1 | 2 | 3 | 4 | 5 | 6 |
|---|---|---|---|---|---|---|
| 1 | — | | | | | |
| 2 | 2.65 | — | | | | |
| 3 | 2.65 | 1.01 | — | | | |
| 4 | 2.65 | 0.65 | 1.01 | — | | |
| 5 | 1.52 | 2.65 | 2.65 | 2.65 | — | |
| 6 | 2.65 | 1.01 | 0.75 | 1.01 | 2.65 | — |

hierarchical representation. The elements of **D*** can be derived from the linkage order or the dendrogram. Taking Figure 6.3(a) as an example, the $d_{ij}^*$ values derived are listed in the lower triangular matrix in Table 6.4. The most widely-used measure of resemblance for comparing the matrices **D** and **D*** is the product-moment correlation coefficient, known in this context as the 'cophenetic correlation coefficient' or $r_c$ (Sokal and Rohlf, 1962). It is calculated in exactly the same way as a normal correlation coefficient with the elements of the strict lower triangles of **D** and **D*** considered as forming linear arrays when read row-wise. Values of $r_c$ range in general from 0.6 to 0.95, and values above 0.75 to 0.8 are intuitively felt to be 'good'.

Besides allowing comparison of **D*** and **D** matrices the cophenetic correlation coefficient can be used to measure the resemblance between **D*** matrices output by different clustering methods, and so provide at least a partial answer to the question: how closely do two solutions resemble each other? Sneath and Sokal (1973, p. 280) prefer to call such measures of resemblance 'matrix correlations'.

The problem of comparing the goodness of fit of two point configurations, such as the observed one as represented by **D** and the predicted or output configuration, measured by **D***, occurs also in ordination techniques (Section 6.3). Measures such as 'stress' in Nonmetric Multidimensional Scaling (Kruskal, 1964A) and 'mapping error' in Nonlinear Mapping (Sammon, 1969) are no more than an attempt to measure the goodness of fit of a low-dimensional representation to the original data. Jardine and Sibson (1968) have proposed a family of measures which are comparable to stress and mapping error. Their measure indicates degree of distortion rather than of resemblance (as with $r_c$). The general form of the coefficient is

$$\Delta_\mu = \frac{\left[\sum_{j<k}^{n} |d_{jk} - d_{jk}^*|^{1/\mu}\right]^\mu}{\left[\sum_{j<k}^{n} d_{jk}^{*\,1/\mu}\right]^\mu} \qquad (6.13)$$

where $\mu$ is an arbitrary value in the range $0 \leqslant \mu \leqslant 1$. Jardine and Sibson (1968) use $\mu = 0.5$ in their example. Other methods have been proposed by Hartigan (1967), Phipps (1971) and Williams and Clifford (1971). Little experience has been gained with any measure other than $r_c$ though it does seem that none of the measures can cope adequately with the situation — which may be a common one — in which one or two individuals are responsible for a considerable part of the discrepancy between **D** and **D***.

Comparisons of the various clustering algorithms using one or other of the measures detailed above have not been numerous. Farris (1969) used $r_c$ to compare many methods of hierarchical clustering and found that the group average method invariably produced higher values of $r_c$ than any other of the techniques discussed earlier, although he also found that even higher values of $r_c$ could be obtained if backward links in the dendrogram were allowed. This casts some doubt on the use of $r_c$ as a measure of the optimality of a clustering strategy, for structures displaying backward links are not generally acceptable. Cunningham and Ogilvie (1972) used two measures one which is equivalent to $r_c$ and a second which is simply Kendall's rank correlation coefficient measured between **D** and **D***. Their study employed six sets of artificial data to assess the relative merits of all the algorithms described in Section 6.1.2 with the exception of the flexible strategy. Their results showed that — for the particular data used — the group average method was always at least as good as any other method. This conclusion matches that of Farris (1969). Both the simple average and furthest neighbour methods performed almost as well as group average, while the behaviour of nearest neighbour was largely dependent on the type of data used; in general, its performance was poor. The variance minimization algorithm of Ward (1963) proved to be significantly prone to chaining. Together with the centroid and median methods, this strategy introduced distortion into one artificial data set that was designed to represent a pure hierarchical structure.

Pritchard and Anderson (1970) applied several hierarchical grouping strategies to a real data set, measured on three varied plant communities. Nearest neighbour and centroid were found to pick outliers and thus to be least useful. Furthest neighbour gave results which were most in agreement with field observations, with group average performing almost as well. This study did not employ any measure of goodness of fit but instead relied on intuitive and background knowledge. Other comparative analyses are those of Anderson (1971A), Boyce (1969) and Gower (1967B).

These limited studies appear to indicate that group average is probably the safest of the grouping strategies to employ initially in an exploratory investigation. This tentative conclusion is indicative of the work that is necessary before any firm guidelines can be laid down. The use of several strategies is recommended, wtih $r_c$ acting as a guide to the goodness of fit of the resulting structures. As with most exploratory work, common sense and a knowledge of the phenomena under investigation are most important; servile reliance on the result of an arbitrary optimality measure is not likely to reveal anything of fundamental importance.

### 6.1.5 Nonhierarchical Clustering Methods

The use of nonhierarchical clustering schemes in physical geography has not been widespread. It seems that in some cases the use of ordination methods has been preferred as an alternative to the hierarchical classification procedures described in the previous section. (See, for example, Whittington and Hughes (1972, Section 2(d)). In the remarks introducing this chapter it was noted that the prevalence of hierarchical classifications may be due as much to the popularity of the methods — and hence availability of programs — as to a belief that the data is structured in a hierarchical fashion. The method described in this section is one of several nonhierarchical clustering algorithms that have been proposed. It has the twin merits of relative simplicity and of the availability of an approximate test for the appropriate number of clusters (Beale, 1969; Sparks, 1973).

The assumptions of the method are that the Euclidean distances **D** separating the $n$ points (representing individuals) in a $p$-dimensional space are proportional to the dissimilarities between the objects, and secondly that no object can belong simultaneously to two clusters. Initially a given number of cluster centres, $c_{max}$, are located in the $p$-space. The positions of these centres can be chosen by inspection of the data matrix or they can be selected randomly. In the latter case, $c_{max}$ should be rather higher than the envisaged final number of clusters. In the accompanying program the coordinate axes of the $p$-space can be optionally standardized by the usual method:

$$z_{ij} = (x_{ij} - \bar{x}_j)/s_j$$

or the data can be converted to deviation form by

$$x_{ij}^* = x_{ij} - \bar{x}_i.$$

The cluster centres, if chosen randomly, are allowed to fall in the range between the maximum and minimum observed values on each variable.

*Step 1* involves the allocation of the points $x_{ij}$ to the cluster having the nearest centre. The usual formula for $d^2$ is used (Equation 6.1) though $d^2$ is employed rather than $d$ as given in (6.1). This formula implies that the $p$ variables are mutually uncorrelated; thus, a preliminary principal components analysis may be necessary or, alternatively, the formula given in Equation (6.3) may be used.

*Step 2.* Once the $n$ individuals have been assigned to clusters using Step 1 the centres of the clusters are redefined as the centroids of the clusters of points by the formula

$$y_{kj} = \frac{1}{n_k} \sum_{i=1}^{n_k} x_{ij} \quad (j = 1,2,\ldots,p; k = 1,2,\ldots,c_{max})$$

in which $n_k$ represents the number of individuals assigned to cluster $k$, and the $y_{kj}$ are the coordinates of the $k$th cluster centre on the $j$th axis (variable).

*Step 3.* At this stage the objects (points) are moved in turn to other clusters to see whether the total squared distance from the points to the cluster centres is reduced when, at the same time, the cluster centres are themselves moved to take account of the reallocation of the point. That is, point $i$ is moved from cluster $j$ to cluster $k$ if

$$\frac{n_k}{n_k + 1} \sum_{m=1}^{p} (x_{im} - y_{km})^2 < \frac{n_j}{n_j - 1} \sum_{m=1}^{p} (x_{im} - y_{jm})^2$$

If this condition is satisfied, the move is made permanent and the values of $y_{jk}$ are recomputed in the manner described at Step 2. The set of points is scanned repeatedly until no further moves take place. The final configuration provides the solution for $c_{max}$ clusters. At this stage, information such as the coordinates of the cluster centres, the cluster membership, distances from the points to the centre of the cluster, and the r.m.s. deviation of the points from the cluster centres can be output. The latter value, $s_c$, is given by

$$s_c = \left( \frac{1}{n - c} \sum_{i=1}^{n} d_{ik}^2 \right)^{1/2}$$

where $c$ is the current number of clusters and $d_{ik}^2$ is the squared distance of the $i$th point from the centre of cluster $k$, to which it has been assigned, i.e.

$$d_{ik}^2 = \sum_{j=1}^{p} (x_{ij} - y_{kj})^2 \quad (i = 1,2,\ldots,n_k; k = 1,2,\ldots,c)$$

The residual sums of squares for the solution involving $c$ clusters can now be defined as

$$\text{RSS}(c) = s_c^2/p(n - c).$$

*Step 4.* The number of clusters, $c$, is reduced by 1 unless $c = c_{min}$, in which case Step 4 is omitted and we proceed directly to Step 5. The pair of clusters to be merged is found by locating that combination of two clusters which minimizes the increase in the squared deviations of the observations from their cluster centres, which is defined as

$$\frac{n_i \cdot n_j}{n_i + n_j} \sum_{k=1}^{p} (y_{ik} - y_{jk})^2 \quad (i = 1, c-1; j = i+1, c)$$

This value is calculated for all $(i,j)$ and the minimum chosen. If the minimum is found when $i = m1$ and $j = m2$ then clusters $m1$ and $m2$ are amalgamated and the centroid of the resulting cluster calculated by the procedures of Step 2.

*Step 5.* At this stage all clusterings have been performed for $c_{min} \leq c \leq c_{max}$ and RSS($c$) values are available for each value of $c$ in this range. These can be used in an $F$-ratio test of the null hypothesis that the solution for $n1$ clusters provides no better fit than the solution for $n2$ clusters, with $n1 \leq n2$. The $F$-ratio is computed from:

$$F(nl, n2) = \frac{\text{RSS}(n2) - \text{RSS}(n1)}{\text{RSS}(n1)} \cdot \frac{n-n2}{n-n1} \cdot \left(\frac{n1}{n2}\right)^{2/p} - 1$$

which has degrees of freedom of $p(n1 - n2)$ and $p(n - n1)$. The null hypothesis is rejected if this $F$-ratio exceeds the tabled value of $F$ at some selected level of significance. This test is an approximation and involves no explicit distributional assumptions. It is necessary, however, to ensure that the Euclidean distance function is an appropriate measure to use in allocating points to clusters. Also, the test is a sequential one which implies that the standard $F$-distribution is not fully appropriate (see below for futher discussion of this point). Nevertheless, the outcome of the $F$ tests should give a rough guide in exploratory analyses. An alternative method is to plot against $c$ and look for a break point. This has been found to be more useful in practice.

## 6.2 Ordination

The methods described in the preceding section require the assumption that the individuals on which observations were recorded fall into one or more classes, which may be arranged either hierarchically or in the form of nonoverlapping clusters. Ordination methods do not require such assumptions. Instead the distance (dissimilarity) relationships among the individuals are represented in a space of reduced dimensionality (usually two or three dimensions are chosen). Any groupings present in the data should then be apparent from visual examination of scatter plots, provided that the distortion introduced by the low-dimensional representation is small and that the number of individuals is not excessive. The pairwise plotting of scores on the first two or three principal components, a procedure suggested in Chapter 4, is an example of ordination. However, several other techniques have been developed specifically to cope with ordination, and these will be described in the following pages. Principal coordinates analysis is an extension of principal components analysis but with several distinct advantages when the data is not purely quantitative. Nonmetric multidimensional scaling (MDS) involves an attempt to solve the problem of ordination by minimizing the distortion accruing from the representation of $n$ individuals in a space of $k$ instead of $p$ dimensions, where $k < p$. Lastly, nonlinear mapping is a technique analogous to multidimensional scaling developed independently by Sammon (1969).

### 6.2.0 Principal Coordinates Analysis

This technique was suggested by Gower (1966). Other useful references are Blackith and Reyment (1971, p. 163), Rohlf (1972) and Gower (1967A). The starting point is the matrix **D** of dissimilarity coefficients (such as Euclidean distances) which is transformed into the matrix **E** by the relationship

$$e_{ij} = -\tfrac{1}{2} d_{ij}^2.$$

Alternatively a matrix of similarity coefficients **E** can be computed. In the accompanying program (Section 6.4) Gower's general similarity coefficient is used. This has the advantages that (i) it can handle quantitative, two-state (presence—absence) or qualitative multistate variables and (ii) the resulting matrix **E** is always symmetrical and positive semi-definite, that is, the $n$ objects can be represented as a set of points in Euclidean space (Sneath and Sokal, 1973, p. 136). The coefficients $e_{ij}$ are computed separately for the three types of variable and are then weighted by the reciprocal of the number of variables involved and the resulting values are summed. That is,

$$e_{ij} = e_{ij}^Q / p^Q + e_{ij}^T / p^T + e_{ij}^M / p^M \qquad (6.14)$$

where $e_{ij}^Q$, $e_{ij}^T$ and $e_{ij}^M$ are the values of the coefficient for quantitative, two-state and multistate variables respectively and $p^Q$, $p^T$ and $p^M$ are the numbers of such variables.

**E** is now modified so that the mean of each row and column is removed, since the mean is unimportant in the determination of the distance between any two points. The modified matrix $\mathbf{F} = (f_{ij})$ is given by

$$f_{ij} = e_{ij} - \bar{e}_i - \bar{e}_j + \bar{e}$$

where $\bar{e}_i$, $\bar{e}_j$ and $\bar{e}$ are the means of the $i$th column and the $j$th row, and the overall mean, respectively. Since **E** and consequently **F** is positive semi-definite, the Householder-QL algorithms can be used to determine a specified number of the largest eigenvalues of **F**, and the corresponding eigenvectors can be determined by the method of inverse iteration. Since the number of eigenvalues required is usually small relative to the order of **F** it is more efficient to use the combination

TRIDI, RATQR, EIGVEC and TRBAK1 rather than HTDQL (see Appendix A for details of these subroutines). The magnitude of the $k$th eigenvalue gives the relative importance of the $k$th dimension in the determination of the variation in interpoint distances. Published results (e.g. Blackith and Reyment, 1971, p. 167) indicate that much of this variation is contained in the first two or three dimensions. The eigenvectors give the coordinates of the $n$ points (representing in each dimension. It is usual to plot these coordinates pairwise and inspect the resulting scatter diagram for pattern and structure. The eigenvectors $Q$ are usually scaled so that the sum of squares of each eigenvector is numerically equal to the corresponding eigenvalue. This is achieved by setting

$$q_{ij}^* = q_{ij}\sqrt{\lambda_j}$$

where $q_{ij}^*$ is a scaled eigenvector element and $\lambda_j$ the $j$th eigenvalue. Gower (1966) provides a more extended discussion of the method.

The magnitude of $n$, the number of objects, is an important consideration. The number of elements in the $(n \times n)$ similarity matrix F increases as $n^2$. This means that, if only a few of the largest eigenvalues are required, the sequence of procedures described in the previous paragraph should be used rather than subroutine HTDQL, which would give all $n$ eigenvalues and associated eigenvectors. A second problem, which is also related to the size of $n$, is the production of scatter diagrams on the lineprinter. The number of overprints, and the difficulty of interpreting the results visually, might limit the use of principal coordinates analysis to situations in which $n$ does not exceed 200.

*6.2.1 Nonmetric Multidimensional Scaling*

Ordination methods share the common aim of representing geometrically the relationships among a set of objects by projecting the original $p$ dimensional scatter onto a space of dimensionality $k (k \ll p)$. In some cases — for example in factor analysis (Chapter 5) — the axes of the $k$-space are thought to have some interpretation in terms of the processes which generated the observed relationships, or as descriptors of the measures used to characterize the objects. Nonmetric multidimensional scaling (MDS) can be used solely as an ordination technique or additionally as a method of searching for 'scales' which summarize or describe common attributes the objects. Unlike factor analysis, MDS is designed to deal with data that is not necessarily measureable in quantitative form; thus, it has been used to analyze political opinions, spatial preferences or the response of consumers to different brands of detergent (Golledge and Rushton, 1972; Press, 1973). Uses in physical geography and geology have been few (Tobler, Mielke and Detwyler, 1970; Whittington and Hughes, 1972) but there is obvious scope for further applications especially in situations in which the degree of similarity or dissimilarity between pairs of objects or places is not readily assessed by quantitative measures. Even where a quantitative measure is available some may still prefer to use MDS rather than factor analysis, principal components or coordinates analysis (e.g. Shepard, 1972). Given the considerable use of MDS algorithms in the behavioural sciences (Shepard, Romney and Nerlove, 1972) it is surprising that comparatively little interest has been shown by geographers.

Two properties of MDS are said by Kruskal (1964A) to give the technique a wider applicability and a more acceptable base in theory than other superficially comparable techniques. These properties are:

(i) any kind of similarity of dissimilarity coefficient can be analysed, even those based on subjective estimation or intuition.

(ii) a measure of the 'goodness of fit' of the $k$-space configuration output by MDS to the original configuration is available. The aim of the technique is to minimize this measure, which is termed *stress*.

Multidimensional Scaling is based essentially on the work of Torgerson (1958) as developed by Shepard (1962) and Kruskal (1964A,B). Other related work is that of Coombs (1964), Guttman (1968) and Lingoes and Guttman (1967). As formulated by Kruskal (1964A) the conceptual basis of the technique is simple. Letting $\delta_{ij}$ be a measure of dissimilarity, proximity, association or correlation between individuals $i$ and $j$, the problem is to determine '... that configuration of $n$ points in the space of smallest possible dimension such that, to an acceptable degree of approximation, the resulting interpoint distances $d_{ij}$ are monotonically related to the given proximity data [$\delta_{ij}$'s] in the sense that $d_{ij} < d_{kl}$ whenever $\delta_{ij} < \delta_{kl}$, (Shepard, 1972, p. 7–8). This relationship constitutes a monotonicity constraint. If the $\delta_{ij}$ represent similarities rather than dissimilarities then as $d_{ij} < d_{kl}$ then $\delta_{ij} > \delta_{kl}$ for the monotonicity constraint to be satisfied. This is the only difference between similarities and dissimilarities made in MDS.

The numerical technique can be briefly summarized as follows. (Kruskal, 1964B, gives a detailed description. $\Delta$ $(=\delta_{ij})$ and $\mathbf{D}$ $(=d_{ij})$ are matrices containing the dissimilarity coefficients measured or assessed from the given data and the interpoint distances computed in the $k$-space, respectively. It is computationally more convenient to consider $\Delta$ and $\mathbf{D}$ to be row vectors rather than lower triangular matrices, so that, for example, the elements of $\mathbf{D}$ are stored in a one-dimensional array in the order $d_{21}, d_{31}, d_{32}, d_{41}, d_{42}, \ldots, d_{n,n-1}$. The coordinates of the $n$ points in the $k$-space are held in the $(n \times k)$ matrix $\mathbf{X}$. The starting values of the $x_{ij}$ are chosen by a method to be described. The elements of $\Delta$ are arranged in ascending order (if they represent measures of dissimilarity) or in descending order (if they

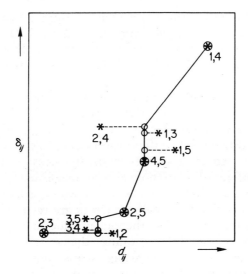

Figure 6.4 Plot of Δ versus **D** to illustrate derivation of $\hat{\mathbf{D}}$ (Reproduced with permission from J. B. Kruskal, *Psychometrika*, 29, 115–129 (1964), Figure 4)

represent similarity measures. A record is kept of the original position of each element. Next, the distances $d_{ij}$ are computed, using Equation (6.1), though other distance or dissimilarity functions could be used (see Kruskal, 1964B). The $d_{ij}$ are calculated in the same order as the $\delta_{ij}$ after ranking. A third matrix $\hat{\mathbf{D}}$ ($= \hat{d}_{ij}$) is now computed. This matrix is also stored as a row vector and it contains the result of replacing those elements of **D** which do not satisfy the monotonicity constraint by values whose calculation is described later. The relationship between **D** and $\hat{\mathbf{D}}$ is best shown diagrammatically (Figure 6.4). In Figure 6.4 the solid line represents a monotonic relationship between Δ and **D**. Points resulting from a scatter plot of Δ against **D** are shown by asterisks (∗). If these lie to the left or right of the solid line they are moved horizontally so that they lie on the line. The new positions are indicated by open circles. If a point lies on the line already it is indicated thus: ⊛. Projecting the ∗ and ⊛ onto the horizontal axis gives the $d_{ij}$ values, which will not always satisfy the monotonicity requirement. The values $\hat{d}_{ij}$ are shown by projecting ○ and ⊛ points on to the horizontal axis. These will satisfy the requirement of monotonicity. 'Raw Stress' ($S^*$) is defined as a the sum of the squares of the horizontal distances that the points have moved (from ∗ to ○ in Figure 6.4) or, more formally,

$$S^* = \sum_{i=1}^{n-1} \sum_{j=2}^{n} (d_{ij} - \hat{d}_{ij})^2$$

$S^*$ ranges upwards from zero, which value would indicate that all the * points needed no adjustment and that $\Delta$ and $\mathbf{D}$ were perfectly monotonically related, or in other words, $\mathbf{D} = \hat{\mathbf{D}}$. Hence, $S^*$ measures the degree to which $\mathbf{D}$ and $\Delta$ depart from a perfect monotone relationship.

Raw stress is dependent upon the scales used to measure the configuration $\mathbf{X}$ of points in the $k$-space. If $S^*$ is the raw stress computed from a configuration $(x_{11}, x_{12}, \ldots, x_{nk})$ then $k^2 S^*$ is the raw stress derived from the configuration $(kx_{11}, kx_{12}, \ldots, kx_{nk})$. It is unreasonable to expect a goodness of fit measure to alter merely because the configuration of points $\mathbf{X}$ is magnified or reduced. Raw stress is therefore normalized by a factor which is influenced by $k$ in the same way as is $S^*$. Two scaling factors are recommended: $\Sigma_{i<j} d_{ij}^2$ and $\Sigma_{i<j}(d_{ij} - \bar{d})^2$, where $\bar{d}$ is the mean value of the $d_{ij}$. Kruskal and Carroll (1969) recommend the second of the two factors. Furthermore, the positive square root of the scaled stress value it taken since, by analogy with conventional regression analysis, raw stress represents residual variance. The square root of scaled stress is analogous to a standard deviation. Stress is therefore defined as

$$S = S^* / \sum_{i<j} d_{ij}^2 \qquad \text{(Formula I)} \qquad (6.15)$$

or

$$S = S^* / \sum_{i<j} (d_{ij} - \bar{d})^2 \qquad \text{(Formula II)} \qquad (6.16)$$

In numerical terms, then, the problem becomes that of minimizing $S$, which is a function of the $d_{ij}$ and $\hat{d}_{ij}$, which are themselves dependent upon the configuration $\mathbf{X}$. This implies that the points representing the $x_{ij}$ should be moved around until $S$ is a minimum. The values of the $x_{ij}$ at the minimum are the coordinates of the points representing individuals in the MDS solution. Kruskal (1964B) makes use of a steepest-descent minimization algorithm (Chapter 3) to find min $(S)$. This is a widely-used method, but more recently-developed algorithms may be computationally more efficient. If the initial (starting) configuration $\mathbf{X}^{(1)}$ is far from the true minimum then convergence will be slow and local minima may intervene. The use of several different starting points is often recommended to check the validity of the solution, but computing demands may be excessive if $n$ is large.

Several techniques of determining a suitable starting configuration have been proposed. If a suitable starting point can be chosen intuitively or on the basis of prior information then convergence may be more rapid. Otherwise an arbitrary starting point must be used. Kruskal (1964B) suggested that the first $n$ $k$-tuples in the series $(1,0,0,\ldots,0), (0,1,0,\ldots,0), \ldots, (0,0,0,\ldots,1), (2,0,0,\ldots,0)$ etc.

could form the rows of $\mathbf{X}^{(1)}$. Alternatively, a pseudorandom number generator could be used to compute an initial value for each $x_{ij}$. A third method is used by Torgerson (1958); this involves transformation of $\Delta$ to a matrix $\mathbf{U}$ of scalar products. The first $k$ eigenvectors of $\mathbf{U}$, scaled by the square roots of the corresponding eigenvalues, are taken as the elements of $\mathbf{X}^{(1)}$. $\mathbf{U}$ is defined by

$$u_{ij} = \left\{\left(\sum_{k=1}^{n}\delta_{ik}^2 + \sum_{k=1}^{n}\delta_{kj}^2\right)\bigg/2n\right\} - \left\{\left(\sum_{k=1}^{n}\sum_{h=1}^{n}\delta_{kh}^2\right)\bigg/2n\right\} - \delta_{ij}^2/2 \tag{6.17}$$

for $i, j = 1, 2, \ldots, n$. If $\mathbf{Q}$ is the matrix of eigenvectors of $\mathbf{U}$ and $\mathbf{\Lambda}$ ($= \lambda_i$) the diagonal matrix of eigenvalues then

$$\mathbf{X}^{(1)} = \mathbf{Q}_k \mathbf{\Lambda}_k^{\frac{1}{2}} \tag{6.18}$$

or

$$x_{ij} = q_{ij} \cdot \lambda_j^{\frac{1}{2}} \quad (i = 1, n; j = 1, k)$$

The notation $\mathbf{Q}_k$, $\mathbf{\Lambda}_k$ means: the first $k$ columns of $\mathbf{Q}$ and the first $k$ diagonal elements of $\mathbf{\Lambda}$. If $\Delta$ and $\mathbf{D}$ are perfectly monotonically related then the rank order of the $\delta_{ij}$ and the $d_{ij}$ will be the same. If the relationship is not monotonic the $d_{ij}$ values must be adjusted to give the values $\hat{d}_{ij}$ expected under the hypothesis of monotonicity (see Figure 6.4). The $\hat{d}_{ij}$ are obtained by a process of checking the (ranked) $d_{ij}$ array to ensure that all values above the current one are larger in magnitude. If not, the set of $d_{ij}$'s which do not increase with position are replaced by their mean value. For example, the sequence 1,2,3,3,2,5 becomes 1,2.5,2.5,2.5,2.5,5. Kruskal (1964B) outlines a method of sorting the $d_{ij}$ into $\hat{d}_{ij}$ which is used in the accompanying program. Stress can now be formed from (6.15) or (6.16). The negative gradient of stress at each point $x_{ij}$ is computed next. This provides an indication of the direction of change in $x_{ij}$ to lower the stress value. The gradient vector is found from

$$g_{kl} = S \sum_{i=1}^{n}\sum_{j=1}^{n}(\Delta^{ki} - \Delta^{kj})\left[\frac{d_{ij} - \hat{d}_{ij}}{S^*} - \frac{d_{ij}}{\sum_{i,j} d_{ij}^2}\right] \cdot \frac{(x_{il} - x_{jl})}{d_{ij}} \tag{6.19}$$

in the case of Euclidean distances $d_{ij}$. The $\Delta^{ki}$ and $\Delta^{kj}$ are Kroekner deltas which take the value 1 if the $i$ and $k$ or $k$ and $j$ indices are equal, otherwise they take the value 0. (If strict lower triangles of $\Delta$ and $\mathbf{D}$ have been used, the term $(\Delta^{ki} - \Delta^{kj})$ can be ignored.) The amount of movement along the direction of the negative gradient is given by the step size, $\alpha$, which should be about 0.2 initially. Thereafter, the step size is computed from

$$\alpha_{present} = \alpha_{previous} \cdot (\text{angle factor}) \cdot (\text{relaxation factor}) \cdot (\text{good luck factor})$$

where

angle factor $= 4.0^{\cos\theta^{3.0}}$

relaxation factor $= 1.3/(1.0 + \text{fivestepratio}^5)$

fivestepratio $= \min\left(1.0, \dfrac{\text{present stress}}{\text{stress 5 iterations ago}}\right)$

good luck factor $= \min\left(1.0, \dfrac{\text{present stress}}{\text{previous stress}}\right)$.

If five iterations have not been claculated then 'stress five iterations ago' becomes 'stress at iteration 1'. A similar procedure can be used for 'previous stress' on iteration 1. The value of $\cos\theta$ is

$$\cos\theta = \frac{\sum_{i,j} g_{ij} g''_{ij}}{\sqrt{\sum_{i,j} g_{ij}^2} \sqrt{\sum_{i,j} g''^{2}_{ij}}} \tag{6.20}$$

where $g''_{ij}$ is the gradient at $x_{ij}$ at the previous iteration.

Iterations proceed until stress becomes acceptably small. There are several rules to indicate when the process have converged: this could be when $S = 0.0$ or when $\text{mag}(g)$ reaches a value of 2–5 per cent of its value for an arbitrary configuration. The value of $\text{mag}(g)$ (the relative magnitude of $g$) is given by

$$\text{mag}(g) = \left(\sum_{i,s} g_{is}^2/n\right)^{1/2} \tag{6.21}$$

Neither rule can ensure that the minimum is global and not local.

In the above summary it has been assumed that $\mathbf{X}$ has been normalized after computation at every iteration. Normalization in this context involves (i) subtracting out the column means of $\mathbf{X}$, and (ii) dividing the resulting deviations $(x_{ij} - \bar{x}_j)$ by $(\Sigma_{ij,j}(x_{ij} - \bar{x}_j)^2)^{1/2}$.

The results of a MDS analysis consist of the value of stress at each iteration and a final configuration, $\mathbf{X}$. Unless there are *a priori* reasons for choosing a particular value of $k$ then the analysis should be repeated for a range of $k$ values. Minimum stress plotted against $k$ may then act as a guide to the possible dimensionality of the system, a break or elbow in the curve, where the curve is approaching zero, indicating a suitable choice of $k$. If minimum stress is suitably low, two-dimensional slices through the chosen $k$-space are then examined visually in order to seek out relationships among the individuals. In some instances it may be thought that the reference axes themselves have a physical interpretation, perhaps in terms of concepts or scales which summarize the way in which the points are arranged. Usually this means that some form of rotation of the axes is desirable (see Chapter 5) since the axes output by MDS are, in general, arbitrary. The varimax rotation has frequently been used in this context.

One of the few applications of MDS in physical geography is reported by Anderson (1971A) whose main purpose was the comparative study of various methods of ordination and hierarchical cluster analysis. He took a set of data from a previous study by Rayner (1966) which related to soil samples of Brown Earths (BE) some of which were described as acidic BE(A) or lessivated BE(L), together with some gley soils and a renzina (Rz) sample. These samples came from Glamorgan, South Wales. The resulting configuration for $k = 2$ is shown in Figure 6.5. No attempt is made by Anderson to interpret the axes which define the 2-space; he is interested only in the relative positions

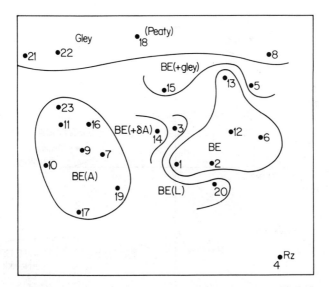

Figure 6.5 Nonmetric multidimensional scaling representation of Glamorgan soils.
Key to abbreviations: BE — brown earth; BE(A) — acid brown earth; BE(L) — lessivated brown earth, and Rz — rendzina (Reproduced with permission from A. J. B. Anderson, *Journ. Intern. Assoc. Mathem. Geol.*, 3, 1–14 (1971), Figure 3, by permission of Plenum Publishing Co.)

of the points. The diagram showed a well-separated group of acidic brown earths, with more dispersed groups of other soil types.

Whittington and Hughes (1972) provide an example of the use of the MDS algorithm in geology, using the technique to aid the definition of Ordovician faunal provinces. Similarities among trilobite faunas found in various Ordovician rocks were assessed by Simpson's coefficient. A three-dimensional representation was found to give acceptably low values of stress, and the points were found to group into recognizable sets which had a distinctive geographical distribution (Figure 6.6). Again, no attempt was made to rotate or interpret the axes of the final configuration.

MDS methods have enjoyed something of a boom in the behavioural and social sciences, where data is often available only in the form of grades, preferences or opinions. Potentially, however, it has wider usage

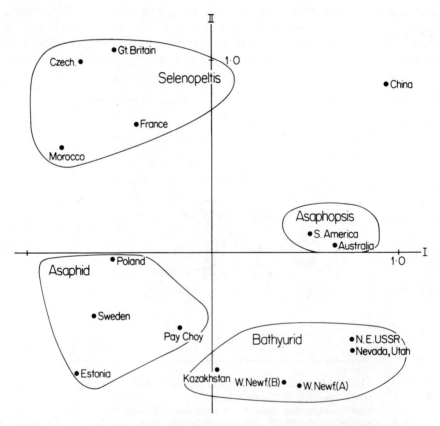

Figure 6.6 Faunal provinces based on nonmetric multidimensional scaling of trilobite genera, Arenig and Llanvirn Series (Reproduced with permission from H. B. Whittington and C. P. Hughes, *Phil. Trans. Roy. Soc., London*, B, 263, 235–273 (1972), Figure 2, by permission of the Royal Society)

than this, and could be fruitfully applied in physical geography. The lack of rigid assumptions makes it an attractive technique, but problems involving minimization of nonlinear functions are always prone to the type of difficulty mentioned in this section.

## 6.2.2 Nonlinear Mapping

Nonlinear mapping, although developed independently by Sammon (1969), is similar both in concept and in execution to MDS methods (Section 6.2.2) in that $n$ points in a $p$-dimensional space are projected onto a $k$-dimensional subspace ($k \ll p$) with a minimum of distortion. Sammon's computational technique is somewhat simpler than that of Kruskal (1964B) though the method has been used only on dissimilarity matrices containing Euclidean distance measures. The output consists of the values of a goodness of fit function, termed mapping error, and a two or three dimensional representation of inter-point relationships. Nonlinear mapping does not aim to ensure monotonicity between observed dissimilarities and calculated distances; rather, the goodness-of-fit function measures the amount of distortion of inter-point distances introduced by mapping onto a $k$-(as opposed to a $p$-) dimensional space. The function to be minimized is

$$E = \frac{1}{\sum_{\substack{i=1 \, j=1 \\ i \neq j}}^{n} \sum^{n} d_{ij}} \sum_{i=1}^{n} \sum_{j=1}^{n} \frac{(\delta_{ij} - d_{ij})^2}{\delta_{ij}^2} \qquad (6.22)$$

where $\delta_{ij}$ is an observed dissimilarity and $d_{ij}$ a distance measured in a $k$-dimensional space. The initial $k$-space representation of the points, $X^{(1)}$, can be chosen randomly or can be taken as the subset of $k$ out of the original $p$ variables that have maximum variance. Alternatively, principle component scores could be employed.

Given the matrix $X^{(1)}$, from which interpoint distances $D$ ($= d_{ij}$) are computed by the normal formula, and $\Delta$ ($= \delta_{ij}$), the matrix of dissimilarities, the method of steepest descent is used to locate a minimum of $E$. Reference should be made to Section 6.2.1 and to Chapter 3 for a brief discussion of nonlinear optimization techniques. The steepest descent method proceeds by computing the $k$-space coordinates $X$ at iteration $(m + 1)$ from:

$$x_{ij}(m + 1) = x_{ij}(m) - MF.\phi_{ij}(m)$$

MF is a parameter termed by Sammon the 'magic factor' — it is in fact a fixed step length which lies in the range 0.3–0.4. A refinement of Sammon's method would include provision for varying the step length, which ideally should shorten as a minimum is approached. The last term, $\phi_{ij}$, is the ratio of the first to the second-order partial derivatives of $E$ with respect to the $x_{ij}$. These derivatives are defined as:

$$\frac{\delta E}{\delta x_{ij}} = -\frac{2}{c} \sum_{\substack{p=1 \\ (p \neq i)}}^{n} \left[ \left( \frac{d_{ij} - \delta_{ij}}{d_{ij} \delta_{ij}} \right) (x_{ij} - x_{pj}) \right]$$

and

$$\frac{\delta^2 E}{\delta x_{ij}^2} = -\frac{2}{c} \sum_{\substack{p=1 \\ (p \neq i)}}^{n} \frac{1}{d_{ij} \delta_{ij}} \left[ (d_{ij} - \delta_{ij}) - \frac{(x_{ij} - x_{pj})^2}{\delta_{ip}} \left( 1 + \frac{d_{ip} - \delta_{ip}}{\delta_{ip}} \right) \right]$$

(6.23)

In order to prevent the calculations from 'blowing up' whenever any $d_{ij} = 0$, all zero values in **D** are set to an arbitrarily small value. The algorithm terminates when a fixed number of iterations have been carried out or whenever $E$ has converged to a suitably small value. Again, different sets of $\mathbf{X}^{(1)}$ should be used as a check against the possibility that the process has reached a local optimum. As with MDS, the question of 'when to stop' is almost as thorny as that of 'how to start'.

Howarth (1973) has investigated the utility of the nonlinear mapping algorithm in a geological context. He points out the virtues of ordination as opposed to clustering techniques when dealing with spatial data, for in geology distributions are continuous rather than discrete. In terms of real data analyzed by Howarth, the nonlinear mapping method appears to give satisfactory results on the basis of visual examination of plots of the final **X** configuration. However, the technique has not been widely used in any of the earth sciences. Further empirical testing along the lines of Howarth's (1973) work is necessary before the method can be accepted as other than experimental.

*6.2.3 Quadratic Loss Functions*

Anderson (1971A,B) introduced the idea of the quadratic loss function. The method has not otherwise been reported in the literature. In brief, it consists of an attempt to minimize the quantity

$$L = \sum_{i=1}^{n} \sum_{j=1}^{n} W_{ij} (d_{ij}^* - d_{ij})^2 \qquad (6.24)$$

where $W_{ij}$ is a weighting factor, $d_{ij}^*$ is the distance between points $i$ and $j$ in a space of $k$ dimensions and $d_{ij}$ is the correponding distance in a $p$-dimensional space representing the dissimilarity between items $i$ and $j$ with respect to $p$ variables or characteristics. One can see immediately

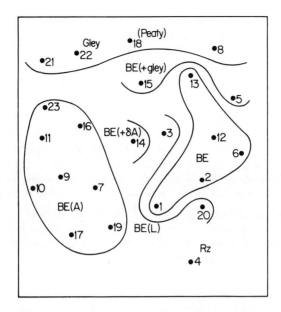

Figure 6.7  Quadratic loss function, Glamorgan soils. The abbreviations are listed in Figure 6.5 (Reproduced with permission from A. J. B. Anderson, *Journ. Assoc. Mathem. Geol.*, 3, 1–14 (1971), Figure 2, by permission of Plenum Publishing Co.)

that (6.23) is yet another form of goodness of fit function, the family of which are represented by Equation (6.13). The weights $W_{ij}$ can be used to increase or decrease the importance attached to large distances. Anderson suggests that the start configuration in the $k$-space should be the scores on the first $r(r > k)$ principal components. The method of steepest descent is again used to locate a minimum of $L$ in $r$ dimensions. This may correspond to a local or a global minimum (Chapter 3). The value of $r$ is reduced by 1 and the process repeated, until $r = k$. The results of applying this technique to the Glamorgan soil example (see Section 6.2.1) are summarized in Figure 6.7. Anderson (1971A) says of this representation that it is preferable to the MDS solution, even though the apparent difference is marginal, because of its relative freedom from computational problems. There has, as yet, been no empirical justification of this proposition.

## 6.3 Examples

In this section several examples of the use of classification and ordination techniques are given. The output in all cases was produced by the programs which are listed and described in Section 6.4.

### 6.3.0 Hierarchical clustering

A selection of 52 sediment samples (Mather, 1969) provides the data for an example of the use of hierarchical procedures. The 52 samples represent differing environmental conditions during the Late Devensian stage of the Pleistocene period in part of North-west England. Each sample was split into 32 half-phi classes in the range $9\phi$ to $-7\phi$, and the weight in each class expressed as a percentage of total sample weight. A 33rd class, percent by weight finer than $9\phi$, was added. The (52 x 33) data matrix was expressed in terms of 13 linearly independent principal components, which accounted for over 97 per cent of the total standardized variance. The principal components analysis allowed the number of variables to be reduced from 33 to 13 with the loss of only 3 per cent of the original information. The original data and the principal components score matrix are shown in Tables 6.5 and 6.6 respectively. The first 13 eigenvalues of the (33 x 33) matrix of correlations among the 33 classes are shown in Table 6.7.

Each of the hierarchical methods of cluster analysis was carried out on the (52 x 13) orthogonal principal components score matrix. As the principal components are uncorrelated the standard Euclidean formula (6.1) was used, though (6.3) could have been used. Also, the scores were not weighted by the size of the corresponding eigenvalue (see Chapter 4). This is often a fundamental decision in classificatory work. If all the original variables are given equal weight then standardized (unit variance) score vectors should not be used. Instead, the vectors of scores should be multiplied by the corresponding eigenvalue so that the contribution of each variable to the score vector is equally represented. If, on the other hand, the principal components analysis represents an attempt to eliminate redundancies in the data then each score vector could be considered to be an independent source of variation and differences between the variances of the principal components, as given by the eigenvalues, would then be viewed as being partly a function of the selection of variables. In other words, each principal component would be given equal importance. Weighting the principal component score vectors by the corresponding eigenvalues indicates that the investigator is giving each variable an equal importance; standardizing the principal component scores is equivalent to giving each principal component equal importance. The latter approach was adopted here.

The results of the cluster analyses are summarized in Table 6.8.

The results obtained from the hierarchical clustering program for the 52 sediment samples are shown in Figures 6.8(a)–(g). The nearest-neighbour algorithm (Figure 6.8(a)) places most of the samples into one large group. Sample 43 joins at a very late stage. The addition of single items to the main stem may indicate the effects of chaining. This dendrogram produced a cophenetic correlation $(r_c)$ of 0.7770. In contrast, the furthest-neighbour algorithm (Figure 6.8(b)) showed several small groups and one main group which can be subdivided into smaller groups. Again, sample 43 is last to join, which it does at a very late stage. For this representation, $r_c$ is 0.7215. The median method gave $r_c$ = 0.847 but backward links were produced. However, sample 43 was again last to join. Ward's error sum of squares algorithm (Figure 6.8(c)) gave $r_c$ = 0.5601. The dendrogram for this method is similar to that produced by the furthest-neighbour algorithm, but $r_c$ is much lower. The group average method (Figure 6.8(d)) gave a dendrogram that was superficially similar to the nearest-neighbour dendrogram. The cophenetic correlation for the group average method was, however, higher ($r_c$ = 0.8225). Samples 1, 24, 10 and 48 are placed in a separate group by GA whereas they show no clear relationship in NN. The one large group in GA is easily subdivided into smaller, compact groups. Several groups with a membership of two or three join the main stem late, and samples 49 and 43 join last. Figures 6.8(f) to 6.8(h) show the results obtained from the flexible strategy method. The cophenetic correlations are 0.0912 ($\beta$ = −0.5), 0.8121 ($\beta$ = 0) and 0.7531 ($\beta$ = 0.5). Thus the results for $\beta$ = −0.5 hardly agree with the observed relationships at all, while the results with $\beta$ = 0 are almost as good as those produced by GA. Comparison of Figures 6.8(a) to 6.8(h) shows that the dendrogram for $\beta$ = −0.5 is very dissimilar to the other dendrograms. One very compact cluster is shown, with a second, larger cluster being divisible into three main groups. Sample 43, which the other methods agree is a 'stray', is placed in one of the smaller subgroups. The dendrograms for $\beta$ = 0 and $\beta$ = 0.5 resemble the GA and NN results, respectively.

The important question is: which of these dendrograms should be used? Earlier discussion led to the conclusion that each method had its own peculiarities and that each method would produce differing conclusions for the same data-set. In other words, one technique may be superior if the data is structured in a particular way, but it may be relatively poor for a different type of data structure. In the present example the group average methods and the flexible strategy with $\beta$ = 0 were roughly comparable. Nearest-neighbour, further neighbour and Ward's error sum of squares methods appear to be less easily interpretable. Both the median and centroid techniques gave backward links, which — although giving high cophenetic correlations — were difficult to incorporate into a hierarchical scheme. It would seem, in this particular instance at least, that the cophenetic correlation is a good guide to the value of a dendrogram.

*6.3.1 Nonhierarchical Cluster Analysis: Nonlinear Mapping*
Tien's data (Table 6.9) is used to illustrate the techniques of non-hierarchical cluster analysis and nonlinear mapping. The results for three cluster centres in the nonhierarchical cluster analysis are shown in

## Table 6.5
### Sediment example — raw data expressed as weight in each phi class

SEDIMENT EXAMPLE -- RAW DATA.                                               SECTION  1

|    | 1      | 2       | 3       | 4       | 5      | 6      | 7      | 8       |
|----|--------|---------|---------|---------|--------|--------|--------|---------|
| 1  | 0.0000 | 0.0000  | 0.0000  | 0.0000  | 0.0000 | 0.0000 | 0.0000 | 0.0000  |
| 2  | 0.0000 | 0.0000  | 0.0000  | 0.0000  | 0.0000 | 0.0000 | 0.0000 | 0.0000  |
| 3  | 0.0000 | 0.0000  | 0.0000  | 0.0000  | 0.0000 | 0.0000 | 0.0000 | 0.0000  |
| 4  | 0.0000 | 0.0000  | 32.1926 | 0.0000  | 0.0000 | 0.9809 | 0.9274 | 2.0511  |
| 5  | 0.0000 | 0.0000  | 0.0000  | 0.0000  | 0.0000 | 0.4950 | 2.4024 | 2.7163  |
| 6  | 0.0000 | 15.4038 | 12.8504 | 7.1183  | 8.6086 | 9.4007 | 1.8812 | 8.5982  |
| 7  | 0.0000 | 0.0000  | 0.0000  | 0.0000  | 0.0000 | 0.0000 | 0.0000 | 0.4708  |
| 8  | 0.0000 | 0.0000  | 0.0000  | 4.4638  | 5.5880 | 8.5464 | 2.7743 | 7.9876  |
| 9  | 0.0000 | 0.0000  | 0.0000  | 1.3592  | 3.7980 | 2.5063 | 1.4459 | 10.2180 |
| 10 | 0.0000 | 0.0000  | 0.0000  | 0.0000  | 0.0000 | 0.0000 | 0.0000 | 0.0000  |
| 11 | 0.0000 | 0.0000  | 0.0000  | 0.0000  | 0.0000 | 0.0000 | 0.0000 | 0.0000  |
| 12 | 0.0000 | 0.0000  | 0.0000  | 0.0000  | 0.9313 | 0.7156 | 1.0578 | 5.4445  |
| 13 | 0.0000 | 0.0000  | 0.0000  | 0.0000  | 0.0000 | 0.0000 | 0.0000 | 0.0000  |
| 14 | 0.0000 | 0.0000  | 0.0000  | 4.8857  | 6.6708 | 2.2683 | 1.0469 | 3.4361  |
| 15 | 0.0000 | 0.0000  | 0.0000  | 0.0000  | 0.0000 | 0.0000 | 0.0000 | 0.2209  |
| 16 | 0.0000 | 0.0000  | 0.0000  | 0.0000  | 0.0000 | 0.0000 | 0.0000 | 0.0000  |
| 17 | 0.0000 | 0.0000  | 0.0000  | 0.0000  | 0.0000 | 0.0000 | 0.0000 | 0.0000  |
| 18 | 0.0000 | 0.0000  | 0.0000  | 0.0000  | 0.0000 | 0.0000 | 0.0000 | 0.0000  |
| 19 | 0.0000 | 0.0000  | 0.0000  | 0.0000  | 0.0000 | 0.0000 | 0.0000 | 0.0000  |
| 20 | 0.0000 | 0.0000  | 0.0000  | 0.0000  | 0.0000 | 0.0000 | 0.0000 | 0.0000  |
| 21 | 0.0000 | 0.0000  | 0.0000  | 0.0000  | 0.0000 | 0.0000 | 0.0000 | 0.0000  |
| 22 | 0.0000 | 0.0000  | 0.0000  | 0.0000  | 0.0000 | 0.0000 | 1.7383 | 0.3200  |
| 23 | 0.0000 | 0.0000  | 0.0000  | 0.0000  | 0.0000 | 0.0000 | 0.0000 | 0.0000  |
| 24 | 0.0000 | 0.0000  | 0.0000  | 0.0000  | 0.0000 | 0.0000 | 0.0000 | 0.0000  |
| 25 | 0.0000 | 0.0000  | 0.0000  | 0.0000  | 0.0000 | 0.0000 | 0.0000 | 0.0000  |
| 26 | 0.0000 | 0.0000  | 0.0000  | 0.0000  | 0.0000 | 0.0000 | 0.0000 | 0.0000  |
| 27 | 0.0000 | 0.0000  | 0.0000  | 0.0000  | 0.0000 | 0.0000 | 0.0000 | 0.0000  |
| 28 | 0.0000 | 0.0000  | 0.0000  | 0.0000  | 0.0000 | 0.0000 | 0.0000 | 0.0000  |
| 29 | 0.0000 | 0.0000  | 0.0000  | 0.0000  | 0.0000 | 0.0000 | 0.0000 | 0.0000  |
| 30 | 0.0000 | 19.1665 | 3.0162  | 11.9643 | 6.2338 | 4.9357 | 2.9156 | 3.6792  |
| 31 | 0.0000 | 0.0000  | 0.0000  | 0.0000  | 5.1414 | 0.7207 | 0.5713 | 0.9668  |
| 32 | 0.0000 | 0.0000  | 11.1188 | 0.0000  | 8.3603 | 8.6277 | 3.5671 | 6.6061  |
| 33 | 0.0000 | 0.0000  | 0.0000  | 0.0000  | 2.2098 | 2.5274 | 0.9660 | 0.8204  |
| 34 | 0.0000 | 0.0000  | 0.0000  | 0.0000  | 0.0000 | 0.0000 | 0.0000 | 0.0000  |
| 35 | 0.0000 | 0.0000  | 0.0000  | 0.0000  | 0.0000 | 0.0000 | 0.0000 | 0.0000  |

|    | 9 | 10 | 11 | 12 | 13 | 14 | 15 | 16 |
|----|---|----|----|----|----|----|----|----|
| 36 | 0.0000 | 0.0000 | 0.0000 | 0.0000 | 0.0000 | 0.0000 | 0.0000 | 0.0284 |
| 37 | 0.0000 | 0.0000 | 0.0000 | 0.0000 | 0.0000 | 0.0000 | 0.9446 | 0.0000 |
| 38 | 0.0000 | 0.0000 | 0.0000 | 0.0000 | 0.5058 | 0.0000 | 0.0000 | 0.1291 |
| 39 | 0.0000 | 0.0000 | 0.0000 | 0.0000 | 0.0000 | 0.0000 | 0.0000 | 0.0911 |
| 40 | 0.0000 | 0.0000 | 0.0000 | 0.0000 | 0.0000 | 0.0000 | 0.0000 | 0.0000 |
| 41 | 0.0000 | 0.0000 | 15.3830 | 1.4730 | 12.5987 | 8.2143 | 5.0430 | 6.4698 |
| 42 | 0.0000 | 18.9352 | 12.7089 | 11.8345 | 8.3068 | 3.6081 | 2.5755 | 3.9993 |
| 43 | 24.4209 | 0.0000 | 9.6720 | 6.7908 | 8.2314 | 6.6765 | 2.5414 | 3.5911 |
| 44 | 0.0000 | 0.0000 | 21.0242 | 17.7679 | 12.6076 | 4.2357 | 2.9780 | 3.8795 |
| 45 | 0.0000 | 0.0000 | 0.0000 | 0.0000 | 0.0000 | 0.0000 | 0.0000 | 0.0000 |
| 46 | 0.0000 | 0.0000 | 0.0000 | 0.0000 | 0.0000 | 0.0000 | 0.0000 | 0.0000 |
| 47 | 0.0000 | 0.0000 | 0.0000 | 0.0000 | 0.0000 | 0.0000 | 0.0000 | 0.0000 |
| 48 | 0.0000 | 0.0000 | 0.0000 | 0.0000 | 0.0000 | 0.0000 | 0.0253 | 0.0000 |
| 49 | 0.0000 | 0.0000 | 0.0000 | 0.0000 | 0.5025 | 0.0000 | 0.0000 | 0.0000 |
| 50 | 0.0000 | 0.0000 | 0.0000 | 0.0000 | 0.0000 | 0.0000 | 0.0000 | 0.0000 |
| 51 | 0.0000 | 0.0000 | 0.0000 | 0.0000 | 0.0000 | 0.0000 | 0.0000 | 0.0000 |
| 52 | 0.0000 | 0.0000 | 0.0000 | 0.0000 | 0.0000 | 0.5692 | 0.0594 | 0.3177 |

SEDIMENT EXAMPLE -- RAW DATA.                                   SECTION 2

|    | 9 | 10 | 11 | 12 | 13 | 14 | 15 | 16 |
|----|---|----|----|----|----|----|----|----|
| 1  | 0.0000 | 0.0000 | 0.0000 | 0.0000 | 0.0000 | 0.0000 | 0.0104 | 0.0242 |
| 2  | 0.0000 | 0.0000 | 0.0000 | 0.0000 | 0.0160 | 0.0157 | 0.0190 | 0.0398 |
| 3  | 0.0173 | 0.0520 | 0.0520 | 0.0520 | 0.0167 | 0.0232 | 0.0662 | 0.1976 |
| 4  | 1.2841 | 1.0701 | 0.9453 | 0.7312 | 1.2662 | 1.3160 | 1.9715 | 2.8417 |
| 5  | 3.6821 | 3.1388 | 3.2897 | 2.9879 | 4.0951 | 3.6691 | 4.8830 | 7.5190 |
| 6  | 8.1813 | 5.7842 | 4.9765 | 3.1944 | 2.1747 | 1.7719 | 1.8320 | 1.7109 |
| 7  | 9.9568 | 9.9872 | 1.0631 | 1.2150 | 1.3791 | 2.4009 | 5.6959 | 13.2781 |
| 8  | 4.0431 | 2.6625 | 3.7473 | 2.5442 | 2.1367 | 1.8570 | 2.3082 | 2.7993 |
| 9  | 7.3261 | 6.5550 | 7.0177 | 7.0080 | 5.3132 | 7.0909 | 8.5072 | 8.4124 |
| 10 | 0.0000 | 0.0000 | 0.0000 | 0.0000 | 0.0164 | 0.0313 | 0.0561 | 0.0811 |
| 11 | 0.0000 | 0.0000 | 0.0000 | 0.0000 | 0.0200 | 0.0363 | 0.0979 | 0.3458 |
| 12 | 4.4075 | 4.6648 | 5.0816 | 5.4964 | 6.1630 | 5.0634 | 6.2446 | 9.1610 |
| 13 | 0.0000 | 0.0000 | 0.0000 | 0.0000 | 0.0000 | 0.0091 | 0.0280 | 0.2861 |
| 14 | 1.0604 | 0.7651 | 0.5637 | 0.4564 | 0.4048 | 0.4405 | 0.6632 | 1.4229 |
| 15 | 0.3787 | 0.3156 | 0.2919 | 0.3156 | 0.2931 | 0.2508 | 0.3924 | 1.1363 |
| 16 | 0.0242 | 0.0605 | 0.2178 | 0.5444 | 0.5651 | 1.3169 | 3.2392 | 10.7306 |
| 17 | 0.0000 | 0.0000 | 0.0000 | 0.0000 | 0.0000 | 0.0000 | 0.0129 | 0.1052 |
| 18 | 0.0000 | 0.0000 | 0.0000 | 0.0000 | 0.0000 | 0.0000 | 0.0126 | 0.0239 |
| 19 | 0.0000 | 0.0000 | 0.0000 | 0.0000 | 0.0000 | 0.0000 | 0.0313 | 0.5476 |
| 20 | 0.0000 | 0.0000 | 0.0000 | 0.0000 | 0.0000 | 0.0000 | 0.0000 | 0.0244 |
| 21 | 6.6659 | 0.5535 | 0.5448 | 0.4756 | 0.5758 | 1.1999 | 2.7012 | 4.7474 |
| 22 | 0.0000 | 0.0000 | 0.0000 | 0.0000 | 0.0000 | 0.0080 | 0.0613 | 0.8362 |

Table 6.5  continued

|  | 17 | 18 | 19 | 20 | 21 | 22 | 23 | 24 |
|---|---|---|---|---|---|---|---|---|
| 23 | 0.0000 | 0.0000 | 0.0000 | 0.0000 | 0.0000 | 0.0000 | 0.0249 | 0.0486 |
| 24 | 0.0000 | 0.0000 | 0.0000 | 0.0000 | 0.0000 | 0.0000 | 0.0000 | 0.0000 |
| 25 | 0.0000 | 0.0000 | 0.0000 | 0.0000 | 0.0021 | 0.0022 | 0.0094 | 0.0155 |
| 26 | 0.0000 | 0.0000 | 0.0000 | 0.0000 | 0.0000 | 0.0000 | 0.0032 | 0.0227 |
| 27 | 0.0000 | 0.0000 | 0.0000 | 0.0000 | 0.0000 | 0.0000 | 0.0213 | 0.0242 |
| 28 | 0.0000 | 0.0000 | 0.0000 | 0.0000 | 0.0141 | 0.1529 | 0.2753 | 0.3438 |
| 29 | 0.0000 | 0.0000 | 0.0000 | 0.0000 | 0.0000 | 0.0000 | 0.0151 | 0.0266 |
| 30 | 3.7486 | 2.7767 | 2.6240 | 2.4435 | 2.4526 | 2.5820 | 2.8052 | 2.6658 |
| 31 | 1.2480 | 1.2216 | 1.0986 | 0.8349 | 1.4324 | 1.3464 | 1.6810 | 2.2239 |
| 32 | 5.9996 | 5.7387 | 3.6845 | 4.1019 | 3.5188 | 3.4129 | 3.9174 | 4.5770 |
| 33 | 2.3157 | 2.3554 | 2.3951 | 1.6342 | 2.4954 | 2.9892 | 3.8343 | 4.1874 |
| 34 | 0.0000 | 0.0000 | 0.0000 | 0.0000 | 0.3608 | 0.3995 | 0.8083 | 1.7865 |
| 35 | 0.0000 | 0.0000 | 0.0000 | 0.0000 | 0.0130 | 0.0226 | 0.0916 | 0.4291 |
| 36 | 0.0852 | 0.1420 | 0.1468 | 0.0852 | 0.0483 | 0.1164 | 0.3063 | 0.6877 |
| 37 | 0.0000 | 0.0228 | 0.0228 | 0.0683 | 0.0116 | 0.0995 | 0.3288 | 1.3488 |
| 38 | 0.2260 | 0.5381 | 0.2691 | 0.1291 | 0.2423 | 0.2174 | 0.4749 | 0.8371 |
| 39 | 0.2049 | 0.2618 | 0.2390 | 0.5691 | 0.3208 | 0.4309 | 0.7839 | 1.5080 |
| 40 | 0.0000 | 0.0000 | 0.0000 | 0.0000 | 0.2265 | 0.3379 | 0.6944 | 1.6823 |
| 41 | 5.7188 | 5.1412 | 4.7946 | 3.7548 | 3.7030 | 1.1502 | 3.6409 | 4.0533 |
| 42 | 3.3067 | 2.4941 | 2.5648 | 2.3482 | 1.9849 | 1.7772 | 2.0831 | 2.7695 |
| 43 | 3.4027 | 2.4193 | 2.1049 | 2.0319 | 1.2956 | 1.5267 | 2.2422 | 3.3385 |
| 44 | 3.2974 | 2.3046 | 2.0885 | 1.8702 | 1.3768 | 1.5113 | 2.1835 | 2.9485 |
| 45 | 0.0000 | 0.0000 | 0.0000 | 0.0000 | 0.0000 | 0.0000 | 0.0234 | 0.0310 |
| 46 | 0.0000 | 0.0000 | 0.0000 | 0.0000 | 0.0000 | 0.0101 | 0.0124 | 0.2477 |
| 47 | 0.0000 | 0.0000 | 0.0000 | 0.0000 | 0.0000 | 0.0000 | 0.0371 | 0.1675 |
| 48 | 0.0000 | 0.0000 | 0.0000 | 0.0000 | 0.0528 | 0.0848 | 0.2866 | 1.1127 |
| 49 | 0.3307 | 0.4524 | 0.8512 | 2.0653 | 0.0576 | 0.0400 | 6.9131 | 5.6582 |
| 50 | 0.0000 | 0.0000 | 0.0000 | 0.0000 | 0.0000 | 0.0000 | 0.0000 | 0.0199 |
| 51 | 0.0000 | 0.0000 | 0.0000 | 0.0000 | 0.0000 | 0.0000 | 0.0223 | 0.0763 |
| 52 | 0.1515 | 0.0382 | 0.0831 | 0.0286 | 0.0232 | 0.0956 | 0.0234 | 0.0654 |

SEDIMENT EXAMPLE -- RAW DATA.                                   SECTION  3

|  | 17 | 18 | 19 | 20 | 21 | 22 | 23 | 24 |
|---|---|---|---|---|---|---|---|---|
| 1 | 0.0935 | 0.3122 | 3.2593 | 10.4056 | 23.7204 | 23.2600 | 10.7145 | 10.7923 |
| 2 | 0.6123 | 4.4071 | 34.1910 | 32.2811 | 18.8216 | 5.9904 | 0.6581 | 0.1371 |
| 3 | 1.0805 | 5.2519 | 24.5625 | 26.9620 | 22.3021 | 9.0048 | 2.0986 | 0.4019 |
| 4 | 5.0325 | 6.2577 | 11.5540 | 6.7991 | 4.9978 | 2.4512 | 0.1606 | 1.9855 |
| 5 | 15.5856 | 18.4773 | 15.3204 | 2.3923 | 0.8764 | 0.4641 | 0.1221 | 0.9592 |
| 6 | 1.3084 | 0.6132 | 0.4492 | 0.2144 | 0.1987 | 0.1352 | 0.2235 | 0.2689 |
| 7 | 25.3471 | 18.2290 | 13.2546 | 3.1029 | 1.1408 | 0.2861 | 1.2302 | 0.5848 |

| | | | | | | | | |
|---|---|---|---|---|---|---|---|---|
| 8  | 4.2965  | 4.7272  | 9.1623  | 7.3481  | 7.5861  | 4.5343  | 2.4773  | 1.4342  |
| 9  | 7.1905  | 3.5812  | 2.6177  | 0.8934  | 0.6107  | 0.4077  | 1.2949  | 0.3290  |
| 10 | 0.1025  | 0.2049  | 1.6402  | 5.6828  | 17.0075 | 21.5597 | 17.0961 | 8.0565  |
| 11 | 1.6792  | 4.7064  | 16.7890 | 16.3479 | 16.6236 | 12.5669 | 7.4299  | 2.9350  |
| 12 | 17.1943 | 13.8686 | 10.9071 | 2.2665  | 0.3982  | 0.1431  | 0.0109  | 0.0000  |
| 13 | 3.0839  | 12.6698 | 40.3452 | 21.9618 | 10.5740 | 4.3376  | 0.4889  | 0.7508  |
| 14 | 0.0862  | 7.2380  | 16.0372 | 12.4546 | 13.8648 | 10.4670 | 3.0756  | 1.2778  |
| 15 | 4.3030  | 8.4282  | 19.4868 | 15.3656 | 16.0036 | 14.4293 | 6.2311  | 1.8855  |
| 16 | 28.2789 | 19.2370 | 17.5317 | 6.8753  | 3.5593  | 1.4679  | 0.5622  | 0.1874  |
| 17 | 2.9455  | 17.1027 | 52.8279 | 18.9413 | 5.0932  | 0.9412  | 0.1146  | 0.1064  |
| 18 | 1.5431  | 13.0535 | 44.4091 | 22.2790 | 10.0358 | 3.3420  | 0.1369  | 0.4383  |
| 19 | 12.0136 | 34.2219 | 40.6419 | 7.6823  | 1.6814  | 0.4705  | 0.1581  | 0.0316  |
| 20 | 0.0630  | 0.5132  | 11.3879 | 27.2817 | 28.0857 | 13.9901 | 4.0035  | 2.2939  |
| 21 | 8.5576  | 9.3582  | 12.9816 | 5.9678  | 4.2796  | 2.6952  | 3.4652  | 2.5472  |
| 22 | 9.5675  | 34.8947 | 42.7251 | 7.7453  | 1.9375  | 0.4680  | 0.2242  | 0.0075  |
| 23 | 0.2150  | 0.2997  | 0.6357  | 7.0407  | 26.8819 | 26.7835 | 16.7917 | 0.7841  |
| 24 | 0.0167  | 0.0142  | 0.0871  | 0.1549  | 3.9970  | 25.0116 | 23.4723 | 11.4578 |
| 25 | 0.0780  | 0.2104  | 2.8621  | 21.1299 | 37.9531 | 21.2301 | 1.9757  | 1.2712  |
| 26 | 0.0960  | 0.5450  | 4.1513  | 14.4433 | 31.0990 | 25.4000 | 7.6755  | 2.5934  |
| 27 | 0.1706  | 1.0841  | 11.1472 | 20.5044 | 26.4184 | 19.7908 | 3.2245  | 1.8529  |
| 28 | 0.3201  | 0.8241  | 13.9569 | 26.7946 | 26.4790 | 14.2643 | 5.4840  | 1.3034  |
| 29 | 0.2273  | 1.4694  | 17.5138 | 26.4475 | 23.6433 | 12.5270 | 4.4511  | 2.0193  |
| 30 | 3.5141  | 3.8933  | 4.5093  | 2.0928  | 1.2368  | 1.2368  | 0.6634  | 0.8376  |
| 31 | 7.8235  | 20.7665 | 31.2075 | 6.7695  | 3.5664  | 1.7605  | 0.7619  | 0.4191  |
| 32 | 0.4876  | 2.8700  | 3.4037  | 2.1197  | 1.6086  | 1.0767  | 1.1462  | 1.4353  |
| 33 | 5.2783  | 4.3602  | 6.3135  | 4.5884  | 4.8362  | 3.7060  | 3.0986  | 2.8474  |
| 34 | 4.0905  | 7.0582  | 16.0107 | 12.3580 | 11.2443 | 6.5822  | 2.3361  | 4.1454  |
| 35 | 2.5717  | 5.1374  | 23.1334 | 27.8441 | 17.6891 | 9.1699  | 2.4705  | 2.2078  |
| 36 | 1.4430  | 2.2511  | 13.4679 | 25.2218 | 29.4604 | 12.9205 | 3.3518  | 4.4336  |
| 37 | 6.6123  | 10.4408 | 19.6716 | 16.5962 | 14.9673 | 10.7642 | 4.4331  | 2.4016  |
| 38 | 2.3671  | 4.8499  | 11.3571 | 8.7023  | 9.1924  | 6.6640  | 4.5861  | 3.2316  |
| 39 | 4.1631  | 7.5483  | 16.8016 | 12.1267 | 11.1557 | 6.9463  | 4.7016  | 2.0381  |
| 40 | 5.3542  | 11.8002 | 27.0964 | 16.3866 | 14.0884 | 10.7031 | 2.0464  | 1.4183  |
| 41 | 4.6293  | 3.1702  | 2.9055  | 1.0022  | 0.5797  | 0.2609  | 0.1800  | 0.2734  |
| 42 | 4.1914  | 3.5680  | 3.8822  | 1.5631  | 0.9924  | 0.4864  | 0.2611  | 0.2550  |
| 43 | 5.0740  | 4.2531  | 4.3910  | 1.5789  | 0.8569  | 0.3655  | 0.2215  | 0.2448  |
| 44 | 4.4224  | 3.7134  | 4.0218  | 1.7076  | 1.0993  | 0.5224  | 0.3844  | 0.2914  |
| 45 | 0.6401  | 3.5036  | 18.6974 | 22.1436 | 25.0927 | 14.0407 | 4.2008  | 0.4778  |
| 46 | 2.9988  | 8.8458  | 28.4089 | 24.5429 | 19.6125 | 7.7989  | 1.0722  | 0.1340  |
| 47 | 1.1824  | 4.9633  | 26.3173 | 27.8109 | 23.1208 | 8.4437  | 0.4902  | 0.7542  |
| 48 | 4.2780  | 6.7507  | 11.2304 | 7.3280  | 11.8125 | 21.0334 | 13.1892 | 4.7167  |
| 49 | 1.9849  | 1.9165  | 2.9432  | 0.4563  | 2.1903  | 0.0143  | 0.6229  | 3.1681  |
| 50 | 0.3071  | 3.9501  | 25.8992 | 27.1083 | 24.5331 | 14.0193 | 2.1915  | 0.1056  |

Table 6.5  *continued*

| | | | | | | |
|---|---|---|---|---|---|---|
| 51 | 2.0532 | 13.0602 | 40.0363 | 23.4372 | 13.2556 | 4.9863 | 0.2447 | 0.1631 |
| 52 | 1.5991 | 10.0699 | 37.3258 | 26.2649 | 14.6502 | 4.9534 | 0.1227 | 0.1091 |

SEDIMENT EXAMPLE -- RAW DATA.                                              SECTION 4

| | 25 | 26 | 27 | 28 | 29 | 30 | 31 | 32 |
|---|---|---|---|---|---|---|---|---|
| 1  | 2.4646 | 5.1108 | 1.7901 | 0.8561 | 0.1038 | 0.4151 | 0.7783 | 0.3632 |
| 2  | 0.3154 | 0.1234 | 0.2605 | 0.0137 | 0.1920 | 0.0548 | 0.0411 | 0.0274 |
| 3  | 0.2233 | 0.3795 | 0.4688 | 0.3572 | 0.1563 | 0.3349 | 0.0447 | 0.6698 |
| 4  | 1.3285 | 0.8760 | 0.7446 | 0.7446 | 0.8614 | 0.9052 | 0.5840 | 0.9636 |
| 5  | 0.0698 | 0.5407 | 0.4011 | 0.4535 | 0.6802 | 0.3314 | 0.0872 | 0.3663 |
| 6  | 0.1327 | 0.2724 | 0.2305 | 0.3004 | 0.2445 | 0.1502 | 0.2410 | 0.0384 |
| 7  | 0.4235 | 0.4235 | 0.3428 | 0.9277 | 0.1008 | 0.9478 | 0.1613 | 0.4235 |
| 8  | 1.0431 | 0.4172 | 0.4172 | 0.3651 | 0.3390 | 0.4303 | 0.3781 | 0.0522 |
| 9  | 0.3503 | 0.7217 | 0.3927 | 0.2441 | 0.6050 | 0.2441 | 0.1910 | 0.1486 |
| 10 | 6.2187 | 3.1414 | 2.9491 | 2.1797 | 2.3934 | 0.8975 | 1.1754 | 0.6625 |
| 11 | 3.0787 | 1.3752 | 1.8267 | 0.9852 | 0.4926 | 0.4310 | 0.7594 | 0.5952 |
| 12 | 0.0109 | 0.0054 | 0.0054 | 0.0054 | 0.0054 | 0.0163 | 0.0272 | 0.0272 |
| 13 | 0.2794 | 0.3143 | 0.0349 | 0.1746 | 0.2095 | 0.1222 | 0.3667 | 0.1222 |
| 14 | 0.7726 | 0.5943 | 0.4755 | 0.4903 | 0.6835 | 0.2526 | 0.2526 | 0.1040 |
| 15 | 2.4063 | 0.7901 | 0.5028 | 1.0595 | 0.3591 | 0.1077 | 0.1975 | 0.4848 |
| 16 | 0.2915 | 0.0416 | 0.1666 | 0.4165 | 0.2499 | 0.0625 | 0.1666 | 0.1249 |
| 17 | 0.1637 | 0.0655 | 0.0082 | 0.0082 | 0.0900 | 0.0409 | 0.0082 | 0.0082 |
| 18 | 0.0411 | 0.3150 | 0.0959 | 0.1781 | 0.0274 | 0.0137 | 0.0137 | 0.0137 |
| 19 | 0.2109 | 0.0527 | 0.0211 | 0.1792 | 0.0105 | 0.0633 | 0.0843 | 0.0527 |
| 20 | 2.0558 | 1.3633 | 0.0649 | 1.1253 | 0.7141 | 0.2813 | 0.1298 | 0.5843 |
| 21 | 8.4449 | 8.0318 | 5.6682 | 3.5799 | 0.9638 | 0.3672 | 1.1474 | 0.5967 |
| 22 | 0.0897 | 0.0523 | 0.0374 | 0.0224 | 0.1495 | 0.3662 | 0.0299 | 0.3887 |
| 23 | 5.2706 | 2.9402 | 2.4828 | 1.3721 | 0.3920 | 0.8276 | 0.7623 | 0.3267 |
| 24 | 7.7700 | 5.3810 | 3.2472 | 2.3426 | 3.0848 | 0.5798 | 1.1829 | 1.4844 |
| 25 | 0.0000 | 0.8886 | 1.0714 | 0.6174 | 0.3269 | 0.2542 | 0.2361 | 0.7627 |
| 26 | 1.7150 | 1.5686 | 1.0457 | 0.9830 | 0.2928 | 0.6902 | 0.7529 | 0.4183 |
| 27 | 2.0213 | 1.5401 | 1.8048 | 0.9385 | 0.4813 | 0.5535 | 0.7460 | 0.6738 |
| 28 | 1.4755 | 1.4017 | 0.7132 | 0.2951 | 0.8853 | 0.0000 | 0.0246 | 0.0000 |
| 29 | 1.4765 | 1.3896 | 0.4560 | 0.8685 | 0.0856 | 0.1954 | 0.4125 | 0.3691 |
| 30 | 1.5093 | 0.7132 | 0.5059 | 0.3317 | 0.4976 | 0.4395 | 0.0746 | 0.0746 |
| 31 | 0.6476 | 0.5143 | 0.2667 | 0.5714 | 0.3810 | 0.4952 | 0.5714 | 0.1143 |
| 32 | 0.3489 | 0.6977 | 0.9569 | 0.3289 | 0.6180 | 0.3887 | 0.2292 | 0.4784 |
| 33 | 0.8793 | 4.7317 | 2.4705 | 2.1355 | 3.0567 | 2.1983 | 1.9890 | 2.0308 |
| 34 | 2.1529 | 2.9544 | 1.8322 | 1.7635 | 2.6109 | 1.8780 | 1.7406 | 2.0154 |
| 35 | 1.2667 | 0.7239 | 0.0905 | 0.0905 | 0.3438 | 0.1629 | 0.0000 | 0.8143 |

| | | | | | | | |
|---|---|---|---|---|---|---|---|
| 36 | 0.8077 | 0.5654 | 0.4038 | 0.2019 | 0.5250 | 0.1615 | 0.0606 | 0.6057 |
| 37 | 1.0477 | 0.7092 | 0.1612 | 0.6770 | 0.1128 | 0.9348 | 0.8220 | 0.2095 |
| 38 | 4.2059 | 2.8514 | 3.2316 | 2.1623 | 2.2336 | 2.5425 | 1.7821 | 1.7584 |
| 39 | 2.3856 | 4.3542 | 1.5054 | 1.9033 | 1.5749 | 1.2044 | 1.7834 | 1.5054 |
| 40 | 0.2837 | 0.9725 | 0.2634 | 0.5673 | 0.0000 | 0.4457 | 0.2431 | 0.1621 |
| 41 | 0.1800 | 0.2067 | 0.2400 | 0.1400 | 0.2200 | 0.1467 | 0.2267 | 0.1933 |
| 42 | 0.2611 | 0.2732 | 0.1154 | 0.2004 | 0.2186 | 0.0243 | 0.2247 | 0.0911 |
| 43 | 0.2682 | 0.0525 | 0.1749 | 0.1224 | 0.2798 | 0.0117 | 0.1691 | 0.1283 |
| 44 | 0.3162 | 0.2480 | 0.3162 | 0.0744 | 0.2046 | 0.2604 | 0.2542 | 0.2294 |
| 45 | 1.1547 | 0.8959 | 1.0353 | 0.3982 | 0.6172 | 0.5973 | 0.4579 | 0.8561 |
| 46 | 1.2565 | 0.3351 | 0.2513 | 0.3686 | 0.1843 | 0.1340 | 0.3518 | 0.3183 |
| 47 | 0.4902 | 0.1508 | 0.2074 | 0.4148 | 0.3583 | 0.3960 | 0.2640 | 0.4337 |
| 48 | 2.8606 | 2.6640 | 0.7861 | 1.2447 | 0.3712 | 1.0263 | 0.3275 | 0.4149 |
| 49 | 6.3184 | 14.9791 | 9.9798 | 10.3182 | 9.5942 | 1.2320 | 1.7568 | 3.1942 |
| 50 | 0.1672 | 0.0176 | 0.0176 | 0.0440 | 0.0528 | 0.0528 | 0.0088 | 0.0440 |
| 51 | 0.0544 | 0.2855 | 0.0272 | 0.0952 | 0.0816 | 0.0136 | 0.2719 | 0.0136 |
| 52 | 0.3136 | 0.0545 | 0.1909 | 0.0409 | 0.0545 | 0.0682 | 0.1909 | 0.0682 |

Table 6.5 *concluded*

```
SEDIMENT EXAMPLE -- RAW DATA.                          SECTION   5
                 33
         1        5.5259
         2        1.7824
         3        5.2242
         4        6.1755
         5        3.9939
         6        1.4913
         7        5.6264
         8        3.5334
         9        3.6193
        10        8.8472
        11       10.8781
        12        0.6745
        13        3.8411
        14        3.7888
        15        4.3636
        16        4.0815
        17        1.4161
        18        3.0272
        19        1.8450
        20        6.0377
        21        7.8253
        22        0.3887
        23        6.1199
        24       10.7156
        25        5.1028
        26        6.5043
        27        7.0025
        28        4.9922
        29        5.4064
        30        3.3668
        31        4.8763
        32        4.1862
        33       16.7492
        34       15.8716
        35        7.7271
        36        5.4719
        37        7.5915
        38       24.7124
        39       14.8924
        40        5.2274
        41        2.3069
        42        2.0948
        43        1.5215
        44        1.8600
        45        5.1365
        46        3.1160
        47        3.9974
        48        8.4288
        49       12.4344
        50        1.4610
        51        1.8217
        52        2.4680
```

Table 6.10. Cluster No. 1 is the most compact, with an average point-to-centre distance of 4.993. Cluster No. 3, with approximately the same membership, is nearly twice as diffuse (average point-to-centre distance of 8.457). This is due mainly to the fact that samples 44, 70 and 71 are relatively distant from the cluster centre. Cluster No. 2 is the largest, with a membership of 42 out of the 74 samples. For its size it is relatively compact (average point-to-centre distance is 5.410) even though some samples are a long way from the centre (e.g. 35 and 36). The coordinates of the cluster centres indicate that clusters 1 and 3 have similar mean values on $X_1$ but differ considerably in terms of the remaining three variables. Clusters 1 and 2 are similar with respect to $X_4$ but again differ in terms of the other variables. Clusters 2 and 3 have nearly equal mean values of $X_2$ but again differ on the other variables.

The nonlinear mapping algorithm was used to project the configuration of 74 points (representing samples) in a four-dimensional space onto two dimensions. The result is shown in Figure 6.9. Table 6.11 shows that the initial mapping error, 0.03131, was quickly reduced until by iteration 10 the mapping error was 0.0005213. Eleven iterations later the mapping error was 0.0001644. This shows that a minimum of the error function has been located, though it could be a local rather than a global minimum. Visual inspection of Figure 6.9 indicates that the groupings picked out by eye differ quite significantly from those produced by the nonhierarchical cluster analysis. Samples 70, 71 and 46 are seen to be distant from other points and so should logically remain unclassified. The reason for their lack of similarity with other samples may be experimental error, contamination or the fact that they are members of some other type of sediment not represented elsewhere in the 74 samples. One should, however, beware of purely visual analyses. The eye is capable of seeing patterns where none exist, or of selecting only what the observer subconsciously expects to see. (This is not the place to discuss the psychology of perception; the interested reader should refer to one of the many books on this subject, for example, Vernon (1962) and Hanson (1958).) Furthermore, the two-dimensional representation given by ordination methods such as nonlinear mapping are only approximations to the relationships between the points in a higher-dimensional space. Mapping error is a measure of average distortion. It does not tell us how the distortion is distributed among the points. Generally, if mapping error is near zero, one can obviously be confident that the maximum distortion at any point will be low.

### 6.3.2 Nonmetric Multidimensional Scaling

The nonmetric multidimensional scaling program was firstly used in an attempt to find the spatial relationships among five points, given only

Table 6.6
Sediment example — principal component scores

PRINCIPAL COMPONENT SCORES, SEDIMENT EXAMPLE

| | 1 | 2 | 3 | 4 | 5 | 6 | SECTION 7 | 8 |
|---|---|---|---|---|---|---|---|---|
| 1 | -0.9331 | 0.5044 | 1.2630 | -1.4460 | -1.2370 | -0.4483 | -0.0506 | 0.4314 |
| 2 | -0.4943 | -1.1300 | 0.0078 | 0.2863 | -1.1163 | -0.5569 | 0.0152 | 0.1389 |
| 3 | -0.5628 | -0.7434 | 0.0896 | 0.1069 | 1.0513 | 0.0088 | 0.1129 | -0.3730 |
| 4 | 0.3525 | 0.2744 | -0.0223 | 0.9453 | -0.2582 | 1.2856 | 0.5294 | -0.4554 |
| 5 | 1.3047 | 0.0816 | -1.6442 | -1.2448 | -0.2322 | -0.0721 | -0.0750 | -0.0075 |
| 6 | 2.1336 | 0.1184 | 1.8254 | 0.8917 | 0.3557 | 0.0162 | -1.7395 | 0.4763 |
| 7 | 0.6305 | 0.0820 | -2.5459 | -0.9083 | -1.5383 | 0.3737 | 0.1359 | -2.6953 |
| 8 | 1.3058 | 0.0962 | 0.8085 | -0.2978 | 0.3579 | 0.1156 | 0.1591 | 1.3581 |
| 9 | 2.5814 | 0.7225 | -0.6853 | -2.8447 | 1.7919 | -0.4597 | -0.0541 | 0.5492 |
| 10 | -1.0514 | 1.2751 | 1.1267 | -1.1225 | -1.5550 | -0.2794 | -0.1868 | 0.7968 |
| 11 | -0.7415 | 0.1475 | 0.3185 | -0.2491 | -0.1939 | 0.2867 | -0.0526 | 0.2308 |
| 12 | 1.9777 | 0.0209 | -1.6985 | -2.4570 | 0.3887 | -0.5802 | -0.1312 | -0.2155 |
| 13 | -0.4205 | -0.9576 | -0.5228 | 0.5975 | 0.2480 | -0.0206 | -0.1748 | 0.8977 |
| 14 | 0.1628 | 0.3913 | 0.5676 | 0.3340 | -0.2906 | -0.2145 | -0.0130 | 0.0271 |
| 15 | -0.4681 | 0.3476 | 0.0757 | -0.4748 | -0.3483 | 0.5531 | -0.0758 | 0.0924 |
| 16 | -0.3166 | 0.4945 | -2.3176 | 0.4696 | -1.9024 | -0.2996 | 0.1183 | -2.3844 |
| 17 | -0.3106 | -1.2450 | -0.8704 | 0.9332 | -0.0266 | -0.3592 | -0.3500 | 1.6513 |
| 18 | -0.3957 | -1.1376 | -0.5008 | 0.6345 | 0.2882 | 0.3564 | -0.2256 | 0.1427 |
| 19 | -0.1481 | -1.1353 | -1.6309 | -0.9369 | -1.4908 | -0.0985 | 0.4879 | 1.5458 |
| 20 | -0.7320 | -0.3682 | 0.6569 | -0.4201 | 0.8848 | 0.3857 | 0.1739 | -0.7380 |
| 21 | -0.2831 | 1.4713 | -0.8190 | 0.3938 | -1.0292 | -1.9729 | -0.2521 | 0.5990 |
| 22 | -0.1711 | -0.0856 | -1.6341 | 1.0464 | -1.2910 | 0.0869 | -0.4810 | 1.6501 |
| 23 | -0.9231 | 0.5150 | 0.1889 | -1.2974 | -0.8558 | -0.4634 | -0.0191 | -0.4425 |
| 24 | -1.1845 | 1.9245 | 1.2912 | -1.3462 | -2.6355 | -0.6420 | 0.3459 | 1.8111 |
| 25 | -0.7591 | -0.3543 | 0.9177 | -0.7843 | 1.0666 | -0.3374 | 0.3610 | -1.5556 |
| 26 | -0.8302 | 0.0155 | 1.0053 | -1.0055 | 0.1523 | 0.2116 | 0.1863 | -0.9243 |
| 27 | -0.7881 | -0.0291 | 0.6297 | -0.3591 | 0.7173 | 0.0731 | 0.1669 | -0.6838 |
| 28 | -0.6420 | 0.5932 | 0.6273 | -0.5516 | 0.7067 | -0.8386 | 0.1333 | -0.6392 |
| 29 | -0.6958 | -0.3927 | 0.4884 | -0.2131 | 0.7827 | 0.3958 | 0.1036 | -0.2940 |
| 30 | 1.4909 | 0.1185 | 0.1598 | 1.1298 | -0.7152 | -0.2550 | -2.6062 | -1.5531 |
| 31 | 0.2957 | -0.3954 | -0.9781 | 0.4113 | 0.5482 | 0.4714 | 0.1889 | -0.1704 |
| 32 | 1.9757 | 0.5202 | 0.6051 | -0.5072 | 0.8167 | 0.3507 | 0.6471 | 1.5755 |
| 33 | 0.2314 | 1.9982 | -0.7574 | 0.0811 | 1.1709 | 2.6177 | 0.2354 | -0.1003 |
| 34 | -0.7720 | 1.2439 | -0.5450 | 0.7102 | 0.4804 | 2.3967 | 0.0624 | -0.1355 |
| 35 | -0.5795 | -0.6157 | 0.0399 | 0.0747 | 0.7486 | 0.0827 | 0.0839 | -0.2383 |

|    |         |         |         |         |         |         |         |         |
|----|---------|---------|---------|---------|---------|---------|---------|---------|
| 36 | -0.5972 | -0.5812 |  0.4564 | -0.4517 |  0.9419 | -0.2603 |  0.2263 | -0.9643 |
| 37 | -0.4725 | -0.2541 | -0.1968 | -0.0611 | -0.3837 |  1.0156 |  0.0138 | -0.1029 |
| 38 | -0.8622 |  1.7544 | -0.3021 |  0.7133 |  0.3914 |  3.3244 | -0.0304 |  0.0878 |
| 39 | -0.6448 |  0.9176 |  0.4245 |  0.4160 |  0.2911 |  1.7539 | -0.0033 |  0.1003 |
| 40 | -0.4029 | -0.6271 | -0.4150 | -0.0266 |  0.1480 |  0.1265 | -0.0612 |  0.0618 |
| 41 |  2.2541 |  0.2501 |  0.9602 |  0.0070 |  0.5094 |  0.2912 |  0.7497 |  1.6068 |
| 42 |  1.5125 | -0.1421 |  1.3088 |  1.6287 | -0.9310 | -0.2218 | -2.3898 | -1.8745 |
| 43 |  1.3991 | -0.2122 |  0.9948 |  1.6207 | -1.5063 | -0.6121 |  5.8056 | -1.6494 |
| 44 |  1.4957 | -0.0876 |  1.3716 |  1.9699 | -1.1489 |  0.2768 | -0.0072 | -1.1089 |
| 45 | -0.6885 | -0.3455 | -0.3431 | -0.0937 |  0.8667 |  0.1844 |  0.1431 | -0.5138 |
| 46 | -0.5112 | -0.8276 | -0.1403 |  0.2452 |  0.5813 | -0.2704 | -0.0115 |  0.0536 |
| 47 | -0.5695 | -0.7803 |  0.0686 |  0.1829 |  1.0830 |  0.0284 |  0.0975 | -0.2964 |
| 48 | -0.7233 |  0.3088 | -0.4433 | -0.9880 | -1.4832 |  0.3657 | -0.1507 |  0.1733 |
| 49 | -0.7145 |  4.2023 | -1.6311 |  2.4843 |  1.9546 | -3.5997 |  0.0408 | -0.3127 |
| 50 | -0.5445 | -1.0802 |  0.3507 | -0.2124 |  0.9021 | -0.5737 |  0.0999 | -0.3522 |
| 51 | -0.4099 | -1.1248 | -0.4418 |  0.5634 |  0.4148 | -0.3297 | -0.1585 |  0.8416 |
| 52 | -0.3929 | -1.0836 | -0.2665 |  0.4878 |  0.6879 | -0.2872 | -0.0787 |  0.6645 |

PRINCIPAL COMPONENT SCORES, SEDIMENT EXAMPLE           SECTION 2

|    | 9       | 10      | 11      | 12      | 13      |
|----|---------|---------|---------|---------|---------|
| 1  |  0.2695 |  1.0510 | -1.9476 | -0.2694 |  2.1644 |
| 2  | -0.0044 |  0.2703 |  0.2738 |  0.6580 |  0.9664 |
| 3  |  0.0166 |  0.1493 |  0.1318 | -0.0507 |  0.1466 |
| 4  |  4.3721 |  3.7530 |  1.8866 |  0.5213 | -0.5744 |
| 5  | -0.0297 |  0.2846 |  0.0489 |  0.3251 |  0.2935 |
| 6  | -1.0084 |  1.0596 |  0.7079 | -3.2620 | -0.4185 |
| 7  |  0.4162 | -0.6502 | -0.3645 | -1.1530 |  0.7546 |
| 8  | -0.4806 | -2.5184 | -0.7116 | -0.9684 |  0.7287 |
| 9  | -0.9282 |  1.2608 |  0.2146 |  2.1337 | -1.2199 |
| 10 | -0.0555 |  0.3836 | -0.0981 |  0.2435 | -0.0780 |
| 11 |  0.1987 | -0.1119 |  0.8671 |  0.5364 |  0.4082 |
| 12 | -0.3427 |  1.2450 |  0.1324 |  1.2354 | -0.7233 |
| 13 |  0.2377 |  0.3713 | -0.0316 |  0.4871 |  0.4962 |
| 14 |  0.0066 | -1.9418 | -1.6358 |  0.4001 | -0.3878 |
| 15 |  0.0757 |  0.1458 |  0.1101 | -0.2152 |  0.5077 |
| 16 |  0.6262 | -1.0578 | -0.3353 | -1.6030 |  1.4063 |
| 17 | -0.2845 |  0.6383 | -0.0470 |  0.6019 |  0.0219 |
| 18 | -0.2196 |  0.5310 |  0.0504 |  0.5654 |  0.3166 |
| 19 | -0.2230 | -0.0510 | -0.3314 | -0.9373 | -1.5502 |
| 20 |  0.1794 |  0.1186 | -0.1337 | -0.1050 |  0.7205 |
| 21 |  0.3534 | -1.8091 |  4.2944 |  1.6015 |  1.9189 |
| 22 | -0.2304 |  0.1944 | -0.7015 | -1.0013 | -2.0515 |

Table 6.6  continued

| | | | | |
|---|---|---|---|---|
| 23 | 0.0148 | -1.3849 | 2.3288 | -0.1182 | -3.5192 |
| 24 | 0.0110 | 1.5567 | -1.3348 | 0.3740 | 0.8675 |
| 25 | 0.3368 | -0.4108 | -0.6264 | -1.1242 | -1.1516 |
| 26 | 0.0702 | -0.5771 | 0.0547 | -0.6389 | -1.5045 |
| 27 | 0.0386 | -0.4453 | 0.5357 | -0.1053 | -0.4321 |
| 28 | 0.1607 | -0.0911 | 0.2112 | 0.1489 | 0.3639 |
| 29 | 0.0446 | 0.1792 | -0.1854 | 0.1699 | 0.6283 |
| 30 | -2.3432 | 0.2999 | 0.4969 | 0.5371 | 0.7429 |
| 31 | -0.3076 | -0.7567 | -0.3016 | 0.3374 | -0.9268 |
| 32 | 1.3117 | -0.7969 | 0.1700 | -1.8393 | 1.0039 |
| 33 | -0.8262 | -0.0850 | -0.9195 | 0.1770 | -0.0803 |
| 34 | -0.5322 | 0.3482 | -0.9137 | 0.0593 | 0.6028 |
| 35 | 0.0498 | 0.5426 | -0.3688 | 0.1921 | 1.7813 |
| 36 | 0.1988 | 0.0606 | -0.3402 | -0.3336 | 0.3232 |
| 37 | -0.0927 | -0.7257 | 0.2317 | 0.1508 | 0.5181 |
| 38 | -1.0195 | -0.5862 | 1.3825 | 0.5150 | 0.0235 |
| 39 | -0.6058 | -0.0681 | 0.4077 | 0.6412 | 0.0436 |
| 40 | -0.0494 | -0.0273 | -0.2716 | -0.2339 | -0.1239 |
| 41 | 2.1666 | -1.7872 | 0.3760 | -0.9425 | 0.9267 |
| 42 | 0.7984 | 1.2267 | 0.3420 | 0.3661 | 0.1492 |
| 43 | -2.6172 | 1.5235 | 0.4658 | 0.0589 | -0.1133 |
| 44 | 2.7003 | -1.5235 | -2.2429 | 3.7932 | -1.1232 |
| 45 | 0.0005 | -0.1739 | 0.1336 | -0.1886 | -0.8887 |
| 46 | -0.0203 | 0.1033 | 0.5238 | 0.2084 | 0.0817 |
| 47 | -0.0093 | 0.0998 | -0.0446 | 0.1123 | 0.4583 |
| 48 | -0.0618 | 0.0642 | -0.4314 | -0.5377 | -0.8844 |
| 49 | 0.3195 | 0.1873 | -1.4790 | -0.9563 | -0.9280 |
| 50 | 0.1149 | -0.0054 | 0.0013 | -0.0076 | -0.2710 |
| 51 | -0.1689 | 0.2905 | 0.0379 | 0.4208 | 0.0921 |
| 52 | -0.1588 | 0.1599 | 0.2720 | 0.3639 | 0.5086 |

Table 6.7
Sediment example — eigenvalues and percentage of variance

|  | 1 | 2 | 3 | 4 | 5 | 6 | 7 |
|---|---|---|---|---|---|---|---|
| Eigenvalue | 12.18 | 7.57 | 3.62 | 2.02 | 1.45 | 1.33 | 1.08 |
| % Variance | 36.90 | 22.95 | 10.96 | 6.12 | 4.39 | 4.30 | 3.26 |

|  | 8 | 9 | 10 | 11 | 12 | 13 |
|---|---|---|---|---|---|---|
| Eigenvalue | 0.83 | 0.68 | 0.44 | 0.33 | 0.31 | 0.28 |
| % Variance | 2.50 | 2.07 | 1.32 | 1.00 | 0.93 | 0.84 |

Table 6.8
Comparison of results of hierarchical clustering algorithms applied to data of Table 6.6

| Method |  | Cophenetic correlation | Backward links |
|---|---|---|---|
| NN |  | 0.7770 | No |
| FN |  | 0.7215 | No |
| Median |  | 0.8470 | Yes |
| Ward's error |  | 0.5600 | No |
| Centroid |  | 0.8651 | Yes |
| Group average |  | 0.8225 | No |
| Flexible: | $\beta = 0.5$ | 0.7531 | No |
|  | $\beta = 0.25$ | 0.7876 | No |
|  | $\beta = 0.1$ | 0.8122 | No |
|  | $\beta = 0.0$ | 0.8121 | No |
|  | $\beta = -0.1$ | 0.7973 | No |
|  | $\beta = -0.25$ | 0.6807 | No |
|  | $\beta = -0.5$ | 0.0912 | No |

the matrix of dissimilarities, in order to illustrate the power of the method. In this artificial example four of the five points were chosen to lie at the corners of a square, while the fifth point lay at the centre. Given that the square has sides of unit length, the distance between each pair from the five points can be measured and used as a dissimilarity measure. The starting coordinates for the multi-dimensional scaling algorithm were chosen randomly from the range $(-2, +2)$; the resulting points are shown in Figure 6.10. Initially, stress was more than 35 per cent but reduced very rapidly so that by iteration 15 it had stabilized at a value not much greater than zero. The final configuration produced by the scaling algorithm is shown in Figure 6.11. The original square is almost perfectly reproduced, even

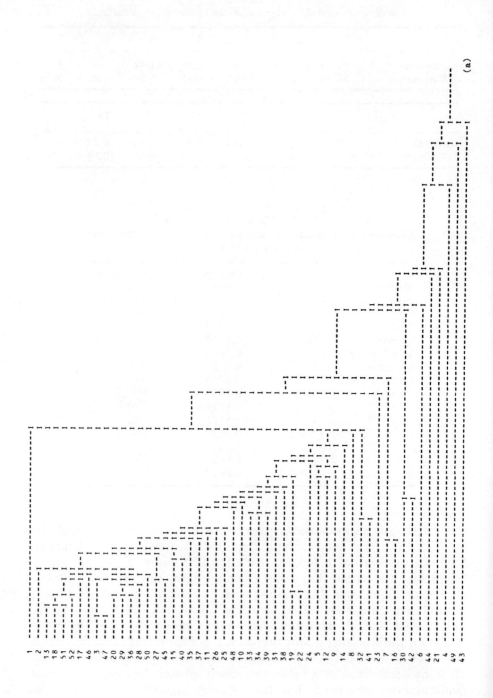

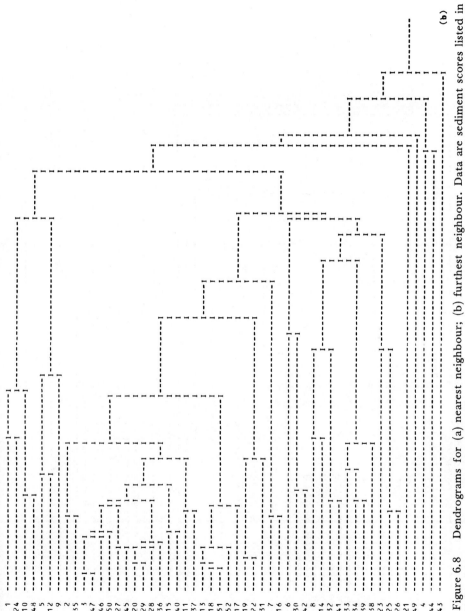

Figure 6.8 Dendrograms for (a) nearest neighbour; (b) furthest neighbour. Data are sediment scores listed in Table 6.5

(c)

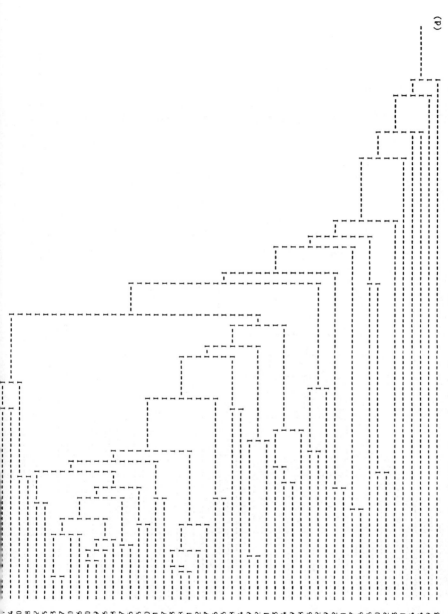

Figure 6.8 *continued* (c) Ward's error sum of squares; (d) group average. Data are sediment scores listed in Table 6.5

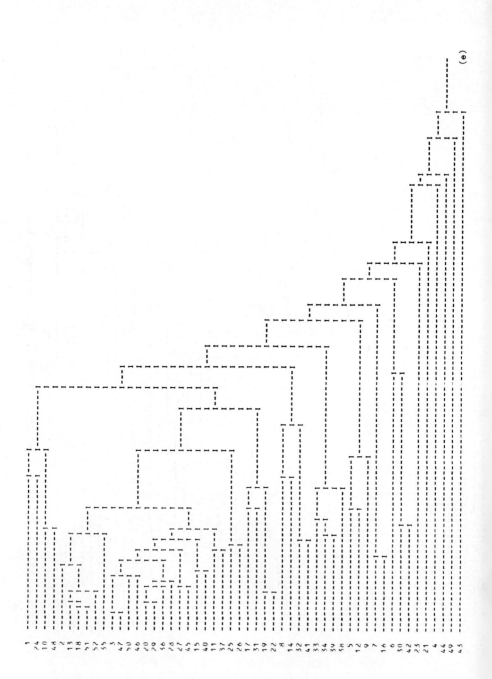

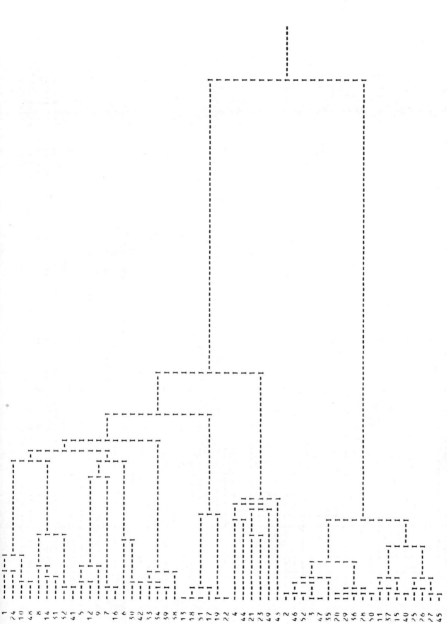

Figure 6.8 *continued* (e) simple average; (f) flexible strategy with $\beta = -0.5$. Data are sediment scores listed in Table 6.5

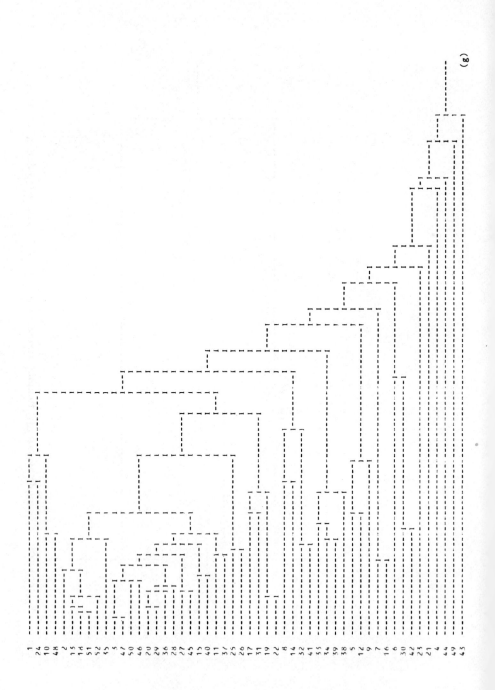

(g)

Figure 6.8 *continued* (g) flexible strategy with $\beta = 0$; (h) flexible strategy with $\beta = 0.5$. Data are sediment scores listed in Table 6.5

## Table 6.9
Tien data — mineralogy of Kansan boulder clays (Reproduced with permission from P. L. Tien, *Clays and Clay Minerals*, **16**, 99−107 (1968))

|    | X1      | X2      | X3      | X4     |
|----|---------|---------|---------|--------|
| 1  | 60.0000 | 13.0000 | 27.0000 | 0.4800 |
| 2  | 55.0000 | 19.0000 | 26.0000 | 0.7500 |
| 3  | 69.0000 | 20.0000 | 11.0000 | 1.8200 |
| 4  | 73.0000 | 17.0000 | 10.0000 | 1.7000 |
| 5  | 72.0000 | 19.0000 | 9.0000  | 2.1100 |
| 6  | 74.0000 | 18.0000 | 8.0000  | 2.2500 |
| 7  | 63.0000 | 25.0000 | 12.0000 | 2.0800 |
| 8  | 68.0000 | 22.0000 | 10.0000 | 2.2000 |
| 9  | 68.0000 | 20.0000 | 12.0000 | 1.6700 |
| 10 | 77.0000 | 13.0000 | 10.0000 | 1.3000 |
| 11 | 72.0000 | 17.0000 | 11.0000 | 1.5500 |
| 12 | 71.0000 | 17.0000 | 12.0000 | 1.4200 |
| 13 | 75.0000 | 14.0000 | 11.0000 | 1.2800 |
| 14 | 76.0000 | 14.0000 | 10.0000 | 1.4000 |
| 15 | 73.0000 | 16.0000 | 11.0000 | 1.4500 |
| 16 | 67.0000 | 21.0000 | 12.0000 | 1.7500 |
| 17 | 70.0000 | 22.0000 | 8.0000  | 2.7500 |
| 18 | 61.0000 | 29.0000 | 10.0000 | 2.9000 |
| 19 | 62.0000 | 27.0000 | 11.0000 | 2.4500 |
| 20 | 68.0000 | 20.0000 | 12.0000 | 1.6700 |
| 21 | 59.0000 | 31.0000 | 10.0000 | 3.1000 |
| 22 | 65.0000 | 26.0000 | 9.0000  | 2.8900 |
| 23 | 62.0000 | 29.0000 | 9.0000  | 3.2200 |
| 24 | 67.0000 | 20.0000 | 13.0000 | 1.5400 |
| 25 | 60.0000 | 17.0000 | 23.0000 | 0.7400 |
| 26 | 58.0000 | 24.0000 | 18.0000 | 1.3500 |
| 27 | 77.0000 | 15.0000 | 8.0000  | 1.8800 |
| 28 | 77.0000 | 15.0000 | 8.0000  | 1.8800 |
| 29 | 71.0000 | 21.0000 | 8.0000  | 2.6500 |
| 30 | 80.0000 | 13.0000 | 7.0000  | 1.8600 |
| 31 | 71.0000 | 21.0000 | 8.0000  | 2.6500 |
| 32 | 75.0000 | 19.0000 | 6.0000  | 3.1700 |
| 33 | 76.0000 | 17.0000 | 7.0000  | 2.4300 |
| 34 | 72.0000 | 20.0000 | 8.0000  | 2.5000 |
| 35 | 84.0000 | 12.0000 | 4.0000  | 3.0000 |
| 36 | 82.0000 | 13.0000 | 5.0000  | 2.6000 |
| 37 | 78.0000 | 18.0000 | 4.0000  | 4.5000 |
| 38 | 57.0000 | 28.0000 | 15.0000 | 1.8700 |
| 39 | 54.0000 | 26.0000 | 20.0000 | 1.3000 |
| 40 | 72.0000 | 16.0000 | 12.0000 | 1.3300 |
| 41 | 74.0000 | 15.0000 | 11.0000 | 1.3500 |
| 42 | 70.0000 | 23.0000 | 7.0000  | 3.2900 |
| 43 | 68.0000 | 24.0000 | 8.0000  | 3.0000 |
| 44 | 66.0000 | 12.0000 | 22.0000 | 0.5500 |
| 45 | 60.0000 | 14.0000 | 26.0000 | 0.5400 |
| 46 | 49.0000 | 21.0000 | 30.0000 | 0.7000 |
| 47 | 56.0000 | 23.0000 | 21.0000 | 1.1000 |
| 48 | 53.0000 | 29.0000 | 16.0000 | 1.6100 |
| 49 | 73.0000 | 22.0000 | 5.0000  | 4.4000 |
| 50 | 59.0000 | 29.0000 | 12.0000 | 2.4200 |
| 51 | 65.0000 | 22.0000 | 13.0000 | 1.6900 |
| 52 | 56.0000 | 18.0000 | 26.0000 | 0.6900 |
| 53 | 56.0000 | 24.0000 | 20.0000 | 1.2000 |
| 54 | 70.0000 | 21.0000 | 9.0000  | 2.3300 |
| 55 | 70.0000 | 21.0000 | 9.0000  | 2.3300 |
| 56 | 67.0000 | 25.0000 | 8.0000  | 3.1300 |
| 57 | 69.0000 | 9.0000  | 22.0000 | 0.4100 |
| 58 | 64.0000 | 15.0000 | 21.0000 | 0.7100 |
| 59 | 59.0000 | 14.0000 | 27.0000 | 0.5200 |
| 60 | 68.0000 | 23.0000 | 9.0000  | 2.5600 |
| 61 | 70.0000 | 21.0000 | 9.0000  | 2.3400 |
| 62 | 76.0000 | 18.0000 | 6.0000  | 1.1500 |

Table 6.9    continued

|    | X1      | X2      | X3      | X4     |
|----|---------|---------|---------|--------|
| 63 | 73.0000 | 22.0000 | 5.0000  | 4.4000 |
| 64 | 77.0000 | 18.0000 | 5.0000  | 3.6000 |
| 65 | 78.0000 | 17.0000 | 5.0000  | 3.4000 |
| 66 | 72.0000 | 15.0000 | 13.0000 | 1.1500 |
| 67 | 80.0000 | 14.0000 | 6.0000  | 2.3300 |
| 68 | 61.0000 | 15.0000 | 24.0000 | 0.6300 |
| 69 | 61.0000 | 15.0000 | 24.0000 | 0.6300 |
| 70 | 41.0000 | 28.0000 | 31.0000 | 0.9000 |
| 71 | 41.0000 | 29.0000 | 30.0000 | 0.9700 |
| 72 | 58.0000 | 25.0000 | 17.0000 | 1.4700 |
| 73 | 64.0000 | 28.0000 | 8.0000  | 3.5000 |
| 74 | 59.0000 | 28.0000 | 13.0000 | 2.1500 |

though a random starting configuration was used. However, the two axes cannot be given any physical meaning. A varimax rotation, while not altering interpoint relationships, may produce a more meaningful set of axes. Comparisons of Figures 6.11 and 6.12 shows that the varimax rotation has caused the square to be moved anti-clockwise around its centre point so that the vertical axis could be considered to represent height and the horizontal axis width. In artificial examples such as this one, the true configuraton is known and the results can therefore be checked. Experiments with the MDS algorithm using the same dissimilarity matrix but different sets of random starting coordinates have shown that, while it is in general very successful in recovering the true shape or configuration of points, on a number of occasions the final configuration was badly distorted. This should encourage users to repeat each run of the program several times, starting with different coordinates each time.

The MDS algorithm was applied to the Tien data (Table 6.9 above) using $(1.0 - g_{ij})$ as a dissimilarity measure, where $g_{ij}$ is Gower's general similarity coefficient. Since $g_{ij}$ lies in the range 0 (no similarity) to 1 (perfect similarity) it follows that $(1.0 - g_{ij})$ lies in the range 0 (no dissimilarity) to 1 (complete dissimilarity). The strict lower triangle of the (74 x 74) matrix of dissimilarity coefficients was read from a disc file where it had been preserved from a previous run of the principal coordinates program.

Since the number of variables is four, and since the fourth variable is linearly dependent upon the first three, the maximum dimensionality of the system is three. For the sake of comparison the program was run with two and one dimension specified, and a maximum of 150 iterations. The result is shown in Figure 6.13 which is seen to be similar to the principal coordinates plot (Figure 6.15) and the nonlinear mapping (Figure 6.9). At the 150th iteration, stress was 4.136 per cent. Stress had, in fact, been oscillating between 4.022 per cent and 4.473 per cent from iteration 80 (Figure 6.14). The plot of the final configuration is, perhaps, slightly easier to interpret than the nonlinear

Table 6.10
Nonhierarchical cluster analysis of Tien data. Three cluster centres are shown

```
RESULTS FOR CURRENT ITERATION WITH   3 CLUSTER CENTRES

CLUSTER   SIZE    DIST FROM GRAND MEAN   COORDINATES

   1       37           1.049            0.7357   -0.4718   -0.5650    0.1350

   2       17           1.629           -0.2541    1.1434   -0.5156    1.0078

   3       20           2.179           -1.1451   -0.0990    1.4835   -1.1064


     DISTANCE MATRIX FOR CLUSTER CENTRES

   2    2.0863
   3    3.0682     3.2869


CLUSTER MEMBERSHIP FOR INDIVIDUALS

CLUSTER NUMBER    1

    3     0.7897    4    0.4322    5    0.3215    6    0.2414    8    1.0694    9    0.9764
   10     1.2623   11    0.6413   12    0.8480   13    1.1180   14    1.0299   15    0.7662
```

```
 16   1.1236    20   0.9764    27   1.1709    28   0.6871    29   0.9153
 30   1.2090    31   0.9153    33   1.2206    34   0.5445    35   1.9840
 36   1.5181    37   2.5969    41   0.9382    51   0.9319    54   0.8069
 55   0.8069    61   0.8100    64   1.0979    65   1.6840    66   1.2179
 67   1.1250

AVERAGE POINT - TO - CENTRE DISTANCE    1.030
CLUSTER NUMBER   2

  7   0.9899    17   1.0072    19   0.7316    21   0.6656    22   1.1845
 23   0.7217    38   1.6543    45   0.9544    49   0.5643    50   1.9904
 56   0.4130    60   0.7862    73   1.9994    74   0.7162        1.2204
                18                                              
                42                                              
                63                                              

AVERAGE POINT - TO - CENTRE DISTANCE    0.9872
CLUSTER NUMBER   3

  1   1.4379     2   0.4228    26   0.6192    39   1.2731    44   1.4612
 45   1.2110    25   1.3005    48   0.8131    52   2.1792    53   0.4804
 57   2.5025    46   1.2537    68   1.2342    69   1.0158    70   1.0158
 71   2.7186    47   1.5454                                   
                58                                            
                59                                            
                72                                            
                                                              1.8227
                                                              1.0684
                                                              2.6416

AVERAGE POINT - TO - CENTRE DISTANCE    1.401

R.M.S. DEVIATION FROM CENTRES =    1.274

CLUSTERS MERGED AT THIS ITERATION:    1  AND   2
```

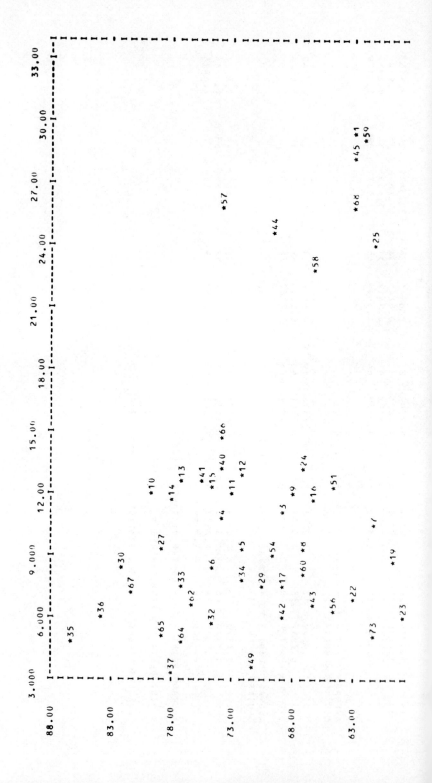

```
58.00   I------I------I------I------I------I------I------I------I------I------I
             *18

              *57
              *2
                                                                          *46
53.00
        *21    *50  *74                *72  *26
                        *38                     *53 *47
                                  *39
                                                                          *71  *70
48.00
                             *48

43.00


38.00   I------I------I------I------I------I------I------I------I------I------I

OVERPRINT TABLE
ITEM NO.   ALSO LOCATION OF ITEM NO.

27           28
49           65
29           31
54           55
9            61
            20
68           69
```

Figure 6.9  Results of nonlinear mapping applied to data of Table 6.9 (see also Table 6.10)

Table 6.11
Output from nonlinear mapping program (Table 6.16) for Tien data shown in Table 6.9. The graphical output from this program is shown in Figure 6.9

```
JOB IDENTIFIER:    TIEN DATA -- NONLINEAR MAPPING
NO OF VARIABLES =    4
NO OF INDIVIDUALS =   74

ACTUAL CORE USED = 3224 WORDS
MAGIC =      0.350
MAX =    100
ETA =    0.10000000E-04

VARIABLES      1 AND    3 HAVE HIGHEST VARIANCES

         ITERATION         MAPPING ERROR

              1            0.313138170E-01
              2            0.136927450E-01
              3            0.768943686E-02
              4            0.456121545E-02
              5            0.283105495E-02
              6            0.183548562E-02
              7            0.124725309E-02
              8            0.889205800E-03
              9            0.665300983E-03
             10            0.521230714E-03
             11            0.425577657E-03
             12            0.359836479E-03
             13            0.312943741E-03
             14            0.278193609E-03
             15            0.251463774E-03
             16            0.230180886E-03
             17            0.212710811E-03
             18            0.197995820E-03
             19            0.185336616E-03
             20            0.174260030E-03
             21            0.164437819E-03
```

mapping in that the groups are more compact, but there is little to choose between the MDS representation and that produced by principal coordinates. However, the computing requirements of the MDS algorithm are considerably more excessive than those of either of the alternative methods. If nonmetric data is used there is justification for employing MDS; when metric data is being analysed principal coordinates analysis appears to be a quick, cheap and efficient alternative.

### 6.3.3 Principal Coordinates Analysis

Tien's data (Table 6.9) is again used. The result is shown in Figure 6.15, in which the first principal axis is represented by the vertical scale. Comparison of Figures 6.15, 6.13 and 6.9 shows that there is a broad agreement between the results of the principal coordinates analysis, the nonmetric multidimensional scaling analysis, and the nonlinear mapping. The advantages of principal coordinates analysis are the speed of computation and the fact that the general similarity coefficient of Gower can be used for binary, multistate and continuously-measureable data.

## 6.4 Computer Programs

The following computer programs are described in this section: (i) hierarchical cluster analysis; (ii) nonhierarchical cluster analysis; (iii) principal coordinates analysis; (iv) nonmetric multidimensional scaling and (v) nonlinear mapping. Comments at the head of each program listing provide instructions regarding control cards and data input. Refer to Appendix A for listings of library subroutines. The following remarks are intended to summarize the characteristics of each program, and to indicate possible modifications.

(i) Hierarchical cluster analysis (Table 6.12). This program is based on Lance and Williams (1967) combinatorial formula which provides for the following clustering strategies (the mnemonics are used in the control card sequence): nearest neighbour (NN), furthest neighbour (FN), group average (GA), centroid (C), median (M), Ward's error sum of squares (WE), simple average (SA) and flexible strategy (FLEX). The programming strategy of subroutine CLUSTER is based on an idea given by Anderberg (1973). In its present form the program operates on a dissimilarity matrix held linearly rowwise in array DIST. If necessary, modifications could be made to subroutine CLUSTER to allow the use of similarity measures so that the highest remaining coefficient is selected at each cycle. Alternatively, a similarity matrix could be converted to a dissimilarity matrix before the program is entered. If mixed-mode data is to be analysed, Gower's general similarity coefficient $g$ may be used. A subroutine (SIMCAL) to calculate the matrix $G$ of similarities is listed with the principal coordinates program (Table 6.14). Since $0 \leq g_{ij} \leq 1$ this coefficient can be converted to a dissimilarity coefficient by defining $f_{ij} = 1.0 - g_{ij}$.

If the available computer system is small (less than 32K memory) the subroutines ARBOR, MINIM and MIN can be overlaid. ARBOR carries out the calculations involved in the production of the dendrogram.

(ii) Nonhierarchical cluster analysis (Table 6.13). The method of Beale (1969) is used. If centre coordinates are to be generated randomly a suitable pseudo-random number generator must be supplied. A generator for ICL 1900 computers is given in Appendix A where details of modifications required for other systems will be found. Most Fortran compiler libraries have an efficient pseudo-random number generator which returns one pseudo-random number from the distribution uniform on the range (0,1) at each call. If random centre coordinates are used the program should be run several times using a different starting point each time. For well-defined clusters the same outcome should result. The starting number of clusters should always

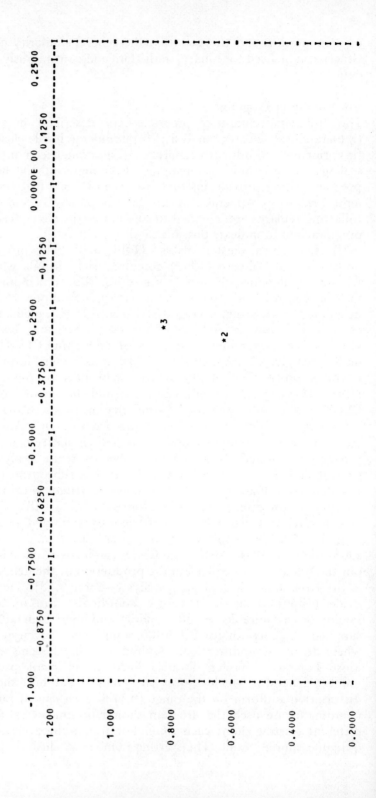

Figure 6.10  Nonmetric multidimensional scaling — starting points for square experiment (see text)

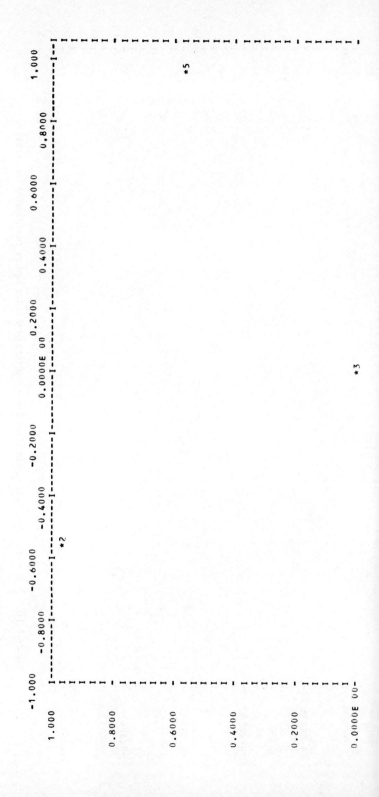

```
-0.2000


-0.4000


-0.6000                *1


-0.8000


-1.000                                                    *4
```

Figure 6.11 Final configuration produced by nonmetric multidimensional scaling program starting from the point shown in Figure 6.10. Final stress — 0.0 per cent

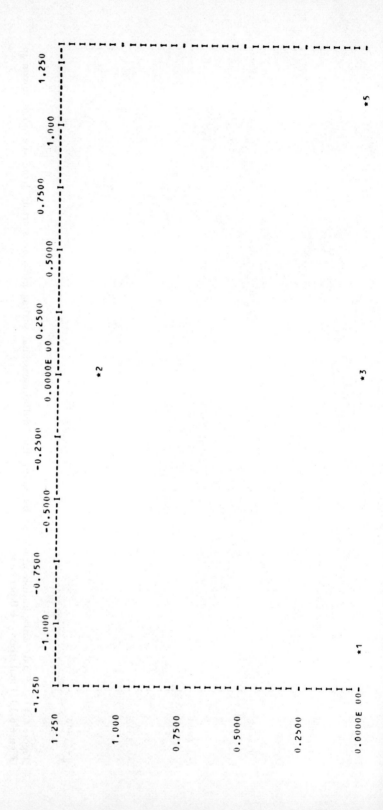

-0.2500
-0.5000
-0.7500
-1.000
-1.250

Figure 6.12   Varimax rotation applied to configuration of Figure 6.11

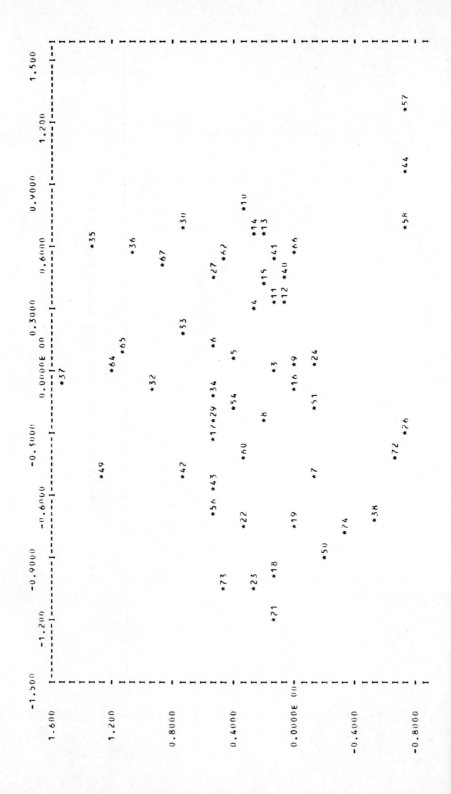

```
                                                                    *45
                                            *25                     *59*1
                                                                        *68

                                                          *2  *52

                                    *46


              *48      *39    *53  *47



                                *71
                                *70
-1.200



-1.600



-2.000



-2.400
```

OVERPRINT TABLE
ITEM NO. ALSO LOCATION OF ITEM NO.

| | |
|---|---|
| 49 | 65 |
| 27 | 28 |
| 29 | 31 |
| 54 | 55 |
| 54 | 61 |
| 9 | 20 |
| 68 | 69 |

Figure 6.13 Nonmetric multidimensional scaling applied to Tien's data (Table 6.9). The initial configuration was randomly selected, and the final stress was 4.136 per cent

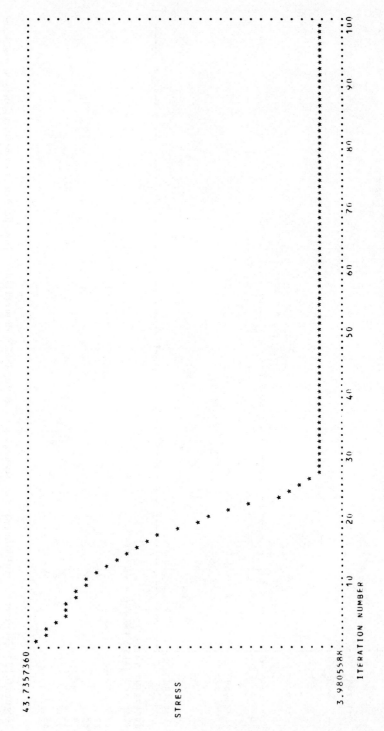

Figure 6.14   Plot of stress v. Iteration number for Tien's data

Table 6.12
Computer program for hierarchical cluster analysis

```
C ******      HIERARCHICAL CLUSTER ANALYSIS      ************
C ******      PROGRAM BY P. M. MATHER             ************
C ******      UNIVERSITY OF NOTTINGHAM            ************
C
C SEE CHAPTER 6 FOR DETAILS
C
C NOTES ON PROGRAM USE
C
C NEEDS I/O UNITS 5(CARDREADER), 6(LINEPRINTER), 3 & 4 (TEMPORARY DISC
C OR MTAPE FILES), 9 (OPTIONAL: DISC FILE CONTAINING LOWER TRIANGLE +
C DIAGONAL OF DISTANCE MATRIX, IF THIS IS TO BE READ -- SEE COL 10 OF
C CARD 1, BELOW --- OR WILL HOLD DISTANCE MATRIX IF THIS IS TO BE
C WRITTEN TO DISC -- SEE COL 10 OF CARD 1 BELOW).
C ADDITIONALLY, AN INPUT DEVICE AS SPECIFIED ON CONTROL CARD 1 (SEE
C BELOW) WILL BE NEEDED -- UNLESS IT CORRESPONDS TO ONE OF THE ABOVE.
C SUBROUTINES NEEDED: MATPR ( SEE APPENDIX A)
C DIMENSION OF X ( AND IX) IN MAIN PROGRAM SHOULD BE AT LEAST NOBS+NOBS*
C (NOBS+1)/2+NOBS-1 AND DIMENSION OF INT AT LEAST 5*NOBS-2, WHERE  NOBS
C IS THE NUMBER OF CASES.
C CHANGE VALUE OF NDI AS NECESSARY
C
C DATA IN ORDER:
C ********* NOTE: THE TERM 'CARD' MAY INCLUDE SEVERAL PHYSICAL CARDS
C ****** ALL CONTROL CARDS -- I.E. EXCEPTING THE DATA MATRIX ALONE --ARE
C READ FROM UNIT NO 5.
C CARD 1:
C COL 1:  INPUT2,THE UNIT NUMBER DEFINING THE INPUT DEVICE FROM WHICH THE
C         DATA MATRIX IS TO BE READ. MUST BE SPECIFIED IN JOB CONTROL!
C COLS 2-4: NOBS, THE NUMBER OF OBSERVATIONS PER VARIABLE.
C COLS 5-7: NV, THE NUMBER OF VARIABLES.
C COL 8 IPR = 1: DATA MATRIX PRINTED OUT;
C             = 0: DATA PRINTOUT SUPPRESSED.
C COL 9: = 1: DATA STANDARDIZED TO UNIT VARIANCE, ZERO MEAN.
C        = 0: RAW DATA USED.
C COL 10: OUTD = 1: WRITE LR. TRIANGLE + DIAGONAL OF MATRIX OF SQUARED
C                  EUCLIDEAN DISTANCES TO UNIT 9 IN UNFORMATTED FORM AS
C                  ONE LOGICAL RECORD.
C              = 2: READ LR. TRIANGLE + DIAGONAL OF SQUARED EUCLIDEAN
C                  DISTANCE MATRIX FROM UNIT 9, ROWWISE, IN UNFORMATTED
C                  FORM, AS ONE LOGICAL RECORD.
C              = 3: DO NEITHER OF THESE.
C CARD 2:
C COLS 1 - 80: JOB NAME.
C CARD 3:
C COLS 1-80: INPUT FORMAT FOR DATA MATRIX. LEAVE BLANK IF DISTANCE
C            MATRIX READ IN. IF DATA IS TO BE READ IN UNFORMATTED FORM
C            FROM INPUT2 THEN PUNCH UNFORMATTED IN COLS 1 TO 11 OF THIS
C            CARD. INPUT2 SHOULD BE A MAGTAPE/DISC FILE IN THIS CASE.
C CARD 4
C ENTER VARI IF DATA IS TO BE READ VARIABLEWISE OR CASE IF DATA TO BE
C            READ CASE BY CASE. IF DISTANCE MATRIX TO BE READ FROM DISC
C            LEAVE THIS CARD BLANK. START PUNCHING IN COLUMN 1.
C CARD 5 ****** INCLUDE ONLY IF DATA IS TO BE READ CASE BY CASE ********
C ********************* OTHERWISE OMIT ******************************
C ENTER DATA CASE BY CASE IN FORMAT SPECIFIED ON CARD 3. THE DEVICE
C            NUMBER INPUT2 (SEE CARD 1, ABOVE) WILL BE USED TO READ IN
C            THE DATA MATRIX. EACH CASE BEGINS A NEW CARD OR A NEW
C            RECORD. IF INPUT2 IS NOT UNIT 5 THEN THE DATA MATRIX WILL
C            BE PRESENTED SEPARATELY IN ACCORDANCE WITH LOCAL PRACTICE.
C            IN THIS CASE, THE CONTROL CARD SEQUENCE, WHICH IS READ FROM
C            UNIT 5, WILL INCLUDE CARD 4 AND OMIT CARD 5.
C CARD 6:
C COLS 1-4: ENTER ONE OF FOLLOWING MNEMONICS: NN,FN,GA,C,M,WE,SA,FLEX
C           STARTING IN COLUMN 1. DON'T INCLUDE THE COMMA.
C CARD 7     ****** NEEDED ONLY IF CODE FLEX PUNCHED ON CARD 6 *********
C ************************ OTHERWISE OMIT ****************************
```

Table 6.12  *continued*

```
C COLS 1 -10: BETA COEFFICIENT FOR FLEXIBLE STRATEGY METHOD.
C CARD 8  ****** NEEDED ONLY IF [1] OUTD (COL.10,CARD 1) IS NOT EQUAL TO
C           2 AND [2] DATA IS TO BE READ VARIABLEWISE *********
C           ENTER DATA MATRIX VARIABLEWISE, EACH VARIABLE STARTING A
C           NEW CARD OR A NEW RECORD.  THE DATA IS READ FROM DEVICE NO.
C           INPUT2 (CARD 1, ABOVE).    SEE REMARKS ON CARD 5, ABOVE,
C           REGARDING PRESENTATION OF DATA MATRIX.  USE FORMAT SPECIFI-
C           ED ON CARD 3 IF DATA IS FORMATTED.
C CARD 9
C COLS 1-4: ENTER LETTERS PRINT OF DONT PRINT, FOR PRINTOUT/NO PRINTOUT
C           OF EUCLIDEAN DISTANCE MATRIX.  IF DONT PRINT IS ENTERED A
C           DUMMY ROUTINE MATPR CAN BE PROVIDED TO SAVE SPACE.
C CARD 10 ****** ONLY IF CARD 9 CONTAINS 'PRINT' ********
C ******************* OTHERWISE OMIT ***********************
C COLS 1-80: FORMAT FOR SUBROUTINE MATPR ( SEE APPENDIX A)
C
      INTEGER OUTD
      DIMENSION X(18000),IX(18000),INT(1000),FMT(20),ALG(8),TITLE(20)
      EQUIVALENCE (X,IX)
      DATA ALG/'NN  ','FN  ','GA  ','C   ','M   ','WE  ','SA  ','FLEX'/
      DATA VBL ,INPUT/'VARI',1/
      REWIND 3
      REWIND 4
C SET NDI TO DIMENSION OF X AND NDI2 TO DIMENSION OF INT.
      NDI=18000
      NDI2=1000
      WRITE(6,900)
  900 FORMAT('1 HIERARCHICAL CLUSTER ANALYSIS'//' PROGRAM BY P.M.MATHER,
     *NOTTINGHAM 1974'///)
C CONTROL CARD 1
      READ(5,1) INPUT2,NOBS,NV,IPR,ISTAN,OUTD
    1 FORMAT(I1,2I3,3I1)
C CHECK THAT DIMENSION OF X IS SUFFICIENT
      N1=NOBS+NOBS*(NOBS-1)/2+NOBS-1
      IF(NDI.LT.N1) GO TO 16
C CHECK DIMENSION OF INT IS SUFFICIENT
      N1=5*NOBS-2
      IF(N1.GT.NDI2) GOTO 17
C CONTROL CARD 2
      READ(5,4) TITLE
      WRITE(6,2) TITLE
    2 FORMAT('0JOB NAME: ',20A4)
      IF(IPR.GT.0) WRITE(6,6)
    6 FORMAT('0                              DATA    MATRIX'///)
      NZ=NOBS-1
C CONTROL CARDS 3 AND 4
      READ(5,4) FMT, TYPE
    4 FORMAT(20A4)
      IF(TYPE.EQ.VBL) GOTO 5
      IF(OUTD.EQ.2) GOTO 5
C CALL SUBROUTINE INPUT IF DATA MATRIX PRESENTED CASE BY CASE
      CALL INPT(X,FMT,NOBS,NV,INPUT2,IPR)
      INPUT = 2
C SET UP BASE ADDRESSES FOR CLUSTERING
    5 N2=NOBS*(NOBS+1)/2
      N=NOBS+NOBS
      N1=N+N2
      N1=N+N2
      DO 10 I=N,N1
   10 X(I)=0.0
      N1=NOBS+1
      N3=N1+NOBS
      N4=N3+NZ
      N5=N4+NOBS
      DO 12 I=1,NOBS
   12 INT(I)=1
      REWIND 3
```

Table 6.12   *continued*

```
C CONTROL CARD 5
      READ(5,4) TYPE
C FIND REQUESTED CLUSTERING STRATEGY
      DO 20 I=1,8
      IF(TYPE.NE.ALG(I)) GO TO 20
      KTYPE=I
      GOTO 30
   20 CONTINUE
C EXIT IF ERROR
      WRITE(6,40) TYPE
   40 FORMAT('OTYPE CODE INCORRECT - SHOULD BE ONE OF NN,FN,GA,C ,M ,WE
     1,SA OR FLEX.'//' YOU HAVE ENTERED',2X,A4)
      STOP
   30 WRITE(6,50) TYPE
   50 FORMAT('OTYPE CODE ENTERED = ',A4)
C CALL MAIN CLUSTERING ROUTINE
      CALL CLUS(X(1),X(NOBS),X(N),INT(1),INT(N1),INT(N3),N2,FMT,NZ,
     1 INT(N4),INPUT2,NOBS,NV,IPR,ISTAN,OUTD,KTYPE,INT(N5),INPUT)
      NN=NOBS+NOBS
C SET UP BASE ADDRESSES FOR DENDROGRAM DRAWING
      N=NOBS+5*NN
      NM=N+3*NN
      MN=110
      IF(NOBS.GT.110) MN=NOBS
C CALL DENDROGRAM ROUTINE
      CALL ARBOR(INT(1),INT(N1),INT(N3),X(1),IX(NOBS),IX(N),IX(NM),NOBS,
     *NZ,NN,TITLE,MN)
      GO TO 7
   16 WRITE(6,8) N1
    8 FORMAT('ODIMENSION OF X TOO SMALL --- INCREASE TO',I6)
      GOTO 7
   17 WRITE(6,18) N1
   18 FORMAT('ODIMENSION OF INT TOO SMALL - INCREASE TO',I6)
    7 STOP
      END



      SUBROUTINE CLUS(DIS,X,DIST,NGRP,RET,DEL,N2,       FMT,NZ,IX,INPUT2,
     *NOBS,NV,IPR,ISTAN,OUTD,KTYPE,IX1,INPUT)
C THIS ROUTINE CONTROLS CLUSTERING PROCEDURE
      INTEGER RET(NZ),DEL(NZ),OUTD,IX(NOBS),IX1(NOBS)
      DIMENSION X(NOBS),DIST(N2),NGRP(NOBS),DIS(NZ),COEFFS(4,8),FMT(20)
      DATA COEFFS/0.5,0.5,0.0,-0.5,0.5,0.5,0.0,0.5,0.0,0.0,0.0,0.0,0.0,0.0,
     +0.0,0.0,0.0,0.5,0.5,-0.25,0.0,0.0,0.0,0.0,0.0,0.0,0.5,0.5,0.0,0.0,
     +0.0,0.0,0.0,0.0/
      DATA HUGE,VAST,PRINT,IDO,UNF,IFM/1.E20,1.E22,'PRIN',0,'UNFO',0/
      IF(FMT(1).EQ.UNF) IFM=1
      IF(IFM.EQ.1) REWIND INPUT2
      IF (KTYPE.LT.8) GOTO 70
C CONTROL CARD 6
      READ(5,76) COEFFS(3,8)
   76 FORMAT(F10.0)
C CALCULATE FLEXIBLE STRATEGY COEFFICIENTS
      S=1.0-COEFFS(3,8)
      COEFFS(2,8)=0.5*S
      COEFFS(1,8) = COEFFS(2,8)
      WRITE(6,71) COEFFS(1,8), COEFFS(3,8)
   71 FORMAT('OFLEXIBLE STRATEGY COEFFICIENTS:'//' ALPHA(J)=ALPHA(K)=',
     +F10.4/' BETA = ',F10.4//)
   70 FN=NV
      OBS=NOBS
      IF(ISTAN.EQ.1) WRITE(6,42)
   42 FORMAT('ODATA STANDARDIZED BY VARIABLE')
      IF(KTYPE.GE.4.AND.KTYPE.LE.6) IDO=1
      GOTO (230,231,230), OUTD
  231 REWIND 9
C READ DISSIMILARITY (DISTANCE) MATRIX FROM DISC/MTAPE FILE, CHANNEL 9
```

**Table 6.12** *continued*

```
C IF REQUESTED ( CARD 1, COL. 10)
      READ(9) DIST
      GOTO 233
C ALTERNATIVELY READ RAW DATA. IF VARIABLEWISE, READ DIRECTLY FROM CARDS.
C IF CASE-BY-CASE IT WILL ALREADY HAVE BEEN READ FROM CARDS BY
C SUBROUTINE INPT AND WRITTEN TO FILE (CHANNEL 3).
   230 DO 13 I=1,NV
       GOTO(40,41), INPUT
    40 IF(IFM.EQ.1) GOTO 440
       READ(INPUT2,FMT) X
       GOTO 441
   440 READ(INPUT2) X
   441 IF(IPR.GT.0) WRITE(6,7711) X
  7711 FORMAT(' ',12F10.4)
       GOTO 43
    41 READ(3) X
C STANDARDIZE DATA TO HAVE MEAN 0 AND VARIANCE 1, IF REQUESTED.
    43 IF(ISTAN.EQ.0) GOTO 72
       S=0.0
       XB=0.
       DO 9 J=1,NOBS
     9 XB=XB+X(J)
       XB=XB/OBS
       DO 10 J=1,NOBS
    10 S=S+(X(J)-XB)*(X(J)-XB)
       S=SQRT(S/OBS)
       DO 12 J=1,NOBS
    12 X(J)=(X(J)-XB)/S
C COMPUTE DISTANCE MATRIX (SQUARED VALUES)
    72 DO 13 L=1,NOBS
       DO 13 J=1,L
       I1=L*(L-1)/2+J
    13 DIST(I1)=DIST(I1)+(X(L)-X(J))*(X(L)-X(J))
C OUTPUT D2 MATRIX TO DISC (CHANNEL 9) IF REQUESTED.
       IF(OUTD.NE.1) GOTO 233
       REWIND 9
       WRITE(9) DIST
       ENDFILE 9
C CONVERT TO DISTANCES UNLESS MEDIAN, CENTROID OR WARD'S ERROR METHODS
C ARE USED ( SQUARED DISTANCES NECESSARY FOR THESE -- SEE TEXT)
   233 IF(IDO.EQ.1) GOTO 341
       DO 18 I=1,N2
    18 DIST(I)=SQRT(DIST(I)/FN)
       GOTO 340
   341 DO 342 I=1,N2
   342 DIST(I)=DIST(I)/FN
   340 REWIND 3
C PREP1 IS USED IN CALCULATION OF COPHENETIC DISTANCES
       CALL PREP1(DIST,N2,   SDIST,NOBS,FN)
C CONTROL CARD 9
       READ(5,78) TYPE
    78 FORMAT(20A4)
       IF(TYPE.EQ.PRINT) GOTO 334
       WRITE(6,335)
   335 FORMAT('0 DISTANCE MATRIX NOT PRINTED.')
       GOTO 333
C CONTROL CARD 10
   334 READ(5,78) FMT
       IP=1
C MATPR IS LISTED IN APPENDIX A
       CALL MATPR(32HDISTANCE COEFFICIENT MATRIX       ,4,FMT,6,DIST,N2,
      *NOBS,NOBS,NOBS,IP,3,8)
   333 WRITE(6,17)
    17 FORMAT('0        PAIRING SEQUENCE'//'  ITEM     JOINS    ITEM   AT  DISTA
      *NCE'///)
C CLUSTR CARRIES OUT CLUSTERING
       CALL CLUSTR(DIST,RET,DEL,DIS,NOBS,N2,NGRP,X,IX,KTYPE,
```

Table 6.12   *continued*

```
      * COEFFS(1,KTYPE),IDO,NZ)
C CALCULATE LINKAGE ORDER
      CALL LINK(RET,DEL,NGRP,NOBS,NZ)
C COMPUTE COPHENETIC CORRELATION
      CALL COPHGN(DIST,N2,RET,DEL,DIS,NOBS,NZ,IX1,IX)
      CALL COPHCR(DIST,N2,FN,SDIST,   NOBS)
      RETURN
      END


      SUBROUTINE PREP1(DIST,N2,    SDIST,NOBS,FN)
      DIMENSION DIST(N2)
      DATA IBUF/1000/
C IBUF IS RECORD LENGTH FOR UNIT 4
      WRITE(4) DIST
      FN=N2-NOBS
      XM=0.0
      DO 1 I=1,N2
    1 XM=XM+DIST(I)
      XM=XM/FN
      SDIST=0.0
      DO 2 I=1,N2
      X=DIST(I)-XM
      SDIST=SDIST+X*X
    2 DIST(I)=X
      SDIST=SQRT(SDIST/FN)
      L=-(IBUF-1)
      K=N2/IBUF
      LEFT=N2-IBUF*K
      IF(K.EQ.0) GOTO 5
      DO 3 J=1,K
      L=L+IBUF
      LL=L+IBUF-1
    3 WRITE(4) (DIST(I),I=L,LL)
      IF(LEFT.LE.0) GOTO 4
    6 L=LL+1
      LL=LL+LEFT
      WRITE(4) (DIST(I),I=L,LL)
    4 ENDFILE 4
      REWIND 4
      READ(4) DIST
      RETURN
    5 LL=0
      GOTO 6
      END


      SUBROUTINE CLUSTR(DIST,IR,ID,DMIN,NA,NO,NGRP,SMIN,IREC,KTYPE,
     + COEFFS,IDO,IZ)
C SUBROUTINE FOR EFFICIENT CLUSTERING
C BASED ON IDEA FROM ANDERBERG (1974)
      DIMENSION DIST(NO),IR(IZ),IREC(NA),ID(IZ),DMIN(IZ),SMIN(NA),
     +NGRP(NA) , COEFFS(4)
      DATA BIG,VAST/1.0E20,1.0E25/
      LEXP(I)=I*(I-1)/2
      DO 3 I=2,NA
C COMPUTE ROW MINIMA IN SMIN( ) AND RECORD THE COLUMN IDENTIFIER IN IREC( )
      I1=I-1
      I2=LEXP(I)
      CALL SMALL(DIST,NO,I1,I2,L,S)
      SMIN(I)=S
    3 IREC(I)=L
      DO 7 I=2,NA
C PERFORM (NA-1) ITERATIONS, FINDING MIN(SMIN) AT EACH ITERATION.
      I1=I-1
      S=BIG
      DO 4 J=2,NA
```

Table 6.12   *continued*

```
C FIND MIN(SMIN(I))
      IF(S.LE.SMIN(J)) GOTO 4
      K=J
      S=SMIN(J)
    4 CONTINUE
C MIN( SMIN(I) ) IS NOW IN DMIN(I1)
      DMIN(I1)=S
C STORE ROW IDENTIFIER IN IR( ) AND COLUMN IDENTIFIER IN ID( )
      L=IREC(K)
      IR(I1)=L
      ID(I1)=K
      I2=LEXP(L)
C RECALCULATE ROW IR( ) OF DISTANCE MATRIX
      FK=NGRP(L)+NGRP(K)
      GOTO (30,30,31,31,30,30,30,30) , KTYPE
   31 COEFFS(1)=NGRP(L)/FK
      COEFFS(2)=NGRP(K)/FK
      IF(KTYPE.EQ.3) GOTO 30
      COEFFS(3)=-(NGRP(L)*NGRP(K))/(FK*FK)
   30 DO 5 J=1,NA
C IGNORING DISTANCE FROM L TO ITSELF OR L TO K
      IF(J.EQ.L) GOTO 5
      I3=IFIND(J,L)
      I4=IFIND(K,J)
      IF(J.EQ.K) GOTO 12
      IF(KTYPE.NE.6) GOTO 33
      GN=NGRP(J)
      HN=FK+GN
      COEFFS(1)=(GN+NGRP(L))/HN
      COEFFS(2)=(GN+NGRP(K))/HN
      COEFFS(3)=-GN/HN
   33 DIST(I3)=COEFFS(1)*DIST(I3)+COEFFS(2)*DIST(I4)+COEFFS(3)*DMIN(I1)+
     + COEFFS(4)*ABS(DIST(I3)-DIST(I4))
      GOTO 5
   12 DIST(I4)=VAST
    5 CONTINUE
      NGRP(L)=NGRP(L)+NGRP(K)
      NGRP(K)=0
      IF(IDO.EQ.1) DMIN(I1)=SQRT(DMIN(I1))
C PRINT CURRENT DETAILS
      WRITE(6,15) L, K, DMIN(I1)
   15 FORMAT(' ',I4,12X,I4,6X,F10.3)
      IF(L.EQ.1) GOTO 20
      I1=L-1
C COMPUTE NEW VALUE FOR SMIN(L)
      CALL SMALL(DIST,NO,I1,I2,LL,S)
      SMIN(L)=S
C RECORD POSITION OF MINIMUM VALUE IN IREC(L)
      IREC(L)=LL
C ENSURE SMIN(K) IS NOT CONSIDERED IN FUTURE ITERATIONS
   20 SMIN(K)=VAST
      IF(L.EQ.NA) GOTO 7
C IF L IS NOT THE LAST ROW OF DIST THEN CHECK ROWS L+1 TO NA TO SEE IF
C NEW DISTANCE FROM L IS SMALLER THAN CURRENT SMIN( ).
      I4=L+1
      DO 6 J=I4,NA
      IF(NGRP(J)) 13,6,13
   13 LL=IREC(J)
      I2=IFIND(J,K)
      DIST(I2)=VAST
      I2=LEXP(J)
      IF(LL.EQ.K.OR.LL.EQ.L) GOTO 14
      IF(DIST(I2+L).GE.DIST(I2+LL)) GOTO 6
      SMIN(J)=DIST(I2+L)
      IREC(J)=L
      GOTO 6
   14 I1=J-1
```

Table 6.12   *continued*

```
      CALL SMALL(DIST,NO,I1,I2,LL,S)
      SMIN(J)=S
      IREC(J)=LL
    6 CONTINUE
    7 CONTINUE
      RETURN
      END


      SUBROUTINE SMALL(DIST,NO,I1,I2,L,S)
      DIMENSION DIST(NO)
       S=DIST(I2+1)
      L=1
      IF(I1-1) 1,1,2
    2 DO 3 J=2,I1
      I3=I2+J
      IF(DIST(I3).GE.S) GOTO 3
      L=J
      S=DIST(I3)
    3 CONTINUE
    1 RETURN
      END


      SUBROUTINE COPHCR(DIST,N2,FN,SDIST,NOBS)
C COMPUTES COPHENETIC CORRELATION COEFFICIENT
      DIMENSION DIST(N2),BUF(1000)
      DATA IBUF/1000/
      XM=0.0
      S2=0.0
      DO 1 I=1,N2
    1 XM=XM+DIST(I)
      XM=XM/FN
      DO 2 I=1,NOBS
      DO 2 J=1,I
      IF(I.EQ.J) GOTO 2
      K=IFIND(I,J)
      DIST(K)=DIST(K)-XM
      S2=S2+DIST(K)*DIST(K)
    2 CONTINUE
      S2=SQRT(S2/FN)
      K=N2/IBUF
      LEFT=N2-K*IBUF
      IF(LEFT.GT.0) K=K+1
      R=0.0
      NN=IBUF
      KT=0
      DO 3 I=1,K
      IF(I.EQ.K.AND.LEFT.GT.0) NN=LEFT
      READ(4)(BUF(J),J=1,NN)
      DO 3 J=1,NN
      KT=KT+1
    3 R=R+BUF(J)*DIST(KT)
      R=(R/FN)/(SDIST*S2)
      WRITE(6,10) R
   10 FORMAT('0COPHENETIC CORRELATION COEFFICIENT = ',F9.5/' ',34('*'))
      RETURN
      END


      SUBROUTINE LINK(NRET,NDEL,K,N,NP)
      DIMENSION NRET(NP),NDEL(NP),K(N)
      DATA ISRCH,KOUNT/1,1/
      K(1)=1
      DO 1 I=1,NP
      IF(ISRCH-NRET(I)) 1,9,1
    9 KOUNT=KOUNT+1
```

Table 6.12　*continued*

```
      K(KOUNT) = NDEL(I)
    1 CONTINUE
    4 ISRCH=ISRCH+1
      IF(ISRCH-N) 10,10,6
   10 IT=-1
      DO 2 I=1,NP
      IF(ISRCH.EQ.NRET(I)) GOTO 3
      IF(ISRCH.EQ.NDEL(I)) GOTO 4
      GOTO 2
    3 IT=IT+1
      J=0
    5 J=J+1
      IF(J.GT.KOUNT) GOTO 2
      IF(ISRCH-K(J)) 5,11,5
   11 IF(J.EQ.KOUNT) GOTO 8
      MN=J+IT
      IF(MN.EQ.KOUNT) GOTO 8
      MN=MN+1
      LL=KOUNT+2
      L=MN-1
    7 L=L+1
      IF(L.GT.KOUNT) GOTO 17
      LL=LL-1
      LJ=LL-1
      K(LL)=K(LJ)
      GOTO 7
   17 KOUNT=KOUNT+1
      K(MN)=NDEL(I)
      GOTO 2
    8 KOUNT=KOUNT+1
      K(KOUNT)=NDEL(I)
    2 CONTINUE
      GOTO 4
    6 RETURN
      END


      SUBROUTINE MINIM(IB,K,IR,N2,M,L1,L2)
      DIMENSION IB(K),IR(N2,5)
      M=1010
      L2=-1010
      DO 1 I=1,K
      J=IB(I)
      IF(J) 1,1,2
    2 IF(IR(J,2)-M) 4,3,3
    4 M=IR(J,2)
      L1=IR(J,1)
    3 IF(IR(J,1).LE.L2) GOTO 1
      L2=IR(J,1)
    1 CONTINUE
      RETURN
      END

      SUBROUTINE MIN(I,L,IP,M)
      DIMENSION I(L)
      N=1010
      DO 1 J=1,L
      IF(I(J)) 1,1,2
    2 IF(I(J)-N) 3,1,1
    3 N=I(J)
      IP=J
    1 CONTINUE
      M=N
      RETURN
      END
```

Table 6.12  *continued*

```
      SUBROUTINE COPHGN(DIS,N2,RET,DEL,DMAX,N,N1,IWORK,JWORK)
C COMPUTES COPHENETIC CORRELATION
      INTEGER RET(N1),DEL(N1),IWORK(N),JWORK(N)
      DIMENSION DIS(N2),DMAX(N1)
      J=N
   10 J=J-1
      IF(J) 30,30,15
   15 X=DMAX(J)
      CALL SEARCH(RET,DEL,N1,J,IWORK,N,1,NO1)
      CALL SEARCH(RET,DEL,N1,J,JWORK,N,2,NO2)
      DO 20 I=1,NO1
      K=IWORK(I)
      DO 20 L=1,NO2
      M=JWORK(L)
      MM=IFIND(K,M)
      IF(K.EQ.M) GOTO 20
      DIS(MM)=X
   20 CONTINUE
      GOTO 10
   30 DO 31 I=1,N
      K=I*(I-1)/2+I
   31 DIS(K)=0.0
      RETURN
      END


      SUBROUTINE SEARCH(RET,DEL,N1,J,IWK,N,IND,MM)
      INTEGER RET(N1),DEL(N1),IWK(N)
      M=1
      MM=1
      GOTO (10,20), IND
   10 IWK(1)=RET(J)
      GOTO 2
   20 IWK(1)=DEL(J)
    2 NN=IWK(MM)
      DO 1 I=1,J
      IF(RET(I)-NN) 1,3,1
    3 M=M+1
      IWK(M)=DEL(I)
    1 CONTINUE
      MM=MM+1
      IF(MM-M) 2,2,4
    4 MM=MM-1
      RETURN
      END


      SUBROUTINE ARBOR(LINK,RET,DEL,DIS,IR,IDN,BUF,N,N1,N2,TITLE)
C SUBROUTINE TO PRINT DENDROGRAM -- FAILS IF BACKWARD LINKS ARE PRESENT
      INTEGER RET,DEL,BUF,BLANK,DOWN
      DIMENSION LINK(N),RET(N1),DEL(N1),DIS(N1),IR(N2,5),IDN(N2,3),
     * BUF(110),TITLE(20)
      DATA LINE,DOWN,BLANK/'---','III','   '/,LI/0/
      DO 96 I=2,N1
      IF(DIS(I).GE.DIS(I-1)) GOTO 96
      WRITE(6,97)
      STOP
   97 FORMAT('OLINKAGE TREE CANNOT BE DRAWN - BACKWARD LINKS PRESENT')
   96 CONTINUE
C SCALE SO THAT 100 CHARACTERS ACROSS = 60 LINES DOWN.
      KT=INT(DIS(N1)+99.99999999999)/100
      KZ=N/60+1
      NSECT=MAXO(KT,KZ)
      SCALE=(100.0/DIS(N1))*NSECT
      DO 1 I=1,N
      IR(I,5)=0
      KL=LINK(I)
```

Table 6.12   *continued*

```
      DO 2 J=1,N1
      IF(RET(J).NE.KL) GOTO 92
      KZ=DEL(J)
      GOTO 3
   92 IF(DEL(J).NE.KL) GOTO 2
      KZ=RET(J)
    3 D=DIS(J)
      M=J
      GOTO 4
    2 CONTINUE
    4 IR(I,4)=INT(D*SCALE+0.5)
      IF(IR(I,4).LT.1) IR(I,4)=1
      IR(I,2)=KL
      IR(I,3)=KZ
    1 IR(I,1)=I
      L=N
      LUN=N
 2000 KON=1
      L1=L-1
      DO 5 I=1,L1
      CALL MIN(IR(1,4),L,IPOS,M)
      BUF(1)=IPOS
      IF(IPOS.EQ.L) GOTO 5
      IPP=IPOS+1
      DO 6 J=IPP,L
      IF(IR(J,4)-M) 6,8,6
    8 KON=KON+1
      BUF(KON)=J
    6 CONTINUE
    9 K=BUF(1)
      KOUNT=KON
      DO 7 J=2,KON
      KK=BUF(J)
      IF(IR(K,2).EQ.IR(KK,2).OR.IR(K,2).EQ.IR(KK,3)) GOTO 7
      IF(IR(K,3).EQ.IR(KK,2).OR.IR(K,3).EQ.IR(KK,3)) GOTO 7
      KOUNT=KOUNT-1
      BUF(J)=-1
    7 CONTINUE
      IF(KOUNT.GT.1) GOTO 12346
      IF(IR(KZ,5).EQ.100) GOTO 12346
      KZ=BUF(1)
      IR(KZ,4)=-IR(KZ,4)
      GOTO 5
12346 CALL MINIM(BUF,KON,IR,N2,M1,L1,L2)
      KL=MINO(L1,L2)
      KZ=IABS(L1-L2)/2+KL
      LUN=LUN+1
      IR(LUN,1)=KZ
      IR(LUN,2)=M1
      IR(LUN,5)=M
      DO 14 J=1,N1
      IF(M1.NE.RET(J)) GOTO 109
      IP=DEL(J)
      GOTO 15
  109 IF(M1.NE.DEL(J)) GOTO 14
      IP=RET(J)
   15 MM=INT(DIS(J)*SCALE+0.5)
      IF(MM.LT.1) MM=1
 1990 IF(MM-M) 14,1995,16
 1995 DO 1992 IJ=1,KON
      IK=BUF(IJ)
      IF(IK) 1992,1992,1996
 1996 IF(IP.EQ.IR(IK,2).OR.IP.EQ.IR(IK,3)) GO TO 14
 1992 CONTINUE
      DO 1997 IK=1,L
      K=IR(IK,2)
      IJ=IR(IK,3)
```

Table 6.12    *continued*

```
      IF(K.EQ.M1.AND.IJ.EQ.IP) GOTO 1998
      IF(K.EQ.IP.AND.IJ.EQ.M1) GOTO 1998
      GOTO 1997
 1998 IF(IR(IK,4)+MM) 16,14,16
 1997 CONTINUE
      GO TO 16
   14 CONTINUE
   16 IR(LUN,4)=MM
      IR(LUN,3)=IP
      DO 19 J=1,KON
      K=BUF(J)
      IF(K) 19,19,20
   20 IR(K,4)=-IR(K,4)
   19 CONTINUE
      IF(IABS(L1-L2).EQ.1) GOTO 22
      LI=LI+1
      IDN(LI,1)=KL+1
      IDN(LI,2)=MAX0(L1,L2)-1
      IDN(LI,3)=M
      GOTO 22
    5 CONTINUE
      GOTO 30
   22 L=L+1
      GOTO 2000
   30 KR1=1
      KR2=100
      L1=L-1
      DO 51 KT=1,NSECT
      WRITE(6,1000) TITLE, KT
 1000 FORMAT('1                         DENDROGRAM FOR ',20A4,2X,'SECTI
     +ON',I4//)
      KZ=(KT-1)*100
      DO 50 J=1,N
   31 DO 33 KL=1,110
   33 BUF(KL)=BLANK
      DO 32 I=1,L1
      MO=0
      IF(J.NE.IR(I,1)) GOTO 32
      M=IR(I,5)+1
      MM=-IR(I,4)
      IF(M.GT.MM) GO TO 1234
      IF(M.GT.KR2.OR.MM.LT.KR1) GOTO 1234
      IF(MM.LE.KR2) GOTO 52
      MM=KR2
      MO=KR2
   52 M=M-KZ
      MM=MM-KZ
      DO 34 KL=M,MM
   34 BUF(KL)=LINE
      BUF(MM)=DOWN
      IF(MO.NE.0) IR(I,5)=MO
 1234 DO 35 KL=1,LI
      IF(J.LT.IDN(KL,1).OR.J.GT.IDN(KL,2)) GOTO 35
      M=IDN(KL,3)
      IF(M.LT.KR1.OR.M.GT.KR2) GOTO 35
      M=M-KZ
      BUF(M)=DOWN
   35 CONTINUE
      IF(KT-NSECT) 32,55,32
   55 IF(J-IR(L,1)) 32,56,32
   56 DO 36 KL=101,110
   36 BUF(KL)=LINE
   32 CONTINUE
   50 WRITE(6,40) IR(J,2), BUF
      KR1=KR2+1
   51 KR2=KR2+100
   40 FORMAT(' ',I4,2X,110A1)
```

Table 6.12  *continued*

```
      RETURN
      END


      SUBROUTINE INPT(X,FMT,N,NV,INPUT2,IPR)
C READS DATA CASE BY CASE AND WRITES TO CHANNEL 3.
      DIMENSION X(N,NV),FMT(20)
      DATA UNF/'UNFO'/
      IF(FMT(1).EQ.UNF) GOTO 10
      DO 1 I=1,N
      READ(INPUT2,FMT) (X(I,J),J=1,NV)
      IF(IPR) 1,1,5
    5 WRITE(6,4) (X(I,J),J=1,NV)
    1 CONTINUE
      GOTO 22
   10 WRITE(6,30) INPUT2
   30 FORMAT('0DATA READ IN UNFORMATTED FORM FROM UNIT',I4//)
      REWIND INPUT2
      DO 11 I=1,N
      READ(INPUT2) (X(I,J),J=1,NV)
      IF(IPR) 11,11,15
   15 WRITE(6,4) (X(I,J),J=1,NV)
   11 CONTINUE
    4 FORMAT(' ',12F10.4)
   22 DO 2 I=1,NV
    2 WRITE(3) (X(J,I),J=1,N)
      ENDFILE 3
      REWIND 3
      RETURN
      END
```

exceed the expected number of clusters if reasonable results are to be obtained.

(iii) Principal coordinates analysis (Table 6.14). This program follows the computing procedure described by Gower (1966) and Blackith and Reyment (1971), though it is thought to be more efficient than other published programs. For large numbers of items the scatter plots become difficult to interpret, and one of the strategies outlines by Ross (1969) should be considered. The principal coordinates analysis program is computationally more efficient than the multidimensional scaling and nonlinear mapping programs since an analytical rather than an iterative procedure is employed. Gower's general coefficient of similarity is used. This makes the program suitable for mixed-mode data.

(iv) Nonmetric multidimensional scaling (Table 6.15). This program is based on Kruskal (1964B) and follows closely the procedure given there. A pseudo-random number generator is needed if the starting point is to be randomly selected. Details of random number generators are given in Appendix A. Several runs of the program should be made if a random starting point is used in order to ensure as far as possible that the minimum of stress is not a local one. The input to the program is the strict lower triangle of a dissimilarity matrix, read rowwise. For a

## Table 6.13
## Computer program for nonhierarchical cluster analysis

```
C ----------------NOTES ON PROGRAM USE --------------------------------
C *** CONTROL CARD 1***
C COLS 1-3     N, THE NUMBER OF POINTS (ITEMS, INDIVIDUALS)
C COLS 4-6     NV, THE NUMBER OF VARIABLES.
C COLS 7-9     MAX, THE STARTING NUMBER OF CLUSTERS.
C *** CONTROL CARD 2 ***
C COLS 1-3     MIN, THE TERMINAL NUMBER OF CLUSTERS.
C COL 4        ISTAN = 1    USE RAW DATA.
C                    = 2    CONVERT DATA TO DEVIATION FORM;
C                    = 3    CONVERT DATA TO STANDARD SCORES.
C COL 5        IREAD1. IF THE VALUE OF IREAD1 IS POSITIVE, THE
C              STARTING VALUES OF THE CENTRE COORDINATES WILL BE READ
C              AS DATA. IF THE VALUE OF IREAD1 IS ZERO, VALUES OF THE
C              CENTRE COORDINATES WILL BE GENERATED RANDOMLY USING A
C              USER-SUPPLIED PSEUDORANDOM NUMBER GENERATOR (SEE APPENDIX
C              A FOR DETAILS OF PSEUDORANDOM NUMBER GENERATORS).
C *** CONTROL CARD 3 ***
C COLS 1-80    DATA INPUT FORMAT, INCLUDING BRACKETS.
C *** CONTROL CARD 4 ***
C COLS 1-7     IX1, A RANDOM, ODD, POSITIVE INTEGER.
C COLS 8-14    IX2, ANOTHER RANDOM, ODD, POSITIVE INTEGER.
C              THESE TWO INTEGERS ARE NEEDED IN THE RANDOM NUMBER
C              GENERATOR USED IN THIS VERSION OF THE PROGRAM.  IF
C              ANOTHER GENERATOR IS USED, YOU WILL NOT NEED THIS CARD,
C              BUT YOU MAY NEED TO REPLACE IT.
C *** DATA DECK ***
C THE DATA IS READ INTO ARRAY X USING THE FORMAT SPECIFIED ON THE 4TH
C CONTROL CARD.  ENTER DATA CASE BY CASE, BEGINNING EACH CASE ON A
C FRESH CARD.
C *** CONTROL CARD 5 ***
C COLS 1-80    FMT1, THE OUTPUT FORMAT FOR SUBROUTINE MATPR. REFER TO
C              APPENDIX A FOR DETAILS.
C *** CENTRE COORDINATES ***
C    **********************************************************
C THE CENTRE COORDINATES ARE NEEDED ONLY IF THE VALUE OF IREAD1 ON
C CONTROL CARD 3 IS POSITIVE.  IF IT IS ZERO, OMIT THIS SECTION OF THE
C DATA DECK COMPLETELY.
C    **********************************************************
C PUNCH CENTRE COORDINATES (IE THE VALUES OF EACH OF THE MAX CENTRES
C FOR EACH OF THE NV VARIABLES) USING THE FORMAT SPECIFIED ON
C CONTROL CARD 3.  START THE DATA FOR EACH CENTRE ON A FRESH CARD.
C *** SUBROUTINES REQUIRED ***
C          MATPR, FROM APPENDIX A.
C
C THE PROGRAM NEEDS A PSEUDO-RANDOM NUMBER GENERATOR. DETAILS OF SUCH
C A GENERATOR ARE GIVEN IN APPENDIX A. MOST COMPUTER MANUFACTURERS
C SUPPLY A RANDOM NUMBER GENERATOR WITH THEIR FORTRAN OPERATING SYSTEM,
C DETAILS OF WHICH WILL BE AVAILABLE FROM YOUR COMPUTING CENTRE.
C
C THE PROGRAM USES INPUT DEVICE 5 (CARD READER) AND OUTPUT DEVICE 6
C (ASSUMED TO BE LINEPRINTER).
C
      DIMENSION X(5000)
      L=5000
C IF DIMENSION OF X IS ALTERED CHANGE L APPROPRIATELY
C THE DIMENSION OF X SHOULD BE AT LEAST (N*P)+(MAX*P)+2*P+3*N+2*MAX+1
C WHERE N IS NUMBER OF OBSERVATIONS,P IS NUMBER OF VARIABLES AND MAX IS
C THE MAXIMUM NUMBER OF CLUSTER CENTRES TO BE USED.
      READ(5,1) N, NV, MAX
    1 FORMAT(3I3)
C SET UP BASE ADDRESSES OF ARRAYS
      N1=N*NV+1
      N2=N1+MAX*NV
      N3=N2+NV
      N4=N3+NV
      N5=N4+N
      N6=N5+MAX
      N7=N6+MAX
      N8=N7+N
      N9=N8+N
      IF(N9.GT.L) GOTO 3
C CALL MASTER ROUTINE
      CALL EUCLID(X(1),X(N1),X(N2),X(N3),X(N4),X(N5),X(N6),X(N7),
     * X(N8),N,NV,MAX)
```

Table 6.13   *continued*

```
      STOP
    3 WRITE(6,2) L, N9
    2 FORMAT('OINCREASE DIMENSION OF X FROM',I4,' TO',I4,' AND RESUBM
     *IT YOUR JOB')
      STOP
      END



      SUBROUTINE EUCLID(X,CENTRE,S,XM,MEMBER,RSS,NUMBER,IWK,WK,N,P,MAX)
      INTEGER P
      DIMENSION X(N,P),CENTRE(MAX,P),S(P),XM(P),MEMBER(N),RSS(MAX),
     * NUMBER(MAX),IWK(N),  WK(N),FMT1(20),FMT2(20)
      IND=1
      DO 300 I=1,MAX
  300 NUMBER(I)=0
      WRITE(6,910)
  910 FORMAT(
     *'1---------------------EUCLIDEAN--------------------------------'//
     *'  ---------------------CLUSTER----------------------------------'//
     *'  ---------------------ANALYSIS---------------------------------'//
     *'  -----------P. M. MATHER--NOTTINGHAM--1974---------------------'//)
      READ(5,1) MIN, ISTAN, IREAD1
    1 FORMAT(I3,2I1)
      PP=N
      READ(5,2) FMT2
C FMT2 HOLDS DATA INPUT FORMAT
    2 FORMAT(20A4)
      READ(5,170) IX1,IX2
  170 FORMAT(2I7)
C IX1 AND IX2 ARE 'SEEDS' FOR RANDOM NUMBER GENERATOR LISTED IN APPENDIX
C A -- SUBROUTINES SETUP AND RANDOM
      CALL SETUP
      READ(5,FMT2) ((X(I,J),J=1,P),I=1,N)
      READ(5,2) FMT1
C FMT1 IS OUTPUT FORMAT USED IN SUBROUTINE MATPR --- CONTROL CARD 2
      NDM=N*P
      IP=0
C SUBROUTINE MATPR IS LISTED IN APPENDIX A
      CALL MATPR(16HDATA AS INPUT   ,2,FMT1,6,X,NDM,N ,N,P,IP,2,8)
      DO 6 I=1,P
      A=0.0
      DO 5 J=1,N
    5 A=A+X(J,I)
    6 XM(I)=A/PP
      DO 8 I=1,P
      B=XM(I)
      F=0.0
      DO 7 J=1,N
      A=(X(J,I)-B)**2
    7 F=F+A
    8 S(I)=SQRT(F/PP)
      WRITE(6,9) (I,XM(I),S(I),I=1,P)
    9 FORMAT('0    MEANS AND STANDARD DEVIATIONS'//
     *('  ',I4,2X,2F12.4))
      GOTO (3,4,10), ISTAN
C CONVERT TO STANDARD SCORES
    4 DO 11 J=1,P
      A=XM(J)
      B=S(J)
      DO 11 I=1,N
   11 X(I,J)=(X(I,J)-A)/B
      GOTO 91
C CONVERT TO DEVIATIONS FROM THE MEAN
   10 DO 13 J=1,P
      A=XM(J)
```

Table 6.13    *continued*

```
      DO 13 I=1,N
   13 X(I,J)=X(I,J)-A
   91 DO 92 J=1,P
   92 XM(J)=0.0
    3 IF (IREAD1) 30,30,21
C READ IN CENTRE COORDINATES IF REQUESTED
   21 READ(5,FMT2) ((CENTRE(I,J),J=1,P),I=1,MAX)
      GOTO 35
C OR GENERATE RANDOMLY IN RANGE (MAX,MIN)
   30 CALL RANGE(X,P,S,WK,N)
      DO 31 J=1,P
      A=S(J)
      B=WK(J)
      DO 31 I=1,MAX
C RANDOM IS THE NAME OF A USER-SUPPLIED FUNCTION WHICH GENERATES
C PSEUDO-RANDOM NUMBERS IN THE RANGE (0,1)
C SEE APPENDIX A, FUNCTION *RANDOM*
   31 CENTRE(I,J)=RANDOM(IX1,IX2)*(A-B)+B
   35 IP=1
      NDM=MAX*P
C SUBROUTINE MATPR IS LISTED IN APPENDIX A
      CALL MATPR(24HINITIAL CLUSTER CENTRES  ,3,FMT1,6,CENTRE,NDM,MAX,
     * MAX,P,IP,2,8)
      L=MAX
      MAXL=MAX
C CALL CLUMP FOR EACH NUMBER OF CLUSTER CENTRES IN RANGE MAX-MIN
  900 CALL CLUMP(X,CENTRE,MAX,MEMBER,NUMBER,IWK,P,N,L,IND)
      IF(L.LT.MAX) GOTO 60
C CHECK THAT NO CLUSTERS HAVE MEMBERSHIP OF 0
      CALL CHEK(NUMBER,CENTRE,MEMBER,N,L,MAX,P,K)
      IF(K) 61,60,61
   61 L=L-K
      MAXL=L
      WRITE(6,920) L,K
  920 FORMAT('0STARTING NUMBER OF CLUSTERS REDUCED TO',I3,' BECAUSE',I3,
     *' CLUSTERS HAD MEMBERSHIP OF 0'//)
   60 CALL OUTPUT(XM,CENTRE,MAX,NUMBER,WK,N,P,L,A,X,MEMBER)
      RSS(L)=A
      IF(L.EQ.MIN) GOTO 901
C MERGE 2 CLUSTERS FOR NEXT ITERATION
      CALL MERGE(CENTRE,MAX,NUMBER,MEMBER,X,P,L,N)
      IND=0
      L=L-1
      IF(L.GE.MIN) GOTO 900
C CALCULATE F-RATIOS
  901 CALL FCALC(RSS,MAX,P,N,MIN,MAXL)
      STOP
      END



      SUBROUTINE CHEK(NUMBER,CENTRE,MEMBER,N,L,MAX,P,K)
C SUBROUTINE TO CHECK THAT NO CLUSTER HAS ZERO MEMBERSHIP
      INTEGER P
      DIMENSION NUMBER(L),CENTRE(MAX,P),MEMBER(N)
      K=0
      I=0
      M=L
    1 I=I+1
      IF(I.GT.M) RETURN
      IF(NUMBER(I)) 1,2,1
    2 K=K+1
      M=M-1
      IF(I.EQ.M) RETURN
      DO 5 JJ=1,N
      IF(MEMBER(JJ).LT.I) GOTO 5
```

Table 6.13 *continued*

```
      MEMBER(JJ)=MEMBER(JJ)-1
    5 CONTINUE
      DO 3 J=I,M
      K1=J+1
      DO 4 JJ=1,P
    4 CENTRE(J,JJ)=CENTRE(K1,JJ)
    3 NUMBER(J)=NUMBER(K1)
      I=I-1
      GOTO 1
      END




      SUBROUTINE RANGE(X,NV,S,XX,N)
C FINDS MAXIMUM AND MINIMUM OF EACH VARIABLE
      DIMENSION X(N ,NV),S(NV),XX(NV)
      DO 2 I=1,NV
      S(I)=X(1,I)
      XX(I)=S(I)
      DO 2 J=2,N
      IF(X(J,I).GT.S(I)) S(I)=X(J,I)
    2 IF(X(J,I).LT.XX(I)) XX(I)=X(J,I)
      RETURN
      END




      SUBROUTINE OUTPUT(XM,CENTRE,MAX,NUMBER,WK,N,P,L,C,X,MEMBER)
C OUTPUTS INFORMATION AT EACH ITERATION
      INTEGER P
      DIMENSION XM(P),CENTRE(MAX,P),NUMBER(L),WK(N),X(N,P),MEMBER(N)
      DIMENSION BUF(6),IBUF(6)
      WRITE(6,1) L
    1 FORMAT('1RESULTS FOR CURRENT ITERATION WITH ',I4,' CLUSTER CENTRES
     *'///' CLUSTER   SIZE    DIST FROM GRAND MEAN    COORDINATES'/)
    2 FORMAT(' ',I7,3X,I4,10X,F12.3/(40X,8F10.4))
      DO 3 I=1,L
      A=0.0
      DO 4 J=1,P
    4 A=A+(CENTRE(I,J)-XM(J))**2
      A=SQRT(A)
    3 WRITE(6,2) I, NUMBER(I), A,(CENTRE(I,J),J=1,P)
      WRITE(6,5)
    5 FORMAT('0      DISTANCE MATRIX FOR CLUSTER CENTRES'///)
      DO 7 I=2,L
      I1=I-1
      DO 8 J=1,I1
      WK(J)=0.0
      DO 8 K=1,P
    8 WK(J)=WK(J)+(CENTRE(I,K)-CENTRE(J,K))**2
      DO 11 J=1,I1
   11 WK(J)=SQRT(WK(J))
    7 WRITE(6,6) I, (WK(J),J=1,I1)
    6 FORMAT(' ',I6,(10F10.4/))
      WRITE(6,9)
    9 FORMAT('0CLUSTER MEMBERSHIP FOR INDIVIDUALS'//)
   10 FORMAT(' CLUSTER NUMBER ',I4/)
      C=0.0
      DO 52 I=1,L
      II=NUMBER(I)
      WRITE(6,10) I
      K=0
      B=0.0
      KN=0
      DO 53 J=1,N
```

Table 6.13   *continued*

```
      IF(MEMBER(J).NE.I) GOTO 53
      K=K+1
      LL=MEMBER(J)
      A=0.0
      DO 54 IJ=1,P
   54 A=A+(X(J,IJ)-CENTRE(LL,IJ))**2
      C=C+A
      A=SQRT(A)
      B=B+A
      KN=KN+1
      IBUF(KN)=J
      BUF(KN)=A
      IF(KN.LT.6) GOTO 59
      KN=0
      WRITE(6,55) (IBUF(M), BUF(M), M=1,6)
   59 IF(K-II) 53,56,53
   53 CONTINUE
   56 IF(KN.EQ.0) GOTO 61
      WRITE(6,55) (IBUF(M), BUF(M), M=1,KN)
   61 B=B/FLOAT(II)
      WRITE(6,57) B
   52 CONTINUE
   55 FORMAT(' ',6(I3,2X,F10.4,2X))
   57 FORMAT('0AVERAGE POINT - TO - CENTRE DISTANCE', G12.4)
      A=N-L
      B=SQRT((1.0/A)*C)
      WRITE(6,60) B
   60 FORMAT('0R.M.S. DEVIATION FROM CENTRES = ',G12.4)
      RETURN
      END



      SUBROUTINE MERGE(CENTRE,MAX,NUMBER,MEMBER,X,P,L,N)
C SUBROUTINE TO DETERMINE WHICH TWO CLUSTERS ARE TO BE COMBINED
      INTEGER P
      DIMENSION CENTRE (MAX,P),NUMBER(L),MEMBER(N),X(N,P)
      DATA BIG/1.E25/
      OLDA=BIG
      L1=L-1
      DO 1 I=1,L1
      Q=NUMBER(I)
      IJ=I+1
      DO 1 J=IJ,L
      QQ=NUMBER(J)
      A=0.0
      DO 2 K=1,P
    2 A=A+(CENTRE(I,K)-CENTRE(J,K))**2
      A=A*Q*QQ/(Q+QQ)
      IF(A.GT.OLDA) GOTO 1
      I1=I
      I2=J
      OLDA=A
    1 CONTINUE
      WRITE(6,8) I1,I2
    8 FORMAT('0CLUSTERS MERGED AT THIS ITERATION:',I4,' AND',I4)
      Q=NUMBER(I1)
      QQ=NUMBER(I2)
      NUMBER(I1)=NUMBER(I1)+NUMBER(I2)
      DO 3 I=1,N
      IF(MEMBER(I).EQ.I2) MEMBER(I)=I1
    3 IF(MEMBER(I).GT.I2) MEMBER(I)=MEMBER(I)-1
      DO 4 J=1,P
    4 CENTRE(I1,J)=(CENTRE(I1,J)*Q+CENTRE(I2,J)*QQ)/(Q+QQ)
      IF(I2.EQ.L) GOTO 90
      DO 6 K=I2,L1
```

Table 6.13  *continued*

```
      K1=K+1
      DO 7 I=1,P
    7 CENTRE(K,I)=CENTRE(K1,I)
    6 NUMBER(K)=NUMBER(K1)
   90 RETURN
      END



      SUBROUTINE FCALC(RSS,MAX,P,N,MIN,MAXL)
C COMPUTES F-RATIOS
      INTEGER P
      DIMENSION RSS(MAX)
      WRITE(6,1)
    1 FORMAT('1   F-RATIO TESTS OF H0: LARGER NUMBER OF CLUSTERS IS UNWAR
     *RANTED'//
     *       '  NUMBER V NUMBER       F         DF1          DF2'/)
      MAX1=MAXL-1
      DO 3 J=MIN,MAX1
      JJ=J+1
      C=(JJ/FLOAT(J))**(2.0/FLOAT(P))
      F=((RSS(J)-RSS(JJ))/RSS(JJ))/(((N-J)/FLOAT(N-JJ))*C-1.0)
      N2=P*(N-JJ)
    3 WRITE(6,4) J,JJ,F,P,N2
    4 FORMAT('  ',I4,3X,I5,2X,F10.4,4X,I3,I11)
      RETURN
      END



      SUBROUTINE CLUMP(X,CENTRE,MAX,MEMBER,NUMBER,IWK,P,N,NO,IND)
C THIS IS THE MAIN CLUSTERING ROUTINE, CALLED ONCE PER ITERATION
      INTEGER P
      DIMENSION X(N,P),CENTRE(MAX,P),MEMBER(N),NUMBER(NO),IWK(N)
      DATA BIG/1.0E20/
      WRITE(6,91)
      IF (IND) 9,9,1
    1 DO 2 I=1,N
      A=BIG
      DO 3 J=1,NO
      B=0.0
      DO 4 K=1,P
    4 B=B+(X(I,K)-CENTRE(J,K))**2
      IF(B-A) 22,3,3
   22 A=B
      M=J
    3 CONTINUE
      MEMBER(I)=M
    2 NUMBER(M)=NUMBER(M)+1
      DO 5 I=1,NO
      DO 5 J=1,P
    5 CENTRE(I,J)=0.0
      DO 6 I=1,N
      J=MEMBER(I)
      DO 6 K=1,P
    6 CENTRE(J,K)=CENTRE(J,K)+X(I,K)
      DO 8 J=1,NO
      IF(NUMBER(J)) 8,8,88
   88 Q=NUMBER(J)
      DO 87 I=1,P
   87 CENTRE(J,I)=CENTRE(J,I)/Q
    8 CONTINUE
    9 KOUNT=0
      DO 10 I=1,N
      K=MEMBER(I)
```

Table 6.13   *continued*

```
        J=K
        IF(NUMBER(K).LE.1) GOTO 20
        C=BIG
        D=0.0
        DO 11 L=1,P
    11  D=D+(X(I,L)-CENTRE(K,L))**2
        DO 12 L=1,NO
        IF(L-K) 13,12,13
    13  E=0.0
        DO 14 IJ=1,P
    14  E=E+(X(I,IJ)-CENTRE(L,IJ))**2
        IF(E-C) 23,12,12
    23  C=E
        IK=L
    12  CONTINUE
        Q=NUMBER(IK)
        QQ=NUMBER(K)
        C=C*Q/(Q+1.0)
        D=D*QQ/(QQ-1.0)
        IF(C-D) 15,20,20
    15  J=IK
        WRITE(6,90) I,K,J
        NUMBER(J)=NUMBER(J)+1
        NUMBER(K)=NUMBER(K)-1
        DO 16 L=1,P
        CENTRE(K,L)=(CENTRE(K,L)*QQ-X(I,L))/(QQ-1.0)
    16  CENTRE(J,L)=(CENTRE(J,L)*Q+X(I,L))/(Q+1.0)
        KOUNT=KOUNT+1
    20  IWK(I)=J
    10  CONTINUE
        IF(KOUNT) 21,21,17
    17  DO 18 I=1,N
    18  MEMBER(I)=IWK(I)
        GOTO 9
    90  FORMAT(' ',3(I4,6X))
    91  FORMAT('0RECORD OF MOVES'///'   ITEM MOVES  FROM',6X,'TO'// ',12X,
       *'CLUSTER   CLUSTER'/)
    21  RETURN
        END
```

large number of points (20 or more) the computing time requirements may become excessive if a large number of iterations is specified.

(v) Nonlinear Mapping (Table 6.16). The nonlinear mapping algorithm was developed by Sammon (1969) at Rome Air Force Base, New York. This program uses Sammon's algorithm as outlined in his 1969 paper. Since expressions for both first and second order derivatives are provided it may be more economical to use a Newton—Raphson type algorithm to locate the minimum mapping error once the region of the minimum has been found by gradient methods (See Chapter 3, Section 5). However, the procedure given here is quite efficient — faster than nonmetric multidimensional scaling but not as rapid as principal coordinates analysis. The step length (Sammon's 'magic factor') has an influence on the rate of convergence; perhaps the step length could be adjusted to reflect the gradient of the mapping error instead of remaining fixed.

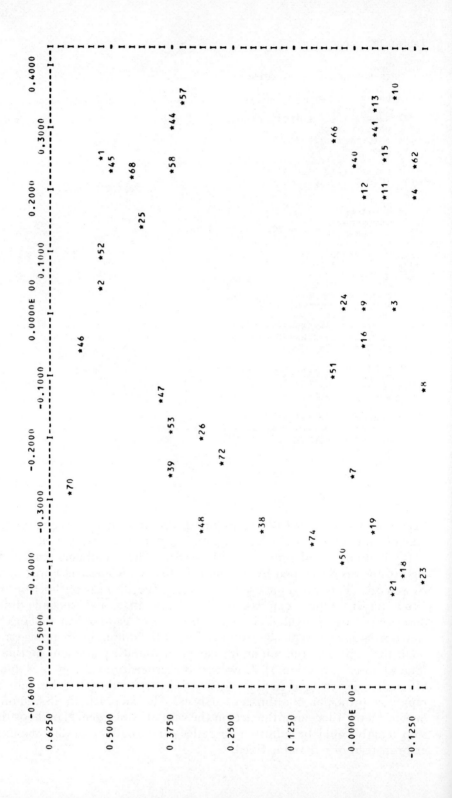

```
                                                                         I   I
                                                                         I   I
                                                                       I-I   I
                                                                         I   I
                                                                         I   I
                                                                       I-I   I
                                                                         I   I
                                                                         I   I
                                                                       I-I   I
                                                                         I   I
                                                   *27  *30               I   I
                                                                       I-I   I
                                              *67                        I   I
                                              *36                        I   I
                                              *35                      I-I   I
                                                                         I   I
                                                                         I   I
                                                                       I-I   I
                                    *33                                  I   I
                                                                         I   I
                              *6                                       I-I   I
                     *64 *65                                             I   I
                     *37                                                 I   I
                                                                       I-I   I
              *5                                                         I   I
                                                                         I   I
        *32                                                            I-I   I
                                                                         I   I
                                                                         I   I
    *54                                                                I-I   I
       *29  *34                                                          I   I
                                                                         I   I
  *17                                                                  I-I   I
                                                                         I   I
*60                                                                      I   I
                                                                       I-I   I
                                                                         I   I
    *42                                                                  I   I
              *49                                                      I-I   I
                                                                         I   I
                                                                         I   I
                                                                       I-I   I
  *56 *43                                                                I   I
                                                                         I   I
                                                                       I-I   I
                                                                         I   I
*22                                                                      I   I
                                                                       I-I   I
*73                                                                      I   I
                                                                         I   I
                                                                       I-I   I
                                                                         I   I
                                                                         I   I
                                                                       I-I---I
-0.2500
-0.3750
-0.5000
-0.6250

OVERPRINT TABLE
ITEM NO.  ALSO LOCATION OF ITEM NO.

70        71
1         59
68        69
9         20
10        14
27        28
54        55
54        61
29        31
49        63
```

Figure 6.15  Results of principal coordinates analysis of Tien's data (Table 6.9). The plot shows the first two dimensions

## Table 6.14
## Computer program for principal coordinates analysis

```
C******************** NOTES ON PROGRAM USE *****************************
C THE FOLLOWING I/O UNITS ARE REQUIRED: CARDREADER (CHANNEL 5),
C LINEPRINTER (CHANNEL 6) AND DISC FILE (CHANNEL 7).
C CORE REQUIRED: N*N+N*NEIG+4*N. IF DIMENSION OF C IN MAIN PROGRAM
C IS ALTERED, DON'T FORGET TO ADJUST VALUE OF NO.
C
C DATA IN ORDER:
C CARD 1: FORMAT 6I3: P N NEIG N1 N2 N3
C P NUMBER OF VARIABLES. PUNCH IN COLS 1-3
C N NUMBER OF CASES. PUNCH IN COLS 4-6
C NEIG NUMBER OF DIMENSIONS. PUNCH IN COLS 7-9
C N1 NUMBER OF CONTINUOUS (QUANTITATIVE) VARIABLES. PUNCH IN COLS 10-12
C N2 NUMBER OF MULTISTATE VARIABLES. PUNCH IN COLS 13-15
C N3 NUMBER OF BINARY (PRESENCE-ABSENCE) VARIABLES. PUNCH IN COLS 16-18
C
C CARD 2: MODE
C MODE SHOULD BE EITHER CASE IF THE DATA MATRIX IS PUNCHED ROWWISE OR
C VARI IF DATA MATRIX TO BE PUNCHED COLUMNWISE. PUNCH EITHER CASE OR
C VARI IN COLUMNS 1-4.
C
C CARD 3: PRINT
C THIS CARD SHOULD CONTAIN THE INSTRUCTION PRINT OR DONT PRINT
C STARTING IN COLUMN 1. THE EFFECT IS TO CAUSE THE SIMILARITY MATRIX TO
C BE PRINTED OR TO SUPPRESS PRINTING.
C
C CARD 4: OQ
C THE INSTRUCTION INCLUDE PUNCHED IN COLUMNS 1-7 WILL CAUSE NEGATIVE
C MATCHES TO BE CONSIDERED IN THE COMPUTATION OF THE SIMILARITY COEFFIC-
C IENTS FOR BINARY AND MULTISTATE DATA. ANY OTHER COMMAND WILL CAUSE
C NEGATIVE MATCHES TO BE EXCLUDED.
C
C CARD 5: FMT
C DATA INPUT FORMAT. USE UP TO 80 COLUMNS.
C
C CARD 6: DATA MATRIX
C PUNCH ROW-WISE OR VARIABLEWISE AS INDICATED ON CARD 2. USE THE
C INPUT FORMAT SPECIFIED ON CARD 5. START EACH CASE OR VARIABLE ON A
C NEW CARD.
C
C SUBROUTINES REQUIRED
C MATPR, TRIDI, RATQR, EIGVEC, TRBAK1, ZERO, PLOTS (AND ASSOCIATED
C ROUTINES). THESE ARE LISTED IN APPENDIX A.
C
      INTEGER P
      REAL MODE
      DIMENSION C(7500),IS(7500),FMT(20)
      EQUIVALENCE(C,IS)
      DATA CASE,VBLE/4HCASE,4HVARI/
C READS PARAMETERS, ALLOCATES BASE ADDRESSES AND CALLS MAIN ROUTINES
      NO=2500
C DON'T FORGET TO ALTER NO IF ARRAY DIMENSIONS ARE CHANGED
      REWIND 7
      INPUT=5
C READ CONTROL CARD 1: P,N,NEIG,N1,N2,N3
      READ(5,1) P,N,NEIG,N1,N2,N3
    1 FORMAT(6I3)
C READ CONTROL CARDS 2,3,AND 4: MODE,OP AND OQ
      READ(5,2) MODE , OP, OQ
    2 FORMAT(A4)
C CARD 5 -- INPUT FORMAT
      READ(5,10) FMT
   10 FORMAT(20A4)
      IF(MODE.EQ.CASE) GOTO 3
      IF(MODE.EQ.VBLE) GOTO 4
      WRITE(6,5) MODE
```

Table 6.14  *continued*

```
    5 FORMAT(' ',A4,' IS NOT A VALID COMMAND.  OPTIONS ARE CASE BY CASE O
     *R VARIABLE BY VARIABLE'/)
C ERROR EXIT IF MODE NOT CORRECTLY SPECIFIED
      STOP
C CALL SUBROUTINE INPUT IF MODE IS 'CASE BY CASE'
    3 CALL INN(C,P,N,FMT)
      INPUT=7
    4 M1=N+1
      M2=M1+N*N
      M3=M2+N
      M4=M3+N
      M5=M4+N
C CALL MAIN ROUTINE
      CALL PCORD(C(1),C(M1),C(M2),C(M3),C(M4),C(M5),P,N,NEIG,INPUT,N1,
     *N2,N3,OP,OQ,FMT)
C CALL PLOTTER ROUTINE -- SEE APPENDIX A
      CALL PLOTS(C(M5),1,NEIG,N,N,NEIG,O,IS,N)
      STOP
      END



      SUBROUTINE INN(C,M,N,FMT)
C IF MODE IS 'CASE BY CASE' THIS SUBROUTINE READS DATA CASEWISE AND
C TRANSFERS IT VARIABLEWISE TO UNIT 7.
      DIMENSION C(N,M), FMT(20)
      DO 1 I=1,N
    1 READ(5,FMT) (C(I,J),J=1,M)
      DO 3 I=1,M
    3 WRITE(7) (C(J,I),J=1,N)
      REWIND 7
      RETURN
      END
      SUBROUTINE PCORD(X,S,EIG,WK,WK2,V,P,N,NEIG,INPUT,N1,N2,N3,OP,OQ,
     1 FMT)
C MAIN ROUTINE
      INTEGER P
C DIMENSION OF INT MUST BE AT LEAST N
      LOGICAL INT(200)
      DIMENSION X(N),S(N,N),EIG(N),WK(N),OUTFMT(20),WK2(N),V(N,NEIG)
      DIMENSION FMT(20)
      DATA OUTFMT(1),OUTFMT(2),OUTFMT(3),OUTFMT(4),OUTFMT(5)/4H(' ',4H,4
     +X,,4HI6,8,4HF12.,4H4)   /,TOT/0.0/
   20 FORMAT('1-------------PRINCIPAL--------------------------------'/
     *          '--------------COORDINATES-------------------------------'/
     *          '---------------ANALYSIS--------------------------------'/
     *          '  -----PROGRAM BY P.M.MATHER,NOTTINGHAM--1974----------'/)
      WRITE(6,20)
C CALCULATE GOWER'S GENERAL COEFFICIENT OF SIMILARITY
      CALL SIMCAL(X,N,S,N,N1,N2,N3,OUTFMT,INPUT,P,OP,OQ,FMT)
      REWIND 7
      FN=N
CALCULATE ROW, COLUMN AND OVERALL MEANS
      DO 2 I=1,N
      S(I,I)=1.0
      EIG(I)=0.0
      DO 2 J=1,I
    2 S(J,I)=S(I,J)
      DO 4 I=1,N
      DO 3 J=1,N
    3 EIG(I)=EIG(I)+S(I,J)
    4 TOT=TOT+EIG(I)
      DO 5 I=1,N
    5 EIG(I)=EIG(I)/FN
      TOT=TOT/(FN*FN)
CONVERT FROM MATRIX E TO MATRIX F (SEE TEXT)
```

Table 6.14    *continued*

```
          DO 6 I=1,N
          DO 6 J=1,I
        6 S(I,J)=S(I,J)-EIG(I)-EIG(J)+TOT
C THE FOLLOWING TWO PARAMETERS ARE MACHINE-DEPENDENT AND
C   MUST BE ALTERED TO SUIT THE PARTICULAR COMPUTER.
C REFER TO APPENDIX A FOR MORE INFORMATION
          ACC=2.0**(-37)
          TOL=2.0**(-255)
          IFL=0
          WRITE(6,8) NEIG
        8 FORMAT('OYOU HAVE REQUESTED',I4,' EIGENVALUES')
C TRIDIAGONALIZE MATRIX F USING ROUTINE FROM APPENDIX A
          CALL TRIDI(S,X,WK,WK2,N,N,TOL)
          DLAM=1.0E-7
          DO 16 I=1,N
       16 EIG(I)=X(I)
C FIND NEIG EIGENVALUES OF MATRIX F USING ROUTINE FROM APPENDIX A
          CALL RATQR(EIG,WK,WK2,N,NEIG,DLAM,ACC,1)
          REWIND 7
          WRITE(7) S
          ENDFILE 7
          REWIND 7
C FIND CORRESPONDING EIGENVECTORS BY INVERSE ITERATION USING ROUTINE
C FROM APPENDIX A
          CALL EIGVEC(X,WK,EIG,V,N,INT,S,ACC,TOL,NEIG,N,IFL)
          IF(IFL) 17,17,22
       17 WRITE(6,7) (I,EIG(I),I=1,NEIG)
        7 FORMAT('OEIGENVALUES OF SIMILARITY MATRIX'//(' ',I4,2X,F12.4))
          READ (7) S
CONVERT EIGENVECTORS OF TRIDIAGONAL MATRIX BACK TO EIGENVECTORS OF THE
C ORIGINAL MATRIX USING ROUTINE FROM APPENDIX A
          CALL TRBAK1(V,N,N,NEIG,S,N,WK)
CONVERT EIGENVECTORS TO PRINCIPAL COORDINATES AND PLOT PAIRWISE USING
C ROUTINE FROM APPENDIX A
          DO 9 I=1,NEIG
          ST=SQRT(EIG(I))
          DO 9 J=1,N
        9 V(J,I)=V(J,I)*ST
          IN=N*NEIG
          IP=1
          CALL MATPR(32HCOORDINATES ON PRINCIPAL AXES    ,8,OUTFMT,6,V,IN,N,
        1 N,NEIG,IP,2,4)
       22 RETURN
          END
          SUBROUTINE SIMCAL(X,N,S,NDIM,N1,N2,N3,OUTFMT,INPUT,P,OP,OQ,FMT)
C SUBROUTINE TO CALCULATE GOWER'S GENERAL COEFFICIENT OF SIMILARITY
          INTEGER P
          DIMENSION X(N),S(NDIM,N),OUTFMT(20),FMT(20)
          DATA PRINT/4HPRIN/
          WRITE(6,100)
      100 FORMAT('ODATA FOR QUANTITATIVE VARIABLES - PRINTED VARIABLEWISE'/)
          FP=P
          IF(N1.EQ.0) GOTO 33
          FN=N1
          CALL ZERO(S,NDIM,N,N,1)
          DO 7 I=1,N1
          IF(INPUT.EQ.7) GOTO 40
          READ(5,FMT) X
          GOTO 11
       40 READ(7) X
       11 WRITE(6,110) X
      110 FORMAT(' ',10F11.3)
          BIG=X(1)
          SMALL=BIG
          DO 5 J=2,N
          IF(X(J).GT.BIG) BIG=X(J)
        5 IF(X(J).LT.SMALL) SMALL=X(J)
```

Table 6.14   *continued*

```
      RANGE=BIG-SMALL
      DO 7 J=1,N
      DO 7 K=1,J
    7 S(K,J)=S(K,J)+(1.0-(ABS(X(J)-X(K))/RANGE))/FN
      DO 50 I=1,N
      DO 50 J=1,I
   50 S(I,J)=S(J,I)
   33 IF(N2.EQ.0) GOTO 34
COMPUTE COEFFICIENTS FOR MULTISTATE VARIABLES, IF ANY
      CALL ZERO(S,NDIM,N,N,2)
      CALL MATCH(X,N,S,NDIM,N2,1,INPUT,IND,P,OQ)
      DO 51 I=1,N
      DO 51 J=1,I
   51 S(I,J)=S(I,J)+S(J,I)
   34 IF(N3.EQ.0) GOTO 43
COMPUTE COEFFICIENTS FOR BINARY VARIABLES, IF ANY
      CALL ZERO(S,NDIM,N,N,2)
      CALL MATCH(X,N,S,NDIM,N3,2,INPUT,IND,P,OQ)
      DO 52 I=1,N
      DO 52 J=1,I
   52 S(I,J)=S(I,J)+S(J,I)
C DECIDE WHETHER TO PRINT SIMILARITY MATRIX, USING ROUTINE FROM
C APPENDIX A
C  43 IF(OP.NE.PRINT) GOTO 45
C
   43 REWIND 9
      WRITE(9) ((S(I,J),J=1,I-1),I=2,N)
      ENDFILE 9
      IF(OP.NE.PRINT) GOTO 45
C
      IP=1
      N4=NDIM*N
      CALL MATPR(56HSIMILARITY MATRIX USING GOWER'S GENERAL COEFFICIENT
     1    ,14,OUTFMT,6,S,N4,NDIM,N,N,IP,1,4)
   45 IF(N2+N3.EQ.0) RETURN
      GOTO (10,20), IND
   10 WRITE(6,310)
      GOTO 30
   20 WRITE(6,300)
   30 RETURN
  300 FORMAT('ONEGATIVE MATCHES CONSIDERED')
  310 FORMAT('ONEGATIVE MATCHES NOT CONSIDERED')
      END
      SUBROUTINE MATCH(X,N,S,N1,NQ,ID,INPUT,IND,P,OQ)
COMPUTES GOWER'S COEFFICIENT FOR MULTISTATE AND BINARY VARIABLES
      INTEGER P
      DIMENSION X(N),S(N1,N)
      DATA OP/4HINCL/
      IND=1
      IF(OP.EQ.OQ) IND = 2
      FN=NQ
      FN=FN/FLOAT(P)
      WRITE(6,11)
      GOTO (13,14), ID
   13 WRITE(6,12)
      GOTO 15
   14 WRITE(6,9)
   15 DO 3 I=1,NQ
      IF(INPUT.EQ.7) GOTO 30
C READ DATA FROM APPROPRIATE SOURCE
      READ(5,1) X
    1 FORMAT(50F0.0)
      GOTO 31
   30 READ(7) X
   31 WRITE(6,10) X
      DO 3 J=1,N
      DO 3 K=1,J
```

Table 6.14 *continued*

```
      IF(IND.EQ.2) GOTO 5
      IF(X(J)+X(K)) 3,5,5
    5 IF(X(J)-X(K)) 4,6,4
    6 S(K,J)=S(K,J)+1.0
    4 S(K,J)=S(K,J)+100.0
    3 CONTINUE
      DO 2 I=1,N
      DO 2 J=1,I
      B=AINT((S(J,I)+0.5)/100.0)
      IF(B) 8,8,7
    7 S(J,I)=((S(J,I)-B*100.0)/B)*FN
      GO TO 2
    8 S(J,I)=0.0
    2 CONTINUE
   10 FORMAT(' ',10F11.3)
   11 FORMAT('0DATA FOR NON-QUANTITATIVE VARIABLES')
    9 FORMAT('+',37X,'(TWO-STATE)'///)
   12 FORMAT('+',37X,'(MULTI-STATE)'///)
      RETURN
      END
```

Table 6.15
Computer program for nonmetric multidimensional scaling

```
C ###### NOTES ON USE OF PROGRAM ######
C REQUIRES I/O UNITS: CARD READER, LINEPRINTER AND DISC FILE.
C CHANNEL NUMBERS ARE: IN , OUT, AND MT RESPECTIVELY. THESE ARE SET IN
C THE DATA STATEMENT. ALSO SET IN THE DATA STATEMENT ARE THE VALUES
C NDIM (FIRST DIMENSION OF ARRAYS X1, X2, X3 AND IX2) AND NSIZE (MAX.
C NUMBER OF ITERATIONS). NDIM2 IS THE DIMENSION OF DOTHER, DHAT, DISSIM,
C DIST, IJ, KEY, AND IWORK, AND IS THE FIRST DIMENSION OF D.
C THESE ARE CURRENTLY SET TO 50 AND 1000 RESPECTIVELY.
C ZERO IS THE CUTOFF POINT: IF STRESS < ZERO ITERATIONS STOP.
C ZERO IS CURRENTLY SET TO 0.0.
C *** DATA IN ORDER:
C
C DATA CARD 1: NRUNS, THE NUMBER OF PROGRAM RUNS (I.E. SEPARATE
C DATA SETS).
C PUNCH IN COLS 1-3 (RIGHT-JUSTIFIED).
C
C DATA CARD 2: NPTS MAXDIM MAXIT INCON ISIZE STEPL IX1 IX3 IFMT
C NPTS    NUMBER OF POINTS ; COLS 1-3.
C MAXDIM STARTING NUMBER OF DIMENSIONS; COLS 4-5.
C MAXIT  MAX. NUMBER OF ITERATIONS; COLS 6-9
C INCON = 0 - START WITH RANDOM COORDINATES; = 1 READ IN STARTING
C POINT. COL 10.
C ISIZE: THE NUMBER OF DIMENSIONS, INITIALLY MAXDIM. IS REDUCED
C BY ISIZE AT THE END OF EACH CYCLE UNTIL THE NUMBER OF
C DIMENSIONS IS LESS THAN ONE. PUNCH IN COLS 11-13.
C STEPL  STEP LENGTH IN STEEPEST-DESCENT ITERATIONS. PUNCH IN
C COLS 20-30, REMEMBERING TO PUNCH DECIMAL POINT.
C IX1    INTEGER STARTING VALUE FOR RANDOM NUMBER GENERATOR.
C        PUNCH IN COLS 31-40,
C        NEEDED ONLY IF INCON = 0; LEAVE BLANK OTHERWISE.
C IX3    AS FOR IX1 EXCEPT PUNCH IN COLS 41-50.
C IFMT   =1 MEANS THAT THE DISSIMILARITY COEFFICIENTS ARE READ
C        FROM UNIT IN, IN FORMATTED FORM. =0 MEANS THAT THE
C        DISSIMILARITY MATRIX IS READ IN UNFORMATTED FORM FROM UNIT 9.
C        PUNCH IN COL. 51.
C
C DATA CARD 3: FMT, THE DATA INPUT FORMAT. UP TO 80 CHARACTERS.
C THIS FORMAT IS USED TO READ IN LOWER AND UPPER LIMITS FOR COORDINATES
C [SEE DATA CARD 4].
C THIS FORMAT IS ALSO USED TO READ THE DISSIMILARITY COEFFICIENTS IF
C IFMT=1 (CARD 2)
C
C DATA CARD 4: IF INCON =0 ENTER LOWER AND UPPER LIMITS OF START
C          COORDINATES FOR EACH OF THE MAXDIM AXES;
C          IF INCON =1 READ START COORDINATES IN SPECIFIED FORMAT.
C NOTE: START COORDINATES FORM A (NPTS*MAXDIM) ARRAY.
C
C DATA CARD 5: DISSIM(J), THE LOWER TRIANGLE (EXCLUDING DIAGONAL)
C OF THE DISSIMILARITY MATRIX PUNCHED ROWWISE IN SPECIFIED FORMAT
C IF IFMT =1, IF IFMT =0 THE LOWER TRIANGLE (EXCLUDING DIAGONAL)
C OF THE DISSIMILARITY COEFFICIENTS MATRIX WILL BE READ IN
C UNFORMATED FORM FROM UNIT 9.
C NOTE: THE DISSIMILARITY MATRIX HAS DIMENSION (NPTS*NPTS).
C
C SUBROUTINES USED: MATPR, SETUP, RANDOM, PLOTS, VARMAX AND RANK -- SEE
C APPENDIX A.
C
C SEE CHAPTER 6 FOR DISCUSSION AND EXAMPLES.
C
      INTEGER OUT, DOTHER(2500)
      DIMENSION X1(50,10), X2(50,10), X3(50,10), DISSIM(2500),
     * DIST(2500),DHAT(2500),IJ(2500),KEY(2500),D(2500,2),FMT(20),
     * STR(1000),IWORK(2500),AMIN(10),AMAX(10),IX2(50,10),WORK(50),
     *WORK2(500)
      EQUIVALENCE(KEY,X2),(WORK,IWORK),(IX2,X3),(D(1,1),WORK2,DIST),
```

Table 6.15  *continued*

```
      1(D(1,2),DISSIM)
       DATA NDIM,IN,OUT,MT/50,5,6,4/,NSIZE/500/,ZERO/0.00000/,NDIM2/2500/
       DATA IBL,IST,IDOT/1H ,1H*,1H./
       READ(IN,1) NRUNS
       DO 10111 IJK=1,NRUNS
       WRITE(6,11111)
C
11111 FORMAT('1NONMETRIC MULTIDIMENSIONAL SCALING'/
      1' PROGRAM BY PAUL M. MATHER, GEOGRAPHY DEPT.,NOTTINGHAM UNIVERSITY'
      2// VERSION DATED JULY,1974'// BASED ON J.B.KRUSKAL,PSYCHOMETRIKA,
      3 1964'//)
C
C READ INPUT PARAMETERS
C
       READ(IN,1) NPTS,MAXDIM,MAXIT,INCON,ISIZE,STEPL,IX1,IX3,IFMT
     1 FORMAT(I3,I2,I4,I1,I3,7X,F10.0,2I10,I1)
       READ(IN,991) FMT
  991 FORMAT(20A4)
       FN=NPTS
       MZ=1
       IF(INCON.EQ.1) GOTO 6000
       WRITE(OUT,5999)
 5999 FORMAT('0START WITH RANDOM COORDINATES')
C SUBROUTINE SETUP (APPENDIX A ) INITIALIZES THE RANDOM NUMBER
C GENERATOR
       CALL SETUP
C
C READ LIMITS FOR RANDOM COORDINATES
C
       READ(IN,FMT) (AMIN(I),AMAX(I),I=1,MAXDIM)
 5998 DO 2 I=1,MAXDIM
       ZM=AMIN(I)
       ZK=AMAX(I)-ZM
       DO 2 J=1,NPTS
C
C GENERATE RANDOM COORDINATES ON UNIFORM DISTRIBUTION (AMAX,AMIN)
C
C *RANDOM* IS A RANDOM NUMBER GENERATOR; LISTING IN APPENDIX A
    2 X1(J,I)=RANDOM(IX1,IX3)*ZK+ZM
       GOTO (6010,1050), MZ
C
C KEY NOW HOLDS THE (I,J) IDENTIFIERS OF THE POINTS STORED IN A SINGLE
C WORD
C ALTERNATIVELY, READ IN START COORDINATES, E.G. FROM PRINCIPAL
C COMPONENTS ANALYSIS.
C
 6000 DO 6101 I=1,NPTS
 6101 READ(IN,FMT) (X1(I,J),J=1,MAXDIM)
       REWIND MT
C
C IF START CONFIGURATION IS USER-SUPPLIED, WRITE TO DISC
C
       WRITE(MT)((X1(I,J),I=1,NPTS),J=1,MAXDIM)
       ENDFILE MT
 6010 N2=NPTS*(NPTS-1)/2
       N21=N2-1
C
C READ MATRIX OF DISSIMILARITY COEFFICIENTS
C
       IF(IFMT.EQ.0) GOTO 8822
       READ(IN,FMT)(DISSIM(I),I=1,N2)
       GOTO 6055
 8822 REWIND 9
       READ(9) (DISSIM(I),I=1,N2)
 6055 L=0
       NPTS1=NPTS-1
       DO 4 I=2,NPTS
```

Table 6.15   *continued*

```
            I1=I-1
            DO 4 J=1,I1
            L=L+1
            K=I*NSIZE+J
         4  KEY(L)=K
C
C
C
C RANK THE DISSIMILARITIES IN ASCENDING ORDER
C
            CALL RANK(DISSIM,IJ,WORK2,N2,1)
C
C PUT THE (I,J) COORDINATES IN THE  SAME ORDER
C
            DO 121 I=1,N2
            J=IJ(I)
       121  IJ(I)=KEY(J)
            REWIND MT
C
C START ITERATIONS
C
      1050  ALPHA=STEPL
            WRITE(OUT,1021) MAXDIM
      1021  FORMAT('1  NUMBER OF DIMENSIONS AT THIS CYCLE = ',I4)
            WRITE(OUT,2000)
      2000  FORMAT('0START COORDINATES'//)
            DO 2001 I=1,NPTS
      2001  WRITE(OUT,2002) (X1(I,J),J=1,MAXDIM)
      2002  FORMAT('  ',10F10.5)
            WRITE(OUT,1020)
      1020  FORMAT('0',2X,'ITER',7X,'STRESS',5X,'STEP LENGTH',5X,'MAG (G)'///)
            DO 1001 ITER=1,MAXIT
            ITS=ITER
C
C NORMALIZE THE CONFIGURATION
C
            CALL NORMLZ(X1,NDIM,NPTS,MAXDIM)
C
C FIND INTERPOINT DISTANCES
C
            CALL DISTPT(X1,NDIM,NPTS,MAXDIM,IJ,DIST,N2,NSIZE)
C
C COMPUTE DHAT
C
            CALL MONOT(DOTHER,DIST,DHAT,DISSIM,N2)
C
C NOW CALCULATE STRESS
C
            S=0.0
            T=0.0
            DO 17 I=1,N2
            S=S+(DIST(I)-DHAT(I))**2
        17  T=T+DIST(I)*DIST(I)
            STR(ITER)=SQRT(S/T)
            STRESS=STR(ITER)*100.
            IF(STR(ITER).LE.ZERO) GOTO 68
            SS=STR(ITER)/S
            ST=STR(ITER)/T
            DO 25 I=1,NPTS
            DO 25 J=1,MAXDIM
        25  X2(I,J)=0.0
        70  DO 19 K=1,N2
            I=IJ(K)/NSIZE
            J=IJ(K)-I*NSIZE
C
C COMPUTE (NEGATIVE) GRADIENTS
C
```

Table 6.15    *continued*

```
   71 DO 19 L=1,MAXDIM
      IF(DIST(K).EQ.0.0) DIST(K)=0.000001
      TERM=(SS*(DIST(K)-DHAT(K))-ST*DIST(K))*(X1(I,L)-X1(J,L))/DIST(K)
      X2(J,L)=X2(J,L)+TERM
      X2(I,L)=X2(I,L)-TERM
   19 CONTINUE
      GSUM=0.0
      DO 29 I=1,NPTS
      DO 29 J=1,MAXDIM
   29 GSUM=GSUM+X2(I,J)*X2(I,J)
      WRITE(OUT,1906) ITER, STRESS, ALPHA, GSUM
 1906 FORMAT(' ',I5,F14.5,2F12.5)
      IF(ITER.EQ.1) GOTO 50
      GLF=AMIN1(1.0,STR(ITER)/STR(ITER-1))
      AF=0.0
      DO 30 I=1,NPTS
      DO 30 J=1,MAXDIM
   30 AF=AF+X2(I,J)*X3(I,J)
      COSTH=AF/SQRT(GSUM*GSUM1)
      AF=4.0**(COSTH*COSTH*COSTH)
      I=MAX0(ITER-5,1)
      FIV=AMIN1(1.0,STR(ITER)/STR(I))
      RF=1.6/(1.0+FIV**5.0)
C NEW ALPHA
   40 ALPHA=ALPHA*AF*RF*GLF
   50 AMAG=SQRT(GSUM/FN)
      IF(AMAG.LE.ZERO) GOTO 90
C
C POINTS MOVED TO NEW CONFIGURATION
C
   56 DO 51 I=1,NPTS
      DO 51 J=1,MAXDIM
   51 X1(I,J)=X1(I,J)+ALPHA/AMAG*X2(I,J)
      GSUM1=GSUM
C
C PRESERVE OLD GRADIENT FOR NEXT ITERATION
C
      DO 52 I=1,NPTS
      DO 52 J=1,MAXDIM
   52 X3(I,J)=X2(I,J)
C
C END OF ITERATION CYCLE
C
 1001 CONTINUE
      GOTO 66
   55 WRITE(OUT,57)
   57 FORMAT('0AMAG ZERO AT THIS CYCLE - PROCEDURE TERMINATES'/)
      GOTO 66
   68 WRITE(OUT,69) ZERO
   69 FORMAT('0*** STRESS NOW LESS THAN CUTOFF = ',E17.11)
      GOTO 66
   90 WRITE(OUT,91)
   91 FORMAT('0 AMAG IS NOW ZERO - ITERATIONS STOPPED')
   66 WRITE(OUT,67)
   67 FORMAT('0FINAL CONFIGURATION FOR THIS CYCLE'///)
      DO 64 I=1,NPTS
   64 WRITE(OUT,63) (X1(I,J),J=1,MAXDIM)
   63 FORMAT(' ',10F11.4)
C
C PLOT DISTANCE V. DISSIMILARITY IF N2 < 1000
C
      IF(N2.GT.1000) GOTO 571
      WRITE(OUT,871)
  871 FORMAT('0 PLOT OF DISTANCE (HORIZONTAL AXIS) V. DISSIMILARITY (VER
     +TICAL AXIS)'/)
      CALL PLOTS(D,1,2,NDIM2,N2,2,0,IX2,N2)
  571 IF(MAXDIM.EQ.1) GOTO 724
```

Table 6.15  *continued*

```
C
C PLOT PAIRS OF DIMENSIONS UNLESS MAXDIM = 1
C
      WRITE(OUT,110)
  110 FORMAT('0 PLOT OF FINAL CONFIGURATION --- 2-DIMENSIONAL SLICES')
      CALL PLOTS(X1,1,MAXDIM,NDIM,NPTS,MAXDIM,0,IX2,NDIM)
C
C PLOT STRESS V. ITERATION
C
      IF(ITS-1) 720,720,724
  724 WRITE(OUT,97)
   97 FORMAT('1       PLOT OF STRESS V. ITERATION NUMBER'//)
      IF(ITS-100) 103,103,105
  103 LEFT=ITS
      NPLOT=1
      GOTO 104
  105 NPLOT=ITS/100
      LEFT=ITS-100*NPLOT
      IF(LEFT.GT.0) NPLOT=NPLOT+1
  104 CALL RANK(STR,DOTHER,WORK2,ITS,-1)
      SMAX=STR(1)+STR(1)*0.01
      SMIN=STR(ITS)-STR(ITS)*0.01
      SSMAX=SMAX*100.0
      SSMIN=SMIN*100.0
      RANGE=SMAX-SMIN
      XINT=RANGE/29.0
      DO 106 L=1,NPLOT
      NO1=102
      IF(L.EQ.NPLOT.AND.LEFT.GT.0) NO1=LEFT+2
      NO12=NO1-2
      L1=(L-1)*100
      L2=L1+100
      IF(L.EQ.NPLOT.AND.LEFT.GT.0) L2=L1+NO1
      WRITE(OUT,108) SSMAX,(IDOT,I=1,NO1)
  108 FORMAT(' ',F16.7,102A1)
      TOP=SMAX+XINT
      BOTTOM=SMAX
      DO 100 I=1,30
      DO 98 J=1,NO12
   98 IWORK(J)=IBL
      TOP=TOP-XINT
      BOTTOM=BOTTOM-XINT
      DO 101 J=1,ITS
      IF(STR(J).LE.BOTTOM.OR.STR(J).GT.TOP) GOTO 101
      IF(DOTHER(J).GT.L2.OR.DOTHER(J).LE.L1) GOTO 101
      K=DOTHER(J)-L1
      IWORK(K)=IST
  101 CONTINUE
      IF(I-15) 111,112,111
  112 WRITE(OUT,113) IDOT,(IWORK(J),J=1,NO12),IDOT
  113 FORMAT(' ',5X,'STRESS',5X,102A1)
      GOTO 100
  111 WRITE(OUT,102) IDOT, (IWORK(J),J=1,NO12),  IDOT
  100 CONTINUE
  102 FORMAT(' ',16X,102A1)
      WRITE(OUT,108) SSMIN,(IDOT,I=1,NO1)
      L21=L1+10
      WRITE(OUT,107) (I,I=L21,L2,10)
  106 WRITE(OUT,308)
  107 FORMAT(' ',18X,10I10)
  308 FORMAT(' ',10X,' ITERATION NUMBER'//)
      IF(MAXDIM.EQ.1) GOTO 720
C
C TRANSFORM COORDINATE SYSTEM USING VARIMAX (NOTE: THE INTERPOINT
C DISTANCES ARE UNALTERED BY THE ROTATION).
C
      CALL VARMAX(X1,X2,X3,DIST,WORK,DHAT,STR,NPTS,MAXDIM,0,50,0.001,
     1NDIM,NDIM,NDIM,MAXDIM,MAXDIM)
```

Table 6.15  *continued*

```
      WRITE(OUT,241)
  241 FORMAT('0       CONFIGURATION AFTER ROTATION'//)
      DO 242 I=1,NPTS
  242 WRITE(OUT,63) (X2(I,J),J=1,MAXDIM)
      CALL PLOTS(X2,1,MAXDIM,NDIM,NPTS,MAXDIM,0,IX2,NDIM)
C
C REDUCE DIMENSIONALITY AND REPEAT ITERATIONS IF NOT ZERO
C
  720 MAXDIM=MAXDIM-ISIZE
      IF(MAXDIM.LE.0) GOTO 10111
C
C READ START COORDINATES FROM BACKING STORE OF FIND NEW RANDOM
C START COORDINATES
C
      MZ=2
      IF(INCON.EQ.0) GOTO 5998
      REWIND MT
      READ(MT) ((X1(I,J),I=1,NPTS),J=1,MAXDIM)
      GO TO 1050
10111 CONTINUE
      STOP
      END




      SUBROUTINE NORMLZ(X1,NDIM,NPTS,MAXDIM)
C
C SUBROUTINE TO NORMALIZE CONFIGURATION
C
      DIMENSION X1(NDIM,MAXDIM)
      FN=NPTS
      DO 8 I=1,MAXDIM
      C=0.0
      DO 9 J=1,NPTS
    9 C=C+X1(J,I)
      C=C/FN
      DO 8 J=1,NPTS
    8 X1(J,I)=X1(J,I)-C
      S=0.0
      DO 10 I=1,NPTS
      DO 10 J=1,MAXDIM
   10 S=S+X1(I,J)**2
      S=SQRT(S/FN)
      DO 11 I=1,NPTS
      DO 11 J=1,MAXDIM
   11 X1(I,J)=X1(I,J)/S
      RETURN
      END




      SUBROUTINE MONOT(DOTHER,DIST,DHAT,DISSIM,N2)
C
C SUBROUTINE TO CHECK FOR TIES IN DISSIM AND CONTROL COMPUTATION OF DHAT
C
      LOGICAL UPSAT,DNSAT
      INTEGER DOTHER(N2)
      DIMENSION DIST(N2),DHAT(N2),DISSIM(N2)
      DO 298 I=1,N2
      DOTHER(I)=0
  298 DHAT(I)=DIST(I)
C FIRST CHECK FOR TIES IN DISSIM & USE KRUSKAL'S STRATEGY 2.
      I=0
  299 I=I+1
      IF(I.GT.N2) GOTO 500
```

Table 6.15   *continued*

```
      J=I+1
      IF(DISSIM(I).NE.DISSIM(J)) GOTO 299
  300 J=J+1
      IF(J.GT.N2) GOTO 297
      IF(DISSIM(I).EQ.DISSIM(J)) GOTO 300
  297 J=J-1
      DOTHER(I)=J-I+1
      DOTHER(J)=I
      S=0.0
      DO 301 K=I,J
  301 S=S+DIST(K)
      G=S/FLOAT(DOTHER(I))
      DHAT(I)=G
      DHAT(I+1)=S
      I=J
      GOTO 299
  500 M=0
C NOW BEGIN FITTING DHAT.
  302 IF(M) 310,310,304
  310 M=1
      GOTO 308
  304 K=DOTHER(M)
      IF(K.EQ.0) K=1
      M=M+K
  308 IF(M-N2) 309,309,305
  309 UPSAT=.FALSE.
      DNSAT=.FALSE.
      IF(M.EQ.1) DNSAT=.TRUE.
      IF(M.EQ.N2) UPSAT=.TRUE.
  303 IF(.NOT.UPSAT) CALL UPSAT1(DHAT,M,UPSAT,DNSAT,DOTHER,N2)
      IF(UPSAT.AND.DNSAT) GOTO 302
      IF(.NOT.DNSAT) CALL DNSAT1(DHAT,M,UPSAT,DNSAT,DOTHER,N2)
      IF(UPSAT.AND.DNSAT) GOTO 302
      GOTO 309
  305 CALL SORT(DHAT,DOTHER,N2)
      RETURN
      END



      SUBROUTINE UPSAT1(DHAT,M,UPSAT,DNSAT,DOTHER,N2)
C
C SUBROUTINE TO CHECK THAT BLOCK IS UPSATISFIED
C
      LOGICAL UPSAT,DNSAT
      INTEGER DOTHER(N2)
      DIMENSION DHAT(N2)
      NO=DOTHER(M)
      IF(NO) 10,20,30
   10 STOP
   30 NO1=M+NO-1
      NO3=NO1+1
      IF(NO3.GT.N2) GOTO 3
      NO4=DOTHER(NO3)
      IF(NO4) 100,110,100
  100 NO5=NO3+NO4-1
      GOTO 120
  110 NO5=NO3
  120 IF(DHAT(M).GE.DHAT(NO3)) GOTO 1
    3 UPSAT=.TRUE.
   22 RETURN
    1 IF(NO4) 70,80,70
   70 S=DHAT(M+1)+DHAT(NO3+1)
      NO2=NO4+NO
      GOTO 90
   80 S=DHAT(M+1)+DHAT(NO3)
```

Table 6.15  *continued*

```
       NO2=NO+1
   90  DHAT(M+1)=S
       DHAT(M)=S/FLOAT(NO2)
       DOTHER(M)=NO2
    5  DOTHER(NO5)=M
       UPSAT=.FALSE.
       IF(M.NE.1) DNSAT=.FALSE.
       GOTO 22
   20  NO5=M+1
       IF(NO5.GT.N2) GOTO 3
       NO3=DOTHER(NO5)
       IF(NO3) 10,50,60
   50  IF(DHAT(M).LT.DHAT(NO5)) GOTO 3
       S=DHAT(M)+DHAT(NO5)
       DHAT(NO5)=S
       DHAT(M)=S*0.5
       DOTHER(M)=2
       GOTO 5
   60  NO5=M+NO3
       NO2=M+1
       IF(DHAT(M).LT.DHAT(NO2)) GOTO 3
       S=DHAT(M)+DHAT(NO5)
       DHAT(M+1)=S
       NO3=DOTHER(NO2)+1
       DHAT(M)=S/FLOAT(NO3)
       DOTHER(M)=NO3
       GOTO 5
       END



       SUBROUTINE DNSAT1(DHAT,M,UPSAT,DNSAT,DOTHER,N2)
C
C SUBROUTINE TO CHECK THAT BLOCK IS DOWNSATISFIED
C
       LOGICAL UPSAT,DNSAT
       INTEGER DOTHER(N2)
       DIMENSION DHAT(N2)
       NO=DOTHER(M)
       IF(NO) 10,20,30
   10  STOP
   30  NO1=DOTHER(M-1)
       IF(NO1) 100,40,100
   40  NO5=M-1
       GOTO 50
  100  NO5=NO1
   50  IF(DHAT(NO5).GE.DHAT(M)) GOTO 41
    3  DNSAT=.TRUE.
   22  RETURN
   41  IF(NO1) 80,90,80
   90  S=DHAT(M-1)+DHAT(M+1)
       NO2=NO+1
       GOTO 101
   80  S=DHAT(NO5+1)+DHAT(M+1)
       NO2=DOTHER(NO5)+NO
  101  DHAT(NO5+1)=S
       S=S/FLOAT(NO2)
       DHAT(NO5)=S
       NO4=M+NO-1
    5  DOTHER(NO5)=NO2
       DOTHER(NO4)=NO5
       M=NO5
       IF(M-1.NE.N2) UPSAT=.FALSE.
       GOTO 22
   20  NO5=M-1
       NO3=DOTHER(NO5)
```

Table 6.15    *continued*

```
        NO4=M
        IF(NO3) 10,60,70
     60 IF(DHAT(M).GT.DHAT(NO5)) GOTO 3
        S=DHAT(M)+DHAT(NO5)
        DHAT(NO5)=0.5*S
        DHAT(M)=S
        NO2=2
        GOTO 5
     70 NO5=NO3
        NO2=DOTHER(NO3)+1
        IF(DHAT(NO3).LT.DHAT(M)) GOTO 3
        S=DHAT(M)+DHAT(NO3+1)
        DHAT(NO3+1)=S
        DHAT(NO3)=S/FLOAT(NO2)
        GOTO 5
        END




        SUBROUTINE SORT(DHAT,DOTHER,N2)
C
C THIS SUBROUTINE GETS DHAT FROM DHAT AND DOTHER LEFT IN SUBROUTINE
C MONOT.
C
        INTEGER DOTHER(N2)
        DIMENSION DHAT(N2)
        I=0
     10 I=I+1
     15 IF(I.GE.N2) RETURN
        J=DOTHER(I)
        IF(J.EQ.0) GOTO 10
        K=I+J-1
        M=I+1
        DO 20 L=M,K
     20 DHAT(L)=DHAT(I)
        I=I+J
        GOTO 15
        END




        SUBROUTINE DISTPT(X1,NDIM,NPTS,MAXDIM,IJ,DIST,N2,NSIZE)
C
C CALCULATES DISTANCES IN SAME ORDER AS THE DISSIMILARITIES
C
        DIMENSION X1(NDIM,MAXDIM), IJ(N2), DIST(N2)
        DO 12 I=1,N2
        J=IJ(I)/NSIZE
        K=IJ(I)-J*NSIZE
        D=0.0
        DO 13 N=1,MAXDIM
     13 D=D+(X1(J,N)-X1(K,N))**2
     12 DIST(I)=SQRT(D)
        RETURN
        END
```

## Table 6.16
### Computer program for nonlinear mapping

```
C NONLINEAR MAPPING PROGRAM.  P. M. MATHER, 1973.
C  BASED ON SAMMON(1969).
C SEE CHAPTER 6 FOR FURTHER DETAILS.
C DATA CARDS:
C 0]: NV  NP
C     NUMBER OF VARIABLES; NUMBER OF POINTS
C PUNCH IN (2I3) FORMAT
C 1]: JOB NAME - UP TO 80 CHARACTERS.
C 2]: FMT
C     DATA INPUT FORMAT, INCLUDING OPENING AND CLOSING BRACKETS
C 3]: X(NP,NV)
C     DATA MATRIX READ ROWISE IN SPECIFIED FORMAT.
C     START EACH ROW ON A NEW CARD.
C 4]: MAGIC MAX ETA IRN IN FORMAT (F10.0,I3,F10.0,I3)
C     MAGIC FACTOR; MAXIMUM NUMBER OF ITERATIONS; CONVERGENCE CRITERION;
C     IRN IS AN INTEGER INDICATOR - IF IRN=0 THE STARTING
C     CONFIGURATION WILL BE COMPUTED.  IF IRN=1 THE STARTING
C     CONFIGURATION WILL BE READ FROM CARDS.
C NEXT CARD ONLY IF IRN = 1.
C 5]: FMT
C     FORMAT CARD FOR STARTING CONFIGURATION
C 6]: Y(NP,2)
C     STARTING CONFIGURATION; COORDINATES OF NP POINTS IN A 2-SPACE.
C     PUNCH Y IN SPECIFIED FORMAT
C
C SUBROUTINES NEEDED:
C PLOTS,SCALE,KALK AND OVCHK FROM APPENDIX A.
C NOTE: A SCRATCH FILE (CHANNEL 3) IS REQUIRED
C
      DIMENSION X(5000)
      NDIM=5000
C IF DIMENSION OF X IS ALTERED CHANGE NDIM ACCORDINGLY
      REWIND 3
      WRITE(6,100)
  100 FORMAT(' NONLINEAR MAPPING FOR DATA STRUCTURE ANALYSIS'/
     *'0 PROGRAMMED BY P. M. MATHER, GEOG. DEPT.,NOTTINGHAM UNIV,1973'/
     *' REFERENCE: IEEE TRANS ON COMPUTERS,VOL C-18,NO. 5,MAY 1969.  AUTH
     *OR JOHN W. SAMMON, JR.'//)
      READ(5,1) NV, NP
    1 FORMAT(2I3)
      NO=NP*(NP+1)/2-NP
      READ(5,30) (X(I),I=1,20)
   30 FORMAT(20A4)
      WRITE(6,5) (X(I),I=1,20), NV, NP
    5 FORMAT('0JOB IDENTIFIER:   ',20A4/' NO OF VARIABLES = ',I4/' NO OF
     +INDIVIDUALS = ',I4/)
C CALL INPUT ROUTINE
      CALL INPT(X,NV,NP)
C SET UP BASE ADDRESSES
      N1=NP+NP+1
      N2=N1+NO
      N3=N2+NP
      N4=N3+NP+NP
      N5=N4+NP
      N6=N5+NP
      N7=N6+NV
C CHECK FOR OVERFLOW
      IF(N7.GT.NDIM) GOTO 11
      WRITE(6,31) N7
   31 FORMAT(' ACTUAL CORE USED = ',I4,' WORDS')
C CALL MAIN ROUTINE
      CALL NONLIN(X(1),X(N1),X(N2),X(N3),X(N4),X(N5),X(N6),NV,NP,NO)
CALL PLOTTER ROUTINE (LISTING --- APPENDIX A)
      CALL PLOTS(X(1),1,2,NP,NP,2,0,X(N1),NP,0)
      WRITE(6,50)
```

Table 6.16  *continued*

```
   50 FORMAT('OPLOT OF TWO-DIMENSIONAL REPRESENTATION')
      GOTO 2
   11 WRITE(6,3) N7
    3 FORMAT('ONOT ENOUGH CORE',I5,' WORDS REQUIRED')
    2 STOP
      END




      SUBROUTINE INPT(X,M,N)
      DIMENSION X(N,M),FMT(20)
      READ(5,3) FMT
    3 FORMAT(20A4)
      DO 1 I=1,N
    1 READ(5,FMT) (X(I,J),J=1,M)
      DO 2 J=1,M
    2 WRITE(3) (X(I,J),I=1,N)
      ENDFILE 3
      REWIND 3
      RETURN
      END




      SUBROUTINE NONLIN(Y,DSTAR,X,YY,WK1,WK2,VAR,NVAR,NPTS,NO)
      REAL MAGIC,DSTAR(NO),X(NPTS),Y(NPTS,2),YY(NPTS,2),WK1(NPTS),
     + WK2(NPTS),VAR(NVAR),FMT(20)
      READ(5,1) MAGIC,MAX,ETA,IRN
    1 FORMAT(2(F10.0,I3))
      WRITE(6,14) MAGIC,MAX,ETA
   14 FORMAT(' MAGIC = ',F8.3//' MAX = ',I4//' ETA = ',E15.8)
      ELAST=1.E20
      IF(IRN.EQ.0) GOTO 21
      READ(5,2) FMT
    2 FORMAT(20A4)
      READ(5,FMT) ((Y(I,J),J=1,2),I=1,NPTS)
   21 CALL DIST(DSTAR,X,NPTS,NVAR,Y,VAR,DEN,NO,IRN)
   22 WRITE(6,11)
   11 FORMAT('OSTARTING Y CONFIGURATION'///)
      WRITE(6,13) ((Y(I,J),J=1,2),I=1,NPTS)
   13 FORMAT(' ',2F12.4)
      WRITE(6,3)
    3 FORMAT('   ITERATION     MAPPING ERROR'///)
      DO 10 ITERS=1,MAX
      CALL DERIVS(Y,DSTAR,NPTS,YY,MAGIC,E,DEN,WK1,WK2,NO)
    6 FORMAT(' ',I6,2X,E20.9)
      WRITE(6,6) ITERS,E
C CHECK FOR CONVERGENCE IF MAPPING ERROR IS REDUCING (IT MAY INCREASE IN
C EARLY ITERATIONS).
      IF(E.GT.ELAST) GOTO 10
      IF((ELAST-E).GT.ETA) GOTO 10
      WRITE(6,20)
      WRITE(6,13) (( Y(I,J),J=1,2),I=1,NPTS)
   20 FORMAT('OFINAL SOLUTION: Y COORDINATES'///)
      RETURN
   10 ELAST=E
      WRITE(6,4) MAX
    4 FORMAT('ONO CONVERGENCE AFTER',I6,2X,'ITERATIONS'///' CURRENT Y CON
     +FIGURATION IS'//)
      WRITE(6,13) ((Y(I,J),J=1,2),I=1,NPTS)
      RETURN
      END
```

Table 6.16    *continued*

```
      SUBROUTINE DIST(D,X,NP,NV,Y,VAR,DEN,N2,IRN)
      DIMENSION D(N2),X(NP),Y(NP,2),VAR(NV),ID(2)
      DEN=0.
      DO 1 I=1,N2
    1 D(I)=0.0
      FN=NP
      DO 31 L=1,NV
      READ(3) X
      K=0
      S=0.0
      B=0.0
      DO 30 I=1,NP
      S=S+X(I)*X(I)
   30 B=B+X(I)
      S=S/FN-(B/FN)*(B/FN)
      VAR(L)=S
      NP1=NP-1
      DO 31 I=1,NP1
      I1=I+1
      DO 31 J=I1,NP
      K=K+1
   31 D(K)=D(K)+(X(I)-X(J))**2
      DO 4 I=1,N2
      D(I)=SQRT(D(I))
      IF(D(I).EQ.0.0) D(I)=1.E-4
    4 DEN=DEN+D(I)
      DEN=2.0*DEN
      IF(IRN.EQ.1) RETURN
      DO 33 L=1,2
      S=VAR(1)
      J=1
      DO 32 I=2,NV
      IF(S.GE.VAR(I)) GOTO 32
      J=I
      S=VAR(I)
   32 CONTINUE
      VAR(J)=0.0
   33 ID(L)=J
      WRITE(6,34) ID(1), ID(2)
   34 FORMAT('OVARIABLES ',I4,' AND',I4,' HAVE HIGHEST VARIANCES'//)
      DO 37 J=1,2
      REWIND 3
      DO 35 I=1,NV
      READ(3)   X
      IF(I.NE.ID(J)) GOTO 35
      DO 36 K=1,NP
   36 Y(K,J)=X(K)
   35 CONTINUE
   37 CONTINUE
      RETURN
      END


      FUNCTION YDIST(I,J,Y,NP)
      DIMENSION Y(NP,2)
      YDIST=0.
      DO 1 K=1,2
    1 YDIST=YDIST+(Y(I,K)-Y(J,K))**2
      YDIST=SQRT(YDIST)
      RETURN
      END
```

Table 6.16    *continued*

```
      SUBROUTINE DERIVS(Y,DSTAR,NP,YY,MAGIC,E,DEN,WK1,WK2,NO)
      INTEGER P,Q
      REAL MAGIC,DSTAR(NO),Y(NP,2),YY(NP,2),WK1(NP),WK2(NP)
      DO 1 I=1,NP
      WK1(I)=0.0
      WK2(I)=0.0
      DO 1 J=1,2
    1 YY(I,J)=Y(I,J)
      NP1=NP-1
      E=0.0
      DO 10 K=1,2
      WK1(NP)=0.
      WK2(NP)=0.
      KT=0
      DO 9 Q=1,NP1
      EP=0.0
      EPP=0.0
      IQ=Q+1
      DO 8 J=IQ,NP
      KT=KT+1
      DQJ=1.0/DSTAR(KT)
      DJQ=YDIST(Q,J,YY,NP)
      IF(DJQ.EQ.0.0) DJQ=1.0E-4
      A=1.0/DJQ
      C=DQJ*DJQ-1.0
      E=E+C*C/DQJ
      Y1=A*(Y(Q,K)-Y(J,K))
      Y2=Y1*Y1
      F=C*Y1
      WK1(J)=WK1(J)+F
      EP=EP+F
      FP=C*A*(1.0-Y2)+DQJ*Y2
      WK2(J)=WK2(J)+FP
    8 EPP=EPP+FP
      EP=EP-WK1(Q)
      EPP=EPP+WK2(Q)
      WK2(Q)=0.0
      WK1(Q)=0.0
    9 Y(Q,K)=Y(Q,K)-MAGIC*EP/ABS(EPP)
   10 Y(NP,K)=Y(NP,K)+WK1(NP)*MAGIC/ABS(WK2(NP))
      E=E/DEN
      RETURN
      END
```

*Chapter 7*
# Discriminant Analysis

## 7.0 Introduction
Discrimination and classification are quite separate, though closely related, concepts. Classification consists of an attempt to discover, usually with no prior information, the number of groups that exist within a given data-set, together with details of group membership. Techniques of classification were discussed in Chapter 6. If the number of groups is known, as well as the characteristics of each group, the problem of assigning unclassified observations to their most likely group becomes logically possible. For example, in Chapter 6 the classification of a set of sediment samples into discrete groups on the basis of their size-distribution characteristics was discussed. *Discriminant analysis*, in this example, would involve the assignment of newly-acquired or previously uncategorized sediment samples to one of these existing classes.

In this chapter the following topics are discussed:

(a) discrimination in the two-group case;
(b) discrimination in the multiple-group case;
(c) discrimination in the two-group case when the variables are of the presence/absence type, and
(d) multivariate analysis of variance.

Since (b) follows naturally from (a), the multivariate extension of simple discriminant analysis is treated before the radically different approach of (c) is dealt with. Multivariate analysis of variance and associated tests are usually carried out prior to multiple discriminant analysis, so (d) will not be treated as a separate topic, but will instead be integrated with (b). The notation used throughout this chapter is shown in Table 7.1.

## 7.1 Two-group discriminant analysis
Fisher (1936) introduced the method of discriminant analysis to deal with the problem of correctly assigning fossil remains to one of two

Table 7.1
Notation for Chapter 7

| | |
|---|---|
| $S$ | variance–covariance matrix for all groups and observations |
| $S_i$ | variance–covariance matrix for group $i$ |
| $S_W$ | pooled within-groups variance-covariance matrix $(S_W = S_1 + S_2 + \ldots + S_k)$ |
| $S_B$ | between-groups variance-covariance matrix $(S_B = S - S_W)$ |
| $\Lambda$ | diagonal matrix of eigenvalues. Not to be confused with $\Lambda$, which is Wilks' Lambda |
| $Q$ | matrix of column eigenvectors |
| $D$ | matrix of discriminant scores. The scores on the first discriminant are stored in column 1 and so on |
| $n$ | total number of individuals |
| $n_i$ | number of individuals in group $i$ |
| $k$ | number of groups |
| $p$ | number of variables |
| $T$ | total sums of squares and cross products matrix |
| $W$ | within-groups pooled sums of squares and cross products matrix |
| $B$ | between-groups sums of squares and cross products matrix |

classes (hominoid and ape) on the basis of measurements on several variables. Since then the topic has been discussed in the context of psychology (Li, 1964), geography (Mather, 1969), and geology (Davis, 1973; Krumbein and Graybill, 1965) and a number of applications in the field of physical geography and geology have been reported in the literature (Griffiths, 1957, 1964, 1966; Hails, 1967; Sahu, 1964, Rhodes, 1969; Middleton, 1962; Potter et al., 1963; Greenwood, 1969, and Greenwood and Davidson-Arnott, 1972).

The simplest case, involving measurements on two variables ($x_1$ and $x_2$) is shown in Figure 7.1. The maximum separation between the centres of the two groups A and B is achieved by projecting each centre point onto the line KK′ rather than onto any other line, for example QQ′. In this example we can also see that groups A and B would appear to overlap if projected on to either axis ($X_1$ or $X_2$) whereas — as can be verified by projecting individual group members on to KK′ — they are in fact quite distinct. This, incidentally is the prime justification of adopting a multivariate approach in that significant differences (or similarities) may well remain hidden if the variables are considered one at a time and not simultaneously.

The aim of discriminant analysis is to find the line which best separates the groups in terms of the projections of the group centroids — see Figure 7.1. The positions of the points along this line can then be used to assign individuals to their most probable class. For

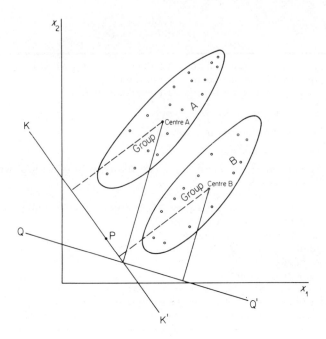

Figure 7.1   Discriminant function for two groups

example, the unclassified individual represented by the point P on Figure 7.1 is more likely to belong to group B than group A.

Graphical methods would be unsuitable if three or more variables were measured. The following algebraic formulation is conceptually identical to the geometrical scheme developed above, but has the advantage that several variables can be included. Furthermore, the relative contributions of the variables to the inter-group difference can be established. This may be of value in practical applications when a small number of diagnostic variables is required.

The main assumptions of discriminant analysis are that the groups have homogeneous variance-covariance matrices, and that the variables are normally distributed. The second assumption is necessary if significance tests are to be applied. Tests for equality of variance–covariance matrices are considered in Section 7.2. In the two-group case the test statistic is Hotelling's T squared, which is used to test the null hypothesis that the multivariate means of the two groups do not differ significantly. In other words, the null hypothesis states that the two individual group means are estimating the same population mean. This is a reasonable hypothesis only when the members of the two groups are random samples drawn from specified populations. The T squared statistic is computed as follows. The two groups are assumed to have sample means $\bar{x}_{i1}$ and $\bar{x}_{i2}$ ($i = 1, 2, \ldots, p$) and the values $d_i$ are the differences between corresponding group means on the $p$ variables, i.e.

$$d_i = \bar{x}_{i1} - \bar{x}_{i2} \quad (i = 1,2,\ldots,p) \tag{7.1}$$

Next, the coefficient known as Mahalanobis' $D$ squared is computed. This is a measure of the overall similarity between the two groups based on all $p$ variables.

$$D^2 = \sum_{i=1}^{p} \sum_{j=1}^{p} s^{ij} d_i d_j \quad (i \neq j)$$

The values $d_i$ and $d_j$ are obtained from Equation (7.1) and $s^{ij}$ is the $(i,j)$th element of $S^{-1}$, the inverse of the pooled variance–covariance matrix for the two groups. (The pooled variance–covariance matrix S is the sum of the variance–covariance matrices of the groups concerned; in this case there are two groups, A and B, so

$$S = S_A + S_B.)$$

If $D^2$ is high then the two groups are similar. Conversely, if is low, they are well-separated. The statistical significance of $D^2$ cannot be tested directly, but an $F$-ratio approximation is available. It is given by:

$$F = \frac{n-p-1}{(n-2)p} \cdot T^2 \tag{7.2}$$

where

$$T^2 = \frac{n_1 n_2}{n} \cdot D^2 \tag{7.3}$$

In these formulae, $T^2$ is Hotelling's $T$ squared, the multivariate analogue of Student's $t$, $n_1$ and $n_2$ are the numbers of individuals in groups A and B, and $n = n_1 + n_2$. As usual, $p$ is the number of variables. If $F$ with $p$ and $(n-p-1)$ degrees of freedom exceeds the tabled value of the $F$-ratio at a chosen significance level then the null hypothesis is not accepted. It is thus probable that the two sample means are estimating different population means. If this is the case, we can proceed to calculate estimates of the coefficients $\alpha_i$ of the equation defining the line KK' (the discriminant function) in Figure 7.1. These coefficients are found from

$$\mathbf{d} = \mathbf{S}\hat{\boldsymbol{\alpha}} \tag{7.4a}$$

S is the pooled variance–covariance matrix, $\hat{\boldsymbol{\alpha}}$ the required vector of estimates and d is defined by Equation (7.1). The comparison between Equation (7.4) and the linear regression equation (Chapter 2) is obvious. Discriminant analysis is equivalent to the regression of inter-group mean differences on the $p$ variables (Kendall, 1965).

Equation 7.4 is solved for $\hat{\alpha}$ by

$$\hat{\alpha} = S^{-1} d \qquad (7.4b)$$

Alternatively $\hat{\alpha}$ can be obtained without explicitly inverting S if an equation-solving algorithm is used. Since S is a symmetric matrix, subroutine SOLVE (Appendix A) could be used. The equation

$$\hat{x}_{i1} = \hat{\alpha}_i x_{i1} + \hat{\alpha}_2 x_{i2} + \ldots + \hat{\alpha}_p x_{ip} \qquad (7.5)$$

gives $\hat{x}_{i1}$, the projection of the $i$th member of the first group on the discriminant axis KK' (Figure 7.1). If the mean values of the $\hat{x}_{i1}$ and $\hat{x}_{i2}$ are calculated, the difference between them will be equal in absolute value to Mahalanobis' $D^2$. Once the coefficients $\hat{\alpha}$ have been found, uncategorized cases can be assigned to their most likely group. If $x_u$ is a vector of measurements on $p$ variables for an unclassified individual then $\hat{x}_u$, the projection of the point representing $x_u$ on to the discriminant axis, is computed from Equation (7.5) as follows:

$$\hat{x}_u = \hat{\alpha}_1 x_{u1} + \hat{\alpha}_2 x_{u2} + \ldots + \hat{\alpha}_p x_{up} \qquad (7.6)$$

If $(\bar{\hat{x}}_1 - \hat{x}_u)$ is less than $(\bar{\hat{x}}_2 - \hat{x}_u)$ then the unclassified individual is assigned to group A rather than group B. Otherwise it is assigned to group B. If either of these values is large, one may suspect that $x_u$ belongs to neither of these groups.

The contributions $c_i$ $(i = 1, 2, \ldots, p)$ of the $p$ variables to the discriminating power of function can be computed from

$$c_i = \frac{\hat{\alpha}_i d_i}{D^2} \cdot 100 \text{ per cent} \qquad (7.7)$$

where $D^2$ is Mahalanobis' $D$ squared, with $d_i$ given by (7.1) and $\hat{\alpha}_i$ by (7.4b). Variables making a small or negative contribution may be eliminated. Sometimes a selection strategy, similar to the backward elimination algorithm described in Chapter 3, is advocated, but the difficulty lies in the fact that variables initially having low $c_j$ values may improve their performance if other variables are eliminated. The $c_i$ values refer only to the relative contributions within the given set of $p$ variables; they are not indicators of absolute importance. A discussion of the choice of variables for use in discriminant analysis is given by Cochran (1964).

Griffiths (1957) gives details of the following analysis of variance design, which is intended to test the null hypothesis that the sample discriminant function accounts for a significant proportion of the intergroup difference. Firstly, the analogue of the squared multiple correlation coefficient, $R^2$, is derived:

$$R^2 = \frac{n_1 n_2}{n + n_1 n_2 D^2}$$

The total sum of squares is

$$\text{TSS} = \frac{n_1 n_2}{n}$$

and the ANOVA layout is:

| Source of variability | d.f. | Sum of squares | Mean square | F |
|---|---|---|---|---|
| Discriminant | df1 $(= p)$ | SS1 $(= \text{TSS} \cdot R^2)$ | MS1 $(= \text{SS1}/\text{df1})$ | $\frac{\text{MS1}}{\text{MS2}}$ |
| Residual | df2 $(= n - p - 1)$ | SS2 $(= \text{TSS} - \text{SS1})$ $(= \text{TSS}(1 - R^2))$ | MS2 $(= \text{SS2}/\text{df2})$ | — |
| Total | $n - 1$ | TSS | — | — |

If the computed $F$-ratio is less than the tabled $F$-value then the null hypothesis is accepted at the given significance level.

The computations in simple discriminant analysis can be summarized in the following algorithm:

(1) Form $S_A$ and $S_B$, the variance–covariance matrices of the two groups, and get $S = S_A + S_B$. (Subroutine SSCP, Appendix A.)
(2) Compute $d$ from (7.1). (The individual variable means are returned by subroutine SSCP.)
(3) Solve the equations $\hat{\alpha} = S^{-1} d$ using subroutine SOLVE, Appendix A.
(4) Compute individual discriminant scores for the members of groups A and B from (7.5) and find the mean values, $\bar{\hat{x}}_1$ and $\bar{\hat{x}}_2$
(5) Mahalanobis' $D$-squared can now be found from

$$D^2 = \bar{\hat{x}}_1 - \bar{\hat{x}}_2$$

Having computed $D^2$, the $F$-ratio test given by (7.2) and (7.3) can be applied.
(6) Compute the elements $c_i$ from (7.7) and the ANOVA tests from (7.8).
(7) If any unclassified cases are to be assigned, use (7.6) to compute discriminant scores.

A computer program to carry out these calculations is listed in Table 7.2.

The distribution of $\frac{1}{2}D$, where $D$ is the square root of Mahalanobis's $D$-squared, is approximately normal. By reference to a table of the

## Table 7.2
### Computer program for two-group discriminant analysis

```
C TWO GROUP DISCRIMINANT ANALYSIS
C PROGRAM BY P.M.MATHER,DEPT OF GEOGRAPHY, THE UNIVERSITY,NOTTINGHAM
C AUGUST,1974
C *********        *********        *********         *********
C USES SUBROUTINES *SSCP*,,*PRNTRI*,*SOLVE* FROM APPENDIX A
C CONTROL CARD 1
C N1       NUMBER OF ITEMS IN GROUP 1
C N2       NUMBER OF ITEMS IN GROUP 2
C NV       NUMBER OF VARIABLES
C PUNCH IN FORMAT(3I3)
C CONTROL CARD 2
C DATA INPUT FORMAT. PUNCH IN COLS 1-80. INCLUDE BRACKETS.
C FOLLOWING THE FORMAT CARD, INPUT THE SCORES FOR GROUP 1 THEN THE
C SCORES FOR GROUP 2, CASE BY CASE, USING FORMAT SPECIFIED ON CARD 2.
C BEGIN DATA FOR EACH CASE ON A FRESH CARD.
C SUBROUTINES FROM APPENDIX A:SSCP,PRNTRI,,SOLVE
C USES I/O UNITS 5 (CARDREADER), 6 (LINEPRINTER) AND 3 (UNFORMATTED
C DISC/MTAPE FILE)
      DOUBLE PRECISION P(10)
      DIMENSION SA(10,10), SB(10,10),S(10,10),X1(10),X2(10),SS(10),
     1 SMT(20),D(10),FMT(20)
      NN1=10
      NN2=100
      READ(5,1) N1, N2, NV
    1 FORMAT(3I3)
      READ(5,2) FMT
    2 FORMAT(20A4)
      N=N1+N2
C *DATIN* READS DATA CASE BY CASE FOR GROUP 1 FOLLOWED BY GROUP 2
C USING SPECIFIED FORMAT (STORED IN *FMT*)
      CALL DATIN(X1,N,NV,FMT)
      IND=-1
C VARIANCE-COVARIANCE MATRIX FOR GROUP 1
      CALL SSCP(X1,SA,SS,NV,N1, 10,IND,3,6,FMT,1)
      IND=-1
C VARIANCE-COVARIANCE MATRIX FOR GROUP 2
      CALL SSCP(X2,SB,SS,NV,N2,10,IND,3,6,FMT,1)
C COMPUTE (SAMPLE) POOLED COVARIANCE MATRIX IN *S*
      DO 3 I=1,NV
      DO 3 J=I,NV
    3 S(I,J)=((SA(I,J)+SB(I,J))*N/FLOAT(N-2))*0.5
      IP=1
      CALL PRNTRI(S,NN2,NN1,24HPOOLED COVARIANCE MATRIX,3,IP,NV,1,8,6)
      DO 4 I=1,NV
C GET MEAN DIFFERENCES IN *D* AND SOLVE LINEAR EQUATIONS USING *SOLVE*
C FROM APPENDIX A.
    4 D(I)=X1(I)-X2(I)
      CALL SOLVE(S,10,NV,DET,P,SS,D)
      WRITE(6,5) (I,SS(I),I=1,NV)
    5 FORMAT('0COEFFICIENTS VECTOR'/10(' ',I4,F12.4)/)
      IP=1
      WRITE(6,6) IP
    6 FORMAT('0DISCRIMINANT SCORES FOR GROUP',I3)
      SUM1=0.0
      REWIND 3
      DO 7 I=1,N1
      XH=0.0
      READ(3) X1
      DO 8 J=1,NV
    8 XH=XH+SS(J)*X1(J)
      WRITE(6,9) I,XH
    9 FORMAT(' ',I4,F12.4)
    7 SUM1=SUM1+XH
      SUM1=SUM1/FLOAT(N1)
      IP=2
```

Table 7.2  *continued*

```
        WRITE(6,6) IP
        SUM2=0.0
        DO 10 I=1,N2
        XH=0.0
        READ(3) X2
        DO 11 J=1,NV
   11   XH=XH+SS(J)*X2(J)
        WRITE(6,9) I,XH
   10   SUM2=SUM2+XH
        SUM2=SUM2/FLOAT(N2)
        DSQUAR=ABS(SUM1-SUM2)
        HALFD=0.5*SQRT(DSQUAR)
        WRITE(6,15) HALFD
   15   FORMAT('OHALF D = ',F10.4)
        T2=DSQUAR*N1*N2/FLOAT(N)
        WRITE(6,12) DSQUAR, T2
   12   FORMAT('OMAHALANOBIS   D SQUARED = ',F10.4/' HOTELLING S T-SQUARED
       1 = ',F12.4/' CONTRIBUTIONS OF VARIABLES TO THE DISCRIMINATING POWE
       1R (EXPRESSED AS PERCENTAGE)'//)
        DO 13 I=1,NV
        C=SS(I)*D(I)*100.0/DSQUAR
   13   WRITE(6,9) I, C
        DD=SQRT(DSQUAR)
        DO 20 I=1,NV
   20   SS(I)=SS(I)/DD
        WRITE(6,21) (I, SS(I), I=1,NV)
   21   FORMAT('OCOEFFICIENTS OF STANDARDIZED DISCRIMINANT FUNCTION'/
       +10(' ',I4,F10.4/))
        SUM1=0.0
        SUM2=0.0
        REWIND 3
        IP=1
        WRITE(6,24) IP
   24   FORMAT('OSCORES ON STANDARDIZED DISCRIMINANT FOR GROUP',I4)
        DO 22 I=1,N1
        XH=0.0
        READ(3) X1
        DO 23 J=1,NV
   23   XH=XH+SS(J)*X1(J)
        WRITE(6,9) I, XH
   22   SUM1=SUM1+XH
        SUM1=SUM1/FLOAT(N1)
        WRITE(6,30) IP, SUM1
   30   FORMAT('OMIDPOINT ON STANDARDIZED AXIS FOR GROUP',I3,' IS',F10.4)
        IP=2
        WRITE(6,24) IP
        DO 25 I=1,N2
        XH=0.0
        READ(3) X2
        DO 26 J=1,NV
   26   XH=XH+SS(J)*X2(J)
        WRITE(6,9) I, XH
   25   SUM2=SUM2+XH
        SUM2=SUM2/FLOAT(N2)
        WRITE(6,30) IP, SUM2
        XH=AMIN1(SUM1,SUM2)
        T2=AMAX1(SUM1,SUM2)
        CUTOFF=(T2-XH)*0.5
        IF(XH.LT.0.0) CUTOFF=(T2+XH)*0.5
        WRITE(6,28) CUTOFF
   28   FORMAT('OMIDPOINT ON STANDARDIZED AXIS = ',F10.4)
        FM=N
        FN=N1*N2
        FV=NV
        R2=(FN*DSQUAR)/(FM*(FM-2.0))
        F=((FM-FV-1.0)*R2)/FV
        NDF1=NV
```

Table 7.2   continued

```
      NDF2=N-NV-1
      WRITE(6,14) F, NDF1, NDF2
   14 FORMAT('0ANOVA TEST OF DISCRIMINANT'///' F = ',F10.4/' NDF1 = ',
     1 I3/' NDF2 = ',I3)
      R2=R2/(1.0+R2)
      WRITE(6,40) R2
   40 FORMAT('0R2 = ',F10.4)
      STOP
      END



      SUBROUTINE DATIN(X,N,NV,FMT)
C *DATIN* READS INPUT DATA CASE BY CASE FOR GROUP 1 FOLLOWED BY GROUP 2
C AND TRANSFERS TO UNIT 3 (DISC OR TAPE).
      DIMENSION X(NV),FMT(20)
      REWIND 3
      WRITE(6,2)
      DO 1 I=1,N
      READ(5,FMT) X
      WRITE(6,3) X
    1 WRITE(3) X
    2 FORMAT('1DATA MATRIX --- GROUP 1 FIRST'///)
    3 FORMAT(' ',12F8.3)
      ENDFILE 3
      REWIND 3
      RETURN
      END
```

standard normal distribution, the probability of misclassification can be found. For example, if $D^2 = 2.30$ and $½D = 1.15$ then $z$, the corresponding standard normal deviate, is 0.8749. The probability of misclassification is thus $(1.0 - 0.8749) = 0.12151$. In other words, an individual that belongs to group A will be placed in group B by the discriminant function on 12.5 per cent of occasions; alternatively, 87.5 per cent of all assignments will, in the long run, be correct.

It is likely that any individual falling close to the half-way point on the discriminant axis will have a high probability of being misclassified. For this reason it is sometimes suggested (e.g. Hope, 1968) that any individual having a probability of 10 per cent or more of belonging to group A should be left unallocated even though it is closer to group B. For example, individual $x_j$ may be closer to the mean discriminant score of group B than to the mean score of group A. If, however, there is a probability of 10 per cent or over that $x_j$ could belong to group A it will not be allocated to either group. The 10 per cent point is found by adding (or subtracting, as the case may be) the 10 per cent point on the standard normal curve to the mean score of the two groups. For instance, if $x_j = 2.8$, $\bar{\bar{x}}_A = 4.04$ and $\bar{\bar{x}}_B = 1.73$ then $x_j$ is closer to the mean of group B. The 10 per cent point of the standard normal distribution is 1.28 so that any individual having a score of $(4.04 - 1.28 =)$ 2.76 or more has at least a 10 per cent chance of belonging to group A. Since 2.8 exceeds 2.76 individual $x_j$ will not be assigned to group B but will be subjected to further examination. Similarly, an individual scoring $(1.73 + 1.28 = )$ 3.01 or less has a 10 per cent chance of belonging to group B, hence an individual scoring 2.9 would not be allocated, even though the midpoint between $\hat{\bar{x}}_A$ and $\hat{\bar{x}}_B$ is 2.89.

In some instances the probability of an individual belonging to group A or group B is not equal, for example if one group occurs more frequently than the other in the population of interest. The cutoff point (which was taken as the midpoint between $\hat{\bar{x}}_A$ and $\hat{\bar{x}}_B$ when the probabilities of membership of group A and B were considered to be equal) is now moved towards the mean of the less frequent group. This should reduce the overall probability of misclassification by increasing the probability that members of the less frequent group will be misclassified and by reducing the probability that members of the more frequent group will be misclassified.

To determine the optimum cutting point, find the ratio of the memberships of the larger group and the smaller group, and denote this ratio by $c$. Then the amount the cutting point is moved is

$$a = \log_e c / D^2.$$

The cutting point is moved a distance $a$ towards the mean of the less frequent group. The probability of misclassifying a member of group A is equal to the probability of the standard normal deviate $z$ where $z$ is the difference between $\hat{x}_A$ and the new cutoff point. The probability of misclassifying a member of group B can be found in a similar way. Hope (1968) discusses this and other related points in some detail.

Thomas (1969) used two-group discriminant analysis in a study of glacial and periglacial sediments taken from slope deposits of the northern uplands of the Isle of Man. Group 1 consisted of solifluxion deposits, and Group 2 was made up of interbedded gravels. The purpose of the analysis was to determine whether there was a statistically significant difference between the two groups in terms of the four variables $X_1$ (mean), $X_2$ (sorting), $X_3$ (skewness) and $X_4$ (kurtosis) computed according to the methods of Folk and Ward. Secondly, if it was found that there was a significant difference between the groups on the basis of these variables, an allocation procedure was required whereby future samples of sediment from this particular area of the Isle of Man could be placed in Group 1 or Group 2 on the basis of their scores on the four variables. The data, which consist of two sets of 22 measurements on the four variables, are listed in Table 7.3, and the sample pooled variance–covariance matrix is shown in Table 7.4. Using the methods outlined above, the discriminant function was found to be:

$$D = -0.6397X_1 + 3.4329X_2 - 4.2759X_3 + 7.5250X_4.$$

Table 7.3

Example data for two-group discriminant analysis (Reproduced with permission from G. S. P. Thomas, *Brit. Geomorph. Res. Group, Occas. Paper 6*, Geoabstracts, Univ. East Anglia, Norwich (1969))

|    | 1       | 2       | 3       | 4      |
|----|---------|---------|---------|--------|
| 1  | 0.4913  | -3.2693 | -0.4213 | 0.9125 |
| 2  | -0.1300 | -3.2490 | -0.5232 | 1.0954 |
| 3  | -0.6481 | -2.7983 | -0.4444 | 1.3912 |
| 4  | -0.8564 | -2.1615 | -0.3953 | 1.3912 |
| 5  | -0.4714 | -2.3650 | -0.2633 | 1.0866 |
| 6  | -0.6286 | -2.4798 | -0.3589 | 1.2991 |
| 7  | -0.4578 | -2.8611 | -0.4953 | 1.1872 |
| 8  | 0.8275  | -4.0372 | -0.5635 | 1.1680 |
| 9  | -1.4238 | -2.3075 | -0.6545 | 1.6537 |
| 10 | -0.0742 | -3.1766 | -0.6580 | 1.0357 |
| 11 | -0.4938 | -3.0081 | -0.5975 | 1.3367 |
| 12 | 0.0052  | -3.3482 | -0.6155 | 1.0704 |
| 13 | 0.0700  | -3.0538 | -0.4518 | 0.9284 |
| 14 | -0.6446 | -2.7040 | -0.5253 | 1.2224 |
| 15 | -1.5146 | -1.6571 | -0.4841 | 1.0469 |
| 16 | 0.3776  | -3.4178 | -0.4896 | 0.9769 |
| 17 | -0.8368 | -2.7783 | -0.6100 | 1.4777 |
| 18 | 0.4913  | -3.2693 | -0.4213 | 0.9125 |
| 19 | -1.0975 | -2.6039 | -0.6964 | 1.5331 |
| 20 | 0.6699  | -3.7452 | -0.5394 | 0.8696 |
| 21 | 0.7162  | -3.5836 | -0.4608 | 0.9028 |
| 22 | -0.1198 | -2.4597 | -0.1513 | 1.0189 |
| 23 | 2.4517  | -4.3949 | -0.1823 | 0.7117 |
| 24 | 1.4428  | -3.8205 | -0.2661 | 0.7213 |
| 25 | 2.2619  | -4.0244 | -0.1576 | 0.7130 |
| 26 | 2.2453  | -4.2391 | -0.1867 | 0.7337 |
| 27 | 1.3328  | -3.9544 | -0.3714 | 0.7344 |
| 28 | 1.6092  | -4.0794 | -0.4027 | 0.8278 |
| 29 | 1.7070  | -3.9824 | -0.2671 | 0.7273 |
| 30 | 1.8808  | -3.8957 | -0.1278 | 0.7841 |
| 31 | 1.5320  | -4.1246 | -0.3959 | 0.7821 |
| 32 | 3.3340  | -4.1482 | -0.1515 | 0.8446 |
| 33 | 0.8023  | -3.9115 | -0.5212 | 0.9712 |
| 34 | 1.2675  | -3.9318 | -0.4164 | 0.7903 |
| 35 | 3.0510  | -4.3519 | 0.0168  | 0.7480 |
| 36 | 1.4040  | -4.0915 | 0.4133  | 0.8150 |
| 37 | 2.7480  | -4.0116 | 0.1939  | 0.7701 |
| 38 | 4.4820  | -4.3326 | 0.1972  | 0.7281 |
| 39 | 1.5329  | -3.6072 | 0.0962  | 0.7253 |
| 40 | -0.0467 | -3.1816 | -0.6163 | 0.8197 |
| 41 | 1.7833  | -4.0487 | -0.3082 | 0.7070 |
| 42 | 1.9980  | -4.0171 | -0.2662 | 0.7837 |
| 43 | 3.5860  | -4.4501 | 0.1046  | 0.8174 |
| 44 | 1.6169  | -4.1473 | -0.3979 | 0.7430 |

Table 7.4

Variance covariance matrix for Thomas data (Table 7.3)

|   | 1      | 2       | 3       | 4       |
|---|--------|---------|---------|---------|
| 1 | 0.7210 | -0.2729 | 0.0880  | -0.0648 |
| 2 |        | 0.1970  | -0.0051 | 0.0315  |
| 3 |        |         | 0.0422  | -0.0073 |
| 4 |        |         |         | 0.0274  |

Table 7.5
Discriminant scores for Thomas data (Table 7.3)

```
DISCRIMINANT SCORES FOR GROUP 1
   1    -2.8695
   2    -0.5903
   3     3.1773
   4     5.2867
   5     1.4853
   6     3.1996
   7     1.5225
   8    -3.1900
   9     8.2320
  10    -0.2503
  11     2.6029
  12    -0.8108
  13    -1.6101
  14     2.5745
  15     5.2281
  16    -2.5299
  17     4.7257
  18    -2.8695
  19     6.2774
  20    -4.4353
  21    -3.9964
  22    -0.0531

DISCRIMINANT SCORES FOR GROUP 2
   1   -10.5205
   2    -7.4727
   3    -9.2231
   4    -9.6693
   5    -7.3132
   6    -7.0825
   7    -8.1481
   8    -8.1299
   9    -7.5612
  10    -9.3697
  11    -4.4041
  12    -6.5808
  13   -11.3345
  14   -10.5782
  15   -10.5634
  16   -13.1047
  17    -8.3172
  18    -2.0888
  19    -8.4015
  20    -8.0328
  21   -11.8670
  22    -7.9791
```

The scores of each set of 22 on this discriminant function are shown in Table 7.5. The first group (solifluxion deposits) have scores in the range $-4.4353$ to $8.2320$, while the interbedded gravels have scores in the range $-2.0888$ to $-13.1047$. There is some overlap, but closer inspection of Table 7.5 shows that only two of the interbedded gravels have scores that lie within the range of the solifluxion deposits, and that six of the solifluxion deposits have scores that lie in the range of the interbedded gravels. Mahalanobis' $D^2$ is 9.4931, and the associated $F$ test results in non-acceptance of the null hypothesis that the two multivariate means are equal ($T^2 = 104.42$, $F = 24.24$ with 39 and 168 degrees of freedom). The contribution of each of the four variables to the power of the discriminant function indicates that sorting is the most useful discriminating variable, accounting for 40.13 per cent of the discriminating power of the function. Kurtosis, accounting for 30.69 per cent, came next followed by the mean (15.24 per cent) and the skewness (13.94 per cent). Finally, the ANOVA test of the discriminant function, which is a test of the null hypothesis that the variance accounted for by the sample function could be due to chance, gives an $F$ value of 1018 with 4 and 39 degrees of freedom. The null hypothesis is therefore rejected conclusively.

## 7.2 Multiple Discriminant Analysis

Although two-group discriminant function analysis (Section 7.1) has been widely used, it is more often the case that several groups are thought to exist in a given set of data. While it is true that each pair of groups could be analyzed separately, using the procedures of Section 7.1, the number of comparisons grows rapidly with $k$, the number of groups. The general formula for $m$, the number of pairwise comparisons of $k$ groups, is $m = \frac{1}{2}(k^2 - k)$. Thus, for $k = 10$, $m = 45$. Multiple discriminant analysis (MDA) provides for the simultaneous comparison of several groups, every member of each group being measured on a number of variables.

The assumptions of MDA are the same as those of two-group discriminant analysis, namely, multivariate normality and equality of the within-group covariance matrices. Tests for multivariate normality are not well developed (but see Ito (1969) and Andrews et al. (1973)), and it is not really known how far departures from normality affect the outcome of the analysis. However, a test for equality of covariance matrices, known as Box's test, is available, though it is also dependent on the normality assumption. Usually it is included in a more general multivariate analysis of variance program, as is the case in this section. A logical pre-requisite of MDA is that the groups are, in fact, separate. Multivariate analysis of variance provides a means of testing the null hypothesis that the (population) means of the $k$ groups are identical. Again, multivariate normality and equality of the variance—covariance matrices of the groups are assumed. The $F$-ratio test of the null hypothesis $H_0: \mu_1 = \mu_2 = \ldots = \mu_k$ (where $\mu_i$ is the multivariate mean of the ith group) is analogous to the Hotelling $T$-squared test used in Section 7.1.

### 7.2.0 Multivariate Analysis of Variance (MANOVA)

Univariate analysis of variance designs have been widely used in many disciplines, notably in the agricultural sciences. Examples of their use in the earth sciences can be found in Griffiths (1967), Schlee (1957),

Doornkamp and King (1971) and Carter and Chorley (1961). A computational guide is provided by Hemmerle (1967).

Given the assumption that the data for each group follow a multivariate normal distribution, and that the groups have equal variances and covariances, it is possible to test the hypothesis that their multivariate means are also equal. If this hypothesis is accepted there is no point in computing discriminant functions. The groups may be postulated a priori (thus, one might divide sediment types into groups on the basis of their supposed environments of deposition) or the existence of the groups may be suggested by the use of one of the classification procedures described in Chapter 6.

The statistical model for MANOVA is:

$$x_{qi} = \mu + (\mu_q - \mu) + (x_{qi} - \mu_q) \tag{7.9}$$

Here, $x_{qi}$ is a vector of measurements on $p$ variables for the $i$th individual (member) of group $q$, $\mu$ is the overall or grand mean vector and $\mu_q$ is the mean vector of the $q$th group. If the $x_{qi}$ are converted to deviation form, i.e.

$$x^*_{qi} = x_{qi} - \mu_q$$

then (7.9) becomes

$$x^*_{qi} = (\mu_q - \mu) + (x_{qi} - \mu_q) \tag{7.10}$$

which states that the scores of individual $i$ in group $q$ over the $p$ variables of interest can be broken down into two components: (i) the deviation of the mean of the $q$th group from the grand mean, and (ii) the deviation of the $i$th individual from the mean of the $q$th group.

If we now find the (corrected) sums of squares and cross-products matrices for the three elements in (7.10), namely:

(i) the $x^*_{qi}$, giving the 'total sums of squares and cross-products matrix', **T**;
(ii) the $(\mu_q - \mu)$, giving the sums of squares and cross-products of the deviations of the group means from the grand mean, or the 'between-groups' matrix, **B**;
(iii) the $(x_{qi} - \mu_q)$, which gives the sums of squares and cross-products matrix for the deviations of the individuals from the means of their groups, or the 'within-groups' matrix, $\mathbf{W}_k$. The sum of the $\mathbf{W}_k$ is **W**, the 'pooled within-groups sums of squares and cross-products matrix',

then we can write:

$$\mathbf{T} = \mathbf{W} + \mathbf{B}, \tag{7.11}$$

which is the fundamental equation of MANOVA.

We now wish to test the null hypothesis that the $\mu_q$ are all equal.

Before this is done, it is necessary to check the assumption that the groups have equal variance—covariance matrices. The variance—covariance matrix for group $q$ is $S_q$ and is derived from

$$S_q = \frac{1}{N_q - 1} W_q \qquad (7.12)$$

where $N_q$ is the number of individuals in group $q$ and $W_q$ the corrected sums of squares and cross-products matrix for the $q$th group. Box (1949) has derived a test statistic $M$ which is exceedingly tedious to compute by hand and hence has rarely been used. It assumes multivariate normality. Very little is known about the effects of heterogeneity of within-group variance-covariance matrices on the outcome of the MANOVA test described below, but given that Box's test is easily programmed there is no reason why we should shelter behind an unverifiable presumption regarding the robustness of MANOVA. Furthermore, it is often valuable to consider in its own right the equality or otherwise of within-group variance—covariance matrices.

The test statistic is

$$M = (N - k)\log_e\{\det(S)\} - \sum_{q=1}^{k}(N_q - 1)\log_e\{\det(S_q)\} \qquad (7.13)$$

Here, $S$ is the pooled variance-covariance matrix, defined as

$$S = \frac{1}{N - k} W$$

$N$ is the total number of individuals, $k$ the number of groups, and $S_q$ the variance—covariance matrix for group $q$ (Equation 7.12). To determine the significance of $M$ it is necessary to transform it into an $F$-ratio. Scalars $a$ and $b$ are defined as:

$$a = \left(\sum_{q=1}^{k}\frac{1}{N_q - 1} - \frac{1}{N - k}\right)\frac{2p^2 + 3p - 1}{6(k - 1)(p + 1)}$$

and

$$b = \left(\sum_{q=1}^{k}\frac{1}{(N_q - 1)^2} - \frac{1}{(N - k)^2}\right)\frac{p(p - 1)(p + 1)}{6(k - 1)}$$

If $b$ is greater than $a^2$ then

$$v_1 = \frac{p(k - 1)(p + 1)}{2},$$

$$v_2 = \frac{v_1 + 2}{b - a^2}$$

and
$$F = \frac{M}{1 - a - \frac{v_1}{v_2}}.$$

If $b$ is less than $a^2$ then $v_1$ is defined as above,

$$v_2 = \frac{v_1 + 2}{a^2 - b}$$

and

$$F = \frac{v_2 M}{v_1 (S - M)}$$

where

$$S = \frac{v_2}{1 - a + \frac{2}{v_2}}$$

If $b = a^2$ the transformation is not defined. The value of $F$ obtained by the above procedure is compared to the tabled value of the $F$ ratio with $v_1$ and $v_2$ degrees of freedom. If the calculated value exceeds the tabled value at a given level of significance then the null hypothesis that the variance—covariance matrices are equal is not accepted.

Hope (1968, p. 28—29) gives details of an alternative procedure which reduces to a chi-square test. The formula is:

$$\chi^2 = -\log_e \prod_{i=1}^{k} \det(S_i)/\det(S)$$

Here, $S_i$ is the variance—covariance matrix for group $i$, $S$ is the pooled variance—covariance matrix, and $k$ the number of groups. The degrees of freedom are $(k - 1)p(p + 1)/2$. If $\chi^2$ is small then the null hypothesis that the within-group variances and covariances are unequal is not accepted. If $\chi^2$ is greater than the tabled value at an appropriate significance level then either or both of the following conclusions apply: (i) the variances and covariances are truly heterogeneous, and (ii) the distributions are non-normal. Hope (1968, p. 29) points out that the $\chi^2$ test is sensitive to non-normality. He also suggests that the MANOVA test, to be described next, is not so sensitive to departures from non-normality nor is it greatly affected by moderate heterogeneity of variances and covariances. Consequently, it is still feasible to carry out the standard MANOVA test even though the $\chi^2$ test or Box's $M$ test indicate heterogeneous variance—covariance matrices. Even so, it is wise to consider the implications of rejecting the null hypothesis.

The basic MANOVA test for equality of group means is known as Wilks' lambda test, and its validity is dependent on the outcome of Box's test as described previously. Wilks' criterion $\Lambda$ is simply:

$$\Lambda = \frac{\det(W)}{\det(T)} \qquad (7.14)$$

and provides a means of testing $H_0 : \mu_k = \mu$. As in the case of Box's $M$, Wilks' $\Lambda$ has to be transformed into an $F$-ratio before it has any practical use. This transformation appears complicated but is easily programmed. For values of $p$ greater than 2 and of $k$ greater than 3 the procedure is:

$$s = \left(\frac{p^2 f^2 - 4}{p^2 + f^2 - 5}\right)^{1/2}$$

where

$$f = k - 1$$

then

$$v_1 = pf$$

and

$$v_2 = s[N - 1 - (p + f + 1)/2] - (v_1 - 2)/2.$$

Letting

$$y = \Lambda^{1/s}$$

then

$$F = \frac{1-y}{y} \frac{v_2}{v_1} \qquad (7.15)$$

This is treated as an $F$ ratio with $v_1$ and $v_2$ degrees of freedom. The null hypothesis is accepted at significance level $\alpha$ if the calculated value of $F$ is less than the tabled value. The special formulae for use when $p \leq 2$ and $q \leq 3$ are given below. (Note: $q = k - 1$).

The computations involved in MANOVA are straigtforward. A computer program is listed in Table 7.7. The basic steps in this program involve:

(i) the calculation of the matrices T, W and B using subroutine SSCP.

(ii) As W is built up by summing the $W_q$ ($q = 1,k$) the corresponding matrices $S_q$ are formed and their determinants evaluated using subroutine DT2.

(iii) Box's criterion $M$ and the corresponding $F$-ratio transformation are now found.

Table 7.6
Formulae for conversion of Wilks' Lambda to an $F$-ratio in special cases
(Reproduced from W. W. Cooley and P. R. Lohnes, *Multivariate Data Analysis*, John Wiley & Sons, Inc., New York, 1971)

| $p$ | $q$ | $F$ | $v_1$ | $v_2$ |
|---|---|---|---|---|
| Any | 1 | $\dfrac{1-\Lambda}{\Lambda}\dfrac{N-p-1}{p}$ | $p$ | $N-p-1$ |
| Any | 2 | $\dfrac{1-\Lambda^{1/2}}{\Lambda^{1/2}}\dfrac{N-p-2}{p}$ | $2p$ | $2(N-p-2)$ |
| 1 | Any | $\dfrac{1-\Lambda}{\Lambda}\dfrac{N-q-1}{q}$ | $q$ | $N-q-1$ |
| 2 | Any | $\dfrac{1-\Lambda^{1/2}}{\Lambda^{1/2}}\dfrac{N-q-2}{q}$ | $2q$ | $2(N-q-2)$ |

(iv) After det (**T**) and det (**W**) have been found, again using subroutine DT2, Wilks' $\Lambda$ and the accompanying $F$-ratio are computed.

As listed, the program will analyse measurements on twenty or fewer variables for twenty or less groups. The number of observations is effectively unlimited. It is necessary only to alter the DIMENSION statement of the main program and the following DATA statement to change these limits. The integer constant N1 is set to the first dimension of the Fortran arrays W, T and WW. MM is equal to the product of the first two dimensions of these arrays. The program uses one backing-store file, which can be tape or disc, and is allocated to channel number 3.

*7.2.1 Multiple Discriminant Analysis*
Two-group discriminant function analysis has been generalized to deal with the $k$-group case by Fisher (1938), Rao (1952) and Rao and Slater (1949). It is assumed that each of the $k$ samples is drawn from a separate population. The $k$ populations will differ in their means (otherwise there would be no point in discriminating between them) but their variance—covariance matrices should be equal and the variables on which measurements are made should be normally distributed. It is considered that moderate departures from these ideal conditions do not have a serious effect on the results.

In the case of two groups, only one discriminant axis was necessary to separate the groups. In the general case of $k$ groups a maximum of $(k-1)$ discriminant axes is necessary. Fewer than $(k-1)$ axes may suffice if they account for a higher proportion of the total inter-group

## Table 7.7
### Computer program for multivariate analysis of variance

```
C  PROGRAM FOR MULTIVARIATE ANALYSIS OF VARIANCE.
C  SEE CHAPTER 7 FOR FURTHER DETAILS
C
C  NEEDS I/O UNITS 5 (CARDREADER),6 (LINEPRINTER) AND 3 (UNFORMATTED DISC
C  FILE).
C
C  DIMENSIONS ALLOW FOR 20 GROUPS, 20 VARIABLES AND AN UNLIMITED NUMBER
C  OF CASES
C
C  DATA IN ORDER:
C  *** CARD 1
C  NV   NUMBER OF VARIABLES (COLS 1-3).
C  NG   NUMBER OF GROUPS (COLS 4-6)
C  *** CARD 2
C  NGR  NUMBER OF ITEMS IN EACH GROUP. PUNCH NUMBER IN GROUP 1 IN COLS
C       1-3,NUMBER IN GROUP 2 IN COLS 4-6 AND SO ON
C  *** CARD 3
C  FMT1  OUTPUT FORMAT FOR USE IN *MATPR* (SEE APPENDIX A).
C  *** CARD 4
C  FMT   DATA INPUT FORMAT. COLS 1-80.
C  *** CARD 5 ON
C  X    DATA, CASE BY CASE, USING FORMAT SPECIFIED IN CARD 4. START EACH
C       CASE ON A FRESH CARD.
C
C  SUBROUTINES REQUIRED (APPENDIX A): MATPR, SSCP, DT2.
C
       DOUBLE PRECISION WORK(20)
       DIMENSION W(20,20),T(20,20),DETS(20),NGR(20),WW(20,20),A(20),AA(20
      +),XCV(20),FMT1(20)
C  IN THE FOLLOWING DATA STATEMENT, N1 IS THE FIRST DIMENSION OF ARRAYS
C  W,T AND WW; MM IS THE PRODUCT OF THE TWO DIMENSIONS OF W,T,AND WW.
C  ALL THREE ARRAYS MUST HAVE A DIMENSION OF AT LEAST (NV*NV).  SINGLE-
C  SUBSCRIPTED ARRAYS NEED A DIMENSION OF AT LEAST NV, EXCEPT NGR WHICH
C  SHOULD HAVE A DIMENSION OF NG.
       DATA N1,MM,WW/20,400,400*0.0/,NIND,IP/0,1/,Y,Z/0.0,0.0/
       WRITE(6,52)
    52 FORMAT('1              MULTIVARIATE ANALYSIS OF VARIANCE'/
      *          '             PROGRAM BY PAUL M MATHER'/
      *          '             UNIVERSITY OF NOTTINGHAM, 1974'/)
       REWIND 3
       READ (5,1) NV, NG
     1 FORMAT(2OI3)
       READ(5,1) (NGR(I),I=1,NG)
       READ(5,3) FMT1
     3 FORMAT(20A4)
       DO 2 I=1,NG
     2 NIND=NIND+NGR(I)
       CALL INPT(NIND,NV,A)
       WRITE(6,51) NIND
    51 FORMAT('OSTATISTICS FOR TOTAL DATA WITH N = ',I4)
C  SUBROUTINE SSCP IS LISTED IN APPENDIX A
       CALL SSCP(A,T,DETS,NV,NIND,N1,3,3,6,FMT1,1)
C  SUBROUTINE MATPR IS LISTED IN APPENDIX A
       CALL MATPR(48HTOTAL SUMS OF SQUARES AND CROSS-PRODUCTS MATRIX ,6 ,
      *FMT1,6,T,MM,N1,NV,NV,IP,1,8)
       CALL SWAP1(T,N1,NV)
       REWIND 3
       DO 4 I=1,NG
       K=NGR(I)
       WRITE(6,50) I, K
    50 FORMAT('OSTATISTICS FOR GROUP',I4,' WITH N = ',I4)
C  SUBROUTINE SSCP IS LISTED IN APPENDIX A
       CALL SSCP(A,W,XCV,NV,K,N1,3,3,6,FMTO,1)
       CALL SWAP1(W,N1,NV)
       DO 20 J=1,NV
```

Table 7.7   *continued*

```
        DO 5 L=1,NV
      5 WW(J,L)=WW(J,L)+W(J,L)
        DO 20 L=1,NV
     20 W(J,L)=W(J,L)*(1.0/FLOAT(K-1))
C SUBROUTINE MATPR IS LISTED IN APPENDIX A
        CALL MATPR(24HWITHIN GROUP SSCP MATRIX,3,FMT1,6,W,MM,N1,NV,NV,
       *IP,1,8)
C SUBROUTINE DT2 IS LISTED IN APPENDIX A
        CALL DT2(W,N1,NV,WORK,DET)
        IF(NV.GT.0) GOTO 10
        WRITE(6,11) I
     11 FORMAT('OVARIANCE-COVARIANCE MATRIX FOR GROUP',I4,' IS SINGULAR')
        STOP
     10 DETS(I)=DET
      4 WRITE(6,6) I,DET
      6 FORMAT('ODETERMINANT OF GROUP',I4,' IS',G14.5)
C SUBROUTINE MATPR IS LISTED IN APPENDIX A
        CALL MATPR(32HPOOLED WITHIN  GROUPS MATRIX      ,4,FMT1,6,WW,MM,N1,
       *NV,NV,IP,1,8)
        XNV=NV
        XNG=NG
        C=NIND
        X=1.0/(C-XNG)
        DO 8 I=1,NV
        DO 8 J=I,NV
      8 W(I,J)=WW(I,J)*X
C SUBROUTINE DT2 IS LISTED IN APPENDIX A
        CALL DT2(W,N1,NV,WORK,D)
        IF(NV.GT.0) GOTO 9
        WRITE(6,12)
     12 FORMAT('OPOOLED WITHIN GROUPS VARIANCE-COVARIANCE MATRIX IS SINGUL
       *AR')
        STOP
      9 WRITE(6,13) D
     13 FORMAT('ODETERMINANT OF POOLED WITHIN GROUPS VARIANCE-COVARIANCE M
       *ATRIX =',G14.5)
        X=(C-XNG)*ALOG(D)
        DO 15 I=1,NG
     15 X=X-(NGR(I)-1.0)*ALOG(DETS(I))
        R=1.0/(C-XNG)**2
        Q=1.0/(C-XNG)
        DO 16 K=1,NG
        P=NGR(K)
        Y=Y+1.0/(P-1.0)
     16 Z=Z+1.0/(P-1.0)**2
        Y=Y-Q
        Z=Z-R
        Y=Y*(2.0*XNV*XNV+3.0*XNV-1.0)/(6.0*(XNG-1.0)*(XNV+1.0))
        Z=Z*((XNV-1.0)*(XNV+2.0))/(6.0*(XNG-1.0))
        IF(Z-Y*Y) 21,22,23
     22 WRITE(6,24)
     24 FORMAT('OA & B ARE EQUAL')
        STOP
     21 P=XNV*(XNG-1.0)*(XNV+1.0)*0.5
        Q=(P+2.0)/(Y*Y-Z)
        B=Q/(1.0-Y+2.0/Q)
        F=Q*X/(P*(B-X))
        GOTO 25
     23 P=(XNG-1.0)*XNV*(XNV+1.0)*0.5
        Q=(P+2.0)/(Z-Y*Y)
        B=P/(1.0-Y-P/Q)
        F=X/B
     25 WRITE(6,26) F ,P, Q
     26 FORMAT('OTEST OF HO: SAMPLE VARIANCE-COVARIANCE MATRICES ARE HOMOG
       *ENEOUS'//' F=',G14.5//' DEGREES OF FREEDOM =',2F8.0)
C SUBROUTINE DT2 IS LISTED IN APPENDIX A
        CALL DT2(T,N1,NV,WORK,DD)
```

Table 7.7   *continued*

```
          IF(NV.GT.0) GOTO 28
          WRITE(6,27)
   27 FORMAT('OTOTAL SSCP MATRIX IS SINGULAR')
          STOP
C SUBROUTINE DT2 IS LISTED IN APPENDIX A
   28 CALL DT2(WW,N1,NV,WORK,D)
          IF(NV.GT.0) GOTO 30
          WRITE(6,29)
   29 FORMAT('OPOOLED WITHIN GROUPS SSCP MATRIX IS SINGULAR')
          STOP
   30 WRITE(6,60) DD
   60 FORMAT('ODETERMINANT OF T = ',E21.11)
   61 FORMAT('ODETERMINANT OF W = ',E21.11)
          WRITE(6,61) D
          WILKS=D/DD
          WRITE(6,31) WILKS
   31 FORMAT('OWILKS LAMBDA= ',G14.5)
          IF(NV.LE.2.OR.NG.LE.3) GOTO 32
          Q=(XNG-1.0)**2
          S=SQRT((XNV*XNV*Q-4.0)/(XNV*XNV+Q-5.0))
          P=XNV*(XNG-1.0)
          Q=S*((NIND-1.0)-(XNV+(XNG-1.0)+1.0)/2.0)-((XNV*(XNG-1.0)-2.0)/2.0)
          S=WILKS**(1.0/S)
          F=(1.0-S)/S*Q/P
          WRITE(6,55) F, P, Q
          STOP
   32 IF(NG-3) 33,34,35
   33 XN=NIND-NV-1.0
          F=((1.0-WILKS)/WILKS)*(XN/XNV)
          GOTO 40
   34 P=SQRT(WILKS)
          XN=NIND-NV-2
          F=((1.0-P)/P)*(XN/XNV)
          XN=XN+XN
          XNV=XNV+XNV
          GOTO 40
   35 IF(NV.GT.1) GOTO 36
          XN=NIND-NG-2
          XNV=NG-1.0
          F=((1.0-WILKS)/WILKS)*(XN/XNV)
          GOTO 40
   36 XN=NIND-NG-3
          XNV=NG-1
          P=SQRT(WILKS)
          S=((1.0-P)/P)*(XN/XNV)
          XNV=XNV+XNV
          XN=XN+XN
   40 WRITE(6,55) F, XNV, XN
   55 FORMAT('OTEST OF HO: MULTIVARIATE MEANS OF GROUPS ARE EQUAL'//' F =
         *',G14.5/' DEGREES OF FREEDOM = ',2F10.3)
          STOP
          END




          SUBROUTINE INPT(N,M,X)
C SUBROUTINE READS DATA MATRIX ROWWISE AND STORES ON UNIT 3
          DIMENSION X(M),FMT(20)
          READ(5,22) FMT
   22 FORMAT(20A4)
          WRITE(6,5000)
 5000 FORMAT('O                    INPUT    DATA'///)
          DO 1 I=1,N
          READ(5,FMT) X
          WRITE(3) X
    1 WRITE(6,2) X
```

Table 7.7 *continued*

```
       ENDFILE 3
       REWIND 3
   2 FORMAT(' ',10G11.5)
       RETURN
       END



       SUBROUTINE SWAP1(S,N1,N)
COPIES LOWER TRIANGLE OF MATRIX INTO UPPER TRIANGLE
       DIMENSION S(N1,N)
       DO 1 I=1,N
       DO 1 J=1,I
   1 S(J,I)=S(I,J)
       RETURN
       END
```

difference, or if $p$, the number of variables, is less than $k$, the number of groups, in which case a maximum of $p$ discriminant axes will be required irrespective of the number of groups. The first discriminant axis can be thought of as the line of closest fit to the means of the $k$ groups in the $p$-dimensional space defined by the variables. The second discriminant axis is orthogonal to the first and to subsequent discriminant axes, and it is the line of closest fit to the $k$ means subject to the constraint of orthogonality. The relationship to the principal axes method used in Chapter 4 is self-evident, for the first principal component is defined as the axis in a $p$-dimensional space such that the sum of the squared deviations of the points representing individuals from the principal axis is minimized.

The coefficients defining the discriminant axes are often termed canonical vectors. These canonical vectors are the eigenvectors of the matrix $W^{-1}A$, where $W^{-1}$ is the inverse of the pooled within-groups sums of squares and cross products matrix of the $k$ groups and $A$ is the between-groups sums of squares and cross-products matrix. The matrix $W^{-1}A$ can be thought of as the ratio of within-group to between-group sums of squares and cross-products. The first canonical axis maximises this ratio, while the second canonical axis maximizes the ratio subject to being at right angles to the first, and so on.

$W^{-1}A$ is nonsymmetric. The determination of the eigenvalues and eigenvectors of a nonsymmetric matrix is more difficult than in the case of a symmetric matrix. However, it is possible to transform $W^{-1}A$ into such a form that the standard procedures for symmetric matrices can be used. The method is described by Martin and Wilkinson, in Wilkinson and Reinsch (eds.) (1971). The necessary subroutines from Appendix A are REDUC, which reduces the eigenproblem to standard form, so that the standard methods (using subroutines HTDQL, or TRIDI, RATQR and E1GVEC) can be applied. The eigenvalues of the reduced form are the same as the eigenvalues of $W^{-1}A$ but the eigenvectors must be back-transformed using subroutine REBAKA. The use of these routines is described in Appendix A.

The eigenvalues $\Lambda$ ($= \lambda_i$) and eigenvectors $Q$ of $W^{-1}A$ give the following information.

(a) The discriminating power of each canonical vector is given by the corresponding eigenvalue, which is usually expressed as a percentage of the sum of the eigenvalues.

(b) The eigenvectors give the coefficients of the discriminant functions (canonical vectors); from these coefficients the scores of the individuals on the discriminant functions can be computed in exactly the same manner as was described in Section 7.1.

(c) The Wilks' Lambda criterion used in MANOVA (Section 7.2.0) is easily computed from

$$\Lambda = \prod_{j=1}^{n} \frac{1}{1 + \lambda_j}$$

where $n$ is the lesser of $p$ and $(k - 1)$.

(d) The following null hypothesis can be tested: the discriminating power of the canonical vectors above the $g$th is due to chance. This is equivalent to stating that the population eigenvalues $\lambda_{g+1}, \lambda_{g+2}, \ldots, \lambda_n$ are zero and that the sample eigenvalues $\hat\lambda_{g+1}, \hat\lambda_{g+2}, \ldots, \hat\lambda_n$ differ from zero only because of sampling fluctuations. The $\chi^2$ test of this null hypothesis is

$$\chi^2 = -N - \tfrac{1}{2}(p + k) - 1 \log_e \Lambda'$$

where

$$\Lambda' = \prod_{j=g+1}^{n} \frac{1}{1 + \lambda_j}$$

and the number of degrees of freedom is $(p - g)(k - g - 1)$. If the calculated $\chi^2$ exceeds the tabled $\chi^2$ for a selected significance level then the null hypothesis is not accepted. The test is then repeated with $g$ incremented by 1 until all $n$ functions have been tested or until $\chi^2$ is not significant.

A computer program for multiple discriminant analysis is given in Table 7.8. The method of use is given in the form of comments within the listing.

Klovan (1966) employed $Q$-mode factor analysis (i.e. analysis of a similarity matrix of individuals) to classify a group of 69 near-shore sediment samples from Barataria Bay, on the Mississippi delta. The 69 samples used by Klovan are a selection from the data matrix published by Krumbein and Aberdeen (1937). Three sources of variability were recognized by Klovan. He termed these current energy, gravitational energy and surf energy factors, and Figure 7.2 shows the locations of

Table 7.8
Computer program for multiple discriminant analysis

```
C ****************   NOTES ON PROGRAM USE  ****************************
C THE FOLLOWING I/O UNITS ARE REQUIRED: 5 (LINEPRINTER), 5 (CARDREADER),
C 4 AND 3 (DISC OR MTAPE SCRATCH FILES).
C DIMENSION OF X IN MAIN PROGRAM SHOULD BE AT LEAST 2(NV*NV)+5*NV+NG+NV
C      *NG OR(KT+2)*NPTS, WHICHEVER IS THE GREATER. KT IS THE LESSER OF
C      NV AND NG-1.
C DIMENSION OF INT SHOULD BE AT LEAST NG.
C IF DIMENSION OF X IS CHANGED ALTER DATA STATEMENT IN MAIN PROGRAM
C      SETTING X TO ZERO.
C
C DATA IN ORDER:
C ***CARD 1:
C COLS 1-3: NV, NUMBER OF VARIABLES
C COLS 4-6: NG, NUMBER OF GROUPS.
C ***CARD 2:
C COLS 1-3: NUMBER OF ITEMS IN GROUP 1;
C COLS 4-6: NUMBER OF ITEMS IN GROUP 2;
C AND SO ON FOR EACH GROUP. START A NEW CARD FOR EACH 20 NUMBERS, I.E.
C      ENTER INFORMATION FOR GROUPS 1-20 ON ONE CARD, FOR GROUPS 21-40 ON
C      A SECOND CARD, AND SO ON. USE AS MANY CARDS AS NECESSARY. NOTE: IF
C      A LARGE NUMBER OF GROUPS IS SPECIFIED THE DIMENSION OF X AND IX
C      IN THE MAIN PROGRAM MAY NEED ALTERATION.
C *** CARD 3:
C COLS 1-4: NAME OF GROUP 1;
C COLS 4-8: NAME OF GROUP 2;
C AND SO ON FOR ALL GROUPS. ENTER 20 GROUP NAMES PER CARD.
C *** CARD 4:
C COLS 1-3: ENTER 000 IF NO UNCLASSIFIED CASES, ALTERNATIVELY ENTER 001.
C *** CARD 5
C COLS 1-80: OUTPUT FORMAT FOR SUBROUTINE MATPR (SEE APPENDIX A).
C *** CARD 6:
C COLS 1-80: INPUT FORMAT FOR DATA.
C *** CARD 7 ON: ENTER DATA CASE BY CASE IN APPROPRIATE FORMAT (AS GIVEN
C ON CARD 5). START EACH CASE ON A FRESH CARD.
C *** CARD 8:                    *** REQUIRED ONLY IF 001 ENTERED ON CARD 4
C COLS 1-3: NUMBER OF UNCLASSIFIED CASES.
C *** CARD 9:                    *** REQUIRED ONLY IF 001 ENTERED ON CARD 4
C ENTER DATA CASE BY CASE FOR UNCLASSIFIED CASES. USE FORMAT ENTERED ON
C CARD 6. START EACH CASE ON A FRESH CARD.
C
C SUBROUTINES REQUIRED FROM APPENDIX A:
C      SSCP, MATPR, PRNTRI, HTDQL, REDUC, REBAKA, PLOTS
C
       DIMENSION X(2500), INT(200), IX(2500)
C DIMENSION OF X AND IX SHOULD BE THE SAME.
       DATA X/2500*0.0/
       EQUIVALENCE(X, IX)
       REWIND 3
       REWIND 4
       READ(5,1) NV, NG
     1 FORMAT(2I3)
C SET UP BASE ADDRESSES
       N=NV*NV
       N1=N+1
       N2=N1+N
       N3=N2+NV
       N4=N3+NV
       N5=N4+NV
       N7=N5+NV
       N8=N7+NV
       N9=N8+NV*NG
C CALL MAIN ROUTINE
       CALL MDA(X(1), X(N1), X(N2), X(N3), X(N4), X(N5), X(N7), X(N8), X(N9),
      1 INT(1), NV, NG, KT, NIND, NUM)
       REWIND 4
```

Table 7.8    *continued*

```
      NIND=NIND+NUM
C CALL PLOTTER ROUTINE
      N1=NIND*KT+1
      CALL PREP(X(1),KT,IX(N1),NIND,NUM)
      STOP
      END
      SUBROUTINE MDA(SSTOT,SSWITH,XM,S,EIG,X,Y,XMEAN,NAMES,NGR,NV,NG,KT,
     1 NIND,NUM)
C MAIN ROUTINE
      INTEGER UNCL
      REAL NAMES
      DIMENSION SSTOT(NV,NV),SSWITH(NV,NV),XM(NV),S(NV),EIG(NV),
     1 NAMES(NG),NGR(NV),X(NV),Y(NV),XMEAN(NV,NG),FMT(20),FMTO(20)
      DATA WILKS,IP,IND,U,W/1.0,1,1,2*0.0/
      MM=NV*NV
      WRITE(6,1000)
 1000 FORMAT('1MULTIPLE DISCRIMINANT ANALYSIS'// PROGRAM BY P.M.MATHER'/
     *' UNIVERSITY OF NOTTINGHAM, 1974'///)
C NOTE EPSI AND ACC ARE MACHINE-DEPENDENT PARAMETERS.
C SEE APPENDIX A FOR DETAILS
      ACC=2.0**(-37)
      EPSI=2.0**(-255)
      TOL=EPSI/ACC
      KT=MINO(NV,NG-1)
      READ(5,120) NGR
  120 FORMAT(20I3)
      READ(5,2) (NAMES(I),I=1,NG)
    2 FORMAT(20A4)
      READ(5,120) UNCL
      READ(5,2) FMTO
      DO 3 I=1,NG
    3 NIND=NIND+NGR(I)
C READ INPUT DATA -- SUBROUTINE INPUT
      CALL INPUT(NIND,NV,X,FMT)
      WRITE(6,51) NIND
   51 FORMAT('0STATISTICS FOR TOTAL DATA WITH N = ',I4)
C SUBROUTINE SSCP IS LISTED IN APPENDIX A
      CALL SSCP(XM,SSTOT,S,NV,NIND,NV,-3,3,6,FMTO,1)
      CALL PRNTRI(SSTOT,MM,NV,48HTOTAL SUMS OF SQUARES AND CROSS-PRODUCT
     *S MATRIX   ,6,IP,NV,1,8,6)
      REWIND 3
      DO 4 I=1,NG
      K=NGR(I)
      WRITE(6,50) I,K
   50 FORMAT(////' STATISTICS FOR GROUP',I4,' WITH N=',I4)
      CALL SSCP(XM,SSWITH,S,NV,K,NV,3,3,6,FMTO,1)
      DO 71 J=1,NV
   71 XMEAN(J,I)=XM(J)
    4 CALL SWAP(SSWITH,NV,NV,EIG)
      REWIND 3
      DO 5 I=1,NV
      SSWITH(I,I)=EIG(I)
    5 S(I)=EIG(I)
      IP=1
      CALL PRNTRI(SSWITH,MM,NV,64HPOOLED WITHIN-GROUPS SUMS OF SQUARES A
     *ND CROSS-PRODUCTS MATRIX   ,8,IP,NV,1,8,6)
      DO 6 I=1,NV
      DO 6 J=1,NV
    6 SSTOT(I,J)=SSTOT(I,J)-SSWITH(I,J)
      CALL PRNTRI(SSTOT,MM,NV,56HBETWEEN-GROUPS SUMS OF SQUARES AND CROS
     *S-PRODUCTS MATRIX ,7,IP,NV,1,8,6)
      CALL REDUC(SSTOT,SSWITH,NV,NV,NV,X,IFL)
      IF(IFL) 20,20,777
   20 CALL HTDQL(NV,NV,NV,SSTOT ,EIG,Y,SSTOT ,ACC,TOL)
      IF(NV.EQ.0) GOTO 779
      DO 40 I=1,KT
   40 U=U+EIG(I)
```

Table 7.8  *continued*

```
   70 U=100.0/U
      WRITE(6,21) (I,EIG(I),I=1,KT)
   21 FORMAT('0EIGENVALUES'/(' ',I4,2X,F12.4))
      CALL REBAKA(SSWITH,NV,X,SSTOT,NV,NV,1,KT)
      WRITE(6,45) KT
   45 FORMAT('0THE MAXIMUM NUMBER OF NONZERO EIGENVALUES = ',I4//)
      CALL MATPR(16HEIGENVECTORS    ,2,FMTO,6,SSTOT,MM,NV,NV,KT,IP,2,8)
      WRITE(6,24)
   24 FORMAT('0PERCENTAGE OF DISCRIMINATING POWER FOR EACH FUNCTION'/)
      DO 43 I=1,KT
      V=EIG(I)*U
   43 WRITE(6,44) I, V
   44 FORMAT(' ',I4,2X,F10.4)
      DO 25 I=1,NV
      EIG(I)=1.0/(1.0+ABS(EIG(I)))
   25 WILKS=WILKS*EIG(I)
   26 WRITE(6,28) WILKS
   28 FORMAT('0WILKS LAMBDA = ',G16.6)
      WRITE(6,30)
   30 FORMAT('0CHI SQUARE TESTS OF SIGNIFICANCE OF RESIDUAL DISCRIMINANT
     *S'// ROOTS ABOVE          CHISQUARE      D.F.'//)
   31 FORMAT(' ',4X,I4,11X,G10.4,4X,I3)
      U=-(NIND-FLOAT(NG+NV)*0.5-1.0)
      J=KT+1
      X(J)=1.0
      DO 33 I=1,KT
      J=J-1
   33 X(J)=X(J+1)*EIG(J)
      DO 34 J=1,KT
   34 Y(J)=U*ALOG(X(J))
      DO 35 J=1,KT
      I=J-1
      IP=(NV-I)*(NG-I-1)
   35 WRITE(6,31) I, Y(J), IP
      LAST=0
      CALL CANON(SSTOT,NV,KT,X,NAMES,NG,NV,NGR,Y,FMTO,LAST,XMEAN)
      IF(UNCL) 80,80,81
   81 CALL PLACE(SSTOT,NV,KT,X,NV,FMT,Y,NUM,LAST,XMEAN,NG,NAMES,S,FMT)
   80 ENDFILE 4
      DO 36 I=1,NV
      SD=SQRT(S(I))
      DO 36 J=1,KT
   36 SSTOT(I,J)=SSTOT(I,J)*SD
      IP=1
      CALL MATPR(64HRELATIVE CONTRIBUTION OF VARIABLES TO DISCRIMINANT F
     1UNCTIONS        ,8,FMTO,6,SSTOT,MM,NV,NV,KT,IP,2,8)
      RETURN
  777 WRITE(6,778)
  778 FORMAT('0FAILURE IN ROUTINE REDUC')
      GOTO 700
  779 WRITE(6,780)
  780 FORMAT('0FAILURE IN ROUTINE HTDQL')
  700 STOP
      END
      SUBROUTINE PLACE(V,N1,KT,X,NV,FMTO,Y,NUM,LAST,XMEAN,NG,NAMES,S,
     1 FMT)
      REAL NAMES
      DIMENSION V(N1,KT),X(NV),FMTO(20),Y(KT),XMEAN(NV,NG),NAMES(NG),
     1 S(NG),FMT(20)
      WRITE(6,1)
    1 FORMAT('1DISCRIMINANT SCORES FOR UNCLASSIFIED ITEMS'// FOLLOWED BY
     1 DISTANCES FROM GROUP CENTRES'///)
      READ(5,120) NUM
  120 FORMAT(I3)
      DO 3 I=1,NUM
      LAST=LAST+1
      READ(5,FMT) X
```

Table 7.8 *continued*

```
      DO 4 K=1,KT
      Y(K)=0.0
      DO 4 L=1,NV
    4 Y(K)=Y(K)+X(L)*V(L,K)
      WRITE(4) Y
      WRITE(6,5) LAST, Y
      DO 8 J=1,NG
      S(J)=0.0
      DO 9 K=1,KT
    9 S(J)=S(J)+(Y(K)-XMEAN(K,J))**2
    8 S(J)=SQRT(S(J))
      WRITE(6,10) (NAMES(J), S(J), J=1,NG)
   10 FORMAT(5('  ',A4,' = ',F11.4)/)
      XV=S(1)
      K=1
      DO 11 J=2,NG
      IF(S(J).GT.XV) GOTO 11
      XV=S(J)
      K=J
   11 CONTINUE
    3 WRITE(6,12) NAMES(K), XV
   12 FORMAT(' NEAREST = ',A4,' AT D = ',F12.4)
    5 FORMAT('  ',I3,5X,10(F9.3,2X))
      RETURN
      END
      SUBROUTINE PREP(X,K,IOVP,N,NUM)
      DIMENSION X(N,K),IOVP(N,4)
      DO 2 I=1,N
    2 READ(4) (X(I,J),J=1,K)
      CALL PLOTS(X,1,K,N,N,K,0,IOVP,N)
      L=N-NUM
      WRITE(6,1) L
    1 FORMAT('0IN THE DIAGRAM IDENTIFIERS HIGHER THAN',I4,'  REFER TO UN
     *CLASSIFIED ITEMS')
      RETURN
      END
      SUBROUTINE CANON(V,N1,KT,X,NAMES,NG,NV,NGR,Y,FMT,LAST,XMEAN)
      REAL NAMES
      DIMENSION V(N1,KT),X(NV),NAMES(NV),NGR(NG),Y(KT),FMT(20),XMEAN(NV,
     1NG)
      DO 3 I=1,NG
      KK=NGR(I)
      WRITE(6,2) NAMES(I)
    2 FORMAT('0DISCRIMINANT SCORES FOR GROUP',2X,A4/' CASE      SCORE'/)
      DO 10 J=1,KK
      READ(3) X
      LAST=LAST+1
      DO 4 K=1,KT
      Y(K)=0.0
      DO 4 L=1,NV
    4 Y(K)=Y(K)+X(L)*V(L,K)
      WRITE(4) Y
   10 WRITE(6,5) LAST, Y
      DO 6 K=1,KT
      Y(K)=0.0
      DO 6 L=1,NV
    6 Y(K)=Y(K)+XMEAN(L,I)*V(L,K)
      WRITE(6,7) NAMES(I), Y
    7 FORMAT('0MEAN SCORE FOR GROUP   ',2X,A4/' ',8X,10(F9.3,2X))
      DO 3 K=1,KT
    3 XMEAN(K,I)=Y(K)
    5 FORMAT('  ',I3,5X,10(F9.3,2X))
      RETURN
      END
      SUBROUTINE INPUT(N,M,X,FMT)
      DIMENSION X(M),FMT(20)
      READ(5,5) FMT
```

Table 7.8   *continued*

```
    5 FORMAT(20A4)
      WRITE(6,5000)
 5000 FORMAT('0                 INPUT   DATA'///)
      DO 1 I=1,N
      READ(5,FMT) X
      WRITE(3) X
    1 WRITE(6,2) X
    2 FORMAT(' ',10G12.5)
      ENDFILE 3
      REWIND 3
      RETURN
      END
      SUBROUTINE SWAP(S,N,M,Z)
      DIMENSION S(N,M),Z(M)
      DO 1 I=1,M
      Z(I)=Z(I)+S(I,I)
      DO 1 J=1,I
    1 S(J,I)=S(J,I)+S(I,J)
      RETURN
      END
```

the 69 samples classified according to these three factors. For the purposes of the present example, representative individuals were selected from each of the three groups — 7 from Group 1 (environment dominated by surf energy); 9 from Group 2 (environment dominated by gravitational energy) and 15 from Group 3 (environment dominated by current energy). The data for these 31 individuals is in the form of scores on the three factors. They are listed in Table 7.9, which also contains the means and standard deviations of the three factors for the 31 individuals, the total sums of squares and cross-products matrix, the means and standard deviations of the three factors computed separately for each group, and the within and between groups sums of squares and cross-products matrices. For Box's test of the null hypothesis that the three sample within-groups variance—covariance matrices come from the same population variance—covariance matrix the determinant of the pooled within-groups variance—covariance matrix is needed. Equation (7.13) is used to compute Box's statistic $M$, which is converted to an $F$-ratio as described in Section 7.2.0. In the present example, det $(S) = 0.75E-6$, det $(S_1) = 0.86E-11$, det $(S_2) = 0.50E-7$ and det $(S_3) = 0.69E-7$. The corresponding $F$ value is 8.388 with 12 and 1770 degrees of freedom. The tabled value of $F$ with these degrees of freedom and at the 5 per cent level is 1.75, so the null hypothesis is not accepted and it is concluded that, at this confidence level, the three variance—covariance matrices are not homogeneous. This in itself is a useful finding, and could lead to questions concerning the causes of the heterogeneity, which could be due to the low sample size, to nonnormality or to differing environmental conditions. In the context of this example, however, it should be concluded that one of the assumptions of MANOVA (and of multiple discriminant analysis) is probably unsatisfied. Bearing in mind

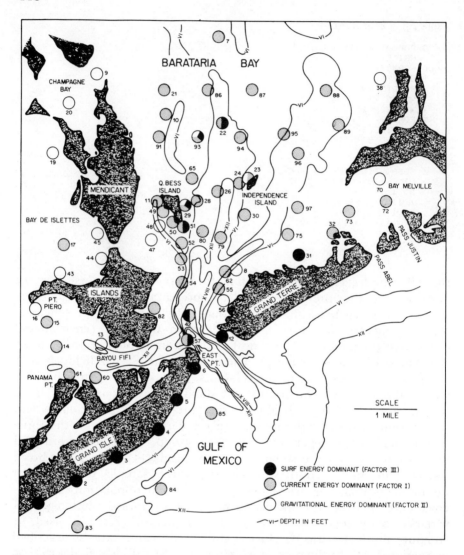

Figure 7.2 Location of sediment samples, Barataria Bay (Reproduced with permission from J. E. Klovan, *Journ. Sedim. Petrol.*, 36, 115–225 (1966), Figure 6, by permission of the Society of Economic Paleontologists and Mineralogists)

the remarks made in Section 7.2.0, we will nevertheless continue (though in practice it would be wise to carry out further investigations, as suggested above).

The primary aim of this MANOVA test is to examine the null hypothesis that the group (population) means are equal. This is achieved by computing Wilks' Lambda (Equation (7.14)) and the associated $F$ statistic. In this example, det $(\mathbf{T}) = 1.89$, det $(\mathbf{W}) = 0.0165$

Table 7.9
(a) Data for multiple discriminant analysis example;
(b) means and standard deviations of the three variables;
(c) total sums of squares and cross-products matrix;
(d) means and standard deviations of the three variables shown separately for each group;
(e) pooled within-groups sums of squares and cross-products matrix;
(f) between-groups sums of squares and cross-products matrix.

(a)
| | | |
|---|---|---|
| 0.27720 | 0.42900E-01 | 0.95870 |
| 0.27750 | 0.43200E-01 | 0.95880 |
| 0.21940 | 0.43000E-01 | 0.97310 |
| 0.19360 | 0.43000E-01 | 0.97830 |
| 0.40470 | 0.42400E-01 | 0.91290 |
| 0.29430 | 0.43200E-01 | 0.95390 |
| 0.57000 | 0.78700E-01 | 0.81620 |
| 0.59090 | 0.77070 | 0.22560 |
| 0.48050 | 0.86400 | 0.10240 |
| 0.36360 | 0.87870 | 0.94000E-02 |
| 0.38070 | 0.88500 | 0.91000E-02 |
| 0.38360 | 0.90640 | 0.61400E-01 |
| 0.20560 | 0.93390 | 0.23780 |
| 0.24650 | 0.94850 | 0.10270 |
| 0.99000E-01 | 0.97760 | 0.42400E-01 |
| 0.17780 | 0.97990 | 0.35600E-01 |
| 0.73990 | 0.24250 | 0.62690 |
| 0.74320 | 0.54900 | 0.56940 |
| 0.79350 | 0.42550 | 0.42680 |
| 0.82280 | 0.43940 | 0.33530 |
| 0.82740 | 0.45340 | 0.32220 |
| 0.82980 | 0.32900 | 0.44780 |
| 0.82990 | 0.19150 | 0.52200 |
| 0.84230 | 0.47010 | 0.22890 |
| 0.85320 | 0.22030 | 0.47010 |
| 0.85390 | 0.29120 | 0.42030 |
| 0.85770 | 0.48330 | 0.14190 |
| 0.86890 | 0.29730 | 0.39160 |
| 0.89150 | 0.39180 | 0.22220 |
| 0.92750 | 0.21790 | 0.29760 |
| 0.96230 | 0.12880 | 0.20500 |

(b) MEANS AND STANDARD DEVIATIONS.

| | | |
|---|---|---|
| 1 | 0.574468E 00 | 0.280514E 00 |
| 2 | 0.439100E 00 | 0.332341E 00 |
| 3 | 0.413106E 00 | 0.321356E 00 |

(c)
| | 1 | 2 | 3 |
|---|---|---|---|
| 1 | 2.4393 | -0.8574 | -0.4986 |
| 2 | | 3.4240 | -2.8085 |
| 3 | | | 3.2014 |

(d) STATISTICS FOR GROUP 1 WITH N= 7

MEANS AND STANDARD DEVIATIONS.

| | | |
|---|---|---|
| 1 | 0.319529E 00 | 0.119564E 00 |
| 2 | 0.480571E-01 | 0.125124E-01 |
| 3 | 0.935986E 00 | 0.526537E-01 |

Table 7.9  *continued*

```
              STATISTICS FOR GROUP    2 WITH N=   9
           MEANS AND STANDARD DEVIATIONS.

                    1   0.325356E 00    0.147600E 00
                    2   0.904967E 00    0.618415E-01
                    3   0.918222E-01    0.813781E-01


              STATISTICS FOR GROUP    3 WITH N=  15
           MEANS AND STANDARD DEVIATIONS.

                    1   0.842907E 00    0.570605E-01
                    2   0.342067E 00    0.122076E 00
                    3   0.361867E 00    0.126529E 00


    (e)               1              2             3
         1         0.3450         -0.1201       -0.0795
         2                         0.2591       -0.1047
         3                                       0.3192


    (f)               1              2             3
         1         2.0944         -0.7373       -0.4191
         2                         3.1549       -2.7038
         3                                       2.8822
```

and $\Delta = \det(\mathbf{W})/\det(\mathbf{T}) = 0.00872$. Using Table 7.6 to find the corresponding $F$-ratio, with $p = 3$ and $q = 3-1 = 2$, we get $F = (1-0.00872)^{\frac{1}{2}}/(0.00872)^{\frac{1}{2}} \times 26/3 = 84.13$, with degrees of freedom $v_1 = 6$ and $v_2 = 2 \times 26 = 52$. The tabled value of $F_{0.05, 12, 52}$ is 1.96, which is easily exceeded. At the 0.01 level, the critical value of $F$ is 2.6, so the null hypothesis is not accepted, and the conclusion is that the population means of the three groups probably differ.

Given this conclusion, it is logical to ask in what respects the environments differ, and to determine an equation which will allow uncategorized samples to be allocated to their most likely group. Multiple discriminant analysis is used to provide this information. Since in this case there are 3 groups and 3 variables, the maximum possible number of non-zero eigenvalues of the matrix $\mathbf{W}^{-1}\mathbf{A}$ is two; these are $\lambda_1 = 15.2027$ and $\lambda_2 = 6.0757$. The eigenvalues give the 'importance' of each discriminant function in explaining between-group differences. Thus, the first function accounts for 71.45 per cent of the discriminating power and the second function takes the remaining 28.55 per cent. Given that the assumptions set out earlier are satisfied a chi-square test of the statistical significance of the functions can be calculated. Normally $n$, the number of individuals, should be large (at least 50) but the results are given in the present case even though $n$ is only 31. The chi-square value computed under the null hypothesis that the discriminating power of both the functions is zero in the population is 128.0 with 6 degrees of freedom. The tabled value of $\chi^2_{0.05,6}$ is 1.64 so it can safely be concluded that at least one discriminant function exists in the population. A second null hypothesis, that the residual discriminating power differs significantly from zero after the effects of the first discriminant functions have been removed, is also tested by chi-square. In this case, the calculated chi-square value is 52.8, with 2 degrees of freedom. The tabled value $\chi^2_{0.05,2}$ is 0.1 so again the null hypothesis is not accepted and we may conclude that, at a significance level of 0.05, both sample functions are present in the population.

The column eigenvectors give the coefficients of the discriminant functions in terms of the three factors. Discriminant function 1 is defined by

$$df1 = 0.20x_1 + 0.87x_2 - 0.41x_3$$

and

$$df2 = 0.71x_1 + 0.03x_2 + 0.02x_3$$

Each of the 31 individuals has a score on the two discriminants (Table 7.10). Individuals in Group 1 (surf energy dominant) have scores on df1 ranging from $-0.327$ to $-0.155$ with a mean of $-0.280$. On df2 the scores range from 0.178 to 0.307 with a mean of 0.248. The individuals in Group 2 (gravitational energy dominant) have a similar mean on df2 $-0.261-$ but have a wider spread of scores, from 0.102 to 0.447. Scores on df1 for Group 2 are considerably different from the Group 1 scores, ranging from 0.693 to 0.871, with a mean of 0.812. Group 3 (current energy dominant) lies between Group 1 and Group 2 on df1, having a spread from 0.099 to 0.530 and a mean of 0.314. Group 3 differs from both Groups 1 and 2 on df2, with a mean score of 0.614 and a range of scores from 0.544 to 0.689. The table of relative contributions of the variables to the discriminant functions (Table 7.11) shows that df1 is mainly characterized by variable (factor) 2, with variable (factor) 3 having a inverse effect. This appears to correspond to an 'importance of gravitation' dimension, with items whose sediment size characteristics are determined almost entirely by gravitational forces having high positive scores, and those items deposited in an environment in which gravity is secondary to forces of turbulence having negative scores. The second discriminant function is characterized almost entirely by variable 1, indicating that it represents a current energy dimension.

The discriminant scores for the 69 sediment samples used by Klovan are plotted using df1 (vertical axis) and df2 (horizontal axis) in Figure 7.3. Group 1 lies at the lower left of the scatter diagram; samples 5 and 7 seem to be more influenced by current energy than do samples

Table 7.10
Discriminant scores for data shown in Table 7.9(a)

```
DISCRIMINANT SCORES FOR GROUP    SURF
CASE     SCORE

  1      -0.302         0.218
  2      -0.302         0.218
  3      -0.319         0.178
  4      -0.327         0.159
  5      -0.259         0.307
  6      -0.297         0.230
  7      -0.155         0.423

MEAN SCORE FOR GROUP    SURF
         -0.280         0.248

DISCRIMINANT SCORES FOR GROUP    GRAV
CASE     SCORE

  8       0.693         0.447
  9       0.803         0.369
 10       0.830         0.285
 11       0.839         0.297
 12       0.837         0.301
 13       0.753         0.180
 14       0.830         0.207
 15       0.851         0.102
 16       0.871         0.158

MEAN SCORE FOR GROUP    GRAV
          0.812         0.261

DISCRIMINANT SCORES FOR GROUP    CURR
CASE     SCORE

 17       0.099         0.544
 18       0.471         0.551
 19       0.350         0.585
 20       0.406         0.603
 21       0.424         0.606
 22       0.265         0.607
 23       0.115         0.604
 24       0.480         0.615
 25       0.166         0.620
 26       0.248         0.622
 27       0.530         0.624
 28       0.268         0.632
 29       0.424         0.647
 30       0.249         0.669
 31       0.217         0.689

MEAN SCORE FOR GROUP    CURR
          0.314         0.614
```

Table 7.11
Relative contributions of the variables to the two discriminant functions

|   | 1 | 2 |
|---|---|---|
| 1 | 0.1574 | 0.5648 |
| 2 | 0.7569 | 0.0277 |
| 3 | -0.1594 | 0.0085 |

1, 2, 3, 4 and 6. Of the unallocated samples, numbers 34 and 35 are assigned to group 1, with distances of 0.346 and 0.269 from the group centroid. Sample 34 is 0.364 units of distance from the centre of group 3, and sample 35 is 0.492 units from the centre of group 3. Thus, sample 34 may be considered to be a transitional deposit. If the sizes of the three groups were taken into account it would be possible to compute probabilities of belonging to each class. (See Section 7.1).

Group 2 is also elongated, with sample 8 lying further from sample 15 (which is placed in the same group) than from sample 18 (which is placed in Group 3). Sample 39 is obviously a candidate for inclusion in group 2, as is sample 37.

The shape of group 3 reflects the arbitrary way in which the representative samples were chosen. It is possible to draw a fairly clear boundary around two possible subgroups; this could lead to a reconsideration of the original classification. It is clear that, by the use of multiple discriminant analysis the original classification of Klovan (1966) can be refined and modified. The use of MANOVA shows that the three groups are not homogeneous in terms of variances and covariances; this could lead to further investigation of the underlying causes.

## 7.3 Discriminant Analysis Based on Binary Attributes

In some situations quantitative measurements of the attributes of interest are not possible. Instead, binary (presence—absence) measures may be considered. With such measures it is not possible to use the methods described earlier in this chapter. Ramsayer and Bonham-Carter (1974) describe a technique which they term 'adaptive pattern-recognition' which allows the discrimination between two groups on the basis of binary attributes alone. The adaptive pattern-recognition model can be represented by a linear equation such as Equation (7.5) except that the $x$'s denote binary variables. The procedure begins with a trial set of coefficients which are progressively modified until maximum discrimination is attained.

Ramsayer and Bonham-Carter illustrate the method by means of a simple example which is summarized here. Eight individuals are known to belong either to Group A or to Group B. Each individual is measured on four binary variables, represented by $x_1$, $x_2$, $x_3$ and $x_4$. An extra (dummy) variable $x_5$ is added. It is set initially to 1 for each individual. Its purpose is to ensure that the average discriminant score $(\bar{d})$ for the combined groups is zero. The discriminant score for individual $i$ is:

$$d_i = \lambda_1 x_{i1} + \lambda_2 x_{i2} + \lambda_3 x_{i3} + \lambda_4 x_{i4} + \lambda_5 x_{i5} \qquad (7.16)$$

The coefficients $\lambda_i$ are designed to ensure that $d_i > \bar{d}$ if the $i$th individual belongs to Group A and that $d_i < \bar{d}$ if individual $i$ belongs to Group B. If any individual is wrongly allocated then the $\lambda_i$ are adjusted

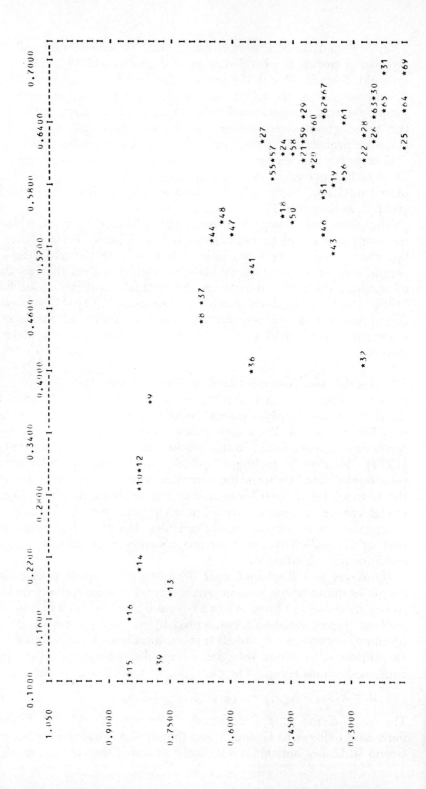

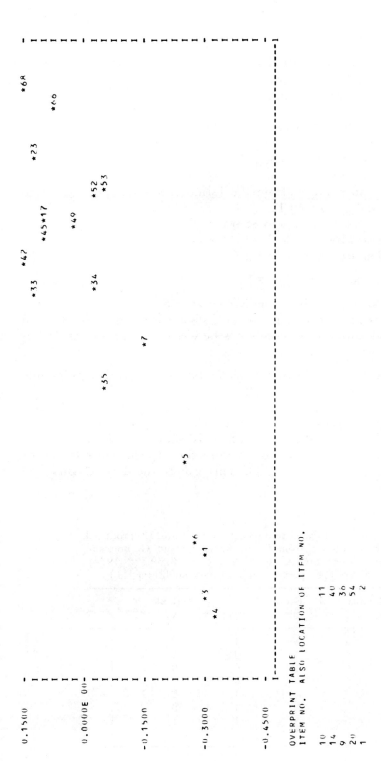

Figure 7.3 Plot of scores of sediment samples on first two discriminant functions

and the calculations repeated. The steps are as follows:

(1) Set each of the coefficients $\lambda_i$ to zero;
(2) Set a counter $k$ to zero;
(3) Compute the discriminant score for individual 1 using Equation (7.16);
(4) If this individual belongs to Group A go to Step 5, otherwise go to Step 8;
(5) If $d_1$ exceeds 0 go to Step 11;
(6) Increase $k$ by 1;
(7) Recompute the $\lambda_i$ from the formula: $\lambda_i = \lambda_i c x_{1i}$ for $i = 1$ to $i = 5$ and go to Step 11;
(8) If $d_1$ is less than 0 go to Step 11;
(9) Increase $k$ by 1;
(10) Recompute the $\lambda_i$ from the formula

$$\lambda_i = \lambda_i + cx_{1i}(i = 1,5)$$

(11) Repeat steps 3–10 for individuals 2 to 8;
(12) If $k$ is zero then none of the individuals are misclassified and (7.16) correctly discriminates between the two groups. If $k$ is positive, go to step 2.

In steps 7 and 10 the value of $c$ is usually taken to be 1, though it could be any value in the range $0 < c < 2$.

The operation of the algorithm can be shown by the following example. The data is listed first (Table 7.12).

The progressive modification of $\lambda$ is shown in Table 7.13. At interation 3 all eight individuals are correctly assigned. The coefficients vector is $(2, -2, -1, -1, 1)$. This can be used to classify and

Table 7.12
Data for adaptive pattern recognition example (Reproduced, with permission from G. R. Ramsayer and G. Bonham-Carter, *Journ. Intern. Assoc. Math. Geol.*, 6, 59–72 (1974), Figure 1, by permission of Plenum Publishing Co.)

| Individual | Group | Variable | | | | |
|---|---|---|---|---|---|---|
| | | $x_1$ | $x_2$ | $x_3$ | $x_4$ | $x_5$ |
| 1 | A | 1 | 1 | 0 | 0 | 1 |
| 2 | B | 0 | 1 | 1 | 1 | 1 |
| 3 | B | 0 | 1 | 0 | 0 | 1 |
| 4 | A | 0 | 0 | 0 | 0 | 1 |
| 5 | B | 1 | 1 | 1 | 1 | 1 |
| 6 | A | 1 | 0 | 0 | 0 | 1 |
| 7 | A | 1 | 0 | 1 | 1 | 1 |
| 8 | B | 0 | 0 | 1 | 1 | 1 |

Table 7.13
Stages in the application of the adaptive pattern recognition algorithm

| Individual | Group | Coefficient | | | | | Discriminant Score ($d_i$) |
|---|---|---|---|---|---|---|---|
| | | 1 | 2 | 3 | 4 | 5 | |
| *Iteration 1* | | | | | | | |
| 1 | A | 0 | 0 | 0 | 0 | 0 | 0 |
| 2 | B | 1 | 1 | 0 | 0 | 1 | 2 |
| 3 | B | 1 | 0 | −1 | −1 | 0 | 0 |
| 4 | A | 1 | −1 | −1 | −1 | −1 | −1 |
| 5 | B | 1 | −1 | −1 | −1 | 0 | −2 |
| 6 | A | 1 | −1 | −1 | −1 | 0 | 1 |
| 7 | A | 1 | −1 | −1 | −1 | 0 | −1 |
| 8 | B | 2 | −1 | 0 | 0 | 1 | 1 |
| *Iteration 2* | | | | | | | |
| 1 | A | 2 | −1 | −1 | −1 | 0 | 1 |
| 2 | B | 2 | −1 | −1 | −1 | 0 | −3 |
| 3 | B | 2 | −1 | −1 | −1 | 0 | −1 |
| 4 | A | 2 | −1 | −1 | −1 | 0 | 0 |
| 5 | B | 2 | −1 | −1 | −1 | 1 | 0 |
| 6 | A | 1 | −2 | −2 | −2 | 0 | 1 |
| 7 | A | 1 | −2 | −2 | −2 | 0 | −3 |
| 8 | B | 2 | −2 | −1 | −1 | 1 | −1 |
| *Iteration 3* | | | | | | | |
| 1 | A | 2 | −2 | −1 | −1 | 1 | 1 |
| 2 | B | 2 | −2 | −1 | −1 | 1 | −3 |
| 3 | B | 2 | −2 | −1 | −1 | 1 | −1 |
| 4 | A | 2 | −2 | −1 | −1 | 1 | 1 |
| 5 | B | 2 | −2 | −1 | −1 | 1 | −1 |
| 6 | A | 2 | −2 | −1 | −1 | 1 | 3 |
| 7 | A | 2 | −2 | −1 | −1 | 1 | 1 |
| 8 | B | 2 | −2 | −1 | −1 | 1 | −1 |

recognize unknown or previously uncategorized individuals in the manner described in Section 7.1.

One drawback of the method is that the order in which the individuals are presented may have an effect on the resulting coefficients. It can be shown that if a linear function exists which can separate the groups then the coefficients vector will converge to a solution. This solution will not necessarily be unique, for if the individuals had been listed in a different order then the vector could have converged to a different and equally feasible solution. For example, a second feasible coefficients vector for the example data set is (4, − 3, − 2, − 1, 1). This is not a major disadvantage, for the coefficients are not given any physical meaning. However, it does imply

## Table 7.14
## Computer program for adaptive pattern recognition

```
C--------------- NOTES ON PROGRAM USE ---------------------------------
C I/O UNITS 5 (CARD READER) AND 6 (LINEPRINTER) REQUIRED.
C WITH PRESENT DIMENSIONS MAXIMUM NUMBER OF INDIVIDUALS IN GROUP 1
C + GROUP 2 IS 100; MAXIMUM NUMBER OF VARIABLES IS 10.
C MAXIMUM NUMBER OF ITERATIONS IS SET TO 1000.
C DATA IN ORDER:
C DATA CARD 1: (2I3): N (NO OF INDUVIDUALS - TOTAL);
C M (NO OF VARIABLES. INCL. DUMMY UNIT COLUMN).
C CARD 2:(20A4): FMT (INPUT FORMAT FOR DATA)
C CARD 3:(FMT): X(I,J) (RAW DATA ROW BY ROW IN SPECIFIED FORMAT).
C START EACH ROW ON A NEW CARD.
C CARD 4:(80L1): CLASS (CLASS MEMBERSHIP INDICATOR; PUNCH T FOR
C CLASS A OR F FOR CLASS B IN SMAE ORDER AS INDIVIDUALS ARE PUNCHED
C IN DATA MATRIX.
C PUNCH 80 PER CARD.
C
C SUBROUTINES USED: NONE.
C ERROR MESSAGES: IF NO. OF ITERATIONS EXCEEDS 1000 A MESSAGE IS PRINTED
C AND THE PROGRAM IS TERMINATED.
C
      LOGICAL CLASS(100)
      DIMENSION X(100,10),XLAM(10),R(10),FMT(20)
      WRITE(6,50)
   50 FORMAT('1ADAPTIVE PATTERN RECOGNITION PROGRAM AFTER BONHAM-CARTER
     *AND RAMASAYER, I. A. M. G. JOURN.,1974'// SEE CHAPTER 7 FOR DETAILS'/
     *20X,'DATA MATRIX'/)
      KOUNT=0
      READ(5,20) N,M
   20 FORMAT(2I3)
      READ(5,30) FMT
   30 FORMAT(20A4)
      DO 1 I=1,N
      READ(5,FMT) (X(I,J),J=1,M)
    1 WRITE(6,2) (X(I,J),J=1,M)
    2 FORMAT(' ',30F3.0)
      DO 3 I=1,M
    3 XLAM(I)=0.0
      READ(5,7) (CLASS(I),I=1,N)
    7 FORMAT(80L1)
      RC=0.0
      C=1.0
    4 K=0
      KOUNT=KOUNT+1
      IF(KOUNT.GT.1000) GOTO 75
      DO 9 I=1,N
      R(I)=0.0
      DO 6 J=1,M
    6 R(I)=R(I)+XLAM(J)*X(I,J)
      IF(.NOT.CLASS(I)) GOTO 10
      IF(R(I).GT.RC) GOTO 9
      K=K+1
      DO 8 J=1,M
    8 XLAM(J)=XLAM(J)+C*X(I,J)
      GOTO 9
   10 IF(R(I).LT.RC) GOTO 9
      K=K+1
      DO 11 J=1,M
   11 XLAM(J)=XLAM(J)-C*X(I,J)
    9 CONTINUE
      IF(K) 4,12,4
   12 WRITE(6,13) (I,XLAM(I),I=1,M)
   13 FORMAT('0COEFFICIENTS OF DISCRIMINANT FUNCTION'/(' ',I4,2X,F10.4/
     *))
      GOTO 100
   75 WRITE(6,76)
   76 FORMAT(' 1000 ITERATIONS EXCEEDED')
  100 STOP
      END
```

that the computed function may not be the most efficient discriminator.

Ramsayer and Bonham-Carter report their experience with the algorithm using actual geological data. For the sedimentological data recorded and analysed by Purdy, they find that the adaptive pattern-recognition algorithm was approximately as successful as the discriminant function method of Section 7.1. In a second application to the study of brachiopod diversity patterns in Permian marine rocks, Ramsayer and Bonham-Carter carried out pair-by-pair analyses for every possible pair out of seven groups. Some difficulty was experienced in that in situations where an individual could not be allocated to any particular group with any degree of certainty there is no clear basis for making probability statements about the likelihood of the individual belonging to any particular group. This is possible with both two-group and multiple group discriminant analysis (Sections 7.1 and 7.2).

The adaptive pattern-recognition algorithm has not yet been applied to geographical problems, but several possibilities are apparent. These include terrain classification and land-systems classification as well as other studies which involve mixed quantitative and qualitative data. Bonham-Carter (1965) outlines a method of converting quantitative data to 0—1 form to allow both types of data to be considered simultaneously.

A computer program for adaptive pattern-recognition is given in Table 7.14. The method of use is given by the comments included in the listing. For large numbers of individuals the number of cycles to attain convergence may be large, so an arbitrary limit of 500 cycles has been set. If this number of cycles is exceeded the program can be restarted by inputting the 500th estimate of the coefficients vector at run 1 to a second run of the program. Note that convergence is guaranteed only if the two groups are, in fact, linearly separable.

*Appendix A*

# Fortran Subroutine Library

The source listings of the Fortran subroutines that are used in the programs listed elsewhere in this book are collected in this Appendix in the form of a subroutine library. There are two reasons for this — to facilitate the location of a particular routine, and to provide a set of routines that the reader may wish to use in his own programs. Two subroutines not specifically used in this book are included for the sake of completeness. These are subroutines CROUT and NSEIG. The subroutines fall into two main groups; the first group (Tables A1 to A19 below) consists of general purpose routines for matrix printout, lineprinter plots and histograms and the computation of correlation and covariance matrices. The second group (Tables A20 to A34 below) provide efficient Fortran subroutines for the evaluation of the eigenvalues and eigenvectors of both symmetric and nonsymmetric matrices, the computation of the determinants of symmetric and nonsymmetric matrices, the determination of matrix inverses and the solution of linear simultaneous equations. Most of the routines in this second set are my own Fortran translations and adaptions of Algol-60 procedures from Wilkinson and Reinsch (eds.) (1971), to whom acknowledgement is due. The other routines were written specifically for this book with the exception of subroutine SCALE, which is reproduced from *Applied Statistics* by kind permission of the Royal Statistical Society.

Comments within each listing provide instructions on the purpose and use of each routine. The following notes (see below) are intended to provide supplementary information. Suitable test data sets can be found in Wilkinson and Reinsch (eds.) (1971) and in Gregory and Karney (1969). The parameters ETA and MACHEP, or their equivalents, occur in several of the linear algebra routine — they are, respectively, the smallest real number that can be represented in the machine (i.e. so that ETA is detectably greater than 0.0) and the smallest real quantity that can be added to 1.0 to produce a detectable change. These values are ETA = $2^{-124}$ (PDP–11), $2^{-255}$ (ICL 1906A) and $2^{-928}$ (CDC 7600) and MACHEP = $2^{-24}$ (PDP–11), $2^{-37}$ (ICL 1906A) and $2^{-47}$ (CDC 7600). Malcolm (1972) gives details of the calculation of these values for other types of computer.

*Section I: General Purpose Routines* (Tables at end of this appendix)
A1: MATPR   Outputs full or lower triangle of array to printer.
A2: PRNTRI  Outputs upper triangle of array to printer.
A3: SSCP    Forms X'X,R or S together with vectors of means and standard deviations.
A4: COPY    Copies array $B$ into array $A$.
A5: TRMULT  Forms A = B'C
A6: MTMULT  Forms A = BC
A7: DPACC   Double-precision accumulation of inner products.
A8: PLOTS   Plots selected columns of array on lineprinter.
A9: SAMESC  Determines equal scale for $X$ and $Y$ axes in PLOTS.
A10: SCALE  Finds sensible scale for each axis in PLOTS.
A11: KALK   Used in PLOTS to calculate character positions.
A12: OVCHK  Checks for overprints. Used in PLOTS.
A13: RANK   Ranks real array into ascending or descending order.
A14: HISTGM Plots histogram on lineprinter.
A15: SETUP  Initiates pseudo-random number generator.
A16: RANDOM Pseudo-random number generator.
A17: DEVIAT Generates normally distributed pseudo-random numbers.
A18: ZERO   Sets elements of specified part of real array to zero.
A19: IFIND  Finds (I,J)th elements of symmetric array stored linearly rowwise.

*Section II: Linear Algebra Routines* (Tables at end of this appendix)
A20: TRIDI   Householder tridiagonalization.
A21: RATQR   Rational QR method for eigenvalues of symmetric matrix.
A22: REDUC   Reduces eigenproblem Ax = λBx to standard form.
A23: REBAKA  Recovers eigenvectors of Ax = λBx.
A24: HTDQL   Eigenvalues and eigenvectors of symmetric matrix by QL method.
A25: EIGVEC  Eigenvectors of tridiagonal matrix by inverse iteration.
A26: TRBAK1  Eigenvectors of original matrix from output of EIGVEC.
A27: DT2     Determinant and Cholesky factorization of symmetric matrix.
A28: SOLVE   Solution of Ax = b where A is symmetric.
A29: CHOLS   Inverse of symmetric matrix by Cholesky method.

A30: ACCINV    Iterative refinement of inverse of symmetric matrix.
A31: GJDEF1    Inverse of symmetric matrix by Gauss-Jordan method.
A32: CROUT     Finds $A^{-1}$, det $(A)$ or solution to $Ax = b$ where $A$ is nonsymmetric.
A33: NSEIG     Eigenvalues and eigenvectors of nonsymmetric matrix using Jacobi method.
A34: GSORTH    Modified Gram-Schmidt orthogonalization.

*Notes*

A1: MATPR,
(i) The parameter TEXT can be either a Hollerith string or a real array containing alphanumeric information. Thus, CALL MATPR(*8HABCDEFGH*,...) and CALL MATPR(*TEXT*,...) are equally acceptable provided that *TEXT* is a real array holding alphanumerics.
(ii) The value of *IWORD* should be set to the number of characters that are held in one store location. This value is 4 (PDP–11), 8 (ICL 1906A) or 10 (CDC 7600) – consult the manufacturer's handbook for other types of computer. The value of *NW* should equal the number of characters in *TEXT* divided by the value in *IWORD*. The number of characters in *TEXT* must be an exact multiple of *IWORD* (if necessary use blank characters to fill the array).
(iii) The real array FMT should hold output format information, including the opening and closing brackets of the format specification. The format should specify (a) an editing character – blank, 0 or 1; (b) one integer, or one integer plus space characters, occupying 12 columns; (c) up to 8 real numbers each occupying 12 columns. Thus, *FMT* could hold (' '16, 6X, 8F 12.4) with the rest of the array being filled with blanks.

A2: PRNTRI
(i) Notes (i) and (ii) under A1 apply also to PRNTRI.
(ii) If the output format is changed, ensure that any replacement allows for the output of one integer and eight real numbers per line. The output format should be stored in the real array FMT at 4 characters per word.

A3: SSCP
(i) Refer to Section 1.6 for details of the computational method used.
(ii) The output values of the means and standard deviations are population values not estimates computed from a sample.

A5: TRMULT
(i) The current version computes both upper and lower triangles of the product matrix. It would be more efficient to have a second routine to

compute the lower (or upper) triangle alone when the product $\mathbf{A} = \mathbf{B'B}$ is being computed, for $\mathbf{A}$ here is necessarily symmetric.
(ii) Arguments $B$ and $C$ in the call statement may refer to the same array, but A must be a separate array; thus, CALL TRMULT-$(S,P,P,\ldots)$ sets $\mathbf{S} = \mathbf{P'P}$.

## A6: MATPR
(i) Note (i) of A5 refers also to A6.

## A7: DPACC
(i) The arrays $A$ and $B$ are stored in one-dimensional form. The use of $I$, $J$, $K$ and $L$ should be obvious when $A$ and $B$ are actually one-dimensional. If $A$ and $B$, or either of $A$ and $B$, is two-dimensional (but stored in one-dimensional form in DPACC) then the actual addresses of each of the first two elements of $A$ and $B$ must be given. For example, if $A$ and $B$ have dimensions $(10,10)$ and it is desired to find the inner product of row 1 of $A$ and column 2 of $B$ then $I = 1$, $J = 11$, $K = 11$ and $L = 12$.

## A8: PLOTS
(i) Output is directed to channel 6.
(ii) A page width of at least 104 characters is necessary, and printing is at 60 lines per page.
(iii) The overprint table is printed in the order: overprints for the top line (line 1), followed by the list of overprints for lines 2, 3 and so on.

## A10: SCALE
(i) This subroutine is included by permission of the Royal Statistical Society.

## A14: HISTGM
(i) The output is scaled so that the largest class is printed across 100 columns and 3 rows.

## A15: SETUP
(i) The values in the listed version refer to ICL 1900 Series computers. Satisfactory results will be obtained only if these values are altered for other makes and classes of computer. The values are as follows:

| | |
|---|---|
| $M$ | largest positive integer representable in the machine |
| $FM$ | floating-point equivalent of $M$. |
| $IA, IA1$ | multipliers in congruential algorithm (see below) |
| $IC, IC1$ | constants in congruential algorithm (see below). |

Note that two congruential generators are used — see below, A16.

Knuth (1969, p. 155) gives the following advice:

(a) if the computer works to base 2 the multiplier ($IA$ and $IA1$) should satisfy the relationship $IA$ mod $8 = 5$, that is, the remainder after dividing $IA$ (or $IA1$) by 8 is 5. If the machine works in base 10, then the relationship is $IA$ mod $200 = 21$.

(b) $IA$ and $IA1$ should be larger than the square root of $M$. Also, both should preferably exceed $M/100$ but be smaller than $M - M^{1/2}$. The binary representations of $IA$ and $IA1$ should not show regular patterns of digits.

(c) The constant ($IC$ and $IC1$) should be an odd positive integer; the value $C/M$ should be approximately $1/2 - \sqrt{3/6}$ ( $\simeq 0.2113248654$).

(ii) Kuo (1972, p. 337) suggests $M = 2^{31} - 1$, $IA = 65539$ and $IC = 0$ for IBM Series/360 computers.

(iii) The generator should be tested before use — see Knuth (1969).

## A16: RANDOM

(i) The values of $IX$ and $IX1$ should not be altered between calls. If the same sequence of pseudo-random numbers is required, $IX$ and $IX1$ can be reset to their original values.

(ii) The initial values of $IX$ and $IX1$ should be chosen arbitrarily. Ideally an odd integer should be chosen.

(iii) The method is described by Neave (1972). Briefly, it involves the generation and combination of two pseudo-random numbers using the congruential algorithm. This makes it possible for the same number to be generated twice in the same sequence.

(iv) The pseudorandom numbers lie in the range $(0,1)$ and are uniformly distributed in this range. To convert to the range $(A,B)$ where $B \geqslant A$, find

$$V = X*(B - A) + A$$

$X$ is a pseudo-random number in the range $(0,1)$ and $V$ is the corresponding number in the range $(A,B)$.

(v) Subroutine OVERFL is an ICL 1900 Series routine to switch off the overflow indicator and it works in TRACE 1 or TRACE 0 mode only. If it is not included an execution error will occur. The corresponding systems routine should be substituted if another computer is used. Not all systems check for integer overflow, so the routine may not be required.

## A17: DEVIAT

(i) Random normal deviates from the standard normal distribution are returned. The method used is the polar method of Knuth (1969, p. 104).

(ii) The parameter $I$ should be set to an odd and then an even number on successive calls — thus, $I$ could be a do-loop index.

(iii) To convert the values returned by DEVIAT to samples from a normal distribution with mean $\bar{X}$ and variance $s^2$ simply multiply each deviate by $s$ and add $\bar{X}$.

A20: TRIDI
(i) The parameter *TOL* is defined as *MACHEP/ETA* where *MACHEP* and *ETA* are discussed above in the introductory paragraphs.
(ii) TRIDI is normally followed by RATQR and optionally by EIGVEC and TRBAK1. The elements of arrays *A,D,E* and *E2* should not be altered between these calls.

A21: RATQR
(i) If more than 25 per cent of the eigenvalues and eigenvectors are required it is more efficient to use HTDQL than TRIDI/RATQR/ EIGVEC/TRBAK1.
(ii) Do not alter arrays *D,E* and *E2*.
(iii) The value of *EPS* is discussed above.

A22: REDUC
(i) This subroutine reduces the eigenproblem $(\mathbf{B}^{-1}\mathbf{A} - \lambda\mathbf{I})\mathbf{x} = 0$ to standard form thus allowing the use of HTDQL or TRIDI/RATQR/ EIGVEC/TRBAK1.
(ii) **B** must be a positive-definite matrix.
(iii) The elements of arrays *A,B* and *DL* must not be altered before subroutine REBAKA is called.
(iv) To determine the eigenvalues and eigenvectors of a matrix of the form $\mathbf{B}^{-1}\mathbf{A}$ firstly use REDUC then one of the routines for the symmetric eigenvalue problem. Finally, call REBAKA to convert the eigenvectors back to the eigenvectors of $\mathbf{B}^{-1}\mathbf{A}$.

A23: REBAKA
(i) See note (iv) of A22.

A24: HTDQL
(i) See above for a definition of *MACHEP*. TOL is defined as *MACHEP/ETA* where *ETA* is defined in the introduction to this Appendix.
(ii) The parameters *A* and *Z* can refer to the same array. This will result in the eigenvectors being overwritten on to the original symmetric matrix.
(iii) A simple modification, described in Wilkinson and Reinsch (eds.) (1971) allows HTDQL to be used when *A* is a symmetric array stored linearly rowwise.
(iv) The value of *N* should be checked after a call to HTDQL. $N = 0$ on return if failure has occurred.

(v) If the elements of the array $A$ differ greatly in magnitude the smaller elements should be moved to the top left and the larger elements to the bottom right of the array by interchanging rows. If required, the column eigenvectors can be moved back to the original order of the rows of $A$.

## A25: EIGVEC
(i) Returns the eigenvectors of the tridiagonalized matrix. TRBAK1 must be called to get the eigenvectors of the original matrix.
(ii) Do not alter arrays $C,B$ and $ROOT$ between TRIDI/RATQR and EIGVEC.
(iii) $ACC$ is equivalent to $MACHEP$ and $EPS$ to $ETA$ in the call statement. $MACHEP$ and $ETA$ are discussed in the introductory paragraphs.
(iv) $IFL$ should be zero on entry. If it contains the value 1 on exit failure has occurred in isolating an eigenvector — perhaps $ACC$ or $EPS$ is wrongly defined.

## A26: TRBAK1
(i) Arrays $A$ and $E$ must not be altered between calls to TRIDI and TRBAK1.

## A27: DT2
(i) If $N = 0$ on return, failure has occurred.

## A28: SOLVE
(i) If the coefficients matrix is not symmetric, use CROUT (A32).
(ii) If $N = 0$ on return, failure has occurred.

## A29: CHOLS
(i) The upper triangle of the matrix to be inverted and the lower triangle of its inverse are stored as follows, where $a_{ij}$ is an element of the original matrix and $d_{ij}$ an element of the inverse matrix:

$$\begin{bmatrix} a_{11} & a_{12} & a_{13} & a_{14} & a_{15} \\ d_{11} & a_{22} & a_{23} & a_{24} & a_{25} \\ d_{21} & d_{22} & a_{33} & a_{34} & a_{35} \\ d_{31} & d_{32} & d_{33} & a_{44} & a_{45} \\ d_{41} & d_{42} & d_{43} & d_{44} & a_{55} \\ d_{51} & d_{52} & d_{53} & d_{54} & d_{55} \end{bmatrix}$$

Figure A1. Method of storage in CHOLS and ACCINV.

A30: ACCINV
(i) The value of *EPS* is critical. It must be set so that 1.0 + EPS is just detectably greater than 1.0. Values for various computers are given in the opening paragraphs under *MACHEP*.
(ii) The use of double precision arithmetic in ACCINV is mandatory.
(iii) The same storage scheme is used as in CHOLS — see Figure A1.
(iv) Check the values of *N* and *N1* on return. Neither should be called by value.

A31: GJDEF1
(i) See subroutine GJDEF3 (Table 5.19) for details of determinant calculation based on the Gauss-Jordan method.
(ii) *A* should be a symmetric matrix.

A32: CROUT
(i) See Wilkinson and Reinsch (eds.) (1971, p. 100) for details of an iterative refinement algorithm based on CROUT.

A33: NSEIG
(i) If the nonsymmetric matrix is of the form $\mathbf{B}^{-1}\mathbf{A}$ then use REDUC/REBAKA instead of NSEIG.
(ii) If eigenvalues only are required, arguments *A* and *T* can refer to the same array, or *T* can be a dummy array.

## Table A1

```
      SUBROUTINE MATPR(TEXT,NW,FMT,IOUT,A,MM,NDIM1,M,N,IP,IND,IWORD)
C
C  PRINTS EITHER THE WHOLE OR THE LOWER TRIANGLE AND PRINCIPAL DIAGONAL
C  OF A TWO-DIMENSIONAL REAL ARRAY OR THE LOWER TRIANGLE PLUS DIAGONAL
C  OF A SYMMETRIC MATRIX STORED LINEARLY ROWWISE.
C
C  ARGUMENTS:
C  TEXT    HOLLERITH STRING OF UP TO 80 CHARACTERS USED TO LABEL SECTIONS
C          OF OUTPUT. THE NUMBER OF CHARACTERS MUST BE A POSITIVE INTEGER
C          MULTIPLE OF IWORD (SEE BELOW).
C  NW      INTEGER: NO OF CHARACTERS IN TEXT DIVIDED BY IWORD.
C  FMT     REAL ARRAY, DIMENSION 20, HOLDING OUTPUT FORMAT FOR DATA. MUST
C          SPECIFY ONE INTEGER FOLLOWED BY UP TO EIGHT REAL NUMBERS
C          PER LINE. FOR EXAMPLE: ('  ',I6,6X,8F12.4).
C  IOUT    INTEGER. UNIT NUMBER OF OUTPUT DEVICE.
C  A       REAL ARRAY CONTAINING MATRIX TO BE PRINTED. MAY BE ONE OR
C          TWO-DIMENSIONAL.
C  MM      INTEGER. THE PRODUCT OF THE TWO DIMENSIONS OF A AS DECLARED IN
C          THE CALLING PROGRAM [IF A IS 2-DIMENSIONAL] OR THE ACTUAL
C          DIMENSION OF A AS DECLARED IN THE CALLING PROGRAM [IF A IS ONE-
C          DIMENSIONAL]
C  NDIM    INTEGER  FIRST DIMENSION OF A AS DECLARED IN THE CALLING
C          PROGRAM IF A IS 2-DIMENSIONAL. VALUE ARBITRARY IF A IS 1-DIMEN-
C          SIONAL ARRAY.
C  M       INTEGER  NUMBER OF ROWS TO BE PRINTED.
C  N       INTEGER. NUMBER OF COLUMNS TO BE PRINTED.
C  IP      INTEGER. CONTROLS START OF PRINTING. IP=1: START ON NEW PAGE.
C          IP=0  START ON NEXT LINE.
C          IP MUST NOT BE CALLED BY VALUE AS IT IS ALTERED BY THE ROUTINE.
C IND      INTEGER. TAKES ONE OF THE FOLLOWING VALUES:
C          1  PRINT LOWER TRIANGLE OF TWO-DIMENSIONAL ARRAY.
C          2  PRINT TWO DIMENSIONAL ARRAY.
C          3  PRINT CONTENTS OF ONE-DIMENSIONAL ARRAY AS LOWER TRIANGLE
C             PLUS DIAGONAL OF SQUARE MATRIX.
C  IWORD   INTEGER. NUMBER OF CHARACTERS STORED IN ONE SINGLE-PRECISION
C          FLOATING-POINT WORD. MUST BE ONE OF 4,8 OR 10, ALTHOUGH OBVIOUS
C          MODIFICATIONS COULD BE INCORPORATED FOR OTHER COMPUTERS.
C          E.G.  IWORD =4 FOR DEC PDP/11, 8 FOR ICL 1900 SERIES AND 10 FOR
C          CDC 7600/6600 SERIES.
C
      DIMENSION A(MM),TEXT(NW),FMT(20)
      IFT=1
      IF(IWORD.EQ.8) IFT=2
      IF(IWORD.EQ.10) IFT=3
      INC=7
      IR=1
      L=1
      LL=1
      IT=N/8
      LEFT=N-IT*8
      IF(LEFT.GT.0) IT=IT+1
      DO 40 KT=1,IT
      IF(KT.EQ.IT.AND.LEFT.GT.0) INC=LEFT-1
      GOTO (51,52,53) ,IFT
   51 WRITE(IOUT,4) IP, TEXT
      GOTO 55
   52 WRITE(IOUT,8) IP,TEXT
      GOTO 55
   53 WRITE(IOUT,10) IP,TEXT
    4 FORMAT(I1,20A4)
    8 FORMAT(I1,10A8)
   10 FORMAT(I1,8A10)
   55 WRITE(IOUT,2) KT
    2 FORMAT('+',90X,'SECTION',I4/)
      IP=0
```

Table A1  *continued*

```
      LL=L+INC
      WRITE(IOUT,5) (I,I=L,LL)
    5 FORMAT(/ /,8X,8I12)
      GOTO (20,20,30), IND
   30 K=(IR-1)*IR/2+IR
      L=0
      DO 11 I=IR,M
      L=L+1
      J=K+L-1
      KK=J
      IF(J-K.GE.8) KK=K+7
      WRITE(IOUT,FMT) I,(A(LP),LP=K,KK)
   11 K=J+IR
      GOTO 12
   20 IPI=1
      L1=(L-1)*NDIM1
      L2=(LL-1)*NDIM1
      J=0
      IF(IND.EQ.1) IPI=IR
      DO 21 I=IPI,M
      LK=L1+I
      GOTO (15,16), IND
   16 K=L2+I
      GOTO 21
   15 J=J+1
      IF(J.GT.8) J=8
      K=L1+NDIM1*(J-1)+I
   21 WRITE(IOUT,FMT) I,(A(L3),L3=LK,K,NDIM1)
   12 IR=IR+8
   40 L=LL+1
      RETURN
      END
```

Table A2

```
      SUBROUTINE PRNTRI(X,N2,NDIM,TEXT,NW,IP,NO,IND,IWORD,IOUT)
C
C SUBROUTINE TO PRINT UPPER TRIANGLE OF SYMMETRIC MATRIX IN NEAT FORM.
C THE SYMMETRIC MATRIX CAN BE HELD IN A ONE OR TWO-DIMENSIONAL ARRAY.
C ARGUMENTS:
C X       REAL ARRAY HOLDING THE UPPER TRIANGLE OF THE MATRIX TO BE
C         PRINTED. MAY BE IN CONVENTIONAL 2-DIMENSIONAL FORM OR MAY BE
C         STORED LINEARLY ROWWISE IN A 1-DIMENSIONAL ARRAY.
C N2      INTEGER. DIMENSION OF A IF MATRIX IS STORED LINEARLY OR THE
C         PRODUCT OF THE 2 DIMENSIONS IF HELD IN A 2-DIMENSIONAL ARRAY.
C TEXT    HOLLERITH STRING OF UP TO 80 CHARACTERS OR ARRAY CONTAINING
C         ALPHANUMERIC TITLE.
C NW      NUMBER OF COMPUTER WORDS OCCUPIED BY TEXT. I.E. NO. OF
C         CHARACTERS IN TEXT DIVIDED BY IWORD (SEE BELOW). THE NUMBER OF
C         CHARACTERS IN TEXT MUST BE AN EXACT INTEGER MULTIPLE OF IWORD.
C IP      INTEGER. IF =0 ON INPUT, PRINTING STARTS ON A NEW LINE. IF = 1
C         ON INPUT, PRINTING BEGINS ON A NEW PAGE. IF = 0,PRINTING STARTS
C         AFTER 1 BLANK LINE. SINCE THE VALUE OF IP IS ALTERED BY THE
C         C ROUTINE, IT SHOULD NOT BE CALLED BY VALUE.
C NO      INTEGER. THE ORDER OF X.
C IND     INTEGER. IND = 1 - MATRIX IS HELD IN TWO DIMENSIONAL FORM.
C         IND = 2 - MATRIX IS HELD LINEARLY ROWWISE, IN ONE-DIMENSIONAL
C         FORM.
C IWORD   INTEGER. NUMBER OF CHARACTERS IN A COMPUTER WORD. 8 FOR ICL
C         1900 SERIES, 4 FOR PDP-11 AND IBM 360/370, AND 10 FOR CDC 6600
C         AND 7600 MACHINES.
C IOUT    INTEGER. OUTPUT DEVICE UNIT NUMBER.
      DIMENSION X(N2),FMT(20),COPY(20),TEXT(NW)
      DATA SP,K/4H6X, ,9/
```

## Table A2 *continued*

```
C IF OUTPUT FORMAT TO BE ALTERED NOTE AND COPY EXACTLY THE LAYOUT OF THE
C EXTANT DATA STATEMENT;ANY REPLACEMENT MUST BE IN 4 CHARACTER UNITS.
      DATA FMT/4H(I7,,4H4X, ,4HF12.,4H4,    ,4HF12.,4H4,    ,4HF12.,4H4,    ,
     14HF12.,4H4,    ,4HF12.,4H4,    ,4HF12.,4H4,    ,4HF12.,4H4,    ,4HF12.,
     24H4,    ,4HF12.,4H4)  /
      IF(IP.NE.1.AND.IP.NE.0) IP=1
      IFT=1
      IF(IWORD.EQ.8) IFT=2
      IF(IWORD.EQ.10) IFT=3
      K2=NO*NO
      II=NO/K
      LEFT=NO-II*K
      IF(LEFT.GT.0) II=II+1
      DO 2 L=1,II
      GOTO (50,51,52),IFT
   50 WRITE(IOUT,4) IP,TEXT
      GOTO 55
   51 WRITE(IOUT,8) IP, TEXT
      GOTO 55
   52 WRITE(IOUT,10) IP,TEXT
    4 FORMAT(I1,20A4)
    8 FORMAT(I1,10A8)
   10 FORMAT(I1,8A10)
   55 IP=0
      WRITE(IOUT,11) L,II
   11 FORMAT('+',80X,'SECTION',I3,' OF',I3/)
      IR=(L-1)*K+1
      IQ=IR+K-1
      IF(L.EQ.II.AND.LEFT.GT.0) IQ=IR+LEFT-1
      KK=IQ-IR
      WRITE(IOUT,12) (I,I=IR,IQ)
   12 FORMAT(' ',5X,9I12)
      DO 1 I=1,20
    1 COPY(I)=FMT(I)
      GOTO (3,13),IND
    3 LIM=IQ*NDIM
      K1=(IR-1)*NDIM+1
      IU=0
      DO 5 I=1,IQ
      WRITE(IOUT,COPY) I, (X(J),J=K1,LIM,NDIM)
      K1=K1+1
      IF(I.LT.IR) GOTO 5
      MM=I-IR+3+IU
      IU=IU+1
      K1=K1+NDIM
      COPY(MM)=SP
      COPY(MM+1)=SP
    5 CONTINUE
      GOTO 2
   13 K1=IR
      K2=IQ
      KT=0
      IU=0
      DO 6 I=1,IQ
      WRITE(IOUT,COPY) I,(X(J),J=K1,K2)
      IF(I.EQ.IQ) GOTO 6
      IF(I.LT.IR) GOTO 7
      K1=K1+NO-I+1
      KT=KT+1
      K2=K1+KK-KT
      MM=I-IR+3+IU
      IU=IU+1
      COPY(MM)=SP
      COPY(MM+1)=SP
      GOTO 6
    7 K1=K1+NO-I
      K2=K1+KK
```

## Table A2  *continued*

```
    6 CONTINUE
    2 CONTINUE
      RETURN
      END
```

## Table A3

```
      SUBROUTINE SSCP(S,SS,X,NV,NOBS,ICOL,IND,INPUT,OUTPUT,FMT,II)
C
C SUBROUTINE SSCP
C FORMS VARIANCE-COVARIANCE MATRIX (S) ,  CORRELATION MATRIX (R)
C OR SUMS OF SQUARES AND CROSS-PRODUCTS MATRIX FROM RAW DATA USING
C THE METHOD OF YOUNGS AND CRAMER, TECHNOMETRICS,1971.
C *** PARAMETERS ***
C  S     VECTOR OF LENGTH NV; UNDEFINED ON ENTRY; CONTAINS THE MEANS
C        OF THE NV VARIABLES ON EXIT.
C  SS    TWO DIMENSIONAL ARRAY. MINIMUM DIMENSION NV*NV. UNDEFINED ON
C        ENTRY. CONTENTS ON EXIT DEPEND ON THE VALUE OF IND (BELOW)
C  X     VECTOR OF LENGTH NV; UNDEFINED ON ENTRY. CONTAINS THE STD.
C        DEVIATIONS OF THE NV VARIABLES ON EXIT.
C  NV    INTEGER. NUMBER OF VARIABLES.
C  NOBS  INTEGER. NUMBER OF OBSERVATIONS (CASES) PER VARIABLE.
C  ICOL  FIRST DIMENSION OF SS AS DECLARED IN CALLING PROGRAM.
C  IND   INTEGER. ON ENTRY MUST BE SET TO 1 (FOR COMPUTATION OF MATRIX S
C        2 (FOR COMPUTATION OF CORRELATION MATRIX) OR 3 (FOR
C        COMPUTATION OF SUMS OF SQUARES &CROSS-PRODUCTS MATRIX).
C        IF IND IS NEGATIVE ON ENTRY THE UPPER TRIANGLE PLUS DIAGONAL
C        OF THE APPROPRIATE MATRIX WILL BE COMPUTED, OTHERWISE THE
C        LOWER TRIANGLE WILL BE COMPUTED.
C  INPUT INTEGER. LOGICAL UNIT NUMBER OF THE DEVICE FROM WHICH THE
C        DATA ARE TO BE READ. MAY BE FORMATTED OR UNFORMATTED DEPENDING
C        ON THE VALUE OF II (SEE BELOW).
C        AND EACH RECORD (CARD, DISC) MUST CONTAIN THE OBSERVATIONS
C        ON THE NV VARIABLES FOR A GIVEN CASE, I.E. THE DATA MATRIX IS
C        READ ROWWISE, EACH ROW BEGINNING A FRESH RECORD.
C  OUTPUT LOGICAL UNIT NO. OF DEVICE ON WHICH RESULTS ARE TO BE OUTPUT.
C  FMT   REAL ARRAY,DIMENSION 20; HOLDS INPUT FORMAT FOR DATA IF II=0.
C        UNUSED WORDS OR PORTIONS OF WORDS MUST BE SET TO THE CHARACTER
C        BLANK.
C  II    INTEGER EQUAL TO 0 IF INPUT DATA IS FORMATTED OR TO 1
C        IF INPUT DATA IS UNFORMATTED.
C *** OUTPUT ***
C        THE MEANS AND STANDARD DEVIATIONS OF THE VARIABLES ARE
C        OUTPUT TO THE DEVICE WITH LOGICAL UNIT NUMBER 'OUTPUT'.
C *** FAILURE MESSAGES ***
C        IF IND IS ZERO ON ENTRY, EXECUTION IS TERMINATED AND A
C        MESSAGE IS SENT TO DEVICE OUTPUT.
C *** OTHER ROUTINES CALLED ***
C        SORT, IABS
C
      INTEGER OUTPUT
      DIMENSION SS(ICOL,NV),X(NV),S(NV),FMT(20)
      B=0.0
      LOGIC=1
      FOBS=NOBS
      IF(IND) 6,100,7
    6 LOGIC=2
      IND=IABS(IND)
    7 DO 1 I=1,NV
      S(I)=0.0
      GOTO (8,9), LOGIC
    8 DO 11 J=1,I
   11 SS(I,J)=0.0
      GOTO 1
```

Table A3   *continued*

```
    9 DO 12 J=I,NV
   12 SS(I,J)=0.0
    1 CONTINUE
      DO 2 K=1,NOBS
      IF(II.EQ.1) GOTO 18
      READ(INPUT,FMT) X
      GOTO 17
   18 READ(INPUT) X
   17 DO 3 I=1,NV
    3 S(I)=S(I)+X(I)
      B=B+1.
      IF (K-1) 2,2,5
    5 BB=B*B-B
      DO 60 I=1,NV
      XI=B*X(I)-S(I)
      GOTO (13,14), LOGIC
   13 DO 4 J=1,I
    4 SS(I,J)=SS(I,J)+(XI*(B*X(J)-S(J)))/BB
      GOTO 60
   14 DO 15 J=I,NV
   15 SS(I,J)=SS(I,J)+(XI*(B*X(J)-S(J)))/BB
   60 CONTINUE
    2 CONTINUE
      DO 21 I=1,NV
      X(I)=SQRT(SS(I,I)/FOBS)
   21 S(I)=S(I)/FOBS
      WRITE(OUTPUT,10) (I,S(I),X(I),I=1,NV)
   10 FORMAT(/OMEANS AND STANDARD DEVIATIONS.'//(' ',I6,2E15.6))
      IF(IND.EQ.2) RETURN
      DO 23 I=1,NV
      GOTO (30,31), LOGIC
   30 DO 22 J=1,I
   22 SS(I,J)=SS(I,J)/FOBS
      GOTO 23
   31 DO 24 J=I,NV
   24 SS(I,J)=SS(I,J)/FOBS
   23 CONTINUE
      IF(IND.EQ.1) RETURN
      DO 26 I=1,NV
      XI=X(I)
      GOTO (32,33),LOGIC
   32 DO 20 J=1,I
   20 SS(I,J)=SS(I,J)/(XI*X(J))
      GOTO 26
   33 DO 27 J=I,NV
   27 SS(I,J)=SS(I,J)/(XI*X(J))
   26 CONTINUE
      RETURN
  100 WRITE(OUTPUT,101)
  101 FORMAT('ERROR - IND SET TO ZERO ON ENTRY')
      STOP
      END
```

Table A4

```
      SUBROUTINE COPY(A,N,B,M,K,L)
C
C COPIES ROWS 1-K AND COLUMNS 1-L OF MATRIX [B] INTO MATRIX [A]
C
C ARGUMENTS:
C A       REAL ARRAY, DIMENSION AT LEAST (K*L)
C N       INTEGER. FIRST DIMENSION OF A AS DECLARED IN CALLING PROGRAM.
C B       REAL ARRAY, MINIMUM DIMENSION (K*L).
```

Table A4    *continued*

```
C  M        INTEGER. THE FIRST DIMENSION OF B IN THE CALLING PROGRAM.
C  K,L      INTEGERS. ROWS 1-5 AND COLUMNS 1-6 OF ARRAY B WILL BE COPIED
C  INTO ARRAY A.
C
      DIMENSION A(N,L),B(M,L)
      DO 1 I=1,K
      DO 1 J=1,L
    1 A(I,J)=B(I,J)
      RETURN
      END
```

Table A5

```
      SUBROUTINE TRMULT(A,B,C,N1,N2,N3,K1,K2,K3)
C
C SETS A=B'C
CC  *** ARGUMENTS ***
C   A,B AND C     REAL ARRAYS DEFINED BY A=B'C.
C   N1,N2,N3      FIRST DIMENSIONS OF A,B AND C RESPECTIVELY AS
C                 DECLARED IN THE CALLING PROGRAM.
C   K1            NUMBER OF ROWS IN A AND COLUMNS IN B.
C   K2            NUMBER OF COLUMNS IN A AND C.
C   K3            NUMBER OF ROWS IN B AND C.
C  *** OTHER SUBROUTINES REQUIRED ***
C NONE.
C  *** ERROR MESSAGES ***
C NONE.
C
      DIMENSION A(N1,K2),B(N2,K1),C(N3,K2)
      DO 1 I=1,K1
      DO 1 J=1,K2
      A(I,J)=0.0
      DO 1 L=1,K3
    1 A(I,J)=A(I,J)+B(L,I)*C(L,J)
      RETURN
      END
```

Table A6

```
      SUBROUTINE MTMULT(A,B,C,L,M,N,I1,I2,I3)
C
C SUBROUTINE MTMLT
C SETS A=BC
C  *** ARGUMENTS ***
C   A      REAL MATRIX, ORDER (I1,I2).
C   B      REAL MATRIX, ORDER (I1,I3).
C   C      REAL MATRIX, ORDER (I3,I2).
C ** A,B AND C DEFINED BY A=BC **
C   L      FIRST DIMENSION OF A AS DECLARED IN CALLING PROGRAM.
C   M      FIRST DIMENSION OF B.
C   N      FIRST DIMENSION OF C.
C   I1     NUMBER OF ROWS IN A AND B.
C   I2     NUMBER OF COLUMNS IN A AND C.
C   I3     NUMBER OF COLUMNS IN B AND ROWS IN C.
C
C  *** OTHER SUBROUTINES NEEDED ***
C NONE.
C  *** ERROR MESSAGES ***
C NONE
      DIMENSION A(L,I2),B(M,I3),C(N,I2)
      DO 1 I=1,I1
```

Table A6  *continued*

```
      DO 1 J=1,I2
      A(I,J)=0.0
      DO 1 K=1,I3
    1 A(I,J)=A(I,J)+B(I,K)*C(K,J)
      RETURN
      END
```

Table A7

```
      SUBROUTINE DPACC(A,I,J,IJ,B,K,L,KL,X,SUM,N,IFL)
C
C     SUBROUTINE DPACC FOR DOUBLE PRECISION ACCUMULATION OF INNER
C     PRODUCTS, I.E. SUM=(+,-)SUM(+,-)AB
C         A       VECTOR ON THE LEFT.
C         I,J     NUMERICAL IDENTIFIERS OF THE FIRST 2 ELEMENTS OF A IN THE
C                 MULTIPLICATION PROCESS.
C         IJ      DIMENSION OF A
C         B,K,L,KL  ANALOGOUS TO A,I,J,IJ - SEE ABOVE.
C         X       QUANTITY TO BE ADDED TO PRODUCT OF SPECIFIED ELEMENTS
C                 OF A AND B.
C         SUM     RESULT.
C         IFL     = 1 FOR SUM = AB+X.
C         IFL     = 2 FOR SUM = X-AB.
C         IFL     = 3 FOR SUM = AB-X.
C         IFL     = 4 FOR SUM = -AB-X = -(AB+X).
C
      DOUBLE PRECISION P,Q,R
      DIMENSION A(IJ),B(KL)
      R=0.D0
      IF(I.GT.IJ.OR.K.GT.KL) GOTO 7
      IF(N.LE.0) GOTO 7
      M=J-I
      M1=L-K
      DO 1 IK=1,N
      I1=I+(IK-1)*M
      P=A(I1)
      I1=K+(IK-1)*M1
      Q=B(I1)
    1 R=R+P*Q
    7 P=DBLE(X)
      GOTO (2,3,4,5), IFL
    2 SUM=SNGL(P+R)
      GOTO 6
    3 SUM=SNGL(P-R)
      GOTO 6
    4 SUM=SNGL(R-P)
      GOTO 6
    5 SUM=SNGL(-P-R)
    6 RETURN
      END
```

Table A8

```
      SUBROUTINE PLOTS(A,N1,N2,NDIM,NO,NDIM2,ISAME,IOVP,NDIM3)
C
C SUBROUTINE TO PLOT PAIRWISE COMBINATIONS OF THE COLUMNS OF INPUT
C MATRIX, A.
C ARGUMENTS:
C   A     MATRIX HOLDING VALUES TO BE PLOTTED. MINIMUM DIMENSION (NO*N2)
C   N1    FIRST COLUMN TO BE PLOTTED. IF NEGATIVE, THIS COLUMN WILL BE
C         HELD CONSTANT (SEE BELOW).
C   N2    UPPER COLUMN TO BE PLOTTED. IF NEGATIVE, WILL BE HELD CONSTANT.
```

Table A8 *continued*

```
C           FOR EXAMPLE: IF N1 = 5 AND N2 = 10 THEN ALL POSSIBLE COMBINAT-
C           IONS OF COLUMNS IN THE RANGE (5-10) WILL BE PLOTTED.
C           IF N1 = 1 AND N2 = -5 THEN COLUMNS 1,2,3 AND 4 WILL BE PLOTTED
C           AGAINST COLUMN 5.   IF N1 = -1 AND N2 = -5 THEN COLUMN 1 WILL B
C           PLOTTED AGAINST COLUMN 5.
C     *** IMPORTANT --- THE PARAMETERS N1 AND N2 MUST NOT BE GIVEN
C     NUMERICAL VALUES IN THE CALL STATEMENT -- THEIR VALUES MUST BE
C     PREVIOUSLY SET IN ASSIGNMENT,READ OR DATA STATEMENTS.
C     THE VALUE OF N1 AND N2 ON RETURN IS ALWAYS POSITIVE.
C NDIM    FIRST DIMENSION OF A AS DECLARED IN CALLING PROGRAM.
C NO      NUMBER OF POINTS (ITEMS) TO BE PLOTTED.
C           IF NO IS POSITIVE,PRINTING WILL START ON A NEW PAGE.  IF NO IS
C           NEGATIVE, PRINTING WILL START ON A NEW LINE.
C     **** DO NOT GIVE NO A NUMERIC VALUE IN THE CALL STATEMENT -- ITS
C     VALUE MUST BE PREVIOUSLY SET. THE VALUE OF NO ON RETURN IS ALWAYS
C     POSITIVE.
C NDIM2   SECOND DIMENSION IF A AS DECLARED IN CALLING PROGRAM.
C ISAME   INTEGER; IF 1 ON ENTRY,X AXIS AND Y AXIS WILL HAVE SAME SCALE.
C           OTHERWISE SCALE FOR EACH AXIS WILL BE SET INDEPENDENTLY BY
C           ROUTINE.
C IOVP    INTEGER ARRAY, MINIMUM DIMENSION (NO*4), USED AS WORKSPACE
C NDIM3   FIRST DIMENSION OF IOVP AS DECLARED IN CALLING PROGRAM.
C
C OTHER ROUTINES NEEDED : SCALE,OVCHK,SAMESC,KALK
C ERROR MESSAGES: NONE.
C
      INTEGER SLINE,TEM,CHAR
      DIMENSION A(NDIM,NDIM2),SLINE(102),CHAR(4),SLIST(11),SLIST2(11),
     *TEM(10),IOVP(NDIM3,4)
      DATA CHAR/1H*,1H ,1H-,1HI/,TEM/1H1,1H2,1H3,1H4,1H5,1H6,1H7,1H8,1H9
     1,1H0/
      IPR=1
      IF(NO) 66,66,67
   66 IPR=0
      NO=-NO
   67 SMALL=1.0E-9
      IF(N1) 60,60,61
   60 N1=-N1
      N21=N1
      GOTO 62
   61 N21=IABS(N2)-1
   62 DO 6 I=N1,N21
      KOUNT=0
C SET UP SCALES FOR X AND Y AXES
      CALL SCALE(NO,10,A(1,I),SLIST,IFL,11)
      IF(N2) 63,63,64
   63 N22=-N2
      I1=-N2
      GOTO 65
   64 I1=I+1
      N22=N2
   65 DO 6 II=I1,N22
      CALL SCALE(NO,10,A(1,II),SLIST2,IFL,11)
      DO 50 J=1,11
      IF(ABS(SLIST(J)).LT.SMALL) SLIST(J)=0.0
   50 IF(ABS(SLIST2(J)).LT.SMALL) SLIST2(J)=0.0
C SET UP EQUAL SCALE INTERVAL IF REQUESTED
      IF(ISAME.EQ.1) CALL SAMESC(SLIST ,SLIST2)
C OUTPUT SCALE INFORMATION FOR HORIZONTAL AXIS
      WRITE(6,150) IPR,(SLIST2(J),J=1,11,2)
      IPR=1
      WRITE(6,160) (SLIST2(J),J=2,10,2)
      RY=(SLIST2(11)-SLIST2(1))/99.0
      DO 12 J=1,102
   12 SLINE(J)=CHAR(3)
      DO 13 J=10,100,10
   13 SLINE(J)=CHAR(4)
```

Table A8  *continued*

```
      WRITE(6,130) SLIST(11),SLINE
      P=(SLIST(11)-SLIST(1))/59.0
      KTR=11
C START MAIN LOOP
      DO 3 J=1,60
      Q=SLIST(11)-P*(J-1)
      Z=Q-P
      DO 4 K=1,102
    4 SLINE(K)=CHAR(2)
      DO 5 K=1,NO
C FIND VALUES IN RANGE FOR THIS LINE
      IF(A(K,I).GT.Q.OR.A(K,I).LE.Z) GOTO 5
      ICHK=0
      M=KALK(A(K,II),SLIST2(1),RY)
      ISET=0
      KK=K
C CHECK FOR OVERPRINTS
      IF(K.LT.100) GOTO 20
      CALL OVCHK(IND,K,KOUNT,M,IOVP,SLINE,3,CHAR   ,ICHK,NDIM3)
      IF(IND.EQ.1) GOTO 5
      SLINE(M)=CHAR(1)
      ISET=1
      M=M+1
      L=KK/100
      KK=KK-L*100
      SLINE(M)=TEM(L)
      M=M+1
   20 IF(K.LT.10) GOTO 21
      IF(ICHK.EQ.0) CALL OVCHK(IND,K,KOUNT,M,IOVP,SLINE,2,CHAR,ICHK,
     * NDIM3)
      IF(IND.EQ.1) GOTO 5
      IF(ISET.EQ.1) GOTO 70
      SLINE(M)=CHAR(1)
      ISET=1
      M=M+1
   70 L=KK/10
      KK=KK-L*10
      IF(L.EQ.0) L=10
      SLINE(M)=TEM(L)
      M=M+1
   21 IF(KK.EQ.0) KK=10
      IF(ICHK.EQ.0) CALL OVCHK(IND,K,KOUNT,M,IOVP,SLINE,1,CHAR,ICHK,
     * NDIM3)
      IF(IND.EQ.1) GOTO 5
      IF(ISET.EQ.1) GOTO 71
      SLINE(M)=CHAR(1)
      ISET=1
      M=M+1
   71 SLINE(M)=TEM(KK)
    5 CONTINUE
      IF(MOD(J,6)) 1,11,1
    1 WRITE(6,100) SLINE
      GOTO 3
   11 KTR=KTR-1
C OUTPUT CURRENT LINE
      WRITE(6,120) SLIST(KTR),SLINE
    3 CONTINUE
      WRITE(6,100) (CHAR(3),J=1,102)
      WRITE(6,110) I,II
C PRINT OVERPRINT TABLE
      WRITE(6,754) KOUNT
  754 FORMAT('0 NUMBER OF OVERPRINTS = ',I6/)
      IF(KOUNT) 30,6,30
   30 WRITE(6,170)
      DO 31 K=1,KOUNT
   31 WRITE(6,32) (IOVP(K,J),J=2,4),IOVP(K,1)
    6 CONTINUE
```

Table A8  *continued*

```
      RETURN
C FORMATS
   32 FORMAT(' ',3A1,16X,I7)
  100 FORMAT(' ',12X,'I',102A1,'I')
  110    FORMAT('0PLOT OF COL.',I3,' (VERTICAL AXIS) AGAINST COL.',I3,
     *' (HORIZONTAL AXIS)'///)
  120 FORMAT(' ',G12.4,'-',102A1,'-')
  130 FORMAT(' ',G12.4,'I',102A1,'I')
  150 FORMAT(I1, 6X,5(G12.4,8X),G12.4)
  160 FORMAT(' ',16X,4(G12.4,8X),G12.4)
  170 FORMAT('0OVERPRINT TABLE'/' ITEM NO.   ALSO LOCATION OF ITEM NO.'/)
      END
```

Table A9

```
      SUBROUTINE SAMESC(S1,S2)
C USED IN SUBROUTINE PLOTS
      DIMENSION S1(11),S2(11)
      A1=S1(2)-S1(1)
      A2=S2(2)-S2(1)
      IF(A1-A2) 10,20,30
   30 DO 31 I=2,11
   31 S2(I)=S2(I-1)+A1
      GOTO 20
   10 DO 11 I=2,11
   11 S1(I)=S1(I-1)+A2
   20 RETURN
      END
```

Table A10

(Reproduced with permission from R. P. Thayer and R. F. Storer, *Applied Statistics*, 18, 206−208 (1969))

```
      SUBROUTINE SCALE(N,INT,X,SLIST,IFAULT,INT1)
C
C        ALGORITHM AS21, JOURN. ROYAL STATISTICAL SOC., SERIES C,
C        (APPLIED STATISTICS), VOL 18, PP. 206 - 208, BY R.P. THAYER AND
C        R.F. STORER.  REPRODUCED BY PERMISSION OF THE ROYAL STATISTICAL
C        SOCIETY.  LATER AMENDMENTS INCORPORATED.
C
C
C USED IN SUBROUTINE PLOTS
      DIMENSION X(N),SLIST(INT1)
      XMIN=X(1)
      XMAX=X(1)
      DO 1 I=2,N
      IF(X(I).LT.XMIN)XMIN=X(I)
      IF(X(I).GT.XMAX)XMAX=X(I)
    1 CONTINUE
      R=XMAX-XMIN
      B=XMIN
      IFAULT=0
      IF(R.EQ.0.0)IFAULT=1
      IF(INT.LT.1)IFAULT=2
      IF(N.LT.2)IFAULT=3
      IF(IFAULT.NE.0)GO TO 5
      KOUNT=0
      KOD=0
   21 IF(R.GT.0) GO TO 20
      KOUNT=KOUNT+1
      R=R*10.
      GO TO 21
```

Table A10  *continued*

```
   20 IF(R.LE.10.0) GO TO 22
      KOUNT=KOUNT-1
      R=R/10.
      GO TO 20
   22 B=B*10.**KOUNT
      IF(B.LE.0.0) B=B-1
      IB=IFIX(B)
      B=FLOAT(IB)
      B=B/10.**KOUNT
      R=XMAX-B
      A=R/FLOAT(INT)
      KOUNT=0
   26 IF(A.GT.1.0) GO TO 25
      KOUNT=KOUNT+1
      A=A*10.
      GO TO 26
   25 IF(A.LE.10.0) GO TO 27
      KOUNT=KOUNT-1
      A=A/10.
      GO TO 25
   27 IA=IFIX(A)
      IF(IA.EQ.6) IA=7
      IF(IA.EQ.8) IA=9
      AA=0.0
      IF(A.LE.1.25) AA=-0.75
      IF(A.GT.1.25.AND.A.LE.1.5.OR.A.GT.2.0.AND.A.LE.2.5.OR.A.GT.6.0.
     *AND.A.LE.7.5) AA=-0.5
      A=FLOAT(IA)
      A=AA+1.0+A
      A=A/10.**KOUNT
      IF(KOD.EQ.1) GO TO 41
      TEST=FLOAT(INT)*A
      TEST1=(XMAX-XMIN)/TEST
      IF(TEST1.GT.0.8) GO TO 41
      KOUNT=1
      KOD=1
      R=XMAX-XMIN
      B=XMIN
      GO TO 21
   41 IAB=IFIX(B/A)
      IF(IAB.LT.0) IAB=IAB-1
      C=A*FLOAT(IAB)
      D=C+FLOAT(INT)*A
      IF(D.LT.XMAX) GOTO 43
      B=C
      GOTO 42
   43 C=C+A
      IF(XMIN.GE.C) B=C
   42 SLIST(1)=B
      DO 30 J=1,INT
   30 SLIST(J+1)=SLIST(J)+A
    5 CONTINUE
      RETURN
      END
```

Table A11

```
      INTEGER FUNCTION KALK(A,S,R)
C USED IN SUBROUTINE PLOTS
      Q=S
      DO 1 I=1,100
      Q=Q+R
      IF(A.GE.Q) GOTO 1
      KALK=I
```

Table A11  *continued*

```
      RETURN
    1 CONTINUE
      WRITE(6,2) A
    2 FORMAT('0FAIL IN KALK WITH A = ',F10.4)
      STOP
      END
```

Table A12

```
      SUBROUTINE OVCHK(IND,K,KT,M,IOVP,SLINE,KN,CHAR,ICHEK,NDIM3)
      INTEGER SLINE(102),CHAR(3),IOVP(NDIM3,4)
      IND=0
      IF(SLINE(M).EQ.CHAR(2)) GOTO 1
C IF SLINE(M) IS BLANK, GOTO LABEL 1 OTHERWISE LOOK FOR '*' AND COPY
C CHARACTERS FOLLOWING * UNTIL BLANK OR ANOTHER * IS MET
      I=M
    8 IF(SLINE(I).EQ.CHAR(1)) GOTO 2
      I=I-1
      GOTO 8
    2 IND=1
      J=2
      KT=KT+1
      IOVP(KT,1)=K
    3 I=I+1
      IF(I.GT.102) GOTO 10
      IF(SLINE(I).EQ.CHAR(2).OR.SLINE(I).EQ.CHAR(1)) GOTO 11
      IOVP(KT,J)=SLINE(I)
      J=J+1
      GOTO 3
   11 IF(J.GT.4) GOTO 10
      DO 12 I=J,4
   12 IOVP(KT,I)=CHAR(2)
      GOTO 10
C NEXT SECTION EXECUTED ONLY IF MTH CHARACTER OF LINE IS BLANK. STRATEGY
C IS TO LOOK FORWARD KN CHARACTERS CHECKING FOR A '*'.
    1 MM=M+KN
      M1=M+1
      DO 4 J=M1,MM
      IF(SLINE(J).EQ.CHAR(2)) GOTO 4
      I=J
      GOTO 2
    4 CONTINUE
   10 ICHEK=1
      RETURN
      END
```

Table A13

```
      SUBROUTINE RANK(S,IDEN,WORK,NO,IND)
C
C SUBROUTINE TO RANK INPUT ARRAY S INTO ASCENDING OR DESCENDING ORDER
C IDEN IS AN INTEGER ARRAY, SAME DIMENSION AS S. ON EXIT IT HOLDS THE
C POSITION OF EACH ELEMENT OF S IN THE NEWLY-ORDERED SEQUENCE, I.E.
C IDEN(I) GIVES THE POSITION OF S(I) IN THE RANKED SEQUENCE.
C WORK IS AN INTEGER ARRAY OF THE SAME DIMENSION AS S; WORKSPACE.
C NO IS THE NUMBER OF ELEMENTS IN S
C IF IND = 1 S IS SORTED INTO ASCENDING ORDER
C IF IND = -1 S IS SORTED INTO DESCENDING ORDER
C EXECUTION TERMINATES IF IND IS NOT EQUAL TO EITHER +1 OR -1
C
      DIMENSION S(NO),IDEN(NO), WORK(NO)
      IF(IND.NE.1.AND.IND.NE.-1) GOTO 20
```

Table A13  *continued*

```
         IF (IND) 11,11,10
    11 DO 12 I=1,NO
    12 S(I)=-S(I)
    10 DO 1 I=1,NO
     1 IDEN(I)=I
       M=NO
     2 M=M/2
       IF (M) 3,8,3
     3 K=NO-M
       J=1
     4 I=J
     5 IM=I+M
       L=IDEN(I)
       LM=IDEN(IM)
       IF(S(L)-S(LM)) 7,7,6
     6 IDEN(I)=LM
       IDEN(IM)=L
       I=I-M
       IF (I-1) 7,5,5
     7 J=J+1
       IF(J-K) 4,4,2
     8 DO 13 I=1,NO
       L=IDEN(I)
    13 WORK(I)=S(L)
       A=IND
       DO 15 I=1,NO
    15 S(I)=WORK(I)*A
       RETURN
    20 WRITE(6,21)
    21 FORMAT('OIND SHOULD BE SET TO 1 OR -1 -- EXECUTION TERMINATED')
       STOP
       END
```

Table A14

```
       SUBROUTINE HISTGM(X,NCLASS,N)
C
C HISTOGRAM DRAWING ROUTINE P.M. MATHER JULY 1974
C ARGUMENTS:
C X       REAL ARRAY, DIMENSION AT LEAST N, HOLDING THE VECTOR OF RAW DAT
C         FOR WHICH A FREQUENCY PLOT IS REQUIRED.
C NCLASS INTEGER. THE NUMBER OF CLASSES TO BE USED IN THE HISTOGRAM.
C N      INTEGER. THE NUMBER OF OBSERVATIONS.
C
C NOTE THAT THE MAXIMUM NUMBER OF ITEMS IN ANY ONE CLASS IS 100. TO INCR
C OR DECREASE THIS, CHANGE THE DIMENSION OF *MEMBER*
C
       DIMENSION X(N),MEMBER(100)
       DATA IBORD/1HI/, ISTAR/1H*/
C
C FIND MINIMUM AND MAXIMUM OF X( ).
C
       S=X(1)
       P=S
       DO 1 I=2,N
       IF(X(I).GT. S) S=X(I)
       IF(X(I).LT. P) P=X(I)
     1 CONTINUE
C
C FIND RANGE AND CLASS INTERVAL FOR NCLASS CLASSES.
C
       RANGE=S-P
       CLASS=RANGE/FLOAT(NCLASS)
C
C SET MEMBERSHIP OF EACH CLASS TO ZERO.
```

Table A14   *continued*

```
C
      DO 100 I=1,NCLASS
  100 MEMBER(I)=0
C
C DETERMINE NUMBER IN EACH CLASS.
C
      BOTTOM=S
      DO 2 I=1,NCLASS
      TOP=BOTTOM
      BOTTOM=TOP-CLASS
      IF(I-NCLASS) 10,11,10
   10 DO 12 J=1,N
      IF((X(J).GT.BOTTOM).AND.(X(J).LE.TOP)) MEMBER(I)=MEMBER(I)+1
   12 CONTINUE
      GOTO 2
C
C LAST CLASS IS SPECIAL CASE.
C
   11 DO 13 J=1,N
      IF((X(J).GE.BOTTOM).AND.(X(J).LE.TOP)) MEMBER(I)=MEMBER(I)+1
   13 CONTINUE
    2 CONTINUE
C
C PRINT HISTOGRAM ACROSS PAGE. FIRST SCALE ELEMENTS OF MEMBER TO GET
C THE BEST FIT IN 100 COLUMNS WIDTH.
C
      M=MEMBER(1)
      DO 3 I=2,NCLASS
      IF(MEMBER(I).GT.M) M=MEMBER(I)
    3 CONTINUE
      SCALE=100.0/FLOAT(M)
      WRITE(6,7) SCALE
      P=S+0.5*CLASS
      DO 5 I=1,NCLASS
      P=P-CLASS
      M=MEMBER(I)
      PERC=M/FLOAT(N)*100.0
      K=INT(M*SCALE+0.5)
C
C CHECK FOR K=0 AND ACT ACCORDINGLY.
C
      IF(K) 23,21,23
   21 WRITE(6,20) IBORD
      WRITE(6,6) M,PERC,IBORD
      WRITE(6,24) P, IBORD
      GOTO 5
   23 WRITE(6,20) IBORD,(ISTAR,J=1,K)
      WRITE(6,6) M,PERC,IBORD,(ISTAR,J=1,K)
      WRITE(6,24) P, IBORD,(ISTAR,J=1,K)
      WRITE(6,20) IBORD,(ISTAR,J=1,K)
    5 CONTINUE
    6 FORMAT(1H ,I4,3X,F10.4,2X,101A1)
    7 FORMAT(18H0SCALING FACTOR = ,F10.4///7H NUMBER,5X,7HPERCENT/1H ,
     1 5X,8HMIDPOINT//)
   20 FORMAT(' ',19X,101A1)
   24 FORMAT(1H ,F10.4,9X,101A1)
      RETURN
      END
```

Table A15

```
      SUBROUTINE SETUP
C SETS UP MACHINE-DEPENDENT CONSTANTS FOR RANDOM-NUMBER GENERATOR.
C THE VALUES GIVEN BELOW ARE APPROPRIATE FOR AN ICL 1900 SERIES MACHINE.
```

Table A15   *continued*

```
      COMMON /RANDS/ IA,M,FM,IA1,IC1,IC
      IA1=61739
      IC1=1075
      IA=131069
      IC=27697
      M=8388607
      FM=8388607.0
      RETURN
      END
```

Table A16

```
      FUNCTION RANDOM(IX1,IX2)
      COMMON /RANDS/ IA,M,FM,IA1,IC1,IC
      IX=IX*IA+IC
      IF(IX) 1,2,2
    1 IX=IX+M+1
    2 IX1=IX1*IA1+IC1
      IF(IX1) 3,4,4
    3 IX1=IX1+M+1
    4 IV=IX+IX1+1
      IF(IV) 5,6,6
    5 IV=IV+M+1
    6 RANDOM=IV/FM
C NEXT STATEMENT NECESSARY TO SWITCH OFF OVERFLOW INDICATOR ON ICL 1900
C SERIES COMPUTERS. THIS MAY NOT BE NECESSARY WITH OTHER MACHINES. SEE
C MANUFACTURER'S HANDBOOK FOR DETAILS OF 'INTEGER OVERFLOW'.
      CALL OVERFL(J)
      RETURN
      END
```

Table A17

```
      SUBROUTINE DEVIAT(IX1,IX2,D,I)
C
C GENERATES AN  INDEPENDENT RANDOM STANDARD NORMAL DEVIATE  BY THE
C POLAR MATHOD.
C IX1,IX2    INTEGER VARIABLES, AS FOR RANDOM.
C D          UNDEFINED ON ENTRY. HOLDS RANDOM DEVIATE ON EXIT.
C I          I IS SET TO 1 AND 2 ON ALTERNATE CALLS
C
      COMMON /SAVE/ D1,D2
      R(IX1,IX2)=RANDOM(IX1,IX2)*2.0-1.0
      IF(MOD(I,2).EQ.0) GOTO 2
    1 R1=R(IX1,IX2)
      R2=R(IX1,IX2)
      S=R1*R1+R2*R2
      IF(S.GE.1.0) GOTO 1
      SLOG=SQRT(-2.0*ALOG(S)/S)
      D1=R1*SLOG
      D2=R2*SLOG
      D=D1
      RETURN
    2 D=D2
      RETURN
      END
```

## Table A18

```
      SUBROUTINE ZERO(S,L,M,N,ID)
C
C ZEROS  SPECIFIED PART OF ARRAY
C ARGUMENTS
C S       REAL ARRAY TO BE ZEROED.
C L       INTEGER, FIRST DIMENSION OF S AS DECLARED IN CALLING PROGRAM.
C IF ID = 1 THEN ROWS 1-M AND COLUMNS 1-NOF ARRAY S ARE SET TO ZERO.
C IF ID = 2 THE UPPER TRIANGLE OF THE ARRAY S IS SET TO ZERO.
C THE UPPER TRIANGLE IS DEFINED BY THE FIRST M ROWS OF S
C
      DIMENSION S(L,N)
      GOTO (10,20),ID
   10 DO 1 I=1,M
      DO 1 J=1,N
    1 S(I,J)=0.0
      RETURN
   20 DO 2 I=1,M
      DO 2 J=1,I
    2 S(J,I)=0.0
      RETURN
      END
```

## Table A19

```
      FUNCTION IFIND(I,J)
C
C ON RETURN IFIND IS SET TO THE INTEGER VALUE OF THE POSITION OF THE
C (I,J)TH ELEMENT OF A SQUARE SYMMETRIC MATRIX THAT HAS BEEN STORED
C LINEARLY ROWWISE IN A 1-DIMENSIONAL ARRAY.
C
      L=MAXO(I,J)
      M=MINO(I,J)
      IFIND=L*(L-1)/2+M
      RETURN
      END
```

## Table A20

```
      SUBROUTINE TRIDI(A,D,E,E2,N,N1,TOL)
C
C TRIDIAGONALIZE A SYMMETRIC MATRIX USING HOUSEHOLDER'S METHOD.
C SEE CONTRIBUTION II/2, WILKINSON AND REINSCH(1971)
C ARGUMENTS:
C A       REAL ARRAY, MINUMUM DIMENSION (N*N) HOLDING THE SYMMETRIC
C         MATRIX. ONLY THE LOWER TRIANGLE + DIAGONAL IS NEEDED.
C         ON EXIT THE LOWER TRIANGLE OF A HOLDS DETAILS OF HOUSEHOLDER
C         TRANSFORMATIONS FOR USE WITH *TRBAK1* IN RECOVERING
C         EIGENVECTORS. THE UPPER TRIANGLE IS UNALTERED.
C D       REAL ARRAY, MINIMUM DIMENSION N. HOLDS DIAGONAL OF
C         TRIDIAGONALIZED MATRIX ON EXIT.
C E       REAL ARRAY, MINIMUM DIMENSION N. HOLDS SUBDIAGONAL OF THE
C         TRIDIAGONAL FORM ON EXIT, WITH E(1) SET TO 0.
C E2      REAL ARRAY, MINIMUM DIMENSION N. HOLDS SQUARES OF THE E(I)
C         ON EXIT.
C N       INTEGER. ORDER OF A.
C N1      INTEGER. FIRST DIMENSION OF A AS DECLARED IN CALLING PROGRAM.
C TOL     REAL MACHINE-DEPENDENT PARAMETER DEFINED BY
C                       TOL = ETA / MACHEP
C         WHERE ETA IS SMALLEST REAL POSITIVE NUMBER THAT CAN BE HELD IN
C         THE COMPUTER AND MACHEP IS DEFINED BY 1.0+MACHEP>1.0, I.E.
C         MACHEP IS THE SMALLEST NUMBER SUCH THAT 1.0+MACHEP EXCEEDS 1.0.
C
      DIMENSION A(N1,N),D(N),E(N),E2(N)
      DO 1 I=1,N
    1 D(I)=A(I,I)
      DO 2 II=1,N
      I=N-II+1
      L=I-1
      H=0.0
      IF(L) 3,3,4
    4 DO 5 K=1,L
    5 H=H+A(I,K)*A(I,K)
    3 IF(H-TOL) 6,6,7
    6 E(I)=0.0
      E2(I)=0.0
      GOTO 14
    7 E2(I)=H
      F=A(I,I-1)
      G=-SQRT(H)
      IF(F.LT.0.0) G=-G
      E(I)=G
      H=H-F*G
      A(I,I-1)=F-G
      F=0.0
      IF(L) 8,8,9
    9 DO 10 J=1,L
      G=0.0
      DO 11 K=1,J
   11 G=G+A(J,K)*A(I,K)
      J1=J+1
      IF(J1.GT.L) GO TO 13
      DO 12 K=J1,L
   12 G=G+A(K,J)*A(I,K)
   13 G=G/H
      E(J)=G
   10 F=F+G*A(I,J)
    8 H=F/(H+H)
      IF(L) 14,14,15
   15 DO 16 J=1,L
      F=A(I,J)
      G=E(J)-H*F
      E(J)=G
      DO 16 K=1,J
```

Table A20    *continued*

```
   16 A(J,K)=A(J,K)-F*E(K)-G*A(I,K)
   14 H=D(I)
      D(I)=A(I,I)
      A(I,I)=H
    2 CONTINUE
      RETURN
      END
```

Table A21

```
      SUBROUTINE RATQR(D,B,B2,N,M,DLAM,EPS,IND)
C
C SUBROUTINE RATQR: FINDS A SPECIFIED NUMBER OF THE LARGEST OR SMALLEST
C EIGENVALUES OF A SYMMETRIC TRIDIAGONALIZED MATRIX USING THE QR METHOD.
C REFERENCE : CONTRIBUTION II/6, WILKINSON AND REINSCH (1971)
C USES MODIFIED QR ALGORITHM WITH NEWTON SHIFT.
C ARGUMENTS
C  D       DIAGONAL ELEMENTS OF TRIDIAGONAL FORM STORED AS A REAL ARRAY
C          OF DIMENSION N.
C  B       REAL ARRAY, MINIMUM DIMENSION N, CONTAINING SUBDIAGONAL OF THE
C          TRIDIAGONAL FORM AS OUTPUT BY SUBROUTINE TRIDI.
C  B2      REAL ARRAY,MINIMUM DIMENSION N, THE SQUARES OF THE SUBDIAGONAL
C          ELEMENTS OF THE TRIDIAGONAL FORM AS OUTPUT BT SUBROUTINE TRIDI
C          MINIMUM DIMENSION N. TYPE REAL.
C  N       INTEGER. ORDER OF TRIDIAGONALIZED MATRIX.
C  M       NUMBER OF EIGENVALUES REQUIRED. INTEGER.
C  DLAM    REAL, TOLERANCE FOR THE THEORETICAL ERROR OF THE COMPUTED
C          EIGENVALUES. DLAM = 0 IS PERMISSIBLE.
C  EPS     SMALLEST POSITIVE REAL NUMBER FOR WHICH 1.0+EPS>1.0.
C  IND     IF IND < 0 THEN THE M SMALLEST EIGENVALUES WILL BE COMPUTED.
C          IF IND >=0 THE M LARGEST EIGENVALUES WILL BE COMPUTED.
C
C OUTPUT
C  D       THE FIRST M EIGENVALUES IN DESCENDING ORDER OF MAGNITUDE OCCUPY
C          D(1) TO D(M). THE REMAINING ELEMENTS ARE UNDEFINED.
C  B2      B2(K) IS A BOUND FOR THE THEORETICAL ERROR OF THE KTH
C          EIGENVALUE, EXCLUDING EFFECTS OF ROUNDING ERROR.
C
C *** IF 25% OR MORE OF THE EIGENVALUES ARE REQUIRED, *HTDQL* IS FASTER.
C
      DIMENSION B(N),B2(N),D(N)
      ERR=0.0
      Q=0.0
      S=0.0
      IF(IND.LT.0) GOTO 42
      DO 40 I=1,N
   40 D(I)=-D(I)
   42 B2(1)=0.0
      TOT=D(1)
      DO 1 II=1,N
      I=N-II+1
      P=Q
      Q=ABS(B(I))
      E=D(I)-P-Q
    1 IF(E.LT.TOT) TOT=E
   44 DO 4 I=1,N
    4 D(I)=D(I)-TOT
      DO 6 K=1,M
    7 TOT=TOT+S
      DELTA=D(N)-S
      I=N
      E=ABS(EPS*TOT)
      IF(DLAM.LT.E) DLAM=E
      IF(DELTA-DLAM) 8,8,9
```

Table A21  *continued*

```
    9 E=B2(N)/DELTA
      QP=DELTA+E
      P=1.0
      N1=N-1
      IF(K.GT.N1) GOTO 51
      DO 10 II=K,N1
      I=N1+K-II
      Q=D(I)-S-E
      R=Q/QP
      P=P*R+1.0
      EP=E*R
      D(I+1)=QP+EP
      DELTA=Q-EP
      IF(DELTA-DLAM) 8,8,11
   11 E=B2(I)/Q
      QP=DELTA+E
   10 B2(I+1)=QP*EP
   51 D(K)=QP
      S=QP/P
      IF(TOT+S.GT.TOT) GO TO 7
      S=0.0
      I=K
      DELTA=QP
      I1=K+1
      IF(I1.GT.N) GOTO 8
      DO 12 J=I1,N
      IF(D(J)-DELTA) 13,12,12
   13 I=J
      DELTA=D(J)
   12 CONTINUE
    8 IF(I.LT.N) B2(I+1)=B2(I+1)*E/QP
      I1=I-1
      IF(K.GT.I1) GOTO 52
      DO 14 II=K,I1
      J=I1+K-II
      D(J+1)=D(J)-S
   14 B2(J+1)=B2(J)
   52 D(K)=TOT
      B2(K)=ERR+ABS(DELTA)
    6 ERR=B2(K)
      IF(IND.LT.0) GOTO 43
      DO 41 I=1,M
   41 D(I)=-D(I)
   43 RETURN
      END
```

Table A22

```
      SUBROUTINE REDUC(A,B,N1,N2,N,DL,IFL)
C
C REDUCE THE EIGENPROBLEM [A].[X]=[LAMBDA].[B].[X] TO STANDARD FORM
C RESTRICTIONS: MATRIX [B] MUST BE POSITIVE-DEFINITE
C METHOD: [B] IS REDUCED TO THE FORM [L].[L']=[B] BY THE CHOLESKY METHOD
C THE SYMMETRIC MATRIX [P]=[L-1][A][L-1]' IS THEN FORMED. THE
C EIGENVALUES OF [P] ARE THE SAME AS THOSE OF [B-1].[A] BUT THE EIGENVEC-
C TORS [Y] OF [P] MUST BE CONVERTED TO THE EIGENVECTORS OF [B-1].[A]
C BY SUBROUTINE REBAKA AND TRBAK1.
C
C ARGUMENTS:
C A      REAL ARRAY, MINIMUM DIMENSION (N*N), DEFINED ABOVE. UPPER
C TRIANGLE + PRINCIPAL DIAGONAL REQUIRED.
C B      REAL ARRAY, MINIMUM DIMENSION (N*N). DEFINED ABOVE. UPPER
C TRIANGLE PLUS PRINCIPAL DIAGONAL REQUIRED. [B] MUST BE POSITIVE-
C DEFINITE.
```

## Table A22 *continued*

```
C N1       INTEGER. FIRST DIMENSION OF A AS DECLARED IN CALLING PROGRAM.
C N2       INTEGER. FIRST DIMENSION OF B AS DECLARED IN CALLING PROGRAM.
C N        INTEGER. ORDER OF A AND B.
C DL       REAL VECTOR, MINIMUM DIMENSION N. WORKSPACE.
C IFL      INTEGER. ERROR INDICATOR. SEE BELOW.
C
C ON EXIT THE LOWER TRIANGLE OF [P] IS HELD IN ARRAY A. ITS EIGENVALUES
C AND EIGENVECTORS CAN BE FOUND BY ROUTINES SUCH AS TRIDI-RATQR-EIGVEC
C OR HTDQL, AS EXPLAINED ABOVE.
C
C IFL IS AN ERROR INDICATOR. ON EXIT IFL=0 FOR A SUCCESSFUL RUN. IF IFL
C =1 ON EXIT, THEN [B] IS NOT POSITIVE-DEFINITE.
C
      DIMENSION A(N1,N),B(N2,N),DL(N)
      DO 1 I=1,N
      DO 1 J=I,N
      X=B(I,J)
      IF(I.EQ.1) GOTO 10
      I1=I-1
      DO 2 L=1,I1
      K=I-L
    2 X=X-B(I,K)*B(J,K)
   10 IF(I.EQ.J) GOTO 3
      B(J,I)=X/Y
      GOTO 1
    3 IF (X) 12,12,4
   12 IFL=1
      RETURN
    4 Y=SQRT(X)
      DL(I)=Y
    1 CONTINUE
      DO 5 I=1,N
      Y=DL(I)
      DO 5 J=I,N
      X=A(I,J)
      IF(I.EQ.1) GOTO 5
      I1=I-1
      DO 6 L=1,I1
      K=I-L
    6 X=X-B(I,K)*A(J,K)
    5 A(J,I)=X/Y
      DO 7 J=1,N
      DO 7 I=J,N
      Y=DL(I)
      X=A(I,J)
      IF(I.LE.J) GOTO 11
      I1=I-1
      DO 8 L=J,I1
      K=I-1+J-L
    8 X=X-A(K,J)*B(I,K)
   11 IF(J.EQ.1) GOTO 7
      J1=J-1
      DO 9 L=1,J1
      K=J-L
    9 X=X-A(J,K)*B(I,K)
    7 A(I,J)=X/Y
      RETURN
      END
```

## Table A23

```
      SUBROUTINE REBAKA(B,N2,DL,Z,N1,N,M1,M2)
C
C TO TRANSFORM EIGENVECTORS AFTER REDUC HAS BEEN CALLED.
```

Table A23    *continued*

```
C SEE CONTRIBUTION II/10, WILKINSON AND REINSCH(1971)
C ARGUMENTS
C B        REAL ARRAY,DIMENSION N*N HOLDING SUBDIAGONAL ELEMENTS OF L.
C          THIS IS OUTPUT BY *REDUC*
C N2       FIRST DIMENSION OF B AS DECLARED IN CALLING PROGRAM
C DL       REAL ARRAY AS OUTPUT BY *REDUC*
C Z        REAL ARRAY, DIMENSION N,(M2-M1+1), HOLDING COLUMN EIGENVECTORS.
C N        INTEGER. ORDER OF B.
C N1       INTEGER. FIRST DIMENSION OF Z AS DECLARED IN CALLING PROGRAM.
C M1,M2    INTEGERS: EIGENVECTORS M1 TO M2 WILL BE TRANSFORMED.
C
      DIMENSION B(N2,N),Z(N1,N),DL(N)
      DO 1 J=M1,M2
      DO 1 L=1,N
      I=N+1-L
      Y=DL(I)
      X=Z(I,J)
      IF(I-N) 2,4,4
    2 I1=I+1
      DO 3 K=I1,N
    3 X=X-B(K,I)*Z(K,J)
    4 Z(I,J)=X/Y
    1 CONTINUE
C NORMALIZE EIGENVECTORS
      DO 5 I=M1,M2
      X=Z(1,I)*Z(1,I)
      DO 6 J=1,N
    6 X=X+Z(J,I)*Z(J,I)
      X=SQRT(X)
      DO 5 J=1,N
    5 Z(J,I)=Z(J,I)/X
      RETURN
      END
```

Table A24

```
      SUBROUTINE HTDQL(N,NN,MM,A,D,E,Z,MACHEP,TOL)
C***********************************************************************
C SUBROUTINE HTDQL BASED ON ALGOL PROCEDURES TRED2 BY MARTIN,
C REINSCH AND WILKINSON (1971) AND TQL2 BY BOWDLER, MARTIN,
C REINSCH AND WILKINSON (1971).   INPUT PARAMETERS ARE:
C N        ORDER OF SYMMETRIC MATRIX
C NN       FIRST DIMENSION OF A AS DECLARED IN CALLING PROGRAM
C MM       FIRST DIMENSION OF Z AS DECLARED IN CALLING PROGRAM
C A        SYMMETRIC MATRIX RETURNED UNALTERED UNLESS A=Z IN CALL
C          STATEMENT (SEE WRITEUP). DIMENSION (NN,N) BUT REFER TO
C          WRITEUP FOR DETAILS OF USE WITH LINEAR ARRAYS.
C D        REAL ARRAY, MINIMUM DIMENSION N, CONTAINS EIGENVALUES
C          IN DESCENDING ORDER ON EXIT.
C E        REAL ARRAY, MINIMUM DIMENSION N, USED AS WORK SPACE.
C Z        CONTAINS COLUMN NORMALISED EIGENVECTORS ON EXIT. MAY
C          BE THE SAME AS MATRIX A IN CALL STATEMENT - SEE THE
C          WRITEUP. MINIMUM DIMENSION (MM,N).
C MACHEP   SMALLEST NUMBER FOR WHICH 1+MACHEP>1.
C          (2.0**(-37) ) FOR ICL 1900 MACHINES.) MUST BE OF TYPE REAL.
C TOL      REAL CONSTANT = ETA/MACHEP WHERE ETA IS THE SMALLEST
C          POSITIVE NUMBER REPRESENTABLE IN THE MACHINE (2.0**(-255))
C          FOR ICL 1900 SERIES COMPUTERS) AND MACHEP IS DEFINED ABOVE.
C***********************************************************************
      REAL MACHEP
      DIMENSION A(NN,N),Z(MM,N),D(N),E(N)
      DO 1 I=1,N
      DO 1 J=1,I
    1 Z(I,J)=A(I,J)
```

Table A24   *continued*

```
      I=N+1
      DO 12 IJ=2,N
      I=I-1
      L=I-2
      F=Z(I,I-1)
      G=0.
      IF(L.EQ.0) GOTO 3
      DO 2 K=1,L
   2  G=G+Z(I,K)*Z(I,K)
   3  H=G+F*F
      IF(G.GT.TOL) GO TO 4
      E(I)=F
      H=0.0
      GO TO 11
   4  L=L+1
      G=SQRT(H)
      IF(F.GE.0.0) G=-G
      E(I)=G
      H=H-F*G
      Z(I,I-1)=F-G
      F=0.0
      DO 9 J=1,L
      Z(J,I)=Z(I,J)/H
      G=0.0
      DO 5 K=1,J
   5  G=G+Z(J,K)*Z(I,K)
      J1=J+1
      IF(L.LT.J1) GO TO 8
      DO 6 K=J1,L
   6  G=G+Z(K,J)*Z(I,K)
   8  E(J)=G/H
   9  F=F+G*Z(J,I)
      HH=F/(H+H)
      DO 10 J=1,L
      F=Z(I,J)
      G=E(J)-HH*F
      E(J)=G
      DO 10 K=1,J
  10  Z(J,K)=Z(J,K)-F*E(K)-G*Z(I,K)
  11  D(I)=H
  12  CONTINUE
      D(1)=0.
      E(1)=0.
      DO 16 I=1,N
      L=I-1
      IF(L.EQ.0.OR.D(I).EQ.0.0) GO TO 14
      DO 13 J=1,L
      G=0.0
      DO 123 K=1,L
 123  G=G+Z(I,K)*Z(K,J)
      DO 13 K=1,L
  13  Z(K,J)=Z(K,J)-G*Z(K,I)
  14  D(I)=Z(I,I)
      Z(I,I)=1.0
      IF(L.EQ.0) GOTO 16
      DO 15 J=1,L
      Z(I,J)=0.
  15  Z(J,I)=0.
  16  CONTINUE
      DO 17 I=2,N
  17  E(I-1)=E(I)
      E(N)=0.
      B=0.
      F=0.
      DO 27 L=1,N
      J=0
      H=MACHEP*(ABS(D(L))+ABS(E(L)))
```

Table A24  *continued*

```
      IF(B.LT.H) B=H
      DO 18 M=L,N
      IF(ABS(E(M)).LE.B) GOTO 19
   18 CONTINUE
   19 IF(M-L) 20,27,20
   20 IF(J.EQ.30) GOTO 31
      J=J+1
      G=D(L)
      P=(D(L+1)-G)/(2.0*E(L))
      R=SQRT(P*P+1.0)
      RR=P+R
      IF(P.LT.0.0) RR=P-R
      D(L)=E(L)/RR
      H=G-D(L)
      L1=L+1
      DO 21 I=L1,N
      IF(L.EQ.N) GOTO 22
   21 D(I)=D(I)-H
   22 F=F+H
      P=D(M)
      C=1.0
      S=0.0
      I=M
      LL=M-1
      IF(LL.LT.L) GOTO 26
      DO 25 LI=L,LL
      I=I-1
      G=C*E(I)
      H=C*P
      IF(ABS(P).LT.ABS(E(I))) GO TO 23
      C=E(I)/P
      R=SQRT(C*C+1.0)
      E(I+1)=S*P*R
      S=C/R
      C=1.0/R
      GOTO 24
   23 C=P/E(I)
      R=SQRT(C*C+1.0)
      E(I+1)=S*E(I)*R
      S=1.0/R
      C=C/R
   24 P=C*D(I)-S*G
      D(I+1)=H+S*(C*G+S*D(I))
      DO 25 K=1,N
      H=Z(K,I+1)
      Z(K,I+1)=S*Z(K,I)+C*H
   25 Z(K,I)=C*Z(K,I)-S*H
   26 E(L)=S*P
      D(L)=C*P
      IF(ABS(E(L)).GT.B) GO TO 20
   27 D(L)=D(L)+F
      L1=N-1
      DO 30 I=1,L1
      K=I
      P=D(I)
      J1=I+1
      DO 28 J=J1,N
      IF(P.GT.D(J)) GOTO 28
      P=D(J)
      K=J
   28 CONTINUE
      IF(K.EQ.I) GOTO 30
      D(K)=D(I)
      D(I)=P
      DO 29 J=1,N
      P=Z(J,I)
      Z(J,I)=Z(J,K)
```

Table A24   *continued*

```
   29 Z(J,K)=P
   30 CONTINUE
      RETURN
   31 WRITE(6,32)
   32 FORMAT('0***EXECUTION TERMINATED AFTER 30 ITERATIONS')
      N=0
      RETURN
      END
```

Table A25

```
      SUBROUTINE EIGVEC(C,B,ROOT,VEC,M1,INT,X,ACC,EPS,M,N,IFL)
C
C TO FIND SELECTED EIGENVECTORS OF A SYMMETRIC TRIDIAGONAL MATRIX USING
C INVERSE ITERATION.
C
C ARGUMENTS:
C  C      REAL ARRAY,DIMENSION AT LEAST N, HOLDING DIAGONAL OF THE
C         TRIDIAGONAL MATRIX.
C  B      REAL ARRAY, MINIMUM DIMENSION N,  HOLDING SUBDIAGONAL OF THE
C         TRIDIAGONAL MATRIX.
C  ROOT   REAL ARRAY, MINIMUM DIMENSION M, HOLDING THE FIRST M EIGEN-
C         VALUES OF THE TRIDIAGONAL MATRIX IN DESCENDING ORDER OF
C         MAGNITUDE.
C  VEC    TWO-DIMENSIONAL REAL ARRAY, MINIMUM DIMENSION N*M.
C         UNDEFINED ON ENTRY.
C         CONTAINS COLUMN EIGENVECTORS ON EXIT. THESE ARE NORMALIZED TO
C         UNIT LENGTH.
C  M1     INTEGER. FIRST DIMENSION OF VEC AS DECLARED IN CALLING PROGRAM.
C  INT    LOGICAL ARRAY OF MINIMUM DIMENSION N USED AS WORKSPACE.
C  X      TWO DIMENSIONAL REAL ARRAY, MINIMUM DIMENSION (N,7). WORKSPACE.
C  ACC    SMALLEST POSITIVE REAL CONSTANT SUCH THAT 1.0+ACC>1.0.
C         THIS IS 2^-37 FOR ICL 1900 COMPUTERS AND 2^-24 FOR PDP-11.
C  EPS    SMALLEST POSITIVE REAL NUMBER THAT CAN BE STORED IN THE
C         MACHINE.  THIS IS 2.0^(-255) FOR ICL 1906A AND 2.0^(-124) FOR
C         PDP -11.
C  M      INTEGER, NUMBER OF EIGENVECTORS TO BE COMPUTED.
C  N      INTEGER. ORDER OF TRIDIAGONAL MATRIX.
C  IFL    INTEGER. IF NONZERO ON EXIT INDICATES FAILURE - AN EIGENVECTOR
C         COULD NOT BE ISOLATED AFETR 5 ITERATIONS.
C
C IF MORE THAN 25% OF THE EIGENVALUES AND EIGENVECTORS ARE REQUIRED,
C USE SUBROUTINE HTDQL.
C
      INTEGER R,S,GROUP
      REAL NORM
      LOGICAL INT(N)
      DIMENSION C(N),B(N),ROOT(M),VEC(M1,M),X(N,7)
      R=1
      NORM=ABS(C(1))
      DO 22 I=2,N
   22 NORM=NORM+ABS(C(I))+ABS(B(I))
   81 EPS2=NORM*1.0E-3
      EPS3=ACC*NORM
      EPS4=FLOAT(N)*EPS3
      GROUP=0
      S=1
      DO 30 K=1,M
      ITS=1
      X1=ROOT(R)
      IF(K.EQ.1) GOTO 82
      IF(X1-X0.GE.EPS2) GOTO 25
   24 GROUP=GROUP+1
      GOTO 82
```

## Table A25  *continued*

```
   25 GROUP=0
   82 U=EPS4/SQRT(FLOAT(N))
      DO 27 I=1,N
   27 X(I,7)=U
      U=C(1)-X1
      V=B(2)
      DO 29 I=2,N
      J=I-1
      BI=B(I)
      INT(I)=ABS(BI).GE.ABS(U)
      IF(INT(I)) GOTO 28
      XU=BI/U
      X(I,6)=XU
      X(J,3)=U
      X(J,4)=V
      X(J,5)=0.0
      U=C(I)-X1-XU*V
      IF(I.NE.N) V=B(I+1)
      GOTO 29
   28 XU=U/BI
      X(I,6)=XU
      X(J,3)=BI
      X(J,4)=C(I)-X1
      IF(I.EQ.N) GOTO 43
      X(J,5)=B(I+1)
      GOTO 40
   43 X(J,5)=0.0
   40 U=V-XU*X(J,4)
      V=-XU*X(J,5)
   29 CONTINUE
      X(N,3)=U
      IF(ABS(U).LE.EPS) X(N,3)=EPS3
      X(N,4)=0.0
      X(N,5)=0.0
   47 DO 45 II=1,N
      I=N+1-II
      X(I,7)=(X(I,7)-U*X(I,4)-V*X(I,5))/X(I,3)
      V=U
   45 U=X(I,7)
      IG=R-GROUP
      IH=R-1
      IF(IG.GT.IH) GOTO 90
      DO 48 J=IG,IH
      XU=0.0
      DO 86 I=1,N
   86 XU=XU+X(I,7)*VEC(I,J)
      DO 48 I=1,N
   48 X(I,7)=X(I,7)-XU*VEC(I,J)
   90 NORM=0.0
      DO 49 I=1,N
   49 NORM=NORM+ABS(X(I,7))
      IF(NORM-1.0) 50,51,51
   50 IF(ITS-5) 52,53,52
   53 IFL=1
      RETURN
   52 IF(NORM) 54,55,54
   55 X(S,7)=EPS4
      IF(S-N) 56,57,56
   57 S=1
      GO TO 58
   56 S=S+1
      GO TO 58
   54 XU=EPS4/NORM
      DO 59 I=1,N
   59 X(I,7)=X(I,7)*XU
   58 DO 60 I=2,N
      J=I-1
```

Table A25 *continued*

```
      IF(INT(I)) GOTO 33
      X(I,7)=X(I,7)-X(I,6)*X(J,7)
      GOTO 60
   33 U=X(J,7)
      X(J,7)=X(I,7)
      X(I,7)=U-X(I,6)*X(J,7)
   60 CONTINUE
      ITS=ITS+1
      GOTO 47
   51 U=0.0
      DO 64 I=1,N
      U=U+X(I,7)**2
   64 XU=1.0/SQRT(U)
      DO 65 I=1,N
   65 VEC(I,R)=X(I,7)*XU
   72 R=R+1
   30 X0=X1
   70 RETURN
      END
```

Table A26

```
      SUBROUTINE TRBAK1(Z,N1,N,M,A,N2,E)
C
C TO RECOVER THE EIGENVECTORS OF THE ORIGINAL MATRIX FROM THE
C EIGENVECTORS OF THE TRIDIAGONAL FORM, AS OUTPUT BY EIGVEC.
C
C ARGUMENTS:
C Z       TWO-DIMENSIONAL REAL ARRAY, MINIMUM DIMENSION N,M. ON ENTRY
C         HOLDS THE EIGENVECTORS OF THE TRIDIAGONAL MATRIX, AS OUTPUT
C         BY EIGVEC.
C N1      INTEGER, FIRST DIMENSION OF Z AS DECLARED IN CALLING PROGRAM.
C N       INTEGER. ORDER OF TRIDIAGONAL MATRIX.
C M       INTEGER. NUMBER OF EIGENVECTORS.
C         DETAILS OF THE HOUSEHOLDER REDUCTION AS PERFORMED BY TRIDI.
C A       TWO-DIMENSIONAL REAL ARRAY, MINIMUM DIMENSION N,N.  HOLDS
C N2      INTEGER. THE FIRST DIMENSION OF A IN THE CALLING PROGRAM.
C E       REAL ARRAY, MINIMUM DIMENSION N, HOLDING THE SUBDIAGONAL OF THE
C         TRIDIAGONAL MATRIX.
C
      DIMENSION Z(N1,M),A(N2,N),E(N)
      DO 1 I=2,N
      IF(E(I)) 2,1,2
    2 L=I-1
      H=E(I)*A(I,L)
      DO 3 J=1,M
      S=0.0
      DO 4 K=1,L
    4 S=S+A(I,K)*Z(K,J)
      S=S/H
      DO 3 K=1,L
    3 Z(K,J)=Z(K,J)+S*A(I,K)
    1 CONTINUE
      RETURN
      END
```

Table A27

```
      SUBROUTINE DT2(A,M,N,P,DET)
C
C SUBROUTINE DT2
```

Table A27 *continued*

```
C DETERMINANT EVALUATION AND CHOLESKY FACTORIZATION OF A
C            NONSINGULAR SYMMETRIC MATRIX.
C BASED ON CHOLDET1, CONTRIBUTION I/1 TO WILKINSON AND REINSCH(1971) BY
C MARTIN, PETERS AND WILKINSON.
C *** INPUT PARAMETERS ***
C    A     SYMMETRIC MATRIX WHOSE DETERMINANT IS REQUIRED. ONLY THE UPPER
C          TRIANGLE + DIAGONAL ARE NEEDED.
C    M     FIRST DIMENSION OF A AS DECLARED IN CALLING ROUTINE.
C    N     ORDER OF A
C    P     DOUBLE PRECISION WORKSPACE VECTOR OF DIMENSION N.
C    DET   UNDEFINED ON ENTRY. REAL VARIABLE.
C *** OUTPUT ***
C    A     UPPER TRIANGLE AND DIAGONAL UNALTERED. THE STRICT LR. TRIANGLE
C          CONTAINS THE MATRIX L IN THE CHOLESKY FACTORIZATION A=LU.
C    M     UNALTERED BY THE ROUTINE.
C    N     UNALTERED BY THE ROUTINE EXCEPT IN CASE OF ERROR (SEE BELOW).
C    P     CONTAINS THE RECIPROCALS OF THE DIAGONAL ELEMENTS OF THE
C          MATRIX L IN THE CHOLESKY FACTORIZATION A=LU.
C    DET   CONTAINS THE VALUE OF THE DETERMINANT OF A UNLESS ERROR HAS
C          OCCURRED.
C *** ERROR MESSAGES ***
C THERE ARE NO ERROR MESSAGES. IF MATRIX A IS SINGULAR TO WORKING
C          ACCURACY N IS SET TO 0 AND CONTROL IS RETURNED TO THE CALLING
C          ROUTINE.
C
      INTEGER D2
      DOUBLE PRECISION D1, X, P, PROD
      DIMENSION A(M, N), P(N)
      PROD(C, D)=DBLE(C)*DBLE(D)
      D1=1. D0
      D2=0
      DO 1 I=1, N
      I1=I-1
      DO 1 J=I, N
      X=A(I, J)
      IF(I. EQ. 1) GOTO 8
      DO 2 K=1, I1
    2 X=X-PROD(A(J, K), A(I, K))
    8 IF(J. NE. I) GOTO 7
      D1=D1*X
      IF(X) 3, 6, 3
    6 N=0
      RETURN
    3 IF(DABS(D1). LT. 1. D0) GOTO 4
      D1=D1*0. 625D-1
      D2=D2+4
      GOTO 3
    4 IF(DABS(D1). GE. 0. 625D-1) GOTO 5
      D1=D1*16. D0
      D2=D2-4
      GOTO 4
    5 IF(X. LT. 0. D0) GOTO 6
      P(I)=1. D0/DSQRT(X)
      GOTO 1
    7 A(J, I)=X*P(I)
    1 CONTINUE
      X=D1*2. D0**D2
      DET=SNGL(X)
      RETURN
      END
```

Table A28

```
      SUBROUTINE SOLVE(A,IA,N,DET,P,X,B)
C
C SUBROUTINE SOLVE
C SOLVES LINEAR SIMULTANEOUS EQUATIONS WHEN COEFFICIENTS MATRIX IS
C SYMMETRIC, USING CHOLESKY METHOD.
C REFERENCE: CONTRIBUTION I/1, WILKINSON AND REINSCH (1971)
C ARGUMENTS:
C A        REAL ARRAY, MINIMUM DIMENSION (N*N) HOLDING COEFFICIENTS MATRIX
C          SUBDIAGONAL ELEMENTS NEED NOT BE SET.
C IA       FIRST DIMENSION OF A IN CALLING PROGRAM. INTEGER.
C N        INTEGER. ORDER OF A.
C DET      REAL VARIABLE. UNDEFINED ON ENTRY. HOLDS DET(A) ON EXIT UNLESS
C          FAILURE HAS OCCURRED.
C P        DOUBLE PRECISION WORKSPACE VECTOR OF DIMENSION N
C X        REAL VECTOR, DIMENSION N, HOLDS SOLUTION ON EXIT.
C B        REAL VECTOR OF RIGHT-HAND SIDES. DIMENSION AT LEAST N.
C ERROR MESSAGES
C          NONE. IF A IS SINGULAR, N IS SET TO 0 AND CONTROL IS RETURNED.
C
C REQUIRES SUBROUTINES DT2 AND DPACC (APPENDIX A).
C
      DOUBLE PRECISION P
      DIMENSION A(IA,N),P(N),X(N),B(N)
      CALL DT2(A,IA,N,P,DET)
      IF(N.EQ.0) RETURN
      L=IA*N
      DO 1 I=1,N
      I1=I+IA
      I2=I-1
      CALL DPACC(A,I,I1,L,X,1,2,N,B(I),Y,I2,2)
    1 X(I)=Y*P(I)
      I=N+1
      DO 2 J=1,N
      I=I-1
      I0=I+1
      I4=N-I
      I1=(I-1)*IA+I+1
      I2=I1+1
      I3=I0+1
      CALL DPACC(A,I1,I2,L,X,I0,I3,N,X(I),Y,I4,2)
    2 X(I)=Y*P(I)
      RETURN
      END
```

Table A29

```
      SUBROUTINE CHOLS(A,N1,N,M)
C
C SUBROUTINE CHOLS.
C *** REQUIRES NO USER-SUPPLIED SUBROUTINES ***
C DOUBLE PRECISION VERSION OF CHOLESKY MATRIX INVERSION.
C BASED ON WILKINSON AND REINSCH, 1971, LINEAR ALGEBRA, SPRINGER-VERLAG.
C *** PARAMETERS ***
C A        MATRIX OF DIMENSION N1 BY N. HOLDS UPPER TRIANGLE OF THE
C          SYMMETRIC MATRIX TO BE INVERTED (WHICH IS RETURNED UNCHANGED)
C          AND THE LOWER TRIANGLE + PRINCIPAL DIAGONAL OF THE INVERSE.
C N1       INTEGER, EQUALS N+1
C N        ORDER OF A
C M        FIRST DIMENSION OF A AS DECLARED IN CALLING PROGRAM.
CC ***     ERROR MESSAGES ***
C THERE ARE NO ERROR MESSAGES.
C IF A IS SINGULAR, N IS SET TO 0 AND CONTROL IS RETURNED TO THE CALLING
C          PROGRAM.
C
```

Table A29  *continued*

```
      DOUBLE PRECISION VAL,Y,PROD
      DIMENSION A(M,N)
      PROD(C,D)=DBLE(C)*DBLE(D)
      DO 1 I=1,N
      DO 1 J=I,N
      VAL=A(I,J)
      IF(I.EQ.1) GOTO 4
      I1=I-1
      DO 2 K=1,I1
    2 VAL=VAL-PROD(A(I+1,K),A(J+1,K))
    4 IF(J.GT.I) GOTO 5
      IF(VAL) 11,11,3
    3 Y=1.D0/DSQRT(VAL)
      A(I+1,I)=SNGL(Y)
      GOTO 1
    5 A(J+1,I)=SNGL(VAL*Y)
    1 CONTINUE
      I1=N-1
      DO 7 I=1,I1
      I2=I+1
      DO 7 J=I2,N
      VAL=0.D0
      J1=J-1
      DO 8 K=I,J1
    8 VAL=VAL+PROD(A(J+1,K),A(K+1,I))
    7 A(J+1,I)=SNGL(VAL*DBLE(-A(J+1,J)))
      DO 9 I=1,N
      DO 9 J=I,N
      VAL=0.D0
      J1=J+1
      DO 10 K=J1,N1
   10 VAL=VAL+PROD(A(K,J),A(K,I))
    9 A(J+1,I)=SNGL(VAL)
      GOTO 12
   11 N=0
   12 RETURN
      END
```

Table A30

```
      SUBROUTINE ACCINV(A,B,Z,N1,N,L,M,M1,EPS)
C SUBROUTINE ACCINV
C THIS SUBROUTINE FINDS THE INVERSE OF A NONSINGULAR SYMMETRIC MATRIX
C         BY THE METHOD OF ITERATIVE REFINEMENT.
C BASED ON WILKINSON AND REINSCH,1971, LINEAR ALGEBRA,SPRINGER-VERLAG.
C FORTRAN VERSION BY P. M. MATHER
C *** INPUT PARAMETERS ***
C    A      MATRIX WITH N1 ROWS AND N COLUMNS. THE UPPER TRIANGLE + PRINCIP
C           -AL DIAGONAL OF THE SYMMETRIC MATRIX TO BE INVERTED ARE
C           PLACED IN THE UPPER TRIANGLE OF A, I.E. LOCATIONS 1-N FOR THE
C           FIRST ROW, 2-N FOR THE SECOND, AND SO ON.
C    B      AN N * N MATRIX. WORKSPACE.
C    Z      A WORKSPACE VECTOR OF DIMENSION N.
C    N1     INTEGER, EQUALS N+1.
C    N      ORDER OF MATRIX TO BE INVERTED.
C    L      INTEGER. UNDEFINED ON ENTRY.
C    M      FIRST DIMENSION OF A.
C    M1     FIRST DIMENSION OF B.
C    EPS    SMALLEST NUMBER SUCH THAT 1+EPS=1. THIS IS 2.0**(-24) ON PDP11
C *** OUTPUT PARAMETERS ***
C THE ONLY INPUT PARAMETERS ALTERED BY THE ROUTINE ARE:
C    A      THE COMPUTED INVERSE IS PLACED IN THE LOWER TRIANGLE, LEAVING
C           THE UPPER TRIANGLE UNCHANGED. THE INVERSE IS HELD IN LOCATIONS
C           A(2,1), A(3,1),A(3,2),A(4,1) ... ... A(N1,N).
```

Table A30   *continued*

```
C   L    NUMBER OF ITERATIONS ACCOMPLISHED.
C *** ERROR MESSAGES ***
C NO ERROR MESSAGES ARE PRINTED BUT THE FOOLWING ERRORS ARE FLAGGED:
C   N1 = 0 ON RETURN INDICATES THAT THE MAX. CORRECTION AT ITERATION L
C          IS MORE THAN HALF OF THE PREVIOUS MAXIMUM. THIS MEANS THAT THE
C          ITERATIVE PROCESS IS DIVERGING (A RARE EVENT!) OR THAT EPS HAS
C          BEEN SET TOO LOW.
C   N = 0 INDICATES THAT THE MATRIX IS SINGULAR.
C *** NEEDS SUBROUTINE CHOLS***
C
      DOUBLE PRECISION VAL,PROD
      DIMENSION A(M,N),B(M1,N),Z(N)
      PROD(C,D)=DBLE(C)*DBLE(D)
      E=1.0
      L=0
      CALL CHOLS(A,N1,N,M)
      IF(N) 21,21,100
  100 DO 12 I=1,N
      DO 12 J=1,N
      VAL=0.D0
      IF(J-I) 7,1,13
    1 VAL=1.D0
   13 DO 2 K=1,I
    2 VAL=VAL-PROD(A(K,I),A(J+1,K))
      IF(I-J) 22,4,4
   22 I1=I+1
      DO 3 K=I1,J
    3 VAL=VAL-PROD(A(I,K),A(J+1,K))
    4 IF(J-N) 23,12,12
   23  J1=J+1
      DO 5 K=J1,N
    5 VAL=VAL-PROD(A(I,K),A(K+1,J))
      GOTO 12
    7 DO 8 K=1,J
    8 VAL=VAL-PROD(A(K,I),A(J+1,K))
      J1=J+1
      DO 9 K=J1,I
    9 VAL=VAL-PROD(A(K,I),A(K+1,J))
      IF(I-N) 24,12,12
   24 I1=I+1
      DO 11 K=I1,N
   11 VAL=VAL-PROD(A(I,K),A(K+1,J))
   12 B(I,J)=SNGL(VAL)
      XMAX=0.0
      ZMAX=0.0
      DO 18 I=1,N
      DO 17 J=1,I
      VAL=0.D0
      DO 15 K=1,I
   15 VAL=VAL+PROD(A(I+1,K),B(K,J))
      IF(I-N) 25,17,17
   25 I1=I+1
      DO 16 K=I1,N
   16 VAL=VAL+PROD(A(K+1,I),B(K,J))
   17 Z(J)=SNGL(VAL)
      DO 18 J=1,I
      C=ABS(A(I+1,J))
      D=ABS(Z(J))
      IF(C.GT.XMAX) XMAX=C
      IF(D.GT.ZMAX) ZMAX=D
   18 A(I+1,J)=A(I+1,J)+Z(J)
      L=L+1
      D=ZMAX/XMAX
      IF(D.GT.E/2.0) GOTO 20
      E=D
      IF(D.GT.2.0*EPS) GOTO 100
      RETURN
```

## Table A30 *continued*

```
   20 N1=0
   21 RETURN
      END
```

## Table A31

```
      SUBROUTINE GJDEF1(N,A,H,IROW)
C
C INVERSION OF POSITIVE-DEFINITE MATRICES BY THE GAUSS-JORDAN METHOD
C BY F. L. BAUER AND C REINSCH. CONTRIBUTION I/3, WILKINSON AND REINSCH,
C 1971. FORTRAN TRANSLATION BY P. M. MATHER.
C SUBROUTINE GJDEF1                       INPUT PARAMETERS:
C   N    ORDER OF A
C   A    MATRIX TO BE INVERTED. LOWER TRIANGLE + DIAGONAL ONLY.
C   H    WORKSPACE VECTOR OF LENGTH N.
C   IROW FIRST DIMENSION OF A AS DECLARED IN CALLING PROGRAM.
C OUTPUT FROM SUBROUTINE:
C   A    INVERSE MATRIX. ELEMENTS ABOVE THE PRINCIPAL DIAGONAL ARE
C        NOT ALTERED FROM INPUT MATRIX
C FAILURE MESSAGES:
C   'FAIL IN GJDEF1 - MATRIX SINGULAR'. THE PARAMETER N IS SET TO 0
C        AND CONTROL IS RETURNED TO THE CALLING PROGRAM.
C        OUTPUT DEVICE ASSUMED TO HAVE LOGICAL NO. 6.
C
      DOUBLE PRECISION P,Q,R,Z
      DIMENSION A(IROW,N),H(N)
      L=N+1
      DO 1 K=1,N
      L=L-1
      P=A(1,1)
      IF(P.LE.0.0) GOTO 10
      DO 2 I=2,N
      Q=A(I,1)
      Z=Q
      IF(I.LE.L) Z=-Q
      H(I)=Z/P
      DO 2 J=2,I
      Z=A(I,J)
      R=H(J)
    2 A(I-1,J-1)=SNGL(Z+Q*R)
      A(N,N)=SNGL(1.D0/P)
      DO 1 I=2,N
    1 A(N,I-1)=H(I)
      RETURN
   10 WRITE(6,11)
   11 FORMAT('0***FAIL IN GJDEF1 - MATRIX SINGULAR')
      N=0
      RETURN
      END
```

## Table A32

```
      SUBROUTINE CROUT(N,A,EPS,DET,FINT,B,R,IA,IB,IFL,ONLY)
C
C SUBROUTINE TO SOLVE NON-SYMMETRIC SYSTEM OF LINEAR EQUATIONS, INVERT A
C NON-SYMMETRIC MATRIX OR FIND DDETERMINANT OF NONSYMMETRIC MATRIX.
C ARGUMENTS
C   N    ORDER OF COEFFICIENTS MATRIX, MATRIX TO BE INVERTED OR WHOSE
C        DETERMINANT IS TO BE EVALUATED.
C   A    COEFFICIENTS MATRIX, OR MATRIX WHOSE INVERSE OR DETERMINANT
C        IS REQUIRED. DIMENSION (N*N).
C   EPS  CONSTANT SUCH THAT 1.0+EPS>1.0
```

Table A32   *continued*

```
C DET       DETERMINANT OF A
C FINT      REAL ARRAY, DIMENSION N, USED AS WORKSPACE.
C B         MATRIX OF RIGHT-HAND SIDES, IF EQUATIONS TO BE SOLVED,
C           OTHERWISE NOT DEFINED. DIMENSION (N*R) IF EQUATIONS TO BE
C           SOLVED, (N*N) IF INVERSE TO BE EVALUATED OR (1*1) IF DETER-
C           MINANT ONLY IS REQUIRED.
C IA        FIRST DIMENSION OF A AS DECLARED IN CALLING PROGRAM.
C IB        FIRST DIMENSION OF B AS DECLARED IN CALLING PROGRAM.
C IFL       SET TO 1 ON ENTRY; RETURNS AS 0 IF ERROR HAS OCCURRED.
C ONLY      INTEGER; ON ENTRY HAS ONE OF THE FOLLOWING VALUES :-
C              -1 IF INVERSE OF A IS TO BE FOUND AND RETURNED IN B;
C               0 IF EQUATIONS ARE TO BE SOLVED AND SOLUTIONS RETURNED IN THE
C                  COLUMNS OF B;
C               1 IF DETERMINANT OF A IS ALL THAT IS REQUIRED.
C OUTPUT    IS DEFINED ABOVE; NOTE B CONTAINS INVERSE OF A OR SOLUTIONS
C             OF EQUATIONS DEPENDING ON VALUE OF *ONLY*. MATRIX A IS OVER-
C             WRITTEN. CHECK VALUE OF IFL ON EXIT.
C
C SUBROUTINES NEEDED: DPACC (APPENDIX A).
      DOUBLE PRECISION D1
      INTEGER D2, ONLY, R
      DIMENSION A(IA,N), B(IB,N), FINT(N)
      IFL=1
      IF(ONLY) 2,20,20
    2 R=N
      DO 21 I=1,N
      DO 22 J=1,N
   22 B(I,J)=0.0
   21 B(I,I)=1.0
   20 KK=IA*N
      D1=1.D0
      D2=0
      DO 1 I=1,N
      I1=I
      I2=I1+IA
      CALL DPACC(A, I1, I2, KK, A, I1, I2, KK, 0.0, Y, N, 1)
    1 FINT(I)=1.0/SQRT(Y)
      DO 9 K=1,N
      LL=K-1
      L=K
      K1=K+1
      X=0.0
      DO 3 I=K,N
      I1=I
      I2=I1+IA
      I3=LL*IA+1
      I4=I3+1
      CALL DPACC(A, I1, I2, KK, A, I3, I4, KK, A(I,K), Y, LL, 2)
      A(I,K)=Y
      Y=ABS(Y*FINT(I))
      IF(Y.LE.X) GOTO 3
      X=Y
      L=I
    3 CONTINUE
      IF(L.EQ.K) GOTO 4
      D1=-D1
      DO 5 J=1,N
      Y=A(K,J)
      A(K,J)=A(L,J)
    5 A(L,J)=Y
      FINT(L)=FINT(K)
    4 FINT(K)=L
      D1=D1*A(K,K)
      IF(X.GE.8.0*EPS) GOTO 6
      IFL=0
      WRITE(6,23)
   23 FORMAT('0*** FAIL IN CROUT - MATRIX SINGULAR')
```

Table A32  *continued*

```
      RETURN
    6 IF(DABS(D1).LT.1.D0) GOTO 7
      D1=D1*.625D-1
      D2=D2+4
      GOTO 6
    7 IF(DABS(D1).GE..625D-4) GOTO 8
      D1=D1*.16D2
      D2=D2-4
      GOTO 7
    8 X=-1.0/A(K,K)
      IF(K.EQ.N) GOTO 9
      DO 10 J=K1,N
      I1=K
      I2=I1+IA
      I3=(J-1)*IA+1
      I4=I3+1
      CALL DPACC(A,I1,I2,KK,A,I3,I4,KK,A(K,J),Y,LL,3)
   10 A(K,J)=X*Y
    9 CONTINUE
      DET=SNGL(D1*2.0D0**D2)
      IF(ONLY.EQ.1) RETURN
      L=IB*N
      DO 11 I=1,N
      J=INT(FINT(I)+0.5)
      IF(J.EQ.I) GOTO 11
      DO 13 K=1,R
      X=B(I,K)
      B(I,K)=B(J,K)
   13 B(J,K)=X
   11 CONTINUE
   12 DO 14 K=1,R
      K1=K-1
      DO 15 I=1,N
      I0=I-1
      I1=I
      I2=I1+IA
      I3=K1*IB+1
      I4=I3+1
      CALL DPACC(A,I1,I2,KK,B,I3,I4,L,B(I,K),X,I0,4)
   15 B(I,K)=X/A(I,I)
      I=N+1
      DO 16 I5=1,N
      I=I-1
      I0=N-I
      I1=I*IA+I
      I2=I1+N
      I3=K1*IB+I+1
      I4=I3+1
      CALL DPACC(A,I1,I2,KK,B,I3,I4,L,B(I,K),X,I0,4)
   16 B(I,K)=X
   14 CONTINUE
      RETURN
      END
```

Table A33

```
      SUBROUTINE NSEIG(A,N,MM,T,NN,IND,NITS)
C SUBROUTINE TO DETERMINE EIGENVALUES AND COLUMN EIGENVECTORS
C OF A NONSYMMETRIC MATRIX USING A JACOBILIKE PROCEDURE.  BASED ON
C ALGOL 60 PROCEDURE EIGEN BY EBERLEIN AND BOOTHROYD (1971).
C A = NONSYMMETRIC MATRIX ON INPUT.  ON EXIT THE REAL EIGENVALUES
C     ARE STORED IN DESCENDING ORDER IN THE PRINCIPAL DIAGONAL.
C     SEE EBERLEIN & BOOTHROYD FOR THE COMPLEX CASE.
C T = UNDEFINED ON ENTRY (NEED NOT BE ZEROED).  ON EXIT HOLDS THE
C     EIGENVECTORS, NORMALIZED TO UNIT LENGTH, BY COLUMN.
```

Table A33   *continued*

```
C      SEE NOTE BELOW RE. VALUE OF IND.
C MM   FIRST DIMENSION OF A AS DECLARED IN CALLING PROGRAM.
C NN FIRST DIMENSION OF T AS DECLARED IN CALLING PROGRAM.
C N = ORDER OF A.
C IND = 1 FOR EIGENVECTORS CALCULATED, 0 IF NOT. IF IND = 0 THEN
C T AND A CAN BE THE SAME ARRAY IN THE CALL STATEMENT.
C IF IND < 0 OR IND = NITS ON EXIT THEN CONVERGENCE HAS NOT OCCURRED.
C NITS = NUMBER OF ITERATIONS TO BE PERFORMED.
       LOGICAL MARK
       DIMENSION A(MM,N),T(NN,N)
       MARK=.FALSE.
       IF (IND) 2,2,22
   22  DO 1 I=1,N
       DO 1 J=1,N
    1  T(I,J)=0.0
       DO 25 I=1,N
   25  T(I,I)=1.0
    2  EP=1.E-10
    3  EPS=SQRT(EP)
       N1=N-1
       DO 20 IT=1,NITS
       IF(.NOT.MARK) GO TO 4
       IND=1-IT
       GOTO 21
    4  DO 5 I=1,N1
       J1=I+1
       DO 5 J=J1,N
       IF(ABS(A(I,J)+A(J,I)).GT.EPS) GO TO 6
       IF(ABS(A(I,J)-A(J,I)).GT.EPS.AND.ABS(A(I,I)-A(J,J)).GT.EPS) GOTO 6
    5  CONTINUE
       IND=IT-1
       GO TO 21
    6  MARK=.TRUE.
       DO 20 K=1,N1
       K1=K+1
       DO 20 M=K1,N
       H=0.0
       G=0.0
       HJ=0.0
       YH=0.0
       DO 8 I=1,N
       TE=A(I,K)*A(I,K)
       TEE=A(I,M)*A(I,M)
       YH=YH+TE-TEE
       IF(I.NE.K.AND.I.NE.M) GOTO 7
       GOTO 8
    7  H=H+A(K,I)*A(M,I)-A(I,K)*A(I,M)
       TEP=TE+A(M,I)*A(M,I)
       TEM=TEE+A(K,I)*A(K,I)
       G=G+TEP+TEM
       HJ=HJ-TEP+TEM
    8  CONTINUE
       H=H+H
       D=A(K,K)-A(M,M)
       C=A(K,M)+A(M,K)
       E=A(K,M)-A(M,K)
       IF(ABS(C)-EP) 9,9,10
    9  CX=1.0
       SX=0.0
       GOTO 11
   10  COT2X=D/C
       SIG=1.0
       IF(COT2X.LT.0.0) SIG=-1.0
       COTX=COT2X+(SIG*SQRT(1.0+COT2X*COT2X))
       SX=SIG/SQRT(1.0+COTX*COTX)
       CX=SX*COTX
   11  IF(YH) 12,13,13
```

Table A33   *continued*

```
   12 TEM=CX
      CX=SX
      SX=-TEM
   13 COS2X=CX*CX-SX*SX
      SIN2X=2.0*SX*CX
      D=D*COS2X+C*SIN2X
      H=H*COS2X-HJ*SIN2X
      DEN=G+2.0*(E*E+D*D)
      TANHY=(E*D-H/2.0)/DEN
      IF(ABS(TANHY)-EP) 14,14,15
   14 CHY=1.0
      SHY=0.0
      GOTO 16
   15 CHY=1.0/SQRT(1.0-TANHY*TANHY)
      SHY=CHY*TANHY
   16 C1=CHY*CX-SHY*SX
      C2=CHY*CX+SHY*SX
      S1=CHY*SX+SHY*CX
      S2=-CHY*SX+SHY*CX
      IF(ABS(S1).LE.EP.OR.ABS(S2).LE.EP) GO TO 20
      MARK=.FALSE.
      DO 17 I=1,N
      AKI=A(K,I)
      AMI=A(M,I)
      A(K,I)=C1*AKI+S1*AMI
   17 A(M,I)=S2*AKI+C2*AMI
   18 DO 19 I=1,N
      AIK=A(I,K)
      AIM=A(I,M)
      A(I,K)=C2*AIK-S2*AIM
      A(I,M)=-S1*AIK+C1*AIM
      IF(IND.EQ.0) GOTO 19
      TIK=T(I,K)
      TIM=T(I,M)
      T(I,K)=C2*TIK-S2*TIM
      T(I,M)=-S1*TIK+C1*TIM
   19 CONTINUE
   20 CONTINUE
      IND=NITS
      RETURN
   21 K=1
      KK=1
      K2=K+1
   26 Z=A(K,K)
      DO 30 I=K2,N
      IF(A(I,I).LE.Z) GOTO 30
      K=I
   30 CONTINUE
      IF(K.EQ.KK) GOTO 40
      DO 31 I=1,N
      Z=T(I,K)
      T(I,K)=T(I,KK)
   31 T(I,KK)=Z
      Z=A(KK,KK)
      A(KK,KK)=A(K,K)
      A(K,K)=Z
   40 K=K2
      IF(K.EQ.N) RETURN
      KK=K
      K2=K+1
      END
```

## Table A34

```
      SUBROUTINE GSORTH(X,V,C,D,N,M,MM,NULL,K1,K2,I2,EPS)
C
C MODIFIED GRAM-SCHMIDT ROUTINE.
C ARGUMENTS:
C X        TWO-DIMENSIONAL ARRAY. MINIMUM DIMENSION (N*M). HOLDS MATRIX
C          TO BE ORTHOGONALIZED. UNALTERED BY THE ROUTINE.
C V        TWO-DIMENSIONAL ARRAY, MINIMUM DIMENSION (N*M). UNDEFINED ON
C          ENTRY. HOLDS ORTHOGONAL VECTORS COLUMNWISE ON EXIT.
C C        ARRAY, DIMENSION NM, UNDEFINED ON ENTRY, HOLDS LOWER TRIANGULAR
C          MATRIX OF COEFFICIENTS ON EXIT.
C D        ARRAY, DIMENSION M, WORKSPACE.
C N        NUMBER OF ROWS IN MATRIX X.
C M        NUMBER OF COLUMNS IN MATRIX X.
C MM       DIMENSION OF ARRAY C, = M*(M+1)/2+M
C NULL     NULL IS INITIALLY SET TO ZERO BY THE ROUTINE AND IS INCREMEN-
C          TED BY 1 FOR EVERY PAIR OF COLLINEAR VECTORS.
C K1,K2    INTEGERS DEFINING THE FIRST AND LAST COLUMNS TO BE ORTHOGONAL-
C          ISED. FOR EXAMPLE, IF K1=1 AND K2=5 THEN COLUMNS 1-5 OF X WILL
C          BE ORTHOGONALISED. ON A SUBSEQUENT CALL (PROVIDED C AND D ARE
C          UNALTERED) K1 CAN BE SET TO 6 AND K2 TO M TO COMPLETE THE
C          GRAM SCHMIDT PROCESS.  SEE PROGRAMS FOR POLYNOMIAL CURVE
C          FITTING AND TREND SURFACE ANALYSIS IN CHAPTER 3 FOR EXAMPLES
C          OF THE SEQUENTIAL USE OF THE ROUTINE.
C I2       INTEGER. SET TO 0 ON ENTRY. DO NOT ALTER BETWEEN CALLS TO THE
C          ROUTINE. MUST BE CALLED BY NAME.
C EPS      CRITERION TO DETECT COLLINEARITY.
C
C OTHER ROUTINES USED: NONE
C ERROR MESSAGES: NONE
C FOR COMPUTATIONAL DETAILS SEE CHAPTER 3.
C
      DOUBLE PRECISION D1,D2,D3,D4,D5
      DIMENSION V(N,M),X(N,M),D(M),C(MM)
      NULL=0
      DO 20 IO=K1,K2
      D5=0.D0
      DO 21 J=1,N
      D1=DBLE(X(J,IO))
      V(J,IO)=SNGL(D1)
   21 D5=D5+D1*D1
      DO 20 I1=1,IO
      D1=0.D0
      DO 22 J=1,N
      D2=DBLE(V(J,IO))
      D3=DBLE(V(J,I1))
   22 D1=D1+D2*D3
      IF(I1.LT.IO) GOTO 23
      RATIO=SNGL(D1/D5)
      IF(ABS(RATIO).GT.EPS) GOTO 30
      DO 27 J=1,N
   27 V(J,IO)=0.0
      NULL=NULL+1
      D1=0.D0
   30 D(IO)=SNGL(D1)
      D2=DBLE(D(I1))
      D4=DSQRT(D2)
      GOTO 26
   23 D3=0.D0
      D2=DBLE(D(I1))
      IF(D2.EQ.0.D0) GOTO 25
      D3=D1/D2
      D4=D3*DSQRT(D2)
   25 DO 24 J=1,N
      D1=DBLE(V(J,IO))
      D2=DBLE(V(J,I1))
      D5=D1-D3*D2
   24 V(J,IO)=SNGL(D5)
   26 I2=I2+1
      C(I2)=SNGL(D4)
   20 CONTINUE
      I2=I2-K2
      RETURN
      END
```

# Bibliography

Abrahams, A. D. (1972). Drainage densities and sediment yeilds in Eastern Australia. *Australian Geog. Studies*, 10, 19–41.
Acton, F. S. (1970). *Numerical methods that work*. Harper and Row, New York.
Agterberg, F. P. (1967). Computer techniques in geology. *Earth-Science Reviews*, 3, 47–77.
Agterberg, F. P. (1974). *Geomathematics*. Elsevier Publishing Co., Amsterdam.
Agterberg, F. P. and P. Cabilio (1969). Two-stage least-squares model for the relationship between mappable geologic variables. *Journ. Intern. Assoc. Mathem. Geol.*, 1, 137–153.
Aigner, D. J. (1970). *Basic econometrics*. Prentice-Hall, Englewood Cliffs, New Jersey.
Anderberg, M. R. (1973). *Cluster analysis for applications*. Academic Press, New York.
Anderson, A. J. B. (1971A). Numeric examination of multivariate soil samples. *Journ. Intern. Assoc. Mathem. Geol.*, 3, 1–14.
Anderson, A. J. B. (1971B). Ordination methods in ecology. *Journ. of Ecology*, 59, 713–726.
Anderson, T. W. and H. Rubin (1956). Statistical inference in factor analysis. In: J. Neyman (ed.), *Proc. Sympos. Mathem. Statist. and Probability*, vol. 5, Berkeley, Univ. California Press, pp. 111–150.
Anderssen, R. S. and M. R. Osborn (eds.) (1969). *Least-squares methods in data analysis*. Australian National Univ., A.N.U. Computer Centre Publication CC2/69, Univ. of Queensland Press.
Anderssen, R. S. and M. R. Osborne (eds.) (1970). *Data representation*. Univ. of Queensland Press.
Andrews, D. F., R. Gnanadesikan and J. L. Warner (1973). Methods for assessing multivariate normality. In: P. K. Krishnaiah (ed.) (1973), pp. 95–116.
Anscombe, F. J. (1960). Rejection of outliers. *Technometrics*, 2, 123–147.
Anscombe, F. J. (1961). Examination of residuals. In: J. Neyman (ed.), *Proc. 4th Berkeley Sympos. on Mathem. Statist. and Probability*. Berkeley, University of California Press.
Anscombe, F. J. and J. W. Tukey (1963). The analysis and examination of residuals. *Technometrics*, 5, 141–160.
Archer, C. O. and R. I. Jennrich (1973). Standard errors for orthogonally rotated factor loadings. *Psychometrika*, 38, 581–592.
Armstrong, J. S. (1967). Derivation of theory by means of factor analysis, or Tom Swift and his electric factor analysis machine. *The American Statistician*, December 1967, 17–21.

Armstrong, J. S. and P. Soelberg (1968). On the interpretation of factor analysis. *Psychol. Bull.*, 70, 361—364.

Bagnold, R. A. (1954). *The physics of blown sand and desert dunes.* Methuen, London.

Bard, Y. (1974). *Nonlinear parameter estimation.* Academic Press, New York.

Barkham J. P. and J. M. Norris (1970). Multivariate procedures in an investigation of vegetation and soil relations of two beech woodlands, Cotswold Hills, England. *Ecology,* 51, 630—639.

Bartlett, M. S. (1953). Factor analysis in psychology as a statistician sees it. *Uppsala Sympos. on Psychol. Factor Analysis.* Almqvist and Wiksell, Uppsala, pp. 23—34.

Beale, E. M. L. (1969). Cluster analysis. *Scientific Control Systems Ltd.*, Berners Street, London.

Bensen, M. A. (1962). Factors influencing the occurrence of floods in a humid region of diverse terrain. *Water Supply Paper 1580-B, U.S. Geological Survey,* Washington, D.C.

Berry, B. J. L. and D. F. Marble (eds.) (1968). *Spatial analysis.* Prentice-Hall Inc., Englewood Cliffs, New Jersey.

Berztiss, A. T. (1964). Least-squares fitting of polynomials to irregularly-spaced data. *Soc. Ind. and Appl. Mathem. Review,* 6, 203—227.

Bjork, A. (1967). Solving linear least-squares problems by Gram—Schmidt orthogonalisation. *BIT,* 7, 1—21.

Blackith, R. E. and R. A. Reyment (1971). *Multivariate morphometrics.* Academic Press, New York.

Bohrnstedt, G. W. and T. M. Carter (1971). Robustness in regression analysis. In: H. L. Costner (ed.), *Sociological Methodology 1971.* Jossey-Bass, San Francisco, pp. 118—145.

Bonham-Carter, G. F. (1965). A numerical method of classification using qualitative and semi-quantitative data, as applied to the facies analysis of limestones. *Canad. Petrol. Geol. Bull.,* 13, 482—502.

Booth, A. D. (1957). *Numerical methods (Second Edition).* Academic Press, New York.

Bottenberg, R. and J. H. Ward (1960). *Applied multiple linear regression analysis.* U.S. Dept. of Commerce, Office of Technical Services.

Box, G. E. P. (1949). A general distribution theory for a class of likelihood criteria. *Biometrika,* 36, 317—346.

Boyce, A. J. (1969). Mapping diversity: a comparative study of some numerical methods. In: A. J. Cole (ed.), (1969), pp. 1—30.

Branfield, J. R. and A. W. Bell (1970). *Matrices and their applications.* Macmillan, London.

Breaux, H. J. (1968). A modification of Efroymson's technique for stepwise regression analysis. *Comm. Assoc. Computer Manufrs.,* 11, 556—557.

Bronson, R. (1969). *Matrix methods: an introduction.* Academic Press, New York.

Browne M. W. (1967). On oblique procrustes rotation. *Psychometrika,* 32, 125—132.

Browne, M. W. (1968). A comparison of factor-analytic techniques. *Psychometrika,* 33, 267—273.

Browne, M. W. (1968). Fitting the factor analysis model. *Psychometrika,* 34, 375—394.

Bruce, J. P. and R. H. Clarke (1966). *Introduction to hydrometeorology.* Pergamon Press, Oxford.

Burton, I. (1963). The quantitative revolution and theoretical geography. *Canad. Geogr.,* 7, 151—162.

Carnahan, B., H. A. Luther and J. O. Wilkes (1969). *Applied numerical methods.* Wiley, New York.

Carroll, J. B. (1961). The nature of the data, or how to choose a correlation coefficient. *Psychometrika*, 26, 347–372.

Carson, M. A. and E. A. Sutton (1971). The hydrological response of the Eaton River Basin, Quebec. *Canad. Journ. Earth Sci.*, 8, 102–115.

Carter, C. S. and R. J. Chorley (1961). Early slope development in an expanding stream system. *Geol. Mag.*, 98, 117–130.

Cattell, R. B. (1965). Factor analysis, in introduction to essentials. *Biometrics*, 21, 190–215 and 405–435.

Cattell, R. B. (ed.) (1966A). *Handbook of multivariate experimental psychology*. Rand McNally, Chicago.

Cattell, R. B. (1966B). The meaning and strategic use of factor analysis. In; R. B. Cattell (ed.) (1966A), pp. 174–243.

Chayes, F. (1970) On deciding whether trend surfaces of progressively higher orders are meaningful. *Bull. Geol. Soc. Amer.*, 81, 1273–1278.

Cheetham, A. H. and J. E. Hazel (1969). Binary (presence–absence) similarity coefficients. *Journ. Paleontol.*, 43, 1130–1136.

Chorley, R. J. (1965). The application of quantitative methods to geomorphology, In: R. J. Chorley and P. Haggett (eds.) *Frontiers in Geographical Teaching*. Methuen, London. pp. 147–163.

Chorley, R. J. (1969). The elevation of the Lower Greensand ridge, South-east England. *Geol. Mag.*, 106, 231–243.

Chorley, R. J. and P. Haggett (1965). Trend surface mapping in geographical research. *Trans. Inst. Brit. Geogrs.*, 37, 47–67.

Chorley, R. J. and B. A. Kennedy (1971). *Physical geography, a systems approach*. Prentice-Hall International, Inc., London.

Chorley, R. J., D. R. Stoddart, P. Haggett and H. O. Slaymaker (1966). Regional and local components in the areal distribution of surface sand facies in the Breckland, eastern England. *Journ. Sedim. Petrol.*, 36, 209–220.

Christiansen, W. I. and R. A. Bryson (1966). An investigation of the potential of component analysis for weather classification. *Monthly Weather Review*, Washington. 94, 697–709.

Clark, P. J. and F. C. Evans (1954). Distance to nearest neighbour as a measure of spatial relationships in populations. *Ecology*, 35, 445–453.

Clarke, M. R. B. (1970). A rapidly convergent method for maximum-likelihood factor analysis. *Brit. Journ. Math. Statist. Psychol.*, 23, 43–52.

Clarke, R. T. (1973). A review of some mathematical models used in hydrology, with observations on their calibration and use. *Journ. of Hydrol.*, 19, 1–20.

Cliff, A. D., P. Haggett, J. K. Ord, K. Bassett and R. Davies (1975). *Elements of spatial structure*. Cambridge University Press, London.

Cliff, A. D. and J. K. Ord (1973). *Spatial autocorrelation*, Pion Press, London.

Cliff, N. and C. D. Hamburger (1967). A study of sampling errors in factor analysis by means of artificial experiments. *Psychol. Bull.*, 68, 430–445.

Cochrane, D. and G. H. Orcutt (1949). Application of least-squares regression to relationships containing autocorrelated error terms. *Journ. Amer. Statist. Assoc.*, 44, 32–61.

Cochran, W. G. (1964). On the performance of the linear discriminant function. *Technometrics*, 6, 179–190.

Cohen, J. (1968). Multiple regression as a general data analytic system. *Psychol. Bull.*, 70, 426–443.

Cole, A. J. (ed.) (1969). *Numerical taxonomy, Proc. Colloq. in Numer. Taxonomy*, St. Andrews. Sept. 1968. Academic Press, London.

Comrey, A. L. (1962). The minimum residual method of factor analysis. *Psychol. Rept.*, 11, 15–18.

Cooley, W. W. and P. R. Lohnes (1971). *Multivariate data analysis*. Wiley, New York.

Coombs, C. H. (1964). *A theory of data*. Wiley, New York.
Cormack, R. M. (1971). A review of classification. *Journ. Royal Statist. Soc.*, A, 134, 321–367.
Craddock, J. M. (1969). *Use of eigenvector analysis in statistical meteorology*. Meteorological Office, Synoptic Climatology Branch, Memoir No. 27.
Craddock, J. M. (1973). Problems and prospects for eigenvector analysis in meteorology. *The Statistician*, 12, 133–145.
Crain, I. K. (1970). Computer interpolation and contouring of two-dimensional data: a review. *Geoexploration*, 8, 71–86.
Crain, I. K. and B. K. Bhattacharyya (1967). The treatment of two-dimensional non-equispaced data with digital computer. *Geoexploration*, 5, 173–194.
Cramer, E. M. (1974). On Browne's solution for oblique procrustes rotation. *Psychometrika*, 39, 159–163.
Crawford, C. B. and G. A. Ferguson (1970). A general rotation criterion and its use in orthogonal rotation. *Psychometrika*, 35, 321–332.
Cronbach, L. J. (1951). Coefficient alpha and the internal structure of tests. *Psychometrika*, 16, 297–334.
Cunningham, K. M. and J. C. Ogilvie (1973). Evaluation of hierarchical grouping techniques: a preliminary study. *Computer Journ.*, 15, 209–213.
Curry, L. (1964). The random spatial economy: an exploration in settlement theory. *Ann. Assoc. Amer. Geogrs.*, 54, 138–146.
Czegledy, P. F. (1972). Efficiency of local polynomials in contour mapping. *Journ. Intern. Assoc. Mathem. Geol.*, 4, 291–306.
Dacey, M. F. (1960). A note on the derivation of nearest-neighbor distances. *Journ. Regional Sci.*, 2, 81–87.
Dacey, M. F. (1962). Analysis of central places and point patterns by nearest-neighbor method. *Lund Studies in Geog.*, Ser. B, 24, 55–75.
Dacey, M. F. (1963). Order neighbor statistics for a class of random patterns in multidimensional space. *Ann. Assoc. Amer. Geogrs.*, 53, 505–515.
Dacey, M. F. (1966). A probability model for central place location. *Ann. Assoc. Amer. Geogrs.*, 56, 550–568.
Daniel, C. and F. S. Wood (with J. W. Gorman) (1971). *Fitting equations to data – computer analysis of multifactor data for scientists and engineers*. Wiley–Interscience, New York.
Darlington, R. B. (1968). Multiple regression in psychological research and practice. *Psychol. Bull.*, 69, 161–182.
Davidon, W. C. (1968). Variance algorithms for minimization. *Computer Journ.*, 10, 406–410.
Davies, M. (1967). Linear approximation using the criterion of least total deviations. *Journ. Royal Statist. Soc.*, B, 29, 101–109.
Davis, J. C. (1970). Information contained in sediment-size analyses. *Journ. Intern. Assoc. Mathem. Geol.*, 2, 105–112.
Davis, J. C. (1973). *Statistics and data analysis in geology, with Fortran programs by Robert J. Sampson*. Wiley, New York.
Davis, J. C. and M. J. McCullagh (eds.) (1975). *Display and analysis of spatial data*. Wiley, London.
Delury, D. B. (1950). *Values and integrals of the orthogonal polynomials up to $n = 26$*. Univ. of Toronto Press, Toronto.
Dickinson, A. W. (1968). Pitfalls in the use of regression analysis. *Proc. SHARE XXX (IBM)*, 3.199–3.218.
Digman, J. M. (1966). Interaction and nonlinearity in multivariate experiments. In: R. B. Cattell (ed.) (1966A), pp. 459–475.
Dixon, L. C. W. (1972). *Nonlinear optimization*. English Universities Press, London.
Dixon, R. (1969). *Orthogonal polynomials as a basis for objective analysis*. Meteorological Office, Sci. Paper No. 30, H.M.S.O., London.

Dixon, R. and E. A. Spackman (1970). *The 3-D analysis of meteorological data.* Meteorological Office, Sci. Paper No. 31, H.M.S.O., London.

Dixon, R., E. A. Spackman, I. Jones and Anne Francis (1972). The global analysis of meteorological data using orthogonal polynomial base functions. *Journ. Atmospher. Sci.,* 29, 609–622.

Doornkamp, J. C. (1972). Trend surface analysis of planation surfaces with an East African case study. In: R. J. Chorley (ed.) (1972) *Spatial analysis in geomorphology.* Methuen, London.

Doornkamp, J. C. and C. A. M. King (1971). *Numerical analysis in geomorphology.* Arnold, London.

Dorn, W. S. and D. D. McCracken (1972). *Numerical methods with Fortran IV case studies.* Wiley, New York.

Douglas, I. (1967). Man, vegetation and the sediment yield of rivers. *Nature,* 215, 925–928.

Doveton, J. H. and A. J. Parsley (1970). Experimental evaluation of trend surface distortions induced by inadequate data point distributions. *Trans. Inst. Min. Metall.,* B, 79, B197–B207.

Draper, N. R. and H. Smith (1966). *Applied regression analysis.* Wiley, New York.

Durbin, J. and G. S. Watson (1950). Testing for serial autocorrelation in least-squares regression, I. *Biometrika,* 37, 409–428.

Durbin, J. and G. S. Watson (1951). Testing for serial autocorrelation in least-squares regression, II. *Biometrika,* 38, 159–178.

Eades, D. C. (1965). The inappropriateness of the correlation coefficient as a measure of taxonomic resemblance. *System. Zoology,* 14, 98–100.

Edwards, A. W. F. and L. L. Cavalli-Sforza (1965). A method for cluster analysis. *Biometrics,* 21, 362–365.

Efroymson, M. A. (1960). Multiple regression analysis. In: A. Ralston and H. S. Wilf (eds.) (1960), pp. 191–203.

Eiselstein, L. M. (1967). A principal component analysis of surface runoff data from a New Zealand watershed. In: *Intern. Hydrology Sympos., Ft. Collins, Colorado,* pp. 479–489.

Ethridge, F. G. and D. K. Davies (1973). Grouped regression analysis – a sedimentologic example. *Journ. Intern. Assoc. Mathem. Geol.,* 5, 377–388.

Farrar, D. E. and D. R. Glauber (1967). Multicollinearity in regression analysis: the problem revisited. *Rev. Econ. Statist.,* 49, 92–107.

Farris, J. S. (1969). On the cophenetic correlation coefficient. *System. Zoology,* 18, 279–285.

Fisher, R. A. (1936). The use of multiple measurements in taxonomic problems. *Ann. of Eugenics,* 7, 179–188.

Fisher, R. A. (1938). The statistical utilization of multiple measurements. *Ann. of Eugenics,* 8, 376–386.

Fletcher, R. (1965). Function evaluation without evaluating derivatives – a review. *Computer Journ.,* 8, 33–41.

Fletcher, R. (1971). *A modified Marquardt subroutine for nonlinear least squares.* Rept. AERE R6799, Atomic Energy Research Establishment, Harwell. H.M.S.O., London.

Fletcher, R. (1972). Conjugate direction methods. In: W. Murray (ed.), *Numerical methods for unconstrained optimization.* Academic Press, London, pp. 73–86.

Fletcher, R. and M. J. D. Powell (1963). A rapidly convergent descent method for minimization. *Computer Journ.,* 6, 163–168.

Forsythe, G. E. (1957). Generation and use of orthogonal polynomials for data fitting with a digital computer, *Soc. Indus. and Appl. Mathem. Journal,* 5, 74–88.

Fournier, F. (1960). *Climat et erosion: la relation entre l'erosion du sol par l'eau et les precipitations atmospherique*. Presses Universitaires de France, Paris.
Fox, L. and D. F. Mayers (1968). *Computing methods for scientists and engineers*. Clarendon Press, Oxford.
Freund, R. J. (1963). A warning of roundoff errors in regression. *The American Statistician*, 17, 13–15.
Fuller, E. L. and W. J. Hemmerle (1966). Robustness of the maximum − likelihood estimation procedure in factor analysis. *Psychometrika*, 31, 255–266.
Furnival, G. M. (1971). All possible regressions with less computation. *Technometrics*, 13, 403–408.
Garside, M. J. (1967). The best subset in multiple regression analysis. *Applied Statist.*, 14, 196–200.
Garside, M. J. (1971A). Some computational procedures for the best subset problem. *Applied Statist.*, 20, 8–15.
Garside, M. J. (1971B). Best subset search. Algorithm AS38, *Applied Statist.*, 20, 112–115.
Geary, R. C. and C. E. V. Leser (1968). Significance tests in multiple regression. *The American Statistician*, 22, 20–21.
Gibson, G. R. and J. Mayatt (1965). *First stages in matrices*. University of London Press, London.
Giggs, J. A. (1973). The distribution of schizophrenics in Nottingham. *Trans. Inst. Brit. Geogrs.*, 59, 55–76.
Golden, J. T. (1965). *Fortran IV programming and computing*. Prentice-Hall, Englewood Cliffs.
Goldfeld, S. M. and R. E. Quandt (1972). *Nonlinear methods in econometrics*. North-Holland Publishing Co., Amsterdam.
Golledge, R. C. and G. Rushton (1972). *Multidimensional scaling: review and applications*. Tech. Paper 10, Assoc. Amer. Geogrs., Commission on College Geography.
Gorman, J. W. and R. J. Thomin (1966). Selection of variables for fitting equations to data. *Technometrics*, 8, 27–51.
Gould, P. R. (1967). On the geographical interpretation of eigenvalues. *Trans. Inst. Brit. Geogrs.*, 42, 53–86.
Gould, P. R. (1970). Is *statistix inferens* the geographical name for a wild goose? *Econ. Geog.*, 46, 439–448.
Gower, J. C. (1966). Some distance properties of latent root and vector methods used in multivariate analysis. *Biometrika*, 53, 325–338.
Gower, J. C. (1967A). Multivariate analysis and multidimensional geometry. *The Statistician*, 17, 13–28.
Gower, J. C. (1967B). A comparison of some methods of cluster analysis. *Biometrics*, 23, 623–637.
Gower, J. C. (1971). A general coefficient of similarity and some of its properties. *Biometrics*, 27, 857–871.
Gower, J. C. and G. J. S. Ross (1969). Minimum spanning trees and single link cluster analysis. *Applied Statist.*, 13, 54–64.
Grant, F. (1957). A problem in the analysis of geophysical data. *Geophysics*, 22, 309–344.
Graybill, F. A. (1969). *Introduction to matrices with applications in statistics*. Wadworth Publishing Co. Inc., Belmont, California.
Greenwood, B. (1969). Sediment parameters and environmental discrimination: an application of multivariate statistics. *Canad. Journ. Earth Sci.*, 6, 1347–1358.
Greenwood, B. and R. G. D. Davidson-Arnott (1972). Textural variation in subenvironments of the shallow-water wave zone, Kouchibouguac Bay, New Brunswick. *Canad. Journ. Earth Sci.*, 9, 679–688.

Greer-Wootten, B. (1972). *A bibliography of statistical applications in geography.* Assoc. Amer. Geogrs., Commission on College Geography, Tech. Paper 9.

Gregory, R. T. and D. L. Karney (1969). *A collection of matrices for testing computational algorithms.* Wiley—Interscience, New York.

Greig-Smith, P. (1964). *Quantitative plant ecology (Second Edition).* Butterworths, London.

Griffiths, J. C. (1957). *Petrographic investigation of the Salt Wash sediments.* Final Rept., U.S. Atomic Energy Commission, RME-3151.

Griffiths, J. C. (1966). Application of discriminant functions as a classification tool in the geosciences. In: D. F. Merriam (ed.) (1966), Colloquium on classification procedures. *Computer Contribution,* 7, State Geological Survey, Lawrence, Kansas, pp. 48—52.

Griffiths, J. C. (1967). *Scientific method in the analysis of sediments.* McGraw-Hill, New York.

Griffiths, J. C. and C. W. Ondrick (1969). Modelling the petrology of detrital sediments. In: D. F. Merriam (ed.) (1969), *Computer Applications in the Earth Sciences,* Plenum Press, New York, pp. 73—97.

Grosenbaugh, L. R. (1967). *REX: Fortran 4 system for combinatorial screening or conventional analysis of multivariate regressions.* Berkeley, California, Pacific S.W. Forest and Range Experimental Sta., U.S. Forest Service Paper PSW-44.

Gruvaeus, G. T. (1970). A general approach to procrustes pattern rotation. *Psychometrika,* 34, 493—505.

Guest, P. G. (1961). *Numerical methods of curve fitting.* Cambridge University Press, London.

Gujatari, D. (1970). Use of dummy variables in testing for equality between sets of coefficients in linear regressions: a generalization. *The American Statistician,* 24, 18—23.

Gustafson, G. C. (1973). Quantitative investigation of the morphology of drainage basins using orthophotography, *Münchener Geographische Abhandlungen, Band 11,* Geographische Institut, München.

Guttman, L. (1953). Image theory for the structure of quantitative variates. *Psychometrika,* 18, 277—296.

Guttman, L. (1955). The determinacy of factor scores matrices with implications for five other basic problems of common factor theory. *Brit. Journ. Statist. Psychol.,* 8, 65—81.

Guttman, L. (1968). A general nonmetric technique for finding the smallest coordinate space for a configuration of points. *Psychometrika,* 33, 469—506.

Haggett, P. (1965). *Locational analysis in human geography.* Arnold, London.

Haggett, P. and K. Bassett (1970). The use of trend surface parameters in inter-urban comparisons. *Environment and Planning,* 2, 225—237.

Hails, J. R. (1967). Significance of statistical parameters for distinguishing sedimentary environments. *Journ. Sedim. Petrol.,* 37, 1059—1069.

Hakstian, A. R. (1971). A comparative evaluation of several prominent methods of oblique factor transformation. *Psychometrika,* 36, 175—193.

Hakstian, A. R. (1972). Optimizing the resolution between salient and non-salient factor pattern coefficients. *Brit. Journ. Mathem. Statist. Psychol.,* 25, 229—245.

Hakstian, A. R. (1973). Formulas for image factor scores. *Educ. and Psychol. Measurement,* 33, 803—810.

Hamaker, H. C. (1962). On multiple regression analysis. *Statistica Neerlandica,* 16, 31—56.

Hanson, N. R. (1958). *Patterns of discovery.* Cambridge Univ. Press, London.

Harbaugh, J. and D. F. Merriam (1968). *Computer applications in stratigraphic analysis.* Wiley, New York.

Harman, H. H. (1967). *Modern factor analysis. Second edition.* Chicago Univ. Press, Chicago.
Harman, H. H. and Y. Fukuda (1966). Resolution of the Heywood case in the minres solution. *Psychometrika,* 31, 563–571.
Harman, H. H. and W. H. Jones (1966). Factor analysis by minimizing residuals (minres). *Psychometrika,* 31 351–368.
Harris, C. W. (1962). Some Rao-Guttman relationships. *Psychometrika,* 27, 247–263.
Harris, C. W. (1964). Some recent developments in factor analysis. *Educ. and Psychol. Measurement,* 24, 193–206.
Harris, C. W. (1967). On factors and factor scores. *Psychometrika,* 32, 363–379.
Hartigan, J. A. (1967). Representation of similarity matrices by trees. *Journ. Amer. Statist. Assoc.,* 62, 1140–1158.
Harvey, D. W. (1969). *Explanation in Geography.* Arnold, London.
Hazel, J. E. (1970). Binary coefficients and clustering in biostratigraphy. *Bull. Geol. Soc. Amer.,* 81, 3237–3252.
Hemmerle, W. J. (1967). *Statistical computations on a digital computer.* Blaisdell Publishing Co., Walton, Massachussetts.
Hendrickson, A. E. and P. O. White (1964). PROMAX: a quick method for rotation to oblique simple structure. *Brit. Journ. Statist. Psychol.,* 17, 65–70.
Hocking, R. R. (1972). Criteria for selection of a subset regression: which one should be used? *Technometrics,* 14, 967–970.
Hocking, R. R. and R. B. Leslie (1967). Selection of the best subset in regression analysis. *Technometrics,* 9, 531–540.
Hodges, S. D. and P. G. Moore (1972). Data uncertainties and least-squares regression. *Applied Statist.,* 21, 185–195.
Hoerl, A. E. (1962). Application of ridge analysis to regression problems. *Chem. Eng. Progr.,* 55, 69–78.
Hoerl, A. E. and R. W. Kennard (1970A). Ridge regression: biased estimation for nonorthogonal problems. *Technometrics,* 12, 56–68.
Hoerl, A. E, and R. W. Kennard (1970B). Ridge regression: application to nonorthogonal problems. *Technometrics,* 12, 69–92.
Hope, K. (1968). *Methods of multivariate analysis.* Univ. of London Press, London.
Horikawa, K. and H. W. Shen (1960). *Sand movement by wind action.* Dept. of the Army, Corps of Engineers, Beach Erosion Board, Tech. Memo. No. 119, Washington D.C.
Horst, P. (1965). *Factor analysis of data matrices.* Holt, Rinehart and Winston, New York.
Hotelling, H. (1933). Analysis of a complex of statistical variables into principal components. *Journ. Experimental Psychol.,* 24, 417–441 and 493–520.
Hotelling, H. (1957). The relations of the newer multivariate statistical methods to factor analysis. *Brit. Journ. Statist. Psychol.,* 10, 69–79.
Howarth, R. J. (1967). Trend surface fitting to random data – an experimental test. *Amer. Journ. Sci.,* 265, 619–625.
Howarth, R. J. (1971). Fortran-IV program for grey-level mapping of spatial data. *Journ. Intern. Assoc. Mathem. Geol.,* 3, 95–121. (See also 4, 276.)
Howarth, R. J. (1973). Preliminary assessment of a nonlinear mapping algorithm in a geological context. *Journ. Intern. Assoc. Mathem. Geol.,* 5, 39–58.
Howe, W. G. (1955). *Some contributions to factor analysis,* Rept. No. ONRL-1919, Oak Ridge Nat. Lab., Oak Ridge, Tennessee.
Huang, D. S. (1970). *Regression and econometric methods.* Wiley, New York.
Imbrie, J. (1963). *Factor and vector analysis programs for analyzing geological data.* Tech. Rept. 6, ONR Task No. 389–135, Office of Naval Research, Geography Branch, Northwestern University, Evanston, Illinois.

Ito, K. (1969). On the effect of heteroscedasticity and non-normality on some multivariate test procedures. In: P. R. Krishnaiah, (ed.) (1969), pp. 87–120.
James, W. R. (1967). Nonlinear models for trend analysis in geology. In: D. F. Merriam and N. C. Cocke (eds.), (1967), pp. 26–30.
James, W. R. (1970). Regression models for faulted structural surfaces. *Bull. Amer. Assoc. Petroleum Geol.*, 54, 638–646.
Jansen, J. M. L. and R. B. Painter (1974). Predicting sediment yield from climate and topography. *Journ. of Hydrol.*, 21, 371–380.
Jardine, N. and R. Sibson (1968). The construction of hierarchic and non-hierarchic classifications. *Computer Journ.*, 11, 177–184.
Jardine, N. and R. Sibson (1971). *Mathematical taxonomy*. Wiley, London.
Jennrich, R. I. (1974). Simplified formulae for standard errors in maximum likelihood factor analysis. *Brit. Journ. Mathem. Statist. Psychol.*, 27, 122–131.
Jennrich, R. I. and D. T. Thayer (1973). A note on Lawley's formulas for standard errors in maximum likelihood factor analysis. *Psychometrika*, 38, 571–604.
Jennrich, R. I. and S. M. Robinson (1969). A Newton-Raphson algorithm for maximum likelihood factor analysis. *Psychometrika*, 34, 111–123.
Johnston, J. (1972). *Econometric methods. Second edition*. McGraw-Hill, New York.
Johnson, L. A. S. (1968). Rainbow's End – the quest for an optimal taxonomy. Reprinted in: *System. Zoology*, 1970, 19, 203–239, from *Proc. Linn. Soc. New South Wales*, 93, 8–45.
Jones, T. A. (1972). Multiple regression with correlated independent variables. *Journ. Intern. Assoc. Mathem. Geol.*, 4, 203–218.
Joreskog, K. G. (1963). *Statistical estimation in factor analysis*. Almqvist and Wiksell, Stockholm.
Joreskog, K. G. (1966). Testing a simple structure hypothesis in factor analysis. *Psychometrika*, 31, 165–178.
Joreskog, K. G. (1967). Some contributions to maximum likelihood factor analysis. *Psychometrika*, 32, 443–482.
Joreskog, K. G. (1969). A general approach to confirmatory maximum likelihood factor analysis. *Psychometrika*, 34, 183–202.
Joreskog, K. G. (1973). Analysis of covariance structures. In: P. K. Krishnaiah (ed.) (1973), pp. 263–285.
Joreskog, K. G. and D. N. Lawley (1968). New methods in maximum likelihood factor analysis. *Brit. Journ. Mathem. Statist. Psychol.*, 21, 85–96.
Joreskog, K. G. and M. Van Thillo (1971). *New rapid algorithms for factor analysis by unweighted least-squares, generalised least squares and maximum likelihood*, Research Memo. RM-71-5, Educational Testing Service, Princeton, New Jersey.
Kaiser, H. F. (1963). Image analysis. In: C. W. Harris (ed.) *Problems in measuring change*. Wisconsin Univ. Press, Madison, Wisconsin, pp. 156–166.
Kaiser, H. F. (1970). A second-generation Little Jiffy. *Psychometrika*, 35, 401–415.
Kaiser, H. F. (1973). The JK method: a procedure for finding the eigenvectors and eigenvalues of a real symmetric matrix. *Computer Journ.*, 15, 271–273.
Kaiser, H. F. and J. Caffrey (1965). Alpha factor analysis. *Psychometrika*, 30, 1–14.
Kaiser, H. F. and J. Rice (1974). Little Jiffy, Mark IV. *Educ. and Psychol. Measurement*, 34, 111–117.
Kendall, M. G. (1954). Review of Uppsala symposium on psychological factor analysis. *Journ. Roy. Statist. Soc.*, A, 119, 83–84.
Kendall, M. G. (1965). *A course in multivariate analysis*. Griffin, London.
Kennard, R. W. (1971). A note on the $C_p$ statistic. *Technometrics*, 13, 899–900.
Kershaw, K. A. (1973). *Quantitative and dynamic plant ecology. Second edition*. Arnold, London.

King, C. A. M. (1969). Trend surface analysis of Central Pennine erosion surfaces. *Trans. Inst. Brit. Geogrs.*, 47, 47–59.
King, L. J. (1969). *Statistical analysis in geography*. Prentice-Hall, Englewood Cliffs.
Kline, M. (1967). *Calculus: an intuitive and physical approach*. Wiley, New York. (Two volumes.)
Klovan, J. E. (1966). The use of factor analysis in determining depositional environments from grain-size distributions. *Journ. Sedim. Petrol.*, 36, 115–125.
Kmenta, J. (1971). *Elements of econometrics*. Macmillan, New York.
Knuth, D. E. (1969). *The art of computer programming*, Volume 2. Addison-Wesley, Reading, Massachusetts.
Koelling, M. E. V. and E. H. T. Whitten (1973). Fortran IV program for spline surface interpolation and contour map production. *Geocom Bulletin*, 6.
Kowalski, C. J. (1972). On the effects of non-normality on the distribution of the sample product-moment correlation coefficient. *Applied Statist.*, 21, 1–12.
Krishnaiah, P. K. (ed.) (1969). *Multivariate analysis II – Proc. 2nd Internat. Sympos. on Multiv. Analysis, Wright State Univ., Dayton, Ohio, 1968*. Academic Press, New York.
Krishnaiah, P. K. (ed.) (1973). *Multivariate analysis III – Proc. 3rd Internat. Sympos. on Multiv. Analysis, Wright State Univ., Dayton, Ohio, 1972*. Academic Press, New York.
Krumbein, W. C. (1959). The sorting out of geological variables illustrated by regression analysis of factors controlling beach firmness. *Journ. Sedim. Petrol.*, 29 575–587.
Krumbein, W. C. and E. J. Aberdeen (1937). The sediments of Barataria Bay, La., *Journ. Sedim. Petrol.*, 7, 3–17.
Krumbein, W. C. and F. A. Graybill (1965). *An introduction to statistical models in geology*. McGraw-Hill, New York.
Kruskal, J. B. (1960). Some remarks on wild observations. *Technometrics*, 2, 1–4.
Kruskal, J. B. (1964A). Multidimensional scaling by optimizing goodness of fit to a nonmetric hypothesis. *Psychometrika*, 29, 1–27.
Kruskal, J. B. (1964B). Nonmetric multidimensional scaling: a numerical method. *Psychometrika*, 29, 115–129.
Kruskal, J. B. and J. D. Carroll (1969). Geometrical models and badness of fit functions. In: P. K. Krisknaiah (ed.), (1969), pp. 639–671.
Kuo, S. S. (1972). *Computer application of numerical methods*. Addison Wesley, Reading, Massachussetts.
Lamotte, L. R. and R. R. Hocking (1970). Computational efficiency in the selection of regression variables. *Technometrics*, 12, 83–93.
Lance, G. N. and W. T. Williams (1967). A general theory of classificatory sorting strategies. I: Hierarchical systems. *Computer Journ.*, 9, 373–380.
Larsen, W. A. and S. J. McCleary (1972). The use of partial residual plots in regression analysis. *Technometrics*, 14, 781–790.
Lavalle, P. (1968). Karst depression morphology in South-Central Kentucky. *Geogr. Annal.*, 50A, 94–108.
Lawley, D. N. and A. E. Maxwell (1971). *Factor analysis as a statistical method*. Second edition. Butterworths, London.
Lee, P. J. (1969A). The theory and application of canonical trend surfaces. *Journ. Geol.*, 77, 303–318.
Lee P. J. (1969B). Fortran IV programs for canonical correlation and canonical trend surface analysis. *Computer Contribution*, 32, State Geological Survey, Lawrence, Kansas.
Li, C. C. (1964). *An introduction to experimental statistics*. McGraw-Hill, New York.
Liebeck, P. (1971). *Vectors and matrices*. Pergamon Press, Oxford.

Lingoes, J. C. and L. Guttman (1967). Nonmetric factor analysis: A rank-reducing alternative to linear factor analysis. *Multiv. Behav. Research*, 2, 485–505.

Linn, R. L. (1968). A monte carlo approach to the number of factors problem. *Psychometrika*, 33, 37–71.

Longley, J. M. (1967). An appraisal of least-squares programs for the electronic computer from the point of view of the user. *Journ. Amer. Statist. Assoc.*, 62, 819–829.

McCammon, R. B. (1966). Principal components analysis and its application in large scale correlation studies. *Journ. Geology*, 74, 721–733.

McCammon, R. B. (1968A). Multiple component analysis and its application in classification of environments. *Amer. Assoc. Petrol. Geol. Bulletin*, 52, 2178–2196.

McCammon, R. B. (1968B). The dendrograph: a new tool for correlation. *Geol. Soc. Amer. Bulletin*, 79, 1663–1670.

McCammon, R. B. (1969). Fortran IV program for nonlinear estimation. *Computer Contribution*, 34, State Geological Survey, Lawrence, Kansas.

McCammon, R. B. (1973). Nonlinear regression for dependent variables. *Journ. Intern. Assoc. Mathem. Geol.*, 5, 365–375.

McCammon, R. B. and G. Wenniger (1970). The dendrograph. *Computer Contribution*, 48, State Geol. Survey, Lawrence, Kansas.

McCracken, D. D. (1965). *A guide to Fortran IV programming*. Wiley, New York.

McDonald, R. P. (1962). A general approach to nonlinear factor analysis. *Psychometrika*, 27, 397–415.

McDonald, R. P. (1967). Nonlinear factor analysis. *Psychometric Monograph No. 15*.

McDonald, R. P. (1974). The measurement of factor indeterminancy. *Psychometrika*, 39, 203–222.

McDonald, R. P. and E. J. Burr (1967). A comparison of four methods of constructing factor scores. *Psychometrika*, 32, 381–401.

Mackay, J. R. (1967). Freeze-up and breakup prediction of the Mackenzie River, N. W. T., Canada. In: W. L. Garrison and D. F. Marble, (eds) (1967), *Quantitative geography, Part II: Physical and cartographic topics. (Studies in Geography No. 14.)* Northwestern University, Evanston, Illinois.

Macomber, L. (1971). Utility of trend surfaces in inter-regional map comparisons. *Journ. Regional Sci.*, 11, 87–90.

Malcolm, M. A. (1971). On accurate floating point summation. *Comm. Assoc. Computer Manuf.*, 14, 731–736.

Malcolm, M. A. (1972). Algorithms to reveal properties of floating point arithmetic. *Comm. Assoc. Computer Manuf.*, 15, 949–951.

Mandelbaum, H. (1966). Comments on D. F. Merriam and P. H. A. Sneath, 'Quantitative comparison of contour maps', *Journ. Geophys. Research*, 71, 4431–4433.

Marquardt, D. W. (1963). An algorithm for least-squares estimation of nonlinear parameters. *Soc. Indus. Appl. Mathem., Journ.*, 11, 431–441.

Marquardt, D. W. (1970). Generalized inverses, ridge regression, biased linear and nonlinear estimation. *Technometrics*, 12, 591–612.

Matalas, N. C. and B. J. Reiher (1965). Some comments on the use of factor analysis. *Water Resources Research*, 3, 213–223.

Mather, P. M. (1969). *Analysis of some Late Pleistocene sediments from South Lancashire and their relation to glacial and fluvioglacial processes*. Unpublished Ph.D. thesis, University of Nottingham.

Mather, P. M. (1970). Principal components and factor analysis. *Computer Applicns. Natural and Soc. Sci.*, Old Series, 10, Dept. Geography, University of Nottingham.

Mather, P. M. (1975A). The use of orthogonal polynomials in least squares problems. *Computer Applicns. Natural and Soc. Sci.*, New Series, 2(1), Dept. Geography, University of Nottingham.

Mather, P. M. (1975B). *Computing in Geography – A practical approach.* Basil Blackwell, Oxford.

Mather, P. M. and J. C. Doornkamp (1970). Multivariate analysis in geography with particular reference to drainage-basin morphometry. *Trans. Inst. Brit. Geogr.*, 51, 163–187.

Melton, M. A. (1957). *An analysis of the relations among elements of climate, surface properties and geomorphology.* Tech. Rept. 11, Project NR 389-042, Dept. of Geology, Columbia University, New York.

Melton, R. S. (1963). Some remarks on failure to meet assumptions in discriminant analysis. *Psychometrika*, 28, 49–53.

Merriam, D. F. and N. C. Cocke (eds.) (1967). Computer applications in the earth sciences — colloquium on trend analysis. *Computer Contribution* 12, State Geol. Survey, Lawrence, Kansas.

Merriam, D. F. and P. H. A. Sneath (1966). Quantitative comparison of contour maps. *Journ. Geophys. Research*, 71, 1105–1115.

Middleton, G. V. (1962). A multivariate statistical technique applied to a study of sandstone composition. *Trans. Royal Soc. Canada*, III, 56, 119–126.

Miesch, A. T. and J. J. Connor (1968). Stepwise regression and nonpolynomial models in trend analysis. *Computer Contribution* 27, State Geol. Survey, Lawrence, Kansas.

Miller, R. L. (1956). Trend surfaces: their application to analysis and description of environments of sedimentation. *Journ. Geology*, 64, 425–446.

Miller, R. L. (1964). Comparison analysis of trend surface maps. *Stanford Univ. Publ. Geol. Sci.*, 9, 669–692.

Miller, R. L. and J. S. Kahn (1962). *Statistical analysis in the geological sciences.* Wiley, New York.

Morgan, J. A. and J. F. Tatar (1972). Calculation of the residual sum of squares for all possible regressions. *Technometrics*, 14, 317–325.

Morisawa, M. E. (1962). Quantitative geomorphology of some watersheds in the Appalachian Plateau. *Bull. Geol. Soc. Amer.*, 73, 1025–1046.

Morrison, D. F. (1967). *Multivariate statistical methods.* McGraw-Hill, New York.

Mukherjee, B. N. (1973). Analysis of covariance structures. *Brit. Journ. Mathem. Statist. Psychol.*, 26, 125–154.

Mulaik, S. A. (1972). *The foundations of factor analysis.* McGraw-Hill, New York.

Mullett, G. M. and T. W. Murray (1971). A new method for examining rounding error in least-squares regression computer programs. *Journ. Amer. Statist. Assoc.*, 66, 496–498.

Nash, J. E. and B. L. Shaw (1966). Flood frequency as a function of catchment characteristics. *Inst. Civil Engrs., Proc. Sympos. on River Flood Hydrol.*, London, 115–136.

National Computing Centre (1972). *Standard Fortran programming handbook.* National Computing Centre, Manchester.

Neave, H. R. (1972). Random number package. *Computer Applicns. Natural and Soc. Sci.*, Old Series, 14, Dept. Geography, University of Nottingham.

Nelder, J. A. and R. Mead (1965). A simplex method for minimization. *Computer Journ.*, 7, 308–313.

Norcliffe, G. B. (1969). On the use and limitations of trend surface models. *Canad. Geogr.*, 13, 338–348.

Oldham, C. H. G. and D. B. Sutherland (1955). Orthogonal polynomials: their use in estimating the regional effect, *Geophysics*, 20, 295–306.

O'Neill, R. (1971). Function minimization using a simplex procedure. *Applied Statist.*, 20, 338—341.
Open University (1971). *An introduction to calculus and algebra: Volume 2, Calculus applied.* Open University Press, Bletchley, Bucks.
Oosterhoff, J. (1963). *On the selection of independent variables in a regression equation.* Stiching Mathematisch Centrum, 2 Boerhaavestraat, 49, Amsterdam. Report s319 (VP23). (In English.)
Overall, J. E. and D. K. Speigel (1969). Concerning least squares analysis of experimental data. *Psychol. Bull.*, 72, 311—322.
Parks, J. M. (1969). Multivariate facies maps. In: D. F. Merriam (ed.) Computer applications in petroleum exploration. *Computer Contribution* 40, State Geological Survey, Lawrence, Kansas, 6—12.
Parsley, A. J. (1971). Application of autocorrelation criteria to the analysis of mapped geologic data from the Coal Measures of Central England. *Journ. Intern. Assoc. Mathem. Geol.*, 3, 281—295.
Parsley, A. J. and J. D. H. Doveton (1969). The role of some statistical and mathematical methods in the interpretation of regional geochemical data. *Econ. Geology*, 64, 830.
Peckham, G. (1970). A new method for minimizing a sum of squares without calculating derivatives. *Computer Journ.*, 13, 418—420.
Pennington, R. H. (1970). *Introductory computer methods and numerical analysis.* Macmillan, New York.
Phipps, J. B. (1971). Dendrogram topology. *System. Zoology*, 20, 306—308.
Poole, M. A. and P. N. O'Farrell (1971). The assumptions of the linear regression model. *Trans. Inst. Brit. Geogr.*, 52, 145—158.
Pope, P. T. and J. T. Webster (1972). The use of an F-statistic in stepwise regression procedures. *Technometrics*, 14, 327—329.
Potter, P. E., N. F. Shimp and J. Witters (1963). Trace elements in marine and freshwater argillaceous sediments. *Geoch. et Cosmoch. Acta*, 27, 669—694.
Potter, W. D. (1953). Rainfall and topographic factors that affect runoff. *Trans. Amer. Geophys. Union*, 34, 67—73.
Powell, M. J. D. (1968). *A Fortran subroutine for solving systems of nonlinear algebraic equations.* Report AERE-R5947, Atomic Energy Research Establishment, Harwell. H.M.S.O., London.
Press, S. J. (1973). *Applied multivariate analysis.* Holt, Rinehart and Winston, New York.
Pritchard, N. M. and A. J. B. Anderson (1970). Observations on the use of cluster analysis in botany with an ecological example. *Journ. of Ecology*, 59, 727—747.
Ralston, A. (1965). *A first course in numerical analysis.* McGraw-Hill, New York.
Ralston, A. and H. S. Wilf (eds.) (1960). *Mathematical methods for digital computers, Vol. 1.* Wiley, New York.
Ralston, A. and H. S. Wilf (eds.) (1967). *Mathematical methods for digital computers, Vol. 2.* Wiley, New York.
Ramsayer, G. R. and G. Bonham-Carter (1974). Numeric classification of geologic patterns characterized by binary variables. *Journ. Intern. Assoc. Mathem. Geol.*, 6, 59—72.
Rao, C. R. (1952). *Statistical methods in biometric research.* Wiley, New York.
Rao, C. R. (1964). The use and interpretation of principal component analysis in applied research. *Sankhya (Indian Journ. of Statistics)*, A, 26, 329—358.
Rao, C. R. and P. Slater (1949). Multivariate analysis applied to the difference between neurotic groups. *Brit. Journ. Psychol. (Statist. Sect.)*, 2, 17—29.
Rao, S. V. L. N. (1971). Correlation between regression surfaces based on direct comparison of matrices. *Modern Geology*, 2, 173—177.

Rao, S. V. L. N. and G. S. Srivastava (1969). Comparison of regression surfaces in geological studies. *Trans. Kansas Acad. Sci.*, 72, 91–97.
Rayner, J. H. (1966). Classification of soils by numerical methods. *Journ. Soil Sci.*, 17, 79–92.
Rayner, J. N. (1971). *An introduction to spectral analysis*. Pion Press, London.
Rees, P. H. (1971). Factorial ecology: an extended definition, survey, and critique of the field. *Econ. Geog.*, 47, Supplement, 220–233.
Rhodes, J. M. (1969). The application of cluster and discriminatory analysis in mapping granite intrusions. *Lithos*, 2, 223–237.
Robinson, A. H. (1962). Mapping the correspondence of isarithmic maps. *Ann. Assoc. Amer. Geogrs.*, 52, 414–425.
Robinson, A. H. and L. Caroe (1967). On the analysis and comparison of statistical surfaces. In: W. L. Garrison (ed.), *Quantitative Geography, I: Economic and Cultural Topics. Northwestern University Studies in Geography*, 13, pp. 252–276. Northwestern Univ., Evanston, Illinois.
Rodda, J. C. (1967). The significance of characteristics of basin rainfall and morphometry in a study of floods in the United Kingdom. *UNESCO Sympos. on Floods and their Computation*, Leningrad.
Rodda, J. C. (1970). A trend surface analysis trial for the planation surfaces of North Cardiganshire. *Trans. Inst. Brit. Geogrs.*, 50, 107–114.
Rohlf, F. J. (1967). Correlated characters in numerical taxonomy. *System. Zoology*, 16, 109–126.
Rohlf, F. J. (1970). Adaptive hierarchical clustering schemes. *System. Zoology*, 19, 58–82.
Rohlf, F. J. (1972). An empirical comparison of three ordination techniques in numerical taxonomy. *System. Zoology*, 21, 271–280.
Rohlf, F. J. (1973). Hierarchical clustering using the minimum spanning tree. *Computer Journ.*, 16, 93–95.
Rohlf, F. J. (1974). Dendrogram plot. Algorithm 81, *Computer Journ.*, 17, 89–91.
Ross, G. J. S. (1969). Classification techniques for large data sets. In: A. J. Cole (ed.) (1969), pp. 224–233.
Rummell, R. J. (1970). *Applied factor analysis*. Northwestern Univ. Press, Evanston, Illinois.
Rushton, S. (1951). On least-squares fitting by orthonormal polynomials using the Choleski method. *Journ. Royal Statist. Soc.*, B, 13, 92–99.
Sahu, B. K. (1964). Depositional mechanisms from size analysis of clastic sediments. *Journ. Sedim. Petrol.*, 34, 73–84.
Sammon, J. W. Jr. (1969). A nonlinear mapping algorithm for data structure analysis. *IEEE Trans. on Computers*, C18, 401–409.
Schatzoff, M., S. Fienberg and R. Tsao (1968). Efficient calculation of all possible regressions. *Technometrics*, 10, 769–779.
Schlee, J. (1957). Upland gravels of South Maryland. *Geol. Soc. Amer. Bull.*, 68, 1371–1410.
Seal, H. L. (1967). The historical development of the Gauss linear model. *Biometrika*, 54, 1–24.
Searle, S. R. (1966). *Matrix algebra for the biological sciences*. Wiley, New York.
Searle, S. R. (1971). *Linear models*. Wiley, New York.
Shepard, R. N. (1962). The analysis of proximities: multidimensional scaling with an unknown distance function. *Psychometrika*, 27, 125–139 and 219–246.
Shepard, R. N. (1972). A taxonomy of some principal types of data and of multidimensional methods for their analysis. In: R. N. Shepard et al. (eds.) (1972), pp. 21–47.
Shepard, R. N., A. K. Romney and S. Nerlove (eds.) (1972). *Multidimensional*

*scaling: theory and applications in the behavioral sciences*. (Two volumes). Seminar Press, New York.

Shideler, G. L. (1973). Textural trend analysis of coastal barrier sediments along the Middle Atlantic bight, North Carolina. *Sedim. Geology*, 9, 195–220. See also 10, 311–316.

Silvey, S. D. (1969). Multicollinearity and imprecise estimation. *Journ. Royal Statist. Soc.*, B.31, 539–552.

Simonett, D. S. (1967). Landslide distribution and earthquakes in the Bewani and Torricelli Mountains, New Guinea; statistical analysis. In: J. N. Jennings and J. A. Mabbutt (eds.) (1967). *Landform studies from Australia and New Guinea.* Cambridge University Press, London, pp. 64–84.

Smith, A. F. M. and M. Goldstein (1975). Ridge regression: some comments on a paper of Stone and Conniffe. *The Statistician*, 24, 61–66.

Sneath, P. H. A. (1966). Estimating concordance between geographic trends. *Systematic Zoology*, 15, 250–252.

Sneath, P. H. A. and R. R. Sokal (1973). *Numerical taxonomy: the principles and practice of numerical classification*. Freeman, San Francisco.

Sokal, R. R. and F. J. Rohlf (1962). The comparison of dendrograms by objective methods. *Taxon*, 11, 33–40.

Sparks, D. N. (1973). Euclidean cluster analysis. Algorithm AS 58, *Applied Statist.*, 22, 127–130.

Spitz, O. T. (1966). Generation of orthogonal polynomials for trend surfacing with a digital computer. *Proc. Sympos. Short Course Computer Operations Research Mineral Ind., Penn. State University*, 3, 2–7.

Spivak, M. (1967). *Calculus*. Benjamin, New York.

Stammers, W. N. (1967). The application of multivariate techniques in hydrology. In: *Statistical methods in hydrology, Proc. Hydrol. Sympos.* 5, 1966, McGill Univ., Ottawa.

Stephenson, R. A. (1971). On the interpretation of suppression variables in a multiple linear regression analysis. *Geogr. Analysis*, 3, 98–99.

Stone, J. and D. Conniffe (1973). A critical view of ridge regression. *The Statistician*, 22, 181–187.

Thayer, R. P. and R. F. Storer (1969). Scale selection for computer plots. *Applied Statist.*, 18, 206–208. See also: 20, p. 118.

Thomas, D. N. and M. A. Benson (1970). Generalization of streamflow characteristics from drainage-basin characteristics. *U.S. Geol. Survey, Water Supply Paper 1975*, Washington, D.C.

Thomas, G. S. P. (1969). Routine sediment analysis – a package of three programs. In: The use of computers in geomorphological research, Symposium held at Nottingham, 1968. *Brit. Geomorph. Res. Group, Occas. Paper 6*, Geoabstracts, Univ. of East Anglia, Norwich.

Thurstone, L. L. (1947). *Multiple factor analysis*. Chicago Univ. Press, Chicago, Illinois.

Tien, P. L. (1968). Differentiation of Pleistocene deposits in Northeastern Kansas by clay minerals. *Clays and Clay Minerals*, 16, 99–107.

Tinkler, K. (1969). Trend surfaces with low 'explanations': the assessment of their significance. *Amer. Journ. Sci.*, 267, 114–123.

Tinkler, K. (1971). Statistical analysis of tectonic patterns in areal volcanism: the Bunyaruguru volcanic field in West Uganda. *Journ. Intern. Assoc. Mathem. Geol.*, 3, 335–355.

Tobler, W. R., H. W. Mielke and T. R. Detwyler (1970). Geobotanical distance between New Zealand and neighbouring islands. *Bio. Sci.*, May 1970, 537–542.

Torgerson, W. S. (1958). *Theory and methods of scaling*. Wiley, New York.

Tucker, L. R. (1971). Relations of factor score estimates to their use. *Psychometrika*, 36, 427—436.
Tucker, L. R. and C. Lewis (1970). *A reliability coefficient for maximum likelihood factor analysis.* Dept. of Psychol., Univ. of Illinois at Urbana-Champagn, ONR Contract USNavy/00014-67-A-0305-003, NR 150-304, Champagn, Illinois.
Tukey, J. W. (1965). The inevitable collision between computation and data analysis. *IBM Scientific Comput. Sympos. on Statist., 1963*, IBM Data Processing Div., White Plains, New York.
Tukey, J. W. (1969). Analyzing data — sanctification or detective work? *Amer. Psychologist*, 24, 83—91.
Unwin, D. J. (1970). Percentage RSS in trend surface analysis. *Area*, 2, 25—28.
Unwin, D. J. (1973). The distribution and orientation of corries in northern Snowdonia. *Trans. Inst. Brit. Geogrs.*, 58, 85—98.
Vernon, M. D. (1962). *The psychology of perception.* Penguin Books, Harmondsworth, Middlesex.
Walls, R. C. and D. L. Weeks (1969). A note on the variance of a predicted response in regression. *The American Statistician*, 223, 24—26.
Walsh, J. (ed.) (1966). *Numerical analysis: an introduction.* (Based on a symposium organised by the Inst. of Mathem. and its Applic., Birmingham, 1965). Academic Press, London.
Wampler, R. H. (1970). A report on the accuracy of some widely-used least squares computer programs. *Journ. Amer. Stat. Assoc.*, 65, 549—565.
Warburton, F. W. (1963). Analytical methods of factor rotation. *Brit. Journ. Statist. Psychol.*, 16, 165—174.
Ward, J. H. Jnr. (1963). Hierarchical grouping to optimise an objective function. *Journ. Amer. Statist. Assoc.*, 58, 236—244.
Ward, R. C. (1967). *Principles of hydrology.* McGraw-Hill, London.
Whitten, E. H. T. (1970). Orthogonal polynomial trend surfaces for irregularly-spaced data. *Journ. Intern. Assoc. Mathem. Geol.,* 2, 141—152. (See also 3, 330.)
Whitten, E. H. T. (1972). More on 'irregularly spaced data and orthogonal polynomial trend surfaces'. *Journ. Intern. Assoc. Mathem. Geol.*, 4, 83.
Whitten, E. H. T. and M. E. V. Koelling (1973). Spline-surface interpolation, spatial filtering and trend surfaces for geologic mapped variables. *Journ. Intern. Assoc. Mathem. Geol.*, 5, 111—126.
Whitten, E. H. T. (1975). The practical use of trend surface analysis in the geological sciences. In: J. C. Davis and M. J. McCullagh (eds.) (1975), pp. 282—297.
Whittington, H. B. and C. P. Hughes (1972). Ordovician geography and faunal provinces deduced from trilobite distribution. *Phil. Trans. Royal Soc., London*, B, 263, 235—273.
Wilkinson, J. H. and C. Reinsch (eds.) (1971). *Linear algebra.* (Handbook for automatic computation, vol. 2.) Springer-Verlag, Berlin.
Williams, P. W. (1972). *Numerical computation.* Nelson, London.
Williams, W. T. and H. T. Clifford (1971). On the comparison of two classifications of the same set of elements. *Taxon*, 20, 519—522.
Williams, W. T. and J. M. Lambert (1959). Multivariate methods in plant ecology. I: Association analysis in plant communities. *Journ. Ecology*, 47, 83—101.
Wishart, D. (1969). Mode analysis, a generalization of nearest neighbour which reduces chaining effects. In: A. J. Cole, (ed.) (1969), pp. 282—311.
Wolberg, J. R. (1967). *Prediction analysis.* Van Nostrand, New York.
Wong, S. T. (1963). A multivariate statistical model for predicting mean annual flood in New England. *Ann. Assoc. Amer. Geogr.*, 53, 298—311.

Wong, S. T. (1971). Effect of stream size and average land slope on the occurrence of floods in New England. *Geogr, Analysis*, 3, 77–83. (See also 4, 194–203.)

Woodruff, J. F. and J. D. Hewlett (1970). Predicting and mapping the average hydrological response for the eastern United States. *Water Resources Research*, 6, 1312–1326.

Yamane, T. (1964). *Statistics: an introductory analysis*. Harper and Row, New York.

Yevyevich, V. C. (1972). *Probability and stastistics in hydrology*. Water Resources Publications, Ft. Collins, Colorado.

Youngs, E. A. and E. M. Cramer (1971). Some results relevant to the choice of sum and sum of product algorithms. *Technometrics*, 13, 657–665.

# Author Index

Aberdeen, E. J.   442
Abrahams, A. D.   103, 109
Acton, F. S.   33
Agterberg, F. P.   3, 83, 145
Aigner, D. J.   40
Anderberg, M. R.   371
Anderson, A. J. B.   326, 327, 340, 341
Anderson, T. W.   245
Anderssen, R. S.   40
Andrews, D. F.   431
Anscombe, F. J.   77
Archer, C. O.   305
Armstrong, J. S.   242, 305

Bagnold, R. A.   207, 211
Baird, A. K.   146
Bard, Y.   202, 203, 205, 206
Barkham, J. P.   237, 238
Bartlett, M. S.   244
Bassett, K.   150
Beale, E. M. L.   327, 371
Bell, A. W.   33
Benson, M. A.   40, 187, 192
Berry, B. J. L.   130
Berztiss, A. T.   141
Bhattacharyya, B. K. 141
Bjork, A.   96, 97, 98, 141
Blackith, R. E.   240, 330, 331, 392
Bohrnstedt, G. W.   52
Booth, A. D.   49
Bottenberg, R.   45
Box, G. E. P.   431, 434
Boyce, A. J.   326
Branfield, J. R.   33
Breaux, H. J.   175
Bronson, R.   10
Browne, M. W.   262, 291

Bruce, J. P.   101
Bryson, R. A.   237
Burr, E. J.   302
Burton, I.   2

Cabilio, P.   145
Caffrey, J.   240, 247, 256
Carnahan, B.   10, 97
Caroe, L.   148
Carrol, J. B.   334
Carson, M. A.   40
Carter, C. S.   432
Carter, T. M.   52
Cattell, R. B.   234, 238, 240, 243
Cavalli-Sforza, A. W. F.   311
Chayes, F.   146
Cheetham, J. E.   315
Chorley, R. J.   1, 117, 119, 129, 130, 432
Christiansen, W. I.   237
Clark, P. J.   122
Clarke, R. H.   101
Clarke, R. T.   40
Cliff, A. D.   84, 87, 93, 150, 160, 169
Clifford, H. T.   326
Cochran, W. G.   424
Cochrane, D.   40
Cohen, J.   194
Comrey, A. L.   261
Conniffe, D.   74
Connor, J. J.   49, 124, 125, 143
Cooley, W. W.   280, 437
Coombs, C. H.   332
Cormack, R. M.   310
Craddock, J. M.   237, 238, 239
Crain, I. K.   141

Cramer, E. M.   9, 33, 34, 53, 262, 471
Crawford, C. B.   279
Cronbach, L. J.   247
Cunningham, K. M.   321, 327
Curry, L.   119, 123
Czegledy, P. F.   141

Dacey, M. F.   123
Daniel, C.   45, 77
Darlington, R. B.   45, 95, 194
Davidon, W. C.   205
Davidson-Arnott, R. G. D.   421
Davies, D. K.   197
Davies, M.   40
Davis, J. C.   33, 117, 123, 131, 142, 220, 237, 279, 280, 421
DeLury, D. B.   117
Detwyler, T. R.   331
Dickinson, A. W.   45
Digman, J. M.   3, 271, 305
Dixon, L. C. W.   205
Dixon, R.   141
Doornkamp, J. C.   51, 117, 147, 237, 432
Dorn, W. S.   10, 32
Douglas, I.   108
Doveton, J. D. H.   120, 121, 128, 144
Draper, N. R.   45, 77, 174, 175
Durbin, J.   84

Eades, D. C.   314
Edwards, A. W. F.   311
Efroymson, M. A.   175
Eiselstein, L. M.   237
Ethridge, F. G.   197
Evans, F. C.   122

Farrar, D. E.   40
Feinberg, S.   178
Ferguson, J. A.   279
Fisher, R. A.   420, 437
Fletcher, R.   204, 205, 209
Folk, R. C.   429
Forsythe, G. E.   98, 141
Fournier, F.   108
Fox, L.   26, 97
Freund, R. J.   53
Fukuda, Y.   262
Fuller, E. L.   270
Furnival, G. M.   178

Garside, M. J.   178
Geary, R. C.   51
Gibson, G. R.   33

Giggs, J. A.   295
Glauber, D. R.   40
Golden, J. T.   10
Goldfeld, S. M.   200
Goldstein, M.   74
Golledge, R. C.   331
Gorman, J. W.   178
Gould, P. R.   28, 120
Gower, J. C.   311, 314, 315, 319, 320, 326, 330, 331, 392
Grant, F. S.   141, 142, 159
Graybill, F. A.   10, 141, 421
Greenwood, B.   421
Greer-Wootten, B.   40
Gregory, R. T.   460
Greig-Smith, P.   123, 124
Griffiths, J. C.   237, 421, 424, 432
Grosenbaugh, L. R.   178, 187, 192
Gruvaeus, G. T.   291
Gujatari, D.   194
Gustafson, G. S.   245, 246, 260, 268, 271, 275, 281, 285, 291
Guttman, L.   273, 302, 332

Haggett, P.   117, 122, 130, 150
Hails, J. R.   421
Hakstian, A. R.   281, 283, 287, 291, 299
Hamaker, H. C.   177
Hanson, N. R.   119
Harbaugh, J.   312
Harman, H. H.   240, 243, 245, 261, 262, 268, 271, 278, 279, 280, 295, 296
Harris, C.   273, 274, 299
Hartigan, J. A.   326
Harvey, D.   122, 124, 310
Hazel, A. H.   314, 315
Hemmerle, W. J.   33, 175, 270, 432
Hendrickson, A. E.   281, 283
Hewlett, J. D.   40
Hocking, R. R.   178
Hodges, S. D.   40
Hoerl, A. E.   73, 74, 75
Hope, K.   428, 429, 435
Horikawa, K.   207, 208
Horst, P.   240, 279, 280
Hotelling, H.   242, 247
Howarth, R. J.   131, 146, 340
Howe, W. G.   270
Huang, D. S.   40
Hughes, C. P.   327, 331, 338

Imbrie, J.   314

Ito, K.   431

James, W. R.   146
Jansen, J. M.   40
Jardine, N.   324, 325, 326
Jennrich, R. I.   270, 305
Johnson, L. A. S.   310
Johnston, J. H.   40, 43, 78, 80, 89, 90, 193, 197, 221
Jones, T. E.   74, 75, 76, 142
Jones, W. H.   261
Joreskog, K. G.   202, 245, 261, 270, 290, 291, 304

Kahn, J. S.   123
Kaiser, H. F.   31, 234, 240, 247, 256, 273, 274, 275, 279
Karney, D. L.   460
Kendall, M. G.   243, 244, 423
Kennard, R. W.   73, 74, 75, 187
Kennedy, B. A.   1
Kershaw, S. A.   123, 124
King, C. A. M.   51, 117, 432
King, L. J.   122, 123, 124
Kline, M.   203
Klovan, J. E.   442, 448, 453
Kmenta, J.   3, 40, 43, 66, 73, 84, 90, 268
Knuth, D. E.   464
Koelling, M. E. V.   143
Krumbein, W. C.   40, 141, 421, 442
Kruskal, J. B.   77, 325, 332, 333, 334, 335, 339, 392
Kuo, S. S.   31, 464

Lamotte, L. R.   178
Lance, G. N.   316, 318, 320, 321, 322, 371
Larsen, W. A.   77
Lavalle, P.   40
Lawley, D. N.   224, 225, 226, 240, 268, 270, 271, 273, 283, 287, 288, 291, 305
Lee, P. J.   142
Leser, C. E. V.   51
Leslie, R. B.   178
Lewis, C.   271
Li, C. C.   421
Liebeck, P.   33
Lingoes, J. C.   332
Linn, R. L.   305
Lohnes, P. R.   280, 437
Longley, J. M.   9, 26, 33, 40, 52, 67, 72

Luther, H. A.   10, 97

McCammon, R. B.   203, 205, 237, 311, 317, 323, 324
McCleary, S. J.   77
McCracken, D. D.   10, 32
McCullagh, M. J.   131
McDonald, R. P.   244, 302, 303
Mackay, J. R.   40
Macomber, L.   150
Malcolm, M. A.   32, 33, 461
Mandelbaum, H.   150
Marble, D. A.   130
Marquardt, D. W.   74, 204
Matalas, N. M.   240
Mather, P. M.   10, 96, 97, 237, 305, 342, 421
Maxwell, A. E.   224, 225, 226, 270, 271, 283, 287, 290, 291, 305
Mayatt, J.   33
Mayers, D. F.   26, 97
Meade, R.   205
Melton, M. A.   40, 232
Merriam, D. F.   117, 149, 312
Middleton, G. V.   421
Mielke, H. W.   331
Miesch, A. T.   49, 124, 125, 143
Miller, R. L.   117, 123, 150
Moore, P. G.   40
Morgan, J. A.   178
Morisawa, M. E.   40
Morrison, D. F.   220, 221, 224
Mukherjee, B. N.   242, 304
Mulaik, S. A.   31, 243, 262, 263, 270, 281, 283, 290, 291, 303
Mullett, G. M.   67
Murray, T. W.   67

Nash, J. E.   60, 66
Neave, H. R.   464
Nelder, J. A.   205.
Nerlove, S.   332
Norcliffe, G. B.   117, 118, 119, 147
Norris, J. M.   237, 238

O'Farrell, P. N.   45
O'Neill, R.   205
Ogilvie, J. C.   321, 327
Oldham, C. H. G.   117, 141, 142
Ondrick, C. W.   237
Oosterhoff, J.   177
Orcutt, G. H.   40
Ord, J. K.   84, 87, 93, 150, 160, 169
Osborne, M. R.   40

523

Painter, R. B.   40
Parks, J. M.   315
Parsley, A. J.   118, 120, 121, 128, 144, 146
Pearson, K.   243
Peckham, G.   205
Pennington, R. H.   10, 32, 33
Phipps, J. B.   326
Poole, M. A.   45
Pope, P. J.   178
Potter, P. E.   421
Potter, W. D.   40
Powell, M. J. D.   205, 209
Press, S. J.   331
Pritchard, N. M.   326

Quandt, R. E.   200

Ralston, A.   10
Ramsayer, G. R.   400, 453, 456, 458, 459
Rao, C. R.   215, 437
Rao, S. V. L. N.   150
Rayner, J. H.   337
Rayner, J. N.   83
Rees, P.   295
Reiher, B. J.   240
Reinsch, C.   6, 10, 26, 31, 437, 460, 465, 484, 485
Reyment, R. A.   240, 330, 331, 392
Rhodes, J. M.   421
Rice, J.   274, 275
Robinson, A. H.   148
Robinson, S. M.   270
Rodda, J. C.   40, 117
Rohlf, F. J.   311, 313, 319, 323, 325, 330
Romney, A. K.   332
Ross, G. J. S.   319, 392
Rubin, H.   245
Rummell, R. J.   240, 243, 278, 279
Rushton, G.   331
Rushton, S.   97, 99, 109, 141

Sahu, B. K.   421
Sammon, J. W.   325, 339, 399
Schatzoff, M.   178
Seal, H. L.   200
Searle, S. R.   10, 35, 95
Shaw, B. L.   60, 66
Shen, H. W.   207, 208
Shepard, R. N.   332
Shideler, G. L.   102, 103, 108
Sibson, R.   324, 325, 326

Silvey, S. D.   40
Simonett, D. S.   197
Slater, P.   437
Slaymaker, H. O.   130
Smith, A. F. M.   74
Smith, H.   45, 77, 174, 175
Sneath, P. H. A.   149, 310, 316, 321, 325, 330
Soelberg, P.   305
Sokal, R. R.   310, 316, 321, 325, 330
Spackman, E. A.   141
Sparks, D. N.   327
Spearman, C.   241, 242, 247
Spitz, O. T.   141
Spivak, M.   203
Srivastava, G. S.   150
Stammers, W. N.   237
Stephenson, R. A.   40
Stoddart, D. A.   130
Stone, J.   74
Storer, R. F.   477
Sutherland, D. B.   117, 141, 142
Sutton, E. A.   40

Tatar, J. F.   178
Thayer, D. T.   305
Thayer, R. P.   477
Thomas, D. N.   40
Thomas, G. S. P.   429
Thomin, R. J.   178
Thurstone, L. L.   241, 275, 279, 295
Tien, P. L.   364, 365
Tinkler, K.   123, 147
Tobler, W. R.   331
Torgerson, W. S.   335
Tsao, R.   178
Tucker, L. R.   271, 302
Tukey, J.   2, 77, 120, 305

Unwin, D.   117, 147

Van Thillo, M.   202, 261, 270
Vernon, M. D.   119

Walls, R. C.   192
Walsh, J.   33
Wampler, R. H.   9, 26, 52, 96, 100
Warburton, F. W.   279
Ward, J. H., Jr.   45, 320, 326
Ward, R. C.   101
Ward, W. C.   429
Watson, G. S.   84
Webster, J. T.   178
Weeks, D. L.   192

Wenniger, G.   324
White, P. O.   281, 283
Whitten, E. H. T.   117, 141, 143
Whittington, H. B.   327, 331, 338
Wilf, H. S.   10
Wilkes, J. O.   10, 97
Wilkinson, J. H.   6, 10, 26, 31, 437, 460, 465, 484, 485
Williams, P. W.   10
Williams, W. T.   316, 318, 320, 321, 322, 326, 371

Wishart, D.   320, 321
Wolberg, J. R.   206
Wong, S. T.   40
Wood, F. S.   45, 77
Woodruffe, J. F.   40

Yamane, T.   3, 268
Yevyevitch, V. C.   40, 83, 101, 102, 237
Youngs, E. A.   9, 33, 34, 53, 471

# Subject Index

(F = Figure, S = Section, T = Table)

Adaptive pattern recognition  453
Agglomerative clustering  310, 311
Alpha coefficient  247
Alpha factor analysis  247–261, 270, 304
Angle factor  336
Autocorrelation  82–93
  spatial  83, 84–88, 150, 160, 168

Backward elimination  173
Basic structure of matrix  222
Binary explanatory variables in regression  193–198
BMD  6
Box's test  434, 447

Canonical vectors  442
Centroid method, cluster analysis  319, 323, 326, 355
Chaining  318, 326, 342
Characteristic equation  28
Characteristic root, see Eigenvalue
Chisquare test for number of factors  270, 271, 304, 205
Cholesky factorization  27, 461
Classification  307–459
  types of  309–310
  defined  309
Cluster analysis  19
  agglomerative  310, 311
  centroid method  319, 323, 326, 355
  combinatorial  317, 318
  combinatorial formula  318
  combinatorial strategy  317, 318
  comparison of methods  324–327
  compatible strategy  318
  computer programs  381–399
  dendrogram  321–324
  divisive  310, 311
  euclidean, see Nonhierarchical cluster analysis
  examples  342–343, 351, 355–363, 366–367
  flexible strategy  321, 323, 343
  furthest neighbour  319, 323, 326, 343, 355
  group average  320, 323, 326, 343, 355
  hierarchical  307, 316–321, 342–343, 355, 371
  median  320, 323, 355
  minimum variance  320, 323, 343
  nearest neighbour  319, 323, 324, 326, 342, 355
  nonhierarchical  308, 327–329, 343–351, 366, 371
  simple average  320, 323
  space distorting properties  318
Cochrane–Orcutt iterative procedure  89–91
Coefficient of determination  49, 173, 178
  corrected for degrees of freedom  49, 178
Column vector  10
Combinatorial approach in regression  178, 182–192
Combinatorial strategy in cluster analysis  317, 318
Common factors  242
Common part, in image analysis  273
Communality  223, 255
Comparison of dendrograms  324–327

Comparison of trend maps 147–150, 168–172
Compatible clustering strategy 318
Complete linkage method, cluster analysis, see Furthest neighbour method
Computation, error in 4, 9, 32–35, 37, 40, 52, 67, 72, 131, 140, 303
Computer programs S0.2, 1
Conditional distribution 45, F2.2
Confirmatory data analysis 1, 119
Confirmatory factor analysis 304
Congruential generator 463
Control area, trend surface analysis 118
Cophenetic correlation 325, 326, 342, 343, 355
Correlation coefficient, computing algorithm 33–34, 52
Cosine theta coefficient 314
$C_p$ statistic 178, 187
Curve fitting 39, 95–116

Dendrogram 321–324
  comparison of 324–327
  cophenetic correlation 325, 326, 342, 343, 355
  monotonicity property 318
Dendrograph 323
Dependent variable 39
Derivative 202–203
  estimation by finite differences 203
Determinant 28, 36, 124, 125, 461, 494, 498
Diagonal matrix 14
  inverse of 35
  square root of 35
  multiplication by 15, 35
Dice coefficient 314, 315
Direction cosine 22
Discriminant analysis 5, 309
  assumptions of 422
  defined 421
  for binary variables 453–459
  multiple group 437–453
  two group 420–432
  uncategorized cases in F7.1, 425
Discriminant function F7.1
Discrimination 308
Disturbance term 4, 39, 82
Divisive clustering 310, 311
Dummy explanatory variable 193–198
Durbin two-stage algorithm 89, 91

Durbin–Watson test 84, 90, 91
Eigenproblem 461
Eigenstructure 222
Eigenvalue 28, 32, 74, 75, 461, 462, 465, 485, 500
Eigenvector 28–32, 461, 462, 465, 466, 487, 491, 493, 500
  normalization of 30, 31
Element of matrix 10
Equamax rotation 279
Error in computation 32–35
Errors in data analysis 2
Estimators, properties of 42, F2.1
Eta 460, 461
Euclidean distance 19, 149, 312, 313, 314, 315, 320, 327, 329, 339
  calculation of 312–313
  with correlated variables 313–14
Euclidean geometry 16, 19
Euclidean space 19
Explanatory variable 39
Exploratory data analysis S0.0, 4, 119, 120

F-ratio test in regression 48
F-ratio test in TSA 143, 145, 158
Factor analysis 5, 19, 213, 240–306, 331
  alpha 247–261, 270, 304
  assumptions of 244
  computer programs 258–259, 264–267, 276–277, 282–283, 288–290, 292–293
  definition 241–243
  image analysis 268, 272–275, 304
  indeterminacy 244, 245
  maximum likelihood estimation 241, 268–272, 304, 305
  minres 261, 269, 281, 284, 293, 304
  nonlinear models 244
  number of factors 304
  pattern 279
  Promax rotation 281–290, 293, 304
  Q mode 442
  reliability coefficient 271
  rotation of factors 275–194, 304
  scores 240
  simple structure 213
  standard errors 305
  statistical model 243–245
  structure 279
  target rotation 279, 290–294

varimax rotation   279–281
Factor interpretation   305–306
Factor loadings   243
  estimation of   245–275
  standard errors of   305
Factor pattern matrix   277, 286
Factor scores   277, 294–303
  Bartlett's method   296
  comparison of methods   299, 302
  computer programs for   297–299, 300–301, 302
  ideal variables   296, 299
  image factors   299
  problems in estimation   303
  regression method   295–296
Factor structure matrix   277, 286, 290
Factors common   241
  indeterminacy   244, 245
  specific   244
Fager coefficient   314
Flexible strategy method   321, 323, 343
Fortran   6, 9, 10
Forward selection   173
Furthest neighbour method   319, 323, 326, 343, 355

Gauss–Jordan algorithm   26, 27
Gauss–Seidel algorithm   261
Gaussian elimination   26, 96, 100, 176, 178
Generalized least squares (GLS)   4, 79, 89
Good luck factor   336
Gower's general coefficient of similarity   315, 316, 330, 365, 370
Gram–Schmidt method   S3.1.1, 140–141
Group average cluster analysis   320, 323, 326, 343, 355

Hessian matrix   204
Heteroscedasticity   77–82
Heywood case   256, 257, 273
  adjustment for   262–263
Hierarchical cluster analysis   307, 316–321, 342–343, 355, 371
Homoscedasticity   44, 77, 80
Hotelling's T2   422, 423
Householder tridiagonalization   461, 484
Hyperplane count   290
Hypothesis matrix   290

Identity matrix   14

Ill-conditioned matrix   24
Image factor analysis   268, 272–275, 304
Image factor scores   299
Independent variable, see Explanatory variable
Induced instability   26
Inherent instability   26
Interpretation of factors   305
Inverse matrix   S1.4
Inversion of nonsymmetric matrix   498
Inversion of symmetric matrix   461, 495, 496
Iterative refinement   462, 496

Jaccard coefficient   314, 315
Jacobean matrix   203
Jacobi algorithm   31

Latent root, see Eigenvalue
Least squares methods   4, 39–211, 261
Linear independence   17
Linear simultaneous equations, solution of   24–28, 495
Linearly dependence   17
Linkage tree see Dendrogram
Local component, trend surface analysis   117, 118

Macheps   460, 461
Magic factor   339
Mahalanobis' D2   423, 424, 425, 431
Map comparison   146–150, 169, 172–173
Mapping error   325, 339, 351
Marquardt's algorithm   146
Matrix   10
  addition   11
  basic structure   222
  characteristic equation   26
  characteristic roots and vectors   28
  condition of   24, 49, 124
  defined   10
  determinant of   26, 36, 124, 125
  diagonal   14
  eigenvalues and eigenvectors of   S1.5
  ill-conditioned   24, 124, F1.9
  inverse   23
  multiplication   11–14, F1.1
  negative definite, semidefinite   31
  order   14
  orthogonal   23, 36

orthonormal 23
positive definite, semidefinite 31
postmultiplication 11
premultiplication 11
rank 17
singular 23
spur 32
square 14
subtraction 11
symmetric 15
trace 32
transpose 15
triangular 16
Matrix inversion, algorithms for 26, 27
Matrix operations, defined 11
Maximum likelihood estimation, factor analysis 268–272, 304, 305
Maximum method of cluster analysis, see Furthest neighbour method
Median method 320, 323, 355
Metric 19
Minimum method of cluster analysis, see Nearest neighbour method
Minimum spanning tree 319
Minimum variance method 320, 323, 343
Minres algorithm, factor analysis 261–269, 281, 284, 293, 304
Monotonicity constraint, nonmetric multidimensional scaling 332, 333, F6.4, 335
Monotonicity of dendrogram 318
Multicollinearity in regression 45, 73–77, 215
  in TSA 142
Multidimensional scaling, see Nonmetric multidimensional scaling
Multiple discriminant analysis 437–453
Multiple regression see Regression model
Multivariate analysis of variance 309, 420, 432–437
  Box's test 434, 447
  Wilks' lambda test 436, 442, 448

N-dimensional space 15
N-space 15
Nearest neighbour analysis of point patterns 122
Nearest neighbour method, cluster analysis 319, 323, 324, 326, 342, 355
Newton–Raphson algorithm 202, 204, 261, 270

Nonhierarchical cluster analysis 302, 327–329, 343–351, 366, 371
Nonlinear least squares 5, 198–211
  conjugate gradients 205
  Gauss–Newton method 204
  Marquardt algorithm 146, 204
  Newton's algorithm 201, 202
  Newton–Raphson method 202, 204
  residual sum of squares function 203
  simplex method 205
  starting point 200, 206
  steepest descent algorithms 200–202
Nonlinear mapping 325, 339–340, 351, 369, 370, 371, 399, 416
  computer program 416–419
  example of 368–369, 370
Nonmetric multidimensional scaling 325 331–339, 341, 351, 365, 370, 371, 392, 399, 407
  computer program 407–415
  example of 337–339, 351, 355, 365, 370, 374–379
  numerical method 335–336
Nonstatistical factor analysis 247–261, 304
Normal varimax 280
Normality assumption in factor analysis 305
Normalized vector 23
Null matrix 14
Number of factors 304
Numerical analysis 9

Oblique rotation 281–294, 304
Order of matrix 14
Ordinary least squares (OLS), see Regression model
Ordination 307, 308, 329–341
Orthogonal matrix 23
Orthogonal polynomials 96–101
Orthogonal rotation 279–81, 304
Orthogonal vectors 23
Orthonormal vectors 23
Otsuka coefficient 314

P-condition number 49, 100, 124
Partial regression coefficient 41
PCA, see Principal components analysis
Phenogram, see Dendrogram
Point pattern, in TSA 120–128
Polynomial curve fitting 95, 110

Positive definite matrix   31
Premultiplication   11
Primary factor pattern   277
Primary factors   278, 279
Primary structure matrix   277
Principal axes   219, 220
Principal component scores, effect of standardization   233, 224
Principal components, geometry of   216–220
Principal components analysis   5, 19, 73, 142, 213, 215–239, 309, 329, 342, 441
   computational method   220–232
   example   232–237
   loadings matrix   216
   scale dependence of   219, 220
   scores   216, 223, 224, 312, 313
   significance tests   225, 226
   use in ordination   309
Principal coordinates analysis   316, 329–331, 370–371, 392, 399
   computer program for   402–406
   example of   370–371
Principal diagonal of square matrix   14
Promax rotation   281–290, 293, 304
Proper root, see Eigenvalue
Pseudo-random number generator   371, 392, 461, 464
Pseudo-random numbers   371

QL algorithm for eigenvalues   31
QR algorithm   461
Quadrat analysis   123–124
Quadratic equation, solution of   29
Quadratic loss function   340–341
Quartimax rotation   279

Random numbers   371
Random variable   44
Rank of matrix   17
Raw stress   339
Raw varimax   279
$r_c$, see Cophenetic correlation
Reference factors   278, 291
Reference pattern   279
Reference structure   279
Regional component, trend surface analysis   118
Regression, combinatorial   4, 178, 182–192
   computer program   54–60
   ridge   4, 73–77

stepwise   4, 174–192
   through origin   66–67
Regression model   14, 19, 39–93, 172–211
   assumptions of   40, 43–45
   combinatorial approach   4, 178, 182–192
   computational aspects   23, 24, 27, 40, S2.3
   estimation   S2.2
   linear model   S2.1
   nonlinear models   77, 95
   selection of explanatory variables   173–193
   stepwise method   4, 174–192
   use of binary explanatory variables   193–198
Relaxation factor   336
Repetition in data analysis   3
Reproduced correlations   245, 261
Residual   39, 77, 82, 118
Residual component, trend surface analysis   118
Residual correlations   261
Ridge methods in TSA   142
Ridge regression   4, 73–77
Rotation of coordinate system   20
Rotation of factors   275–293, 304
Rotation to specified target   279, 290–294
Row vector   10

Scores on image factors   299
Selection of variables in regression   172–193
Serial correlation   83
Similarity measurement of   311–316
Similarity measures   311–316
   $\cos\theta$   314
   Dice coefficient   314, 315
   Euclidean distance   312, 313, 314, 315, 320, 327, 329, 339
   Fager coefficient   314
   Gower's general coefficient   315, 316, 330, 365, 370
   Jaccard coefficient   314, 315
   Otsuka coefficient   314
   Simpson coefficient   314, 315, 328
Simple average method   320, 323
Simple structure   213, 275, 277, 279
Simpson coefficient   314, 315, 328
Simultaneous linear equations, solution of   24–28, 495
Simultaneous varimax   280

Single linkage method, cluster analysis, *see* Nearest neighbour method
Solution of simultaneous linear equations   24–28, 495
Space conserving clustering strategy   318
Space distorting clustering strategy   318
Spatial autocorrelation   83, 84–88, 160, 168
SPSS   6
Square matrix   14
Standard error of regression coefficient   49
Standardized regression coefficient   41, 48
Standardized variable, geometry of   22
Statistical factor analysis   261–272, 304
Steepest descent algorithm   200–202, 334, 339
Step length   336
Stepwise regression   4, 172–192
Stress   325, 334, 336
Symmetric matrix   15
Systems approach   1

T-test in regression, 49
Target matrix   281, 283
Target rotation   279, 290–94
Transpose of matrix   15
Trend coefficients   119, 139–140
Trend component   117, 118
Trend surface analysis   4, 94, 116–172
   assumptions   117, 118, 120
   canonical model   142
   computational aspects   130–142
   effect of data point distributions   119, 120–128
   estimation of coefficients   139
   F-ratio test   143, 145, 158
   four-dimensional applications   142
   generation of polynomial terms   130–131

   lineprinter maps   131–132
   nonlinear models   146
   orthogonal polynomials   140, 141
   significance tests   119, 142, 143, 145, 158
   spatial autocorrelation   145, 160–168
   $Z^2$ array   141–143, 150, 159
Trend surface, significance of   146–147
Triangular matrix   16
TSA, *see* Trend surface analysis

Ultrametric   20
Unique variance   296
Unweighted pair group method using arithmetic averages   320

Varimax rotation   279–281, 365
Vector   10, S1.3
   addition   17
   column   10
   geometric representation   16
   linearly dependent sets   17
   multiplication   17
   normalized   23
   orthogonal   23
   orthonormal   23
   row   10
   length   19, 21
Vector representation of variables   21
Vectors collinear   19

Ward's error sum of squares, cluster analysis   320, 323, 343
Weight matrix, spatial autocorrelation   86–87
Weighted pair group method, cluster analysis   320
Wilks' lambda test   436, 442, 448

$Z^2$ array   141–143, 150, 159